Machine and Deep Learning Using MATLAB

Machine and Deep Learning Using MATLAB

Algorithms and Tools for Scientists and Engineers

Kamal I. M. Al-Malah

This edition first published 2024
© 2024 Kamal I. M. Al-Malah

The right of Kamal I. M. Al-Malah to be identified as the author of this work has been asserted in accordance with law.

Registered Office
John Wiley & Sons, Inc., 111 River Street, Hoboken, NJ 07030, USA

For details of our global editorial offices, customer services, and more information about Wiley products visit us at www.wiley.com.

Wiley also publishes its books in a variety of electronic formats and by print-on-demand. Some content that appears in standard print versions of this book may not be available in other formats.

Limit of Liability/Disclaimer of Warranty
MATLAB® is a trademark of The MathWorks, Inc. and is used with permission. The MathWorks does not warrant the accuracy of the text or exercises in this book. This work's use or discussion of MATLAB® software or related products does not constitute endorsement or sponsorship by The MathWorks of a particular pedagogical approach or particular use of the MATLAB® software. While the publisher and authors have used their best efforts in preparing this work, they make no representations or warranties with respect to the accuracy or completeness of the contents of this work and specifically disclaim all warranties, including without limitation any implied warranties of merchantability or fitness for a particular purpose. No warranty may be created or extended by sales representatives, written sales materials or promotional statements for this work. This work is sold with the understanding that the publisher is not engaged in rendering professional services. The advice and strategies contained herein may not be suitable for your situation. You should consult with a specialist where appropriate. The fact that an organization, website, or product is referred to in this work as a citation and/or potential source of further information does not mean that the publisher and authors endorse the information or services the organization, website, or product may provide or recommendations it may make. Further, readers should be aware that websites listed in this work may have changed or disappeared between when this work was written and when it is read. Neither the publisher nor authors shall be liable for any loss of profit or any other commercial damages, including but not limited to special, incidental, consequential, or other damages.

A catalogue record for this book is available from the Library of Congress

Hardback ISBN: 9781394209088; ePub ISBN: 9781394209101; ePDF ISBN: 9781394209095; oBook ISBN: 9781394209118

Cover Image: © Yuichiro Chino/Getty Images
Cover Design: Wiley

Set in 9.5/12.5pt STIXTwoText by Integra Software Services Pvt. Ltd, Pondicherry, India
Printed and bound by CPI Group (UK) Ltd, Croydon, CR0 4YY

C9781394209088_061023

Contents

Preface *xiii*
About the Companion Website *xvii*

1 Unsupervised Machine Learning (ML) Techniques *1*
Introduction *1*
Selection of the Right Algorithm in ML *2*
Classical Multidimensional Scaling of Predictors Data *2*
Principal Component Analysis (PCA) *6*
k-Means Clustering *13*
  Distance Metrics: Locations of Cluster Centroids *13*
  Replications *14*
Gaussian Mixture Model (GMM) Clustering *15*
  Optimum Number of GMM Clusters *17*
Observations and Clusters Visualization *18*
Evaluating Cluster Quality *21*
  Silhouette Plots *22*
Hierarchical Clustering *23*
  Step 1 – Determine Hierarchical Structure *23*
  Step 2 – Divide Hierarchical Tree into Clusters *25*
PCA and Clustering: Wine Quality *27*
Feature Selection Using Laplacian (fsulaplacian) for Unsupervised Learning *35*
CHW 1.1 The Iris Flower Features Data *37*
CHW 1.2 The Ionosphere Data Features *38*
CHW 1.3 The Small Car Data *39*
CHW 1.4 Seeds Features Data *40*

2 ML Supervised Learning: Classification Models *42*
Fitting Data Using Different Classification Models *42*
  Customizing a Model *43*
  Creating Training and Test Datasets *43*
  Predicting the Response *45*
  Evaluating the Classification Model *45*
KNN Model for All Categorical or All Numeric Data Type *47*
  KNN Model: Heart Disease Numeric Data *48*

Viewing the Fitting Model Properties   *50*
The Fitting Model: Number of Neighbors and Weighting Factor   *51*
The Cost Penalty of the Fitting Model   *52*
KNN Model: Red Wine Data   *55*
Using MATLAB Classification Learner   *57*
Binary Decision Tree Model for Multiclass Classification of All Data Types   *68*
Classification Tree Model: Heart Disease Numeric Data Types   *70*
Classification Tree Model: Heart Disease All Predictor Data Types   *72*
Naïve Bayes Classification Model for All Data Types   *74*
Fitting Heart Disease Numeric Data to Naïve Bayes Model   *75*
Fitting Heart Disease All Data Types to Naïve Bayes Model   *77*
Discriminant Analysis (DA) Classifier for Numeric Predictors Only   *79*
Discriminant Analysis (DA): Heart Disease Numeric Predictors   *82*
Support Vector Machine (SVM) Classification Model for All Data Types   *84*
Properties of SVM Model   *85*
SVM Classification Model: Heart Disease Numeric Data Types   *87*
SVM Classification Model: Heart Disease All Data Types   *90*
Multiclass Support Vector Machine (fitcecoc) Model   *92*
Multiclass Support Vector Machines Model: Red Wine Data   *95*
Binary Linear Classifier (fitclinear) to High-Dimensional Data   *98*
CHW 2.1   Mushroom Edibility Data   *100*
CHW 2.2   1994 Adult Census Income Data   *100*
CHW 2.3   White Wine Classification   *101*
CHW 2.4   Cardiac Arrhythmia Data   *102*
CHW 2.5   Breast Cancer Diagnosis   *102*

3   Methods of Improving ML Predictive Models   *103*
Accuracy and Robustness of Predictive Models   *103*
Evaluating a Model: Cross-Validation   *104*
Cross-Validation Tune-up Parameters   *105*
Partition with K-Fold: Heart Disease Data Classification   *106*
Reducing Predictors: Feature Transformation and Selection   *108*
Factor Analysis   *110*
Feature Transformation and Factor Analysis: Heart Disease Data   *113*
Feature Selection   *115*
Feature Selection Using predictorImportance Function: Health Disease Data   *116*
Sequential Feature Selection (SFS): sequentialfs Function with Model Error Handler   *118*
Accommodating Categorical Data: Creating Dummy Variables   *121*
Feature Selection with Categorical Heart Disease Data   *122*
Ensemble Learning   *126*
Creating Ensembles: Heart Disease Data   *130*
Ensemble Learning: Wine Quality Classification   *131*
Improving fitcensemble Predictive Model: Abalone Age Prediction   *132*
Improving fitctree Predictive Model with Feature Selection (FS): Credit Ratings Data   *134*

Improving fitctree Predictive Model with Feature Transformation (FT): Credit Ratings Data   *135*

Using MATLAB Regression Learner   *136*
   Feature Selection and Feature Transformation Using Regression Learner App   *145*
Feature Selection Using Neighborhood Component Analysis (NCA) for Regression: Big Car Data   *146*
CHW 3.1   The Ionosphere Data   *148*
CHW 3.2   Sonar Dataset   *149*
CHW 3.3   White Wine Classification   *150*
CHW 3.4   Small Car Data (Regression Case)   *152*

4   Methods of ML Linear Regression   *153*
Introduction   *153*
Linear Regression Models   *154*
   Fitting Linear Regression Models Using fitlm Function   *155*
   How to Organize the Data?   *155*
   Results Visualization: Big Car Data   *162*
   Fitting Linear Regression Models Using fitglm Function   *164*
Nonparametric Regression Models   *166*
   fitrtree Nonparametric Regression Model: Big Car Data   *167*
   Support Vector Machine, fitrsvm, Nonparametric Regression Model: Big Car Data   *170*
   Nonparametric Regression Model: Gaussian Process Regression (GPR)   *172*
Regularized Parametric Linear Regression   *176*
   Ridge Linear Regression: The Penalty Term   *176*
   Fitting Ridge Regression Models   *177*
   Predicting Response Using Ridge Regression Models   *178*
   Determining Ridge Regression Parameter, λ *179*
   The Ridge Regression Model: Big Car Data   *179*
   The Ridge Regression Model with Optimum λ: Big Car Data   *181*
   Regularized Parametric Linear Regression Model: Lasso   *183*
Stepwise Parametric Linear Regression   *186*
   Fitting Stepwise Linear Regression   *187*
   How to Specify stepwiselm Model?   *187*
   Stepwise Linear Regression Model: Big Car Data   *188*
CHW 4.1   Boston House Price   *192*
CHW 4.2   The Forest Fires Data   *193*
CHW 4.3   The Parkinson's Disease Telemonitoring Data   *194*
CHW 4.4   The Car Fuel Economy Data   *195*

5   Neural Networks   *197*
Introduction   *197*
Feed-Forward Neural Networks   *198*
Feed-Forward Neural Network Classification   *199*
Feed-Forward Neural Network Regression   *200*
   Numeric Data: Dummy Variables   *200*
Neural Network Pattern Recognition (nprtool) Application   *201*
Command-Based Feed-Forward Neural Network Classification: Heart Data   *210*

Neural Network Regression (nftool)   *214*
Command-Based Feed-Forward Neural Network Regression: Big Car Data   *223*
Training the Neural Network Regression Model Using fitrnet Function: Big Car Data   *226*
Finding the Optimum Regularization Strength for Neural Network Using Cross-Validation: Big Car Data   *229*
Custom Hyperparameter Optimization in Neural Network Regression: Big Car Data   *231*
CHW 5.1   Mushroom Edibility Data   *233*
CHW 5.2   1994 Adult Census Income Data   *233*
CHW 5.3   Breast Cancer Diagnosis   *234*
CHW 5.4   Small Car Data (Regression Case)   *234*
CHW 5.5   Boston House Price   *235*

6   Pretrained Neural Networks: Transfer Learning   *237*
Deep Learning: Image Networks   *237*
Data Stores in MATLAB   *241*
Image and Augmented Image Datastores   *243*
Accessing an Image File   *246*
Retraining: Transfer Learning for Image Recognition   *247*
Convolutional Neural Network (CNN) Layers: Channels and Activations   *256*
    Convolution 2-D Layer Features via Activations   *258*
    Extraction and Visualization of Activations   *261*
    A 2-D (or 2-D Grouped) Convolutional Layer   *264*
Features Extraction for Machine Learning   *267*
    Image Features in Pretrained Convolutional Neural Networks (CNNs)   *268*
    Classification with Machine Learning   *268*
    Feature Extraction for Machine Learning: Flowers   *269*
    Pattern Recognition Network Generation   *271*
    Machine Learning Feature Extraction: Spectrograms   *275*
Network Object Prediction Explainers   *278*
    Occlusion Sensitivity   *278*
    imageLIME Features Explainer   *282*
    gradCAM Features Explainer   *284*
HCW 6.1   CNN Retraining for Round Worms Alive or Dead Prediction   *286*
HCW 6.2   CNN Retraining for Food Images Prediction   *286*
HCW 6.3   CNN Retraining for Merchandise Data Prediction   *287*
HCW 6.4   CNN Retraining for Musical Instrument Spectrograms Prediction   *288*
HCW 6.5   CNN Retraining for Fruit/Vegetable Varieties Prediction   *289*

7   A Convolutional Neural Network (CNN) Architecture and Training   *290*
A Simple CNN Architecture: The Land Satellite Images   *291*
    Displaying Satellite Images   *291*
Training Options   *294*
    Mini Batches   *295*
    Learning Rates   *296*
    Gradient Clipping   *297*
    Algorithms   *298*

Training a CNN for Landcover Dataset   *299*

Layers and Filters   *302*

Filters in Convolution Layers   *307*

Viewing Filters: AlexNet Filters   *308*

Validation Data   *311*

    Using shuffle Function   *316*

Improving Network Performance   *319*

    Training Algorithm Options   *319*

    Training Data   *319*

    Architecture   *320*

Image Augmentation: The Flowers Dataset   *322*

Directed Acyclic Graphs Networks   *329*

Deep Network Designer (DND)   *333*

Semantic Segmentation   *342*

    Analyze Training Data for Semantic Segmentation   *343*

    Create a Semantic Segmentation Network   *345*

    Train and Test the Semantic Segmentation Network   *350*

HCW 7.1   CNN Creation for Round Worms Alive or Dead Prediction   *356*

HCW 7.2   CNN Creation for Food Images Prediction   *357*

HCW 7.3   CNN Creation for Merchandise Data Prediction   *358*

HCW 7.4   CNN Creation for Musical Instrument Spectrograms Prediction   *358*

HCW 7.5   CNN Creation for Chest X-ray Prediction   *359*

HCW 7.6   Semantic Segmentation Network for CamVid Dataset   *359*

8   Regression Classification: Object Detection   *361*

Preparing Data for Regression   *361*

    Modification of CNN Architecture from Classification to Regression   *361*

    Root-Mean-Square Error   *364*

    AlexNet-Like CNN for Regression: Hand-Written Synthetic Digit Images   *364*

    A New CNN for Regression: Hand-Written Synthetic Digit Images   *370*

Deep Network Designer (DND) for Regression   *374*

    Loading Image Data   *375*

    Generating Training Data   *375*

    Creating a Network Architecture   *376*

    Importing Data   *378*

    Training the Network   *378*

    Test Network   *383*

YOLO Object Detectors   *384*

    Object Detection Using YOLO v4   *386*

    COCO-Based Creation of a Pretrained YOLO v4 Object Detector   *387*

    Fine-Tuning of a Pretrained YOLO v4 Object Detector   *389*

    Evaluating an Object Detector   *394*

Object Detection Using R-CNN Algorithms   *396*

    R-CNN   *397*

    Fast R-CNN   *397*

    Faster R-CNN   *398*

Transfer Learning (Re-Training)   *399*

R-CNN Creation and Training   *399*

Fast R-CNN Creation and Training   *403*

Faster R-CNN Creation and Training   *408*

evaluateDetectionPrecision Function for Precision Metric   *413*

evaluateDetectionMissRate for Miss Rate Metric   *417*

HCW 8.1   Testing yolov4ObjectDetector and fasterRCNN Object Detector   *424*

HCW 8.2   Creation of Two CNN-based yolov4ObjectDetectors   *424*

HCW 8.3   Creation of GoogleNet-Based Fast R-CNN Object Detector   *425*

HCW 8.4   Creation of a GoogleNet-Based Faster R-CNN Object Detector   *426*

HCW 8.5   Calculation of Average Precision and Miss Rate Using GoogleNet-Based Faster R-CNN Object Detector   *427*

HCW 8.6   Calculation of Average Precision and Miss Rate Using GoogleNet-Based yolov4 Object Detector   *427*

HCW 8.7   Faster RCNN-based Car Objects Prediction and Calculation of Average Precision for Training and Test Data   *427*

9   Recurrent Neural Network (RNN)   *430*

Long Short-Term Memory (LSTM) and BiLSTM Network   *430*

Train LSTM RNN Network for Sequence Classification   *437*

Improving LSTM RNN Performance   *441*

Sequence Length   *441*

Classifying Categorical Sequences   *445*

Sequence-to-Sequence Regression Using Deep Learning: Turbo Fan Data   *446*

Classify Text Data Using Deep Learning: Factory Equipment Failure Text Analysis – 1   *453*

Classify Text Data Using Deep Learning: Factory Equipment Failure Text Analysis – 2   *462*

Word-by-Word Text Generation Using Deep Learning – 1   *465*

Word-by-Word Text Generation Using Deep Learning – 2   *473*

Train Network for Time Series Forecasting Using Deep Network Designer (DND)   *475*

Train Network with Numeric Features   *486*

HCW 9.1   Text Classification: Factory Equipment Failure Text Analysis   *491*

HCW 9.2   Text Classification: Sentiment Labeled Sentences Data Set   *492*

HCW 9.3   Text Classification: Netflix Titles Data Set   *492*

HCW 9.4   Text Regression: Video Game Titles Data Set   *492*

HCW 9.5   Multivariate Classification: Mill Data Set   *493*

HCW 9.6   Word-by-Word Text Generation Using Deep Learning   *494*

10   Image/Video-Based Apps   *495*

Image Labeler (IL) App   *495*

Creating ROI Labels   *498*

Creating Scene Labels   *499*

Label Ground Truth   *500*

Export Labeled Ground Truth   *501*

Video Labeler (VL) App: Ground Truth Data Creation, Training, and Prediction   *502*

Ground Truth Labeler (GTL) App   *513*

Running/Walking Classification with Video Clips using LSTM   *520*

Experiment Manager (EM) App   *526*

Image Batch Processor (IBP) App   *533*

HCW 10.1   Cat Dog Video Labeling, Training, and Prediction – 1   *537*
HCW 10.2   Cat Dog Video Labeling, Training, and Prediction – 2   *537*
HCW 10.3   EM Hyperparameters of CNN Retraining for Merchandise Data Prediction   *538*
HCW 10.4   EM Hyperparameters of CNN Retraining for Round Worms Alive or Dead Prediction   *539*
HCW 10.5   EM Hyperparameters of CNN Retraining for Food Images Prediction   *540*

Appendix A   Useful MATLAB Functions   *543*
A.1   Data Transfer from an External Source into MATLAB   *543*
A.2   Data Import Wizard   *543*
A.3   Table Operations   *544*
A.4   Table Statistical Analysis   *547*
A.5   Access to Table Variables (Column Titles)   *547*
A.6   Merging Tables with Mixed Columns and Rows   *547*
A.7   Data Plotting   *548*
A.8   Data Normalization   *549*
A.9   How to Scale Numeric Data Columns to Vary Between 0 and 1   *549*
A.10   Random Split of a Matrix into a Training and Test Set   *550*
A.11   Removal of NaN Values from a Matrix   *550*
A.12   How to Calculate the Percent of Truly Judged Class Type Cases for a Binary Class Response   *550*
A.13   Error Function m-file   *551*
A.14   Conversion of Categorical into Numeric Dummy Matrix   *552*
A.15   evaluateFit2 Function   *553*
A.16   showActivationsForChannel Function   *554*
A.17   upsampLowRes Function   *555*
A.18A   preprocessData function   *555*
A.18B   preprocessData2 function   *555*
A.19   processTurboFanDataTrain function   *556*
A.20   processTurboFanDataTest Function   *556*
A.21   preprocessText Function   *557*
A.22   documentGenerationDatastore Function   *557*
A.23   subset Function for an Image Data Store Partition   *560*

Index   *561*

Preface

Welcome to "Machine and Deep Learning Using MATLAB Algorithms and Tools for Scientists and Engineers." In today's data-driven world, machine learning and deep learning have become indispensable tools for scientists and engineers across various disciplines. This book aims to provide a comprehensive guide to understanding and applying these techniques using MATLAB algorithms and tools. Divided into ten chapters, "Machine and Deep Learning Using MATLAB Algorithms and Tools for Scientists and Engineers" offers a comprehensive coverage of both machine learning and deep learning techniques. The book takes a step-by-step approach, guiding readers through the process of acquiring, analyzing, and predicting patterns in both numeric and image data.

The first five chapters provide a solid foundation in machine learning, covering unsupervised learning, classification, predictive model improvement, linear regression, and neural networks. Through clear explanations, practical examples, and hands-on case studies, readers will develop the knowledge and skills necessary to apply these techniques to their own scientific and engineering endeavors. Readers will delve into various techniques that are widely used in the field, including clustering, classification, regression, and feature selection. These five chapters provide a solid foundation in machine learning concepts and methods, allowing readers to gain a deep understanding of how to apply these techniques to real-world problems.

Chapter One focuses on unsupervised machine learning techniques. It explores methods such as Classical Multidimensional Scaling, Principal Component Analysis (PCA), k-Means Clustering, Gaussian Mixture Model (GMM) Clustering, and Feature Selection Using Laplacian. The chapter provides tools for visualization and observation of clusters, as well as Hierarchical Clustering. Readers are guided through practical case studies, including analyzing Iris Flower Features Data, Ionosphere Data Features, Small Car Data, and Seeds Features Data.

Chapter Two delves into fitting data using different classification models. It introduces classification techniques like K-Nearest Neighbors (KNN), Binary Decision Tree, Naïve Bayes, Discriminant Analysis (DA), Support Vector Machine (SVM), Multiclass SVM, and Binary Linear Classifier. The chapter also showcases the MATLAB Classification Learner app, which allows users to explore these techniques without writing code. Through case studies involving Mushroom Edibility Data, Adult Census Income Data, White Wine Classification, Cardiac Arrhythmia Data, and Breast Cancer Diagnosis Data, readers gain practical experience in implementing the discussed methods.

Chapter Three covers methods for improving predictive models in machine learning. It explores topics such as Cross Validation, Feature Transformation and Selection, Factor Analysis, Sequential Feature Selection (SFS), Dummy Variables, and Ensemble Learning. The chapter also introduces Feature Selection Using Neighborhood Component Analysis (NCA) for regression problems. Readers will learn how to utilize the Regression Learner app to perform feature selection and transformation. Through case studies involving Ionosphere Data, Sonar Dataset, White Wine Classification, and Small Car Data (Regression Case), readers gain hands-on experience in implementing these techniques.

Chapter Four focuses on ML linear regression methods. It covers various approaches, including fitting Linear Regression Models using the fitlm/fitglm function, Non-Parametric Regression Models, Gaussian Process Regression (GPR), Regularized Parametric Linear Regression, Lasso, and Stepwise Parametric Linear Regression. Practical case studies, such as Boston House Price, The Forest Fires Data, The Parkinson's Disease Telemonitoring Data, and The Car Fuel Economy Data, enable readers to apply these methods to real-world datasets.

Chapter Five explores neural networks for classification and regression tasks. It introduces Feed-Forward Neural Networks and the Neural Network Pattern Recognition (nprtool) tool for classification problems, as well as Feed-Forward Neural Network Regression and the Neural Network Regression (nftool) for data fitting. The chapter covers training the Neural Network Regression Model using the fitrnet function, finding the optimum regularization strength through cross-validation, and custom hyperparameter optimization in neural network regression. Case studies involving Mushroom Edibility Data, 1994 Adult Census Income Data, Breast Cancer Diagnosis, Small Car Data (Regression Case), and Boston House Price enable readers to implement these techniques effectively.

The remaining five chapters focus on the exciting world of deep learning. Deep learning has gained significant popularity due to its ability to analyze complex patterns in large datasets, particularly in the field of image analysis. In these chapters, readers will explore topics such as neural networks, transfer learning, convolutional neural networks (CNNs), object detection, and recurrent neural networks (RNNs). Additionally, the book introduces image/video-based MATLAB tools that facilitate the implementation of deep learning algorithms and enable the analysis of images and videos. These remaining five chapters build upon the foundation established in the earlier chapters, delving into more advanced topics and techniques. Through practical examples, case studies, and hands-on implementation, readers acquire the knowledge and skills necessary to apply transfer learning, CNN architecture, object detection, RNNs, and image/video-based MATLAB tools to a variety of real-world problems.

Chapter Six focuses on transfer learning of pre-trained neural networks. It covers topics such as Data Stores in MATLAB, Image and Augmented Image Datastores, Retraining for Image Recognition, Convolutional Neural Network (CNN) Layers, Channels and Activations, and Feature Extraction for Machine Learning. Additionally, the chapter explores network object prediction explainers like Occlusion Sensitivity, imageLIME Features Explainer, and gradCAM Features Explainer. Through case studies involving CNN retraining for various predictions, such as Round Worms Alive or Dead, Food Images, Merchandise Data, Musical Instrument Spectrograms, and Fruit/Vegetable Varieties, readers gain practical experience in applying these techniques.

Chapter Seven dives into the architecture and training of Convolutional Neural Networks (CNNs). It covers topics like Training Options, Filters in Convolution Layers, Validation Data, Improving Network Performance, Image Augmentation using the Flowers Dataset, Directed Acyclic Graphs Networks, Deep Network Designer (DND), and Semantic Segmentation. Readers are presented with case studies that involve creating CNNs for predictions such as Round Worms Alive or Dead, Food Images, Merchandise Data, Musical Instrument Spectrograms, and Chest X-ray. By following these examples, readers develop a deeper understanding of CNN architecture and training methodologies.

Chapter Eight explores regression classification and object detection. It covers topics like Preparing Data for Regression, Deep Network Designer (DND) for Regression, YOLO Object Detectors, Object Detection Using Regions with Convolutional Neural Networks (R-CNN), R-CNN Transfer Learning (Re-Training), evaluateDetectionPrecision Function for Precision Metric, and evaluateDetectionMissRate for Miss Rate Metric. Through case studies involving testing object detectors, creating CNN-based object detectors, and creating GoogleNet-Based Fast R-CNN object detectors, readers gain hands-on experience in regression classification and object detection tasks.

Chapter Nine focuses on Recurrent Neural Networks (RNNs). It covers topics such as Long Short-Term Memory (LSTM) and BiLSTM Networks, Classifying Categorical Sequences, Sequence-to-Sequence Regression using Deep Learning, Classifying Text Data for Factory Equipment Failure Analysis, Word-By-Word Text Generation, Training Networks for Time Series Forecasting using Deep Network Designer (DND), and Network Training with Numeric Features. The chapter provides case studies involving text classification, text regression, and multivariate classification, allowing readers to apply RNN techniques to real-world datasets.

Chapter Ten covers image/video-based MATLAB tools. It introduces tools such as Image Labeler (IL), Video Labeler (VL), Ground Truth Labeler (GTL), Experiment Manager (EM), and Image Batch Processor (IBP). Readers are guided through case studies involving tasks like video labeling, training, and prediction, as well as hyperparameter tuning for CNN retraining. By working on these examples, readers gain hands-on experience with the image/video-based MATLAB tools and learn how to apply them effectively.

By covering both machine learning and deep learning, this book provides readers with a comprehensive understanding of these powerful techniques. The step-by-step approach ensures that readers can gradually build their knowledge and skills, starting with the fundamentals of machine learning and progressing to more advanced topics in deep learning. Throughout the book, readers will gain hands-on experience through practical examples and case studies, enabling them to apply the learned techniques to their own projects and research.

Each chapter follows a consistent method of approach to ensure clarity and ease of understanding. Firstly, the MATLAB built-in functions relevant to the topic are introduced, along with their properties, limitations, and applicability. Complete running examples are provided, along with the corresponding MATLAB code. Results in the form of figures and tables are presented alongside the code, enabling a better grasp of the concepts. Additionally, MATLAB tools and apps are explained and utilized where applicable, providing an alternative approach to achieving results. Quizzes are included in each chapter to test your understanding, and important notes are highlighted to draw your attention to critical points and potential pitfalls. Finally, end-of-chapter problems are tailored to reinforce the concepts covered, serving as opportunities to apply the methods to real-world case studies.

Throughout the book, a wide range of datasets are used to illustrate the concepts and techniques discussed. From the Iris Flower Features Data to the Boston House Price dataset, from Mushroom Edibility Data to Factory Equipment Failure Text Analysis, you will gain hands-on experience in solving various problems using MATLAB algorithms and tools.

It is worth mentioning that the solution manual for the end-of-chapter problems will be available exclusively for instructors, allowing them to guide students through the learning process. However, data sets, images, or videos for all running examples and for those of end of chapter problems will be available at Wiley's Web companion for all book users and readers. The Companion Website address is: www.wiley.com/go/al-malah/machinelearningmatlab

This book aims to be a valuable resource for anyone seeking to harness the power of machine learning and deep learning in their research or professional endeavors. Whether you are a scientist, engineer, or student, this book equips you with the knowledge and tools necessary to tackle complex data analysis and prediction tasks. The comprehensive coverage and practical approach make this book an invaluable resource for anyone looking to leverage the power of machine learning and deep learning in their work.

We hope you find this book enlightening and empowering as you embark on your journey into the world of machine and deep learning using MATLAB algorithms and tools.

Happy learning!
Kamal Al-Malah

About the Companion Website

This book is accompanied by a companion website which includes a number of resources created by author for students and instructors that you will find helpful.

www.wiley.com/go/al-malah/machinelearningmatlab

The Instructor website includes:

- End of Chapter Solutions

The website includes the following resources for each chapter:

- Auxiliary Data

Please note that the resources in instructor website are password protected and can only be accessed by instructors who buy the book.

1

Unsupervised Machine Learning (ML) Techniques

Introduction

As quoted by MATLAB R2021b built-in help, machine learning (ML) teaches computers to do what comes naturally to humans: Learn from experience. Machine learning algorithms utilize computational methods to directly learn (or extract) information from data without relying on a deterministic model. The set of algorithms adaptively improve their performance as the number of samples available for learning increases. ML uses two types of learning techniques: Unsupervised and supervised.

The *unsupervised* ML techniques can be used to weigh the importance of predictor variables relative to each other, without the influence of the response variable. Remember, birds of feather (feature) flock (cluster) together. Unsupervised learning finds hidden patterns or intrinsic structures in data. It is used to draw inferences from datasets consisting of input data without labeled responses. *Clustering* is the most common unsupervised learning technique. It is used for exploratory data analysis to find hidden patterns or groupings in data. For example, applications where clustering can be used include gene sequence analysis, market research, and object recognition.

Supervised ML, which trains a model on known input (predictor) and output (response) data so that it can predict future outputs. The aim of supervised machine learning is to build a model that makes predictions based on evidence in the presence of uncertainty. A supervised learning algorithm takes a known set of input data and known responses (output) and trains a model to generate reasonable predictions for the response to new data. Supervised learning uses classification and regression techniques to develop predictive models.

On the one hand, classification techniques predict categorical binary (off/on, 0/1, or no/yes) response cases; for example, whether an email is genuine or spam, or whether a tumor is cancerous or benign. In addition, classification models can also handle categorical multi-response cases, like medical imaging, image and speech recognition, score of a food/product quality, and credit scoring.

On the other hand, regression techniques predict continuous responses, for example, changes in temperature or fluctuations in power demand. Typical applications include electricity load

Machine and Deep Learning Using MATLAB: Algorithms and Tools for Scientists and Engineers, First Edition. Kamal I. M. Al-Malah.
Companion Website: www.wiley.com/go/al-malah/machinelearningmatlab

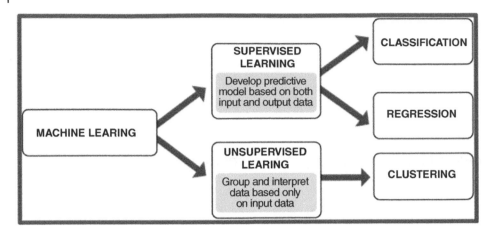

Figure 1.1 The scheme for the two types of ML.

forecasting and algorithmic trading. ML algorithms enable the data analyst to prioritize the input variables based on their impact on or contribution to the response (output) variable. In other words, the investigator can prioritize the list of variables as far as their importance or contribution to the response. Figure 1.1 shows a flowchart for the two ML techniques.

Selection of the Right Algorithm in ML

There are dozens of supervised and unsupervised machine learning algorithms, and each takes a different approach to learning. There is no best method or one-size-fits-all approach. Finding the right algorithm is partly based on trial and error; even a highly experienced data scientist cannot figure out whether an algorithm will work without trying it out. There is a trade-off between highly flexible models tending to overfit data by modeling minor variations that could be noise and simple models that are easier to interpret but might have lower accuracy. Therefore, choosing the right algorithm requires a compromise between choosing one benefit against another, including model speed, accuracy, and complexity. The trial-and-error approach is at the core of machine learning: If one approach or algorithm does not work, we ought to try another. MATLAB® provides tools to help you try out a variety of machine learning models and choose the best. Figure 1.2 shows a summary of some available algorithms that can be used for the prescribed case, namely, classification, regression, or clustering.

Table 1.1 shows some MATLAB apps and functions, which facilitate our tasks in ML. We can directly invoke the app or use command-line features.

Figure 1.3 shows the systematic ML workflow which can help tackle ML challenges. We can complete the entire workflow in MATLAB.

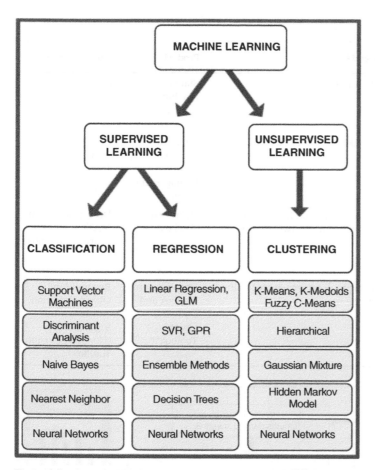

Figure 1.2 Some MATLAB built-in algorithms that can be used in ML scenarios.

Table 1.1 MATLAB apps and functions specifically made for ML.

Task	MATLAB Apps and Functions	Product	More Information
Classification to predict categorical responses	Use the Classification Learner app to automatically train a selection of models and help you choose the best. You can generate MATLAB code to work with scripts. For more options, you can use the command-line interface.	Statistics and Machine Learning Toolbox™	Train Classification Models in Classification Learner App Classification Functions
Regression to predict continuous responses	Use the Regression Learner app to automatically train a selection of models and help you choose the best. You can generate MATLAB code to work with scripts and other function options. For more options, you can use the command-line interface.	Statistics and Machine Learning Toolbox	Train Regression Models in Regression Learner App Regression Functions

(Continued)

Table 1.1   (Continued)

Task	MATLAB Apps and Functions	Product	More Information
Clustering	Use cluster analysis functions.	Statistics and Machine Learning Toolbox	Cluster Analysis
Computational finance tasks such as credit scoring	Use tools for modeling credit risk analysis.	Financial Toolbox™ and Risk Management Toolbox™	Credit Risk (Financial Toolbox)
Deep learning with neural networks for classification and regression	Use pretrained networks and functions to train convolutional neural networks.	Deep Learning Toolbox™	Deep Learning in MATLAB (Deep Learning Toolbox)
Facial recognition, motion detection, and object detection	Use deep learning tools for image processing and computer vision.	Deep Learning Toolbox and Computer Vision Toolbox™	Recognition, Object Detection, and Semantic Segmentation (Computer Vision Toolbox)

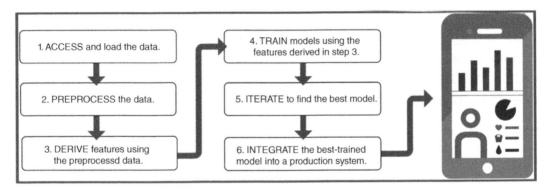

Figure 1.3   The workflow for ML steps, starting from data collection down to a predicting model.

Classical Multidimensional Scaling of Predictors Data

Classical multidimensional scaling can be split into three steps. The first step is to calculate pairwise distance among observations; we use the function **pdist**. The input should be numeric.

>>dist=pdist(Xnum, distance)

where **dist** is an output which represents the Euclidean distance between pairs of observations in **Xnum**, or is a dissimilarity vector containing the distance between each pair of observations. **Xnum** is a numeric matrix containing the data such that each row is considered as an observation. **distance** is an optional input that indicates the method of calculating the dissimilarity or distance. Commonly used methods are **'euclidean'**, the default, **'cityblock'**, and **'correlation'**.

The second step is to use the dissimilarity vector, **dist**, as an input to the function **cmdscale**.

>>[ConfMat, eigen] = cmdscale (dist)

Where **ConfMat** is an **n** × **p** configuration matrix of the reconstructed coordinates in **p**-dimensional space. **n** is the number of observations and **p** is the minimum number of dimensions needed to achieve the given pairwise distances. **eigen** is the Eigenvalues of the matrix **ConfMat*ConfMat'**. We can use Eigenvalues **eigen** to determine if a low-dimensional approximation to the points in **ConfMat** provides a reasonable representation of the data. So, rows of **ConfMat** are the coordinates of **n** points in **p**-dimensional space for some **p** < **n**. When **dist** is a Euclidean distance matrix, the distances between those points are given by **dist**. **p** is the dimension of the smallest space in which the **n** points whose inter-point distances are given by **dist** can be embedded.

When **dist** is Euclidean, by default, the first **p** elements of **eigen** are positive, the rest zero. If the first **k** elements of **eigen** are much larger than the remaining (**n-k**), then we can use the first **k** columns of **ConfMat** as **k**-dimensional points whose inter-point distances approximate **dist**. This can provide a useful dimension reduction for visualization, e.g., for **k** = *2*. The command

>>[ConfMat,eigen] = cmdscale(dist,m)

accepts a positive integer **m** between 1 and **n**. **m** specifies the dimensionality of the desired embedding **ConfMat**. If an **m** dimensional embedding is possible, then **ConfMat** will be of size **n** × **m** and **eigen** will be of size **m** × 1. If only a **q** dimensional embedding with **q** < **m** is possible, then **ConfMat** will be of size **n** × **q** and **eigen** will be of size **m** × 1. Specifying **m** may reduce the computational burden when **n** is very large.

The third step is to use the pareto function to visualize relative magnitudes of a vector in descending order.

>> pareto(eigen)

Figure 1.4 shows the code for generating pareto plot of basketball data after being normalized per game played (**GP**).

```
%% Import & initialize data
bbdata = readtable('basketball.xlsx');
bbdata.pos = categorical(bbdata.pos);
%% Calculate the per-game statistics
temp=bbdata{:,7:end};
normtemp=temp./bbdata.GP;
dist = pdist(normtemp);
[configMat,eigenVal] = cmdscale(dist);
figure(5)
pareto(eigenVal)
```

Figure 1.4 The code for generating Pareto plot of the normalized data per game.

Figure 1.5 shows the Pareto chart. We can see that 100% of the distribution is described in the first 3 variables, out of 14 variables, found in the configuration matrix, **configMat**. Notice that there is no column-wise mapping between **configMat** and the original **normtemp** matrix. What we learn here is simply that transformation into a new world will enable us to describe events or relationships, for example, in a 3-dimension new world rather than 14-dimension old world.

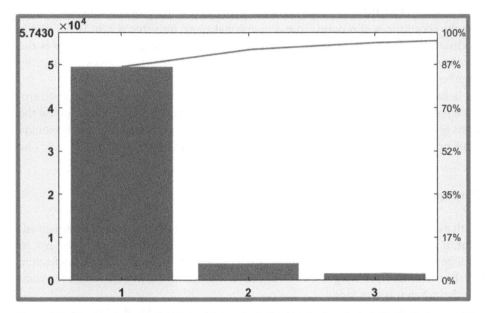

Figure 1.5 Pareto chart for the Eigenvalues where the distribution is describable in terms of the first three variables.

Figure 1.6 shows the workspace variables as created by running Figure 1.4 code.

Name ▲	Value	Size	Class	Min	Max
bbdata	1112x20 table	1112x20	table		
configMat	1112x14 double	1112x14	double	-11.1794	26.9768
dist	1x617716 double	1x617716	double	<Too m...	<Too many ...
eigenVal	1112x1 double	1112x1	double	-1.9708e...	4.9468e+04
normtemp	1112x14 double	1112x14	double	0	30.2242
temp	1112x14 double	1112x14	double	0	29484

Figure 1.6 The workspace variables created by Figure 1.4 code.

Principal Component Analysis (PCA)

Another method for dimensionality reduction is principal component analysis (PCA). The function **pca** is used to perform principal component analysis.

>>[pcs,scrs,latent,tsquared,pexp,mu]= pca(*measurements*)

Or, with a smaller number of outputs:

>>[pcs,scrs, ~,~,pexp,~]=pca(*measurements*)

Where **measurements** are a numeric matrix containing **n** columns corresponding to the observed variables. Each row corresponds to an observation. **pcs** is a **p × p** matrix of principal components. **p** is the number of predictor variables or columns. **scrs** is an **n × p** matrix containing the transformed data using the linear coordinate transformation matrix, **pcs**. **pexp** is a **p × 1** column vector of the percentage of variance, explained by each principal component.

Let us explain a little bit about the principal components. Suppose that the input numeric matrix contains only two variables x1 and x2. After performing PCA, the first column of the output matrix **pcs** contains coefficients of first principal component and the second column those of the second principal component, as shown in Figure 1.7.

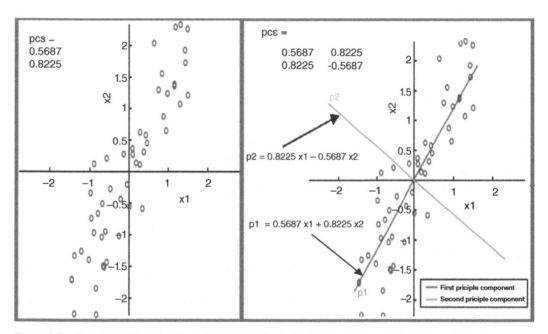

Figure 1.7 Transformation of original data into two principal components with known coefficients.

The second output is a matrix containing the data expressed in the coordinate space of the principal components. Figure 1.8, for example, shows entry (i.e., row) 42 where the original x1 and x2 values were transformed into new coordinate pc1 and pc2 values.

Finally, the last output contains the vector of percent variance explained by each principal component. In this case, most of the variance is explained by the first principal component.

>> pexp

Ans = 95.67   4.33

Since most of the variance is explained by the first principal component, we can only use the first column of the transformed data. Hence, the dimension of the data is reduced from two to one.

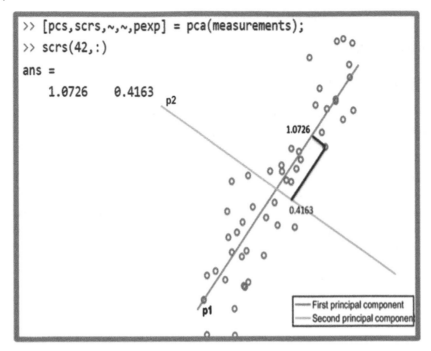

Figure 1.8 The second output is a matrix containing the transformed numeric x1 and x2.

The principal component algorithm that **pca** uses to perform the principal component analysis, specified as the comma-separated pair consisting of '**Algorithm**' and one of the following, as shown in Table 1.2:

Action to take for **NaN** values in the data matrix **X**, specified as the comma-separated pair consisting of '**Rows**' and one of the following, as shown in Table 1.3:

Table 1.2 Algorithms which can be used in PCA.

Algorithm	Description
'**svd**' (default)	Singular value decomposition (SVD) of X.
'**eig**'	Eigenvalue decomposition (EIG) of the covariance matrix. The EIG algorithm is faster than SVD when the number of observations, **n**, exceeds the number of variables, **p**, but is *less accurate* because the condition number of the covariance is the square of the condition number of X.
'**als**'	Alternating least squares (ALS) algorithm. This algorithm finds the best rank-k approximation by factoring **X** into an **n × k** left factor matrix, **L**, and a **p × k** right factor matrix, **R**, where **k** is the number of principal components. The factorization uses an iterative method starting with random initial values.
	ALS is designed to better handle missing values. It is preferable to pairwise deletion ('**Rows**','**pairwise**') and deals with missing values without listwise deletion ('**Rows**','**complete**'). It can work well for data sets with a small percentage of missing data at random, but might not perform well on sparse data sets.

Table 1.3 How to handle NaN values using PCA.

Value	Description
'complete'	Default. Observations with NaN values are removed before calculation. Rows of NaNs are reinserted into **scrs** and **tsquared** at the corresponding locations.
'pairwise'	This option only applies when the algorithm is '**eig**'. If you don't specify the algorithm along with '**pairwise**', then pca sets it to '**eig**'. If you specify '**svd**' as the algorithm, along with the option '**Rows**','**pairwise**', then **pca** returns a warning message, sets the algorithm to '**eig**' and continues.
	When you specify the '**Rows**','**pairwise**' option, **pca** computes the (i,j) element of the covariance matrix using the rows with no NaN values in the columns i or j of X. Note that the resulting covariance matrix might not be positive definite. In that case, **pca** terminates with an error message
'all'	X is expected to have no missing values. **pca** uses all of the data and terminates if any NaN value is found.

Table 1.4 shows the option of column data centering, specified as the comma-separated pair consisting of '**Centered**' and one of these logical expressions.

Table 1.4 The column data centering options upon using PCA.

Value	Description
true (Default)	**pca** centers X by subtracting column means before computing singular value decomposition or eigenvalue decomposition. If X contains NaN missing values, mean(X,'omitnan') is used to find the mean with any available data. You can reconstruct the centered data using **scrs*pcs**'.
false	In this case, **pca** does not center the data. You can reconstruct the original data using **scrs*pcs**'.

Let us calculate the principal component coefficients, **pcs**, principal component scores, **scrs**, and the percentage of total variance explained by each principal component of each variable, **pexp**, of **X**, the predictors' numeric data. **X** will be normalized per game played (**GP**) by the player. Keep in mind that the sixth column named **Minutes** also reflects games played. In other words, the larger number of games played the larger the operation time, in minutes, will be. Thus, those two columns are strongly correlated and both give basically the same information. We can normalize with respect to either but not to both.

Figure 1.9 shows the complete code for reading data, carrying out normalization, implementing PCA to **X** data, and generating Pareto plot.

```
data = readtable('basketball.xlsx');
data.pos = categorical(data.pos);
temp=data{:,7:end};
%%Normalization by division by ./data.GP
normtemp=temp./data.GP;
data{:,7:end}=normtemp;
[pcs,scrs,~,~,pexp,~] = pca(normtemp);
figure(10)
pareto(pexp)
```

Figure 1.9 Preparation of **X** data to PCA implementation and Pareto plot creation.

Figure 1.10 shows Pareto plot for the first three principal components. Notice that 95.8% of the total variance can be explained by three principal components. In this case, 86.1% of the total variance is explained by the first principal component.

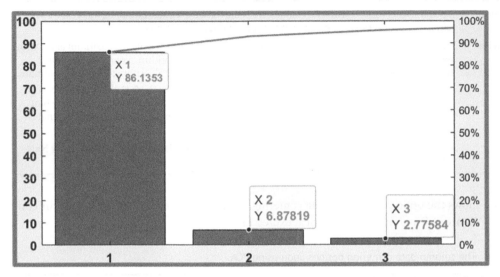

Figure 1.10 The Pareto diagram for the first three principal components showing their contributions to the total variation in **X**.

Figure 1.11 shows the workspace variables where the user can see dimensions of **pcs**, **pexp**, and **scrs** matrix.

Workspace			
Name ▲	Value	Size	Class
data	*1112x20 table*	1112x20	table
normtemp	*1112x14 double*	1112x14	double
pcs	*14x14 double*	14x14	double
pexp	*14x1 double*	14x1	double
scrs	*1112x14 double*	1112x14	double
temp	*1112x14 double*	1112x14	double

Figure 1.11 The workspace variables, including **pcs**, **pexp**, and **scrs** matrix.

On the other hand, if we normalize all the 14 columns, starting from #7 up to the last column (#20) without scaling the value per game played, then Figure 1.12 shows the code without dividing each entry by games played by each player.

Figure 1.13 shows the Pareto diagram where the variation is explained in terms of four principal components. The first four principal components can explain (or contribute) about 95.5% of the total variation in the numeric observed data matrix **X**. In words, when we scale the data by games played by the player, then we found that only three principal components

```
data = readtable('basketball.xlsx');
data.pos = categorical(data.pos);
temp=data{:,7:end};
%%Normalization with mean 0 and sigma 1.
normtemp= normalize(temp);
data{:,7:end}=normtemp;
[pcs,scrs,~,~,pexp] = pca(normtemp);
figure(13)
pareto(pexp)
```

Figure 1.12 The code for creating Pareto plot using the normalized data, without scaling them by the games played.

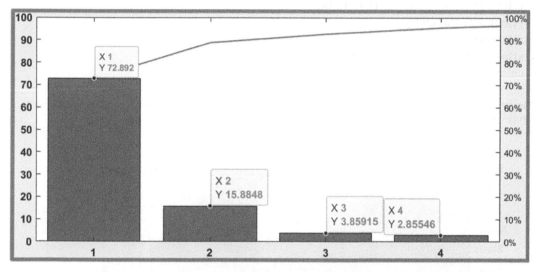

Figure 1.13 The Pareto diagram for the principal components showing their contributions to the total variation in **X**.

were needed to explain the total variance of **X**. If we use the raw data without scaling by the games played or the operation time spent, we found that we need four, not three, principal components to describe the picture; nevertheless, the marginal accuracy is less than 3% between the two cases.

Let us show the contribution of each individual variable to the total variation. Figure 1.14 shows the code for generating a populated pcs-squared plot. The pcs-squared is the square of the principal component coefficients, **pcs**, where they add up to unity, or 100% basis.

Figure 1.15 shows the image screen of pcs-squared values where the total in each column adds up to 100%. For PC#1, the highest contributor is **Points** followed by **FGAttempt** data. You can see the colored bar for other PCs. Notice that the color bar which values the contribution of each color band, is shown to the right of the figure.

```
data = readtable('basketball.xlsx');
data.pos = categorical(data.pos);
temp=data{:,7:end};
%%Normalization by division by ./data.GP
normtemp=temp./data.GP;
data{:,7:end}=normtemp;
[pcs,scrs,~,~,pexp] = pca(normtemp);
pcssqrd=pcs.^2;
Vars=data.Properties.VariableNames;
figure(15)
imagesc(pcssqrd(:,1:3));
labels={Vars{7:20}};
yticks([1 2 3 4 5 6 7 8 9 10 11 12 13 14]);
yticklabels(labels);
xticks([1 2 3]);
xticklabels({'PC#1', 'PC#2', 'PC#3'});
colorbar;
```

Figure 1.14 Code for generation of pc-squared plot for each PC, where the sum of coefficients adds up to unity, or 100%.

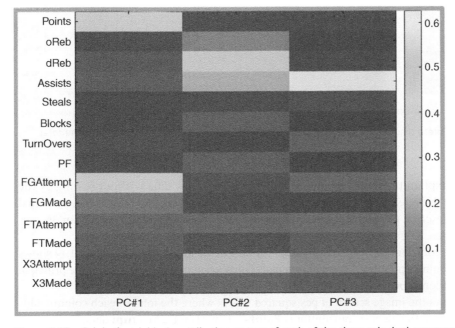

Figure 1.15 Original variables' contribution as part of each of the three principal components: PC#1→PC#3.

k-Means Clustering

The **kmeans** function is used to perform clustering

>>idx=kmeans(X,k)

Where **idx** is the cluster indices, and is returned as a numeric column vector, **X** is the numeric data matrix, and **k** is the number of clusters. There are also several options inputs and outputs that we can supply to tune up the process of clustering.

Distance Metrics: Locations of Cluster Centroids

By default, the Euclidean distance is used to access the similarity between two observations. We can use other metrics such as correlation

>>g = kmeans(X, 2, 'Distance', 'correlation');

Let us say that we have two clusters adjacent to one another and each cluster (a scatter distribution of data) will practically have its centroid and the radius of the circle/or the dimension of a square that will enclose a substantial portion of the distributed data. As a result of close neighborhood, some data from the large-size cluster can be treated as a subset of the neighboring small-size cluster, as shown in Figure 1.16.

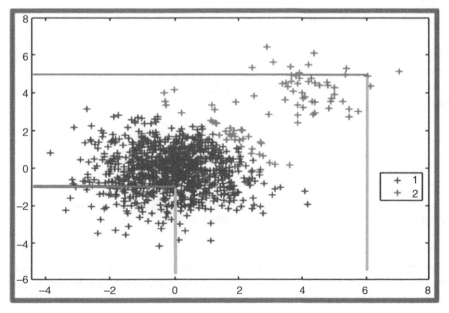

Figure 1.16 Two neighboring clusters affect the inclusion/exclusion of the interfacial data.

Thus, specifying the starting centroids of the two clusters will help improve the results. For example, in Figure 1.16, the starting centroids can be [0 −1] and [6 5]. Doing so, will mitigate the situation by minimizing the overlap between the two zones

>> **g = kmeans(X,2, 'Start', [0 –1; 6 5]);**

Moreover, the following code will return the **k** cluster centroid locations in the **k × p** matrix **C. p** is the number of predictor variables.

opts = statset('Display','final');

[idx,C] = kmeans(X,2,'Distance','cityblock','Replicates',5,'Options',opts);

Replications

Another way to get optimum clustering is to perform the analysis multiple times with different starting centroids and then choose the clustering scheme which will minimize the sum of distances between the observations and their centroids. This can be achieved by using the '**Replicates**' property. Figure 1.17 shows the code for using **kmeans** command which repeats the clustering attempts five times and returns the clusters with the lowest sum of distances.

```
data = readtable('basketball.xlsx');
temp=data{:,7:end};
%%Normalization by division by ./data.GP
normtemp=temp./data.GP;
XNorm = zscore(normtemp);
% Reconstruct coordinates
[pcs,scrs] = pca(XNorm);
% Group data using k-means clustering
try
 grp = kmeans(XNorm,2,"Start","cluster","Replicates",5);
 catch exception
   disp('There was an error fitting the kmeans model')
   error = exception.message
end
figure(18)
scatter3(scrs(:,1),scrs(:,2),scrs(:,3),10,grp)
view(110,40)
```

Figure 1.17 The code for creating kmeans-based two clusters with five cycles to minimize the sum of distances among datapoints and their centroid, within each cluster.

Figure 1.18 shows the scatter plot of the transformed (**scrs**) data from the normalized **X** data. The data are represented using **scrs** matrix which expresses the transformed data using the linear coordinate transformation matrix, **_pcs_**.

Figure 1.19 shows the workspace variables associated with Figure 1.17 code. Notice that **grp** column vector is 1112 × 1 double matrix where it contains either a value of *1* or *2* for each entry, indicating that **kmeans** function classifies the datapoints into two groups or clusters.

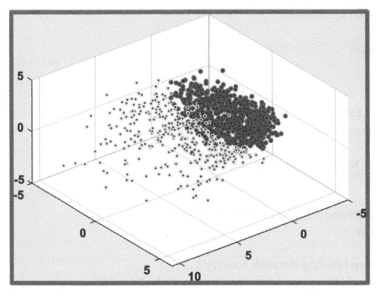

Figure 1.18   **kmeans** function which creates two clusters (*k*=2), labeled by yellow and blue color, with five trials to minimize the sum of distances.

Workspace			
Name ▲	Value	Size	Class
data	*1112x20 table*	1112x20	table
grp	*1112x1 double*	1112x1	double
normtemp	*1112x14 double*	1112x14	double
pcs	*14x14 double*	14x14	double
pexp	*14x1 double*	14x1	double
scrs	*1112x14 double*	1112x14	double
temp	*1112x14 double*	1112x14	double
XNorm	*1112x14 double*	1112x14	double

Figure 1.19   Workspace variables. The **grp** column vector divides X-data into 2 groups.

Gaussian Mixture Model (GMM) Clustering

We can use the function **fitgmdist** to fit several multidimensional gaussian (normal) distributions. The command

>>**gmm = fitgmdist(X,k);**

returns a Gaussian mixture distribution model with **k** components fitted to data (X). We can use the **fitgmdist** function to fit Gaussian mixture models to the data in *k* categories in what is called Gaussian Mixture Models (GMM) clustering.

Figure 1.20 shows the code for using GMM on **XNorm** to group the data into two sets in **grp** and find the corresponding probabilities used to determine the clusters, **gprob**. We set the number of 'Replicates' to five, and give it a 'RegularizationValue' of 0.02. Notice that **options = statset('MaxIter',400)** was used to elevate the maximum number of iterations from 100 (default) to 400. Of course, we can increase it as large as we wish. Moreover, **gmm** was inserted with a try

```
data = readtable('basketball.xlsx');
temp=data{:,7:end};
%%Normalization by division by ./data.GP
normtemp=temp./data.GP;
XNorm = zscore(normtemp);
% Reconstruct coordinates
[pcs,scrs] = pca(XNorm);
% Group data using GMM & return probabilities
options = statset('MaxIter',400);
rng(1); % Reset seed for common start values
try
 gmm=fitgmdist(XNorm,2,"Replicates",5,"RegularizationValue",0.02,
'Options',options);
 catch exception
    disp('There was an error fitting the Gaussian mixture model')
    error = exception.message
end
% grp has entries 1 or 2 group labeling for each X entry
% gprob has the probability of either 0 or 1 for each X entry to be found in one group
[grp,~,gprob] = cluster(gmm,XNorm);
%% View data and groups
figure(21)
scatter3(scrs(:,1),scrs(:,2),scrs(:,3),10,grp)
view(110,40)
```

Figure 1.20 Splitting **XNorm** data into two clusters (categories) in light of implementing multidimensional gaussian (normal) distributions.

catch interception procedure so that the user is warned about potential errors that might result in a divergence issue. Things to consider are additional parameters tailored to **gmm** statement.

Figure 1.21 shows that the entire **XNorm** was split into two clusters, dictated by GMM clustering approach to split the data into 2 clusters or groups.

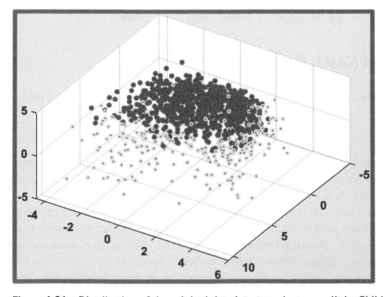

Figure 1.21 Distribution of the original data into two clusters, split by GMM approach.

Optimum Number of GMM Clusters

We can specify optional comma-separated pairs of **Name,Value** arguments. **Name** is the argument name and Value is the corresponding value. **Name** must appear inside quotes. We can specify several name and value pair arguments in any order. For example,

```
fitgmdist(XNorm,k,'Options',options,…
   'CovarianceType', 'diagonal', 'RegularizationValue',0.1);
```

Notice that the default **'CovarianceType'** for GMM is **'full'**.

Figure 1.22 shows the code to examine the AIC over varying numbers of clusters or components. Notice that AIC stands for Akaike's Information Criterion. According to Akaike's theory, the most accurate model has the smallest or minimum AIC. See MATLAB built-in help on **aic**, Akaike's Information Criterion for estimated model.

```
data = readtable('basketball.xlsx');
temp=data{:,7:end};
%%Normalization by division by ./data.GP
normtemp=temp./data.GP;
XNorm = zscore(normtemp);
% Reconstruct coordinates
[pcs,scrs] = pca(XNorm);
% Group data using GMM & return probabilities
options = statset('MaxIter',400);
AIC = zeros(1,4);
GMModels = cell(1,4);
rng(1); % Reset seed for common start values
try
for k = 1:4
    GMModels{k} = fitgmdist(XNorm,k,'Options',options,…
    'CovarianceType', 'diagonal', 'RegularizationValue',0.1);
    AIC(k)= GMModels{k}.AIC;
end
catch exception
    disp('There was an error fitting the Gaussian mixture model')
    error = exception.message
end
[minAIC,numComponents] = min(AIC);
numComponents
```

Figure 1.22 Finding the best cluster number which minimizes AIC.

The result of simulation shows that **numComponents** = **4** has the lowest **AIC** value.

Observations and Clusters Visualization

The following command will create a parallel coordinates plot of the multivariate data in the matrix **X**. We use a parallel coordinates plot to visualize high dimensional data, where each observation is represented by the sequence of its coordinate values plotted against their coordinate indices. Such a plot is particularly helpful when exploring multivariate datasets, as they allow us to quickly identify patterns and relationships between variables.

Some useful information that can be obtained from a parallel coordinates plot includes:

- *Correlation between variables*: Allows us to see how variables are related to one another. When two variables are highly correlated, they will appear to follow a similar pattern in the plot.
- *Outliers*: Outliers can be identified as lines that do not follow the general trend of the data. These points may be indicative of data errors or interesting observations.
- *Clusters*: Parallel coordinate plots can also reveal clusters of data points that share similar values across multiple variables. These clusters can provide insights into underlying patterns or relationships within the data.
- *Data distribution*: The density of lines in different regions of the plot can give an indication of how data is distributed across the variables. Dense areas may indicate regions where many data points have similar values, while sparse areas may indicate regions with greater variability.

>>**parallelcoords (X, 'Groups', g)**

Where **X** is the numeric data matrix, **'Group'** is the property name, and **g** is a vector containing the cluster identifiers.

Figure 1.23 shows the complete code with comments inserted in between.

Figure 1.24 shows the plot of all **X** data using **parallelcoords** function to visualize each datapoint (by the sequence of its coordinate values) lumped into three kmeans-classified groups for the 14 data columns. Column or predictor #5 has a noticeable outlier projection and to a lesser extent can be seen with predictor #2. It is worth-checking the reliability of the source for such raw data.

Figure 1.25 shows the plot of each centroid, stored in variable **C**, while specifying the group property value with a vector, 1:3.

In brief, the coordination value of the centroid represents the average value of each variable for that group. The plot of the coordination value of the centroid can be used to compare the average values of each variable for different groups. This can be helpful in identifying patterns and differences between groups. By comparing the coordination values of the centroids across different variables, we can identify which variables are most important in distinguishing between different groups. By examining the coordination values of the centroids over time or across different experimental conditions, we can identify trends and changes in the values of different variables. The coordination values of the centroid for group #1 (blue curve) and #3 (orange curve) behave in opposite fashion. If the plot of the coordination value of the centroid shows that centroids exhibit opposite behavior between two groups, it means that there is a significant difference in the average values of the variables between the two groups. This does not necessarily mean that there is a negative correlation between the two groups. The opposite behavior of centroids indicates that the two groups have different values and patterns across the variables being analyzed. This could mean that there is a positive or negative correlation between the variables within each group, or that the variables are independent of one another

```
data = readtable('basketball.xlsx');
temp=data{:,7:end};
%%Normalization by division by ./data.GP
normtemp=temp./data.GP;
XNorm = zscore(normtemp);
% Use kmeans function to group X into three groups. Name the output grp and save a
% second output, the centroids of each of 3 groups for each of 14 predictors, named C.
[grp,C] = kmeans(XNorm,3);
% Use the parallelcoords function to visualize the results.
figure (24)
parallelcoords(XNorm,"Group",grp);
% When X size is large, use parallelcoords to plot each centroid stored in variable C.
% Specify the group property value with a vector, 1:3.
figure (25)
parallelcoords(C,'Group',1:3);
% Another option is to plot the median values for each group with bounding lines. In
% addition, to specifying the Group property, you can use the Quantile property with
% a % value in the range (0,1).
% Use parallelcoords to create a bounding plot on the data with a quantile value of 0.25.
figure (26)
parallelcoords (XNorm, 'Group', grp, "Quantile", 0.25);
figure (27)
gscatter (XNorm(:,1),XNorm(:,2),grp,'br','x+')
data.pos = categorical(data.pos);
catag1=categories(data.pos);
count1 = crosstab(grp, data.pos);
figure (28)
bar(count1,'stacked');
legend(catag1);
```

Figure 1.23 The code for creating different parallelcoords, crosstab, and bar plots.

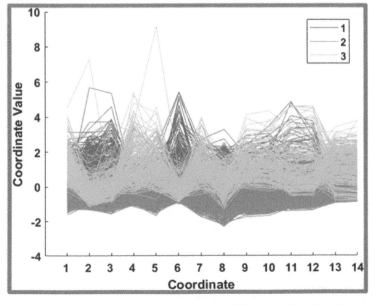

Figure 1.24 kmeans-classified three groups of **X** made of the 14 column predictors.

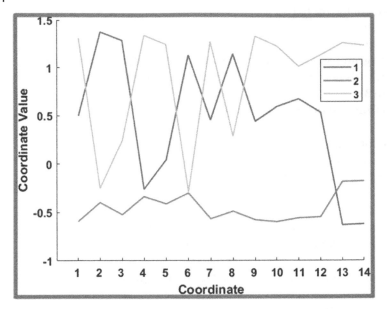

Figure 1.25 The plot of the coordination value of the centroid of each group over the 14 column predictors.

within each group. The nature of the correlation, if any, would depend on the specific variables being analyzed and the nature of the data. To determine the nature of the correlation between the two groups, additional statistical analyses would be needed. Correlation analysis, regression analysis, and other techniques could be used to explore the relationships among the variables and between the two groups.

Figure 1.26 shows the plot of the median values for each group with bounding lines, specifying the **Quantile** property with a value of 0.25 in the range [0,1]. The dotted line below the solid line shows the 25th percentile of measurements for each variable. The dotted line above the solid line shows the 75th percentile of measurements for each variable.

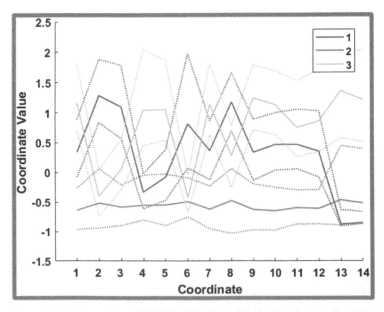

Figure 1.26 The plot of median values for each group with bounding 25th and 75th% lines.

Figure 1.27 (*left*) shows the **gscatter** plot of the first two columns of **XNorm** data, lumped by **grp**. One can roughly say that the predictor data patterns for the first two columns can be divided or allocated into three groups or clusters.

Figure 1.27 (*right*) differentiates the three groups by the category **data.pos** which is further divided into seven sub-categories. The contribution of each of the sub-categories (C up to G) to each of the three groups can be evidenced.

See **crosstab** built-in help. For example, you can use **crosstab** to tell whether or not smoking habit has to deal with the gender of smoker. See the **hospital** data (load hospital) in this regard.

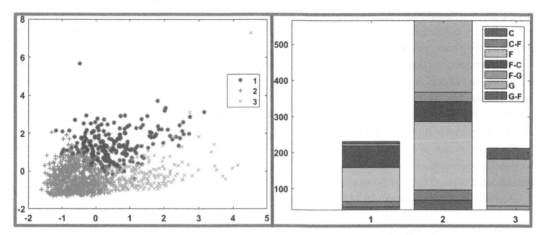

Figure 1.27   Scatter plot for the first two columns lumped by three groups (*left*) and plot of the three groups differentiated by **data.pos** subcategories (*right*).

Figure 1.28 shows the workspace variables where the user can have a look at the dimensions of the variables created by the written code of Figure 1.23.

Name ▲	Value	Size	Class	Min	Max
C	3x14 double	3x14	double	-0.6367	1.3796
catag1	7x1 cell	7x1	cell		
count1	3x7 double	3x7	double	0	234
data	1112x20 table	1112x20	table		
grp	1112x1 double	1112x1	double	1	3
normtemp	1112x14 double	1112x14	double	0	30.2242
temp	1112x14 double	1112x14	double	0	29484
XNorm	1112x14 double	1112x14	double	-2.3093	9.1741

Figure 1.28   The workspace variables created by code of Figure 1.23.

Evaluating Cluster Quality

When using clustering techniques such as k-means and Gaussian mixture model, we have to specify the number of clusters. However, for high dimensional data, it is difficult to determine the optimum number of clusters. We can use the **silhouette** value and plots to judge the quality of the clusters. An observation's silhouette value is a normalized (between −1 and +1) measure of how close that observation is to others in the same cluster, compared to the observations in different clusters.

Silhouette Plots

A silhouette plot shows the silhouette value of each observation, grouped by cluster. Clustering schemes in which most of the observations have high silhouette value are desirable. A silhouette plot is a graphical representation of the silhouette value of each observation in a clustering analysis, grouped by cluster. The silhouette value is a measure of how similar an observation is to its own cluster compared to other clusters. Clustering schemes that result in high silhouette values are generally considered to be more desirable, as they indicate that the clustering is well-defined and the groups are clearly separated.

The silhouette value for each observation can range from −1 to +1. A silhouette value of +1 indicates that the observation is well-matched to its own cluster and poorly matched to other clusters. This is desirable, as it suggests that the observation is clearly assigned to the correct cluster. A silhouette value of 0 indicates that the observation is equally similar to two or more clusters. This is less desirable, as it suggests that the observation could potentially belong to more than one cluster. A silhouette value of −1 indicates that the observation is poorly matched to its own cluster and well-matched to other clusters. This is highly undesirable, as it suggests that the observation may have been assigned to the wrong cluster.

Use the silhouette function in MATLAB to create the silhouette plots.

```
>> [grp,c] = kmeans(X,2);
```

```
>> silhouette(X,grp)
```

The **silhouette** plot will produce, in general, a distribution curve for each group, around the zero reference line. The minimum value is −1 and maximum is +1. Instead of manually experimenting with silhouette plots with different number of clusters, we can automate the process with **evalclusters** function.

The following function call creates 2, 3, 4, and 5 clusters using the k-means clustering, and calculates the silhouette value for each clustering scheme. Any **evalclusters** call can be used.

```
>> clustev = evalclusters(X,'kmeans','silhouette','KList',2:5);
```

```
>> clustev = evalclusters(X,'linkage','silhouette','KList',2:5);
```

```
>> clustev = evalclusters(X,'gmdistribution','silhouette','KList',2:4);
```

The output variable, **clustev**, contains detailed information about the evaluation including the optimum number of clusters.

```
>> kbest = clustev.OptimalK;
```

In place of 'silhouette', we can use other evaluation criteria such as 'CalinskiHarabasz', 'DaviesBouldin', and 'gap'. Refer to the **evalclusters** built-in documentation for further details.

Figure 1.29 shows the code for using automatic silhouette inspection for finding the best number of clusters. If you run the code, you will find out that **kbest** has a value of 2.

```
data = readtable('basketball.xlsx');
temp=data{:,7:end};
%%Normalization by division by ./data.GP
normtemp=temp./data.GP;
XNorm = zscore(normtemp);
options = statset('MaxIter',400);
rng(1); % Reset seed for common start values
try
clustev = evalclusters(XNorm,'kmeans','silhouette','KList',2:5);
%clustev = evalclusters(XNorm,'linkage','silhouette','KList',2:5);
%clustev = evalclusters(XNorm,'gmdistribution','silhouette','KList',2:4);
%clustev = evalclusters(XNorm,'gmdistribution','DaviesBouldin','KList',2:4);
catch exception
    disp('Oops! Something wrong went on!')
    error = exception.message
end
kbest = clustev.OptimalK % k=2.
```

Figure 1.29 Automatic estimation of the best number of clusters using one of classification models.

Copied from Command window: kbest = 2.

Quiz
Which of the following functions will plot each observation as a line against the variable number?
a) bar   b) crosstab   c) kmeans   d) parallelcoords   e) silhouette

Hierarchical Clustering

With hierarchical clustering, we can explore the *subclusters* that were grouped together to form bigger clusters.

Step 1 – Determine Hierarchical Structure

The linkage function calculates the distance between each pair of points and, using these distances, determines the tree hierarchy by linking together pairs of "neighboring" points:

>> Z = linkage(X,'single','euclidean');

Where the optional second and third input argument represent the method of finding the distance between the clusters (default: **'single'**) and to specify the distance metric (default: **'euclidean'**). We can use the dendrogram function to visualize the hierarchy.

For the *second input argument*, one of the following methods can be used, as given in Table 1.5.

Table 1.5 Method of finding the distance between clusters.

Method	Description
'single'	shortest distance (*default*)
'average'	unweighted average distance
'centroid'	centroid distance, appropriate for Euclidean distances only
'complete'	farthest distance
'median'	weighted center of mass distance, appropriate; Euclidean distances only
'ward'	inner squared distance (minimum variance algorithm); Euclidean distances only
'weighted'	weighted average distance

For the *third input argument*, one of the following values can be used, as given in Table 1.6.

Table 1.6 Distance metric, specified as the comma-separated pair consisting of '**Distance**' and a character vector, string scalar, or function handle.

Value	Description
'euclidean'	Euclidean distance (*default*)
'seuclidean'	Standardized Euclidean distance. Each coordinate difference between observations is scaled by dividing by the corresponding element of the standard deviation computed from X. Use the Scale name-value pair argument to specify a different scaling factor.
'mahalanobis'	Mahalanobis distance using the sample covariance of X, C = cov(X,'omitrows'). Use the Cov name-value pair argument to specify a different covariance matrix.
'cityblock'	City block distance.
'minkowski'	Minkowski distance. The default exponent is 2. Use the P name-value pair argument to specify a different exponent, where P is a positive scalar value.
'chebychev'	Chebychev distance (maximum coordinate difference).
'cosine'	One minus the cosine of the included angle between observations (treated as vectors).
'correlation'	One minus the sample correlation between observations (treated as sequences of values).
'hamming'	Hamming distance, which is the percentage of coordinates that differ.
'jaccard'	One minus the Jaccard coefficient, which is the percentage of nonzero coordinates that differ.
'spearman'	One minus the sample Spearman's rank correlation between observations (treated as sequences of values).

@distfun	Custom distance function handle. A distance function has the form:
	 function D2 = **distfun**(ZI,ZJ)
	 ageDifference = abs(ZI(1)-ZJ(:,1));
	 weightDifference = abs(ZI(2)-ZJ(:,2));
	 D2 = ageDifference + 0.2*weightDifference;
	 end
	 ZI is a 1 × n vector containing a single observation.
	 ZJ is an m2 × n matrix containing multiple observations. **distfun** must accept a matrix ZJ with an arbitrary number of observations.
	 D2 is an m2 × 1 vector of distances, and D2(k) is the distance between observations ZI and ZJ(k,:). If your data is not sparse, you can generally compute distance more quickly by using a built-in distance instead of a function handle.

Step 2 – Divide Hierarchical Tree into Clusters

We can use the cluster function to assign observations into groups, according to the linkage distances **Z**.

```
>> grp = cluster(Z,'maxclust',3)
```

Figure 1.30 shows the complete code for importing data, selecting a category, taking a subset out of the entire data, finding the distance between each pair of points, building the tree structure, clustering observations into groups according to the linkage distances Z, and finding the cophenetic correlation coefficient which quantifies how accurately the tree represents the distances between observations. In other words, the cophenetic correlation coefficient is a measure of how faithfully a dendrogram (tree diagram) represents the pairwise distances between the data points. It is computed as the correlation between the pairwise distances among the data points and the corresponding cophenetic distances, which are the distances between the clusters that result from cutting the dendrogram at a certain level.

Thus, the cophenetic correlation coefficient measures how well the dendrogram captures the true distances between the data points. A higher value indicates that the dendrogram is a better representation of the pairwise distances among the data points.

NOTE

You may redefine **guardstats** matrix with different constraints, or even remove the constraint in the first place. For example, if I remove the constraint that **idx=data.pos=='G'**, then the cophenetic correlation coefficient, c, will be 0.7071 compared with c=0.5737 for holding the constraint. For enforcing **idx=data.height>=80**, c=0.8034.

```
%% Import & initialize data
data = readtable('basketball.xlsx');
data.pos = categorical(data.pos);
% Get guard positions
%idx = data.pos == 'G';
idx=data.height>=80;
% Get numeric columns and normalize them
stats = data{:,[ 5 6 11:end ]};
labels = data.Properties.VariableNames([ 5 6 11:end]);
guardstats = zscore(stats(idx,:));
%guardstats = zscore(stats);
%Cluster groups using hierarchical clustering techniques
% The 'ward' method computes the inner squared distance using Ward's minimum
% variance algorithm. Z is a tree structure.
Z = linkage(guardstats,"ward");
% Keep figure 31 active then apply dendrogram to Z.
figure (31)
dendrogram(Z);
```

Figure 1.30 The code for creating a dendrogram, clustering observations into groups, and finding the cophenetic correlation coefficient.

```
% Use the cluster function to assign observations into groups according to the linkage
% distances Z.
gc2 = cluster(Z,'maxclust',2);
gc3 = cluster(Z,'maxclust',3);
%% Visualize the clusters using parallelcoords
figure(32)
parallelcoords(guardstats,'Group',gc2,'Quantile',0.25)
xticklabels(labels);
figure(33)
parallelcoords(guardstats,'Group',gc3,'Quantile',0.25)
xticklabels(labels);
% Use pdist to compute the pairwise distance of the original data, X.
Y=pdist(guardstats);
%The Cophenetic correlation coefficient quantifies how accurately the tree
% represents the distances between observations. Values close to 1 indicate a high-
% quality solution.
c=cophenet(Z,Y);
```

Figure 1.30 (Cont'd)

Figure 1.31 shows the dendrogram, **Z**, for plotting the data (*left*) and the coordinate value plot of the data made of 12 predictor columns and grouped by **gc2** (*right*).

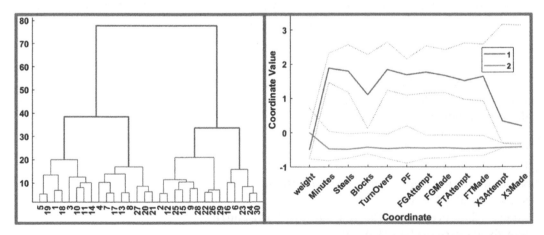

Figure 1.31 The dendrogram, **Z**, for plotting the data (*left*) and the coordinate plot of the data made of 12 predictor columns and grouped by **gc2** (*right*).

Figure 1.32 shows the coordinate value plot of the data made of 12 predictor columns and grouped by **gc3** (*left*) and the workspace variables (*right*), showing the cophenetic correlation coefficient, **c.**

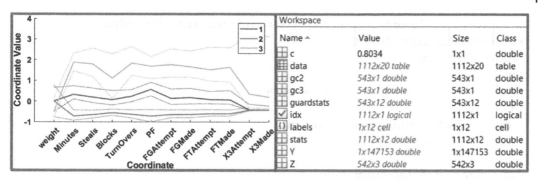

Figure 1.32 The coordinate value plot of the data made of 12 predictor columns and grouped by **gc3** (*left*) and the workspace variables (*right*).

PCA and Clustering: Wine Quality

I will start with the wine example so that the user gets drunk and forgets about machine learning. Figure 1.33 shows the entire code for carrying pca, clustering, and plotting the clustered data.

```
load('winered.mat')
WineRed.quality = categorical(WineRed.quality);
% Extract numeric data
numData = WineRed{:,1:end-1};
%% Perform PCA to transform numeric data. And save to scrs.
[pcs,scrs,~,~,explained]=pca(numData);
figure (34)
pareto(explained);
pcssqrd=pcs.^2;
Vars=WineRed.Properties.VariableNames;
figure(35)
pcs100=pcssqrd*100.0;
imagesc(pcs100(:,1:2));
labels={Vars{1:end-1}};
%yticks([1 2 3 4 5 6 7 8 9 10 11]);
yticklabels(labels);
xticks([1 2]);
xticklabels({'PC#1', 'PC#2'});
colorbar;
%% Cluster into two groups using k-means and/or GMM.
try
g = kmeans(numData,2,'Replicates',5);
catch exception
   disp('Oops! Something wrong went on!')
   error = exception.message
end
figure(36)
```

Figure 1.33 The code for transforming red wine data, showing important **X** variables, clustering them into groups, and presenting representative plots.

```
% scatter plot, colored by g. g, an n×1 vector, will serve as a color matrix.
scatter(scrs(:,1),scrs(:,2),24,g)
% gmm clustering
try
gmmModel = fitgmdist(numData,2,'Replicates',5);
catch exception
   disp('Oops! Something wrong went on!')
   error = exception.message
end
g2= cluster(gmmModel,numData);
figure (37)
% scatter plot by group. The data x,y will be grouped by g2 which is an n×1 vector.
gscatter(scrs(:,1),scrs(:,2),g2, 'br','*d');
%% Cross tabulate grouping and wine quality. Create a stacked bar chart where
% groups 1 and 2 are along the x-axis and wine quality data are represented by %dif-
ferent colors.
counts = crosstab(g,WineRed.quality);
figure (38);
bar(counts,'stacked')
legend(categories(WineRed.quality));
```

Figure 1.33 (Cont'd)

Figure 1.34 (*left*) shows the Pareto plot for principal components where almost 95% of total variance can be explained by the first principal component and Figure 1.34 (*right*) shows the image map for the contribution of individual predictor variables to both PC#1 and PC#2.

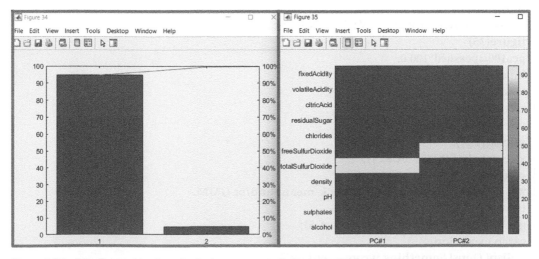

Figure 1.34 The Pareto plot for principal components (*left*) and individual predictor variable contribution to PCs, without predictors normalization (*right*).

NOTE

It is worth mentioning here that without carrying out *normalization* or *scaling* of the original data, then the two columns named "**freeSulfurDioxide**" and "**totalSulfurDioxide**" will have the highest impact on both principal components PC#1 and PC#2. This has to deal with the fact that both have the highest magnitude of numerical values among all columns. On the other hand, if we carry out data normalization or scaling prior to any data treatment step, then the role of each player (predictor) will change. See subsequent figures how the order of influence of predictors changes with the type of initial data being handled.

Figure 1.35 (*left*) shows a plot for the first versus the second column of transformed data (**scrs**) and are colored by the two **kmeans**-based clusters and of the same by **GMM**-based clusters (*right*). Notice that **kmeans**-based are split better than **GMM**-based clusters.

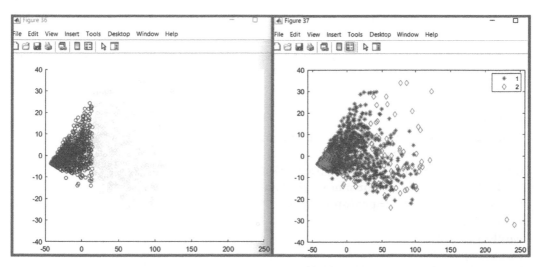

Figure 1.35   Scatter plots for the first two columns of transformed data by **kmeans**-based cluster (*left*) and by **GMM**-based clusters (*right*).

Figure 1.36 (left) shows the stacked bar of the two **kmeans**-based clusters, proportioned by the wine quality (different colored straps) and the workspace variables (right).

Figure 1.37 shows the code to generate the Pareto plot and % pcs-squared for the contribution of predictors but this time **numData** matrix is **normalized**.

Figure 1.38 shows the pareto plot (*left*) and the % pcs-squared for the contribution of predictors (*right*) but this time **numData** matrix is *normalized* prior to carrying out any data analysis. Notice that, however, we need more than two principal components to explain variations in the predictors' data and the contribution magnitude for each predictor is less than that shown earlier in Figure 1.34, without data normalization.

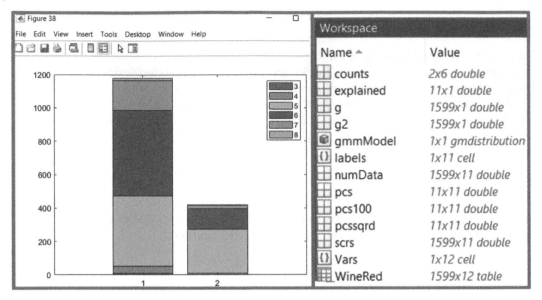

Figure 1.36 (*left*) The stacked bar plot for cross tabulation of the two **kmeans**-based clusters proportioned by the wine quality and (*right*) the workspace variables.

```
load('winered.mat')
WineRed.quality = categorical(WineRed.quality);
% Extract numeric data
numData = normalize(WineRed{:,1:end-1});
%% Perform PCA to transform numeric data. And save to scrs.
[pcs,scrs,~,~,explained]=pca(numData);
figure (38)
pareto(explained);
pcssqrd=pcs.^2;
Vars=WineRed.Properties.VariableNames;
figure(39)
pcs100=pcssqrd*100.0;
imagesc(pcs100(:,1:4));
labels={Vars{1:end-1}};
%yticks([1 2 3 4 5 6 7 8 9 10 11]);
yticklabels(labels);
xticks([1 2 3 4]);
xticklabels({'PC#1', 'PC#2','PC#3', 'PC#4'});
colorbar;
```

Figure 1.37 The code for handling the **normalization** of the original data prior to any data analysis.

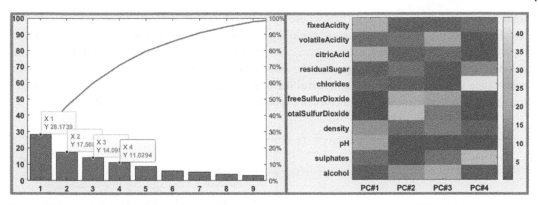

Figure 1.38 The Pareto plot (left) and the % pcs-squared predictors' contribution to PCs (*right*), for **normalized** data.

Figure 1.39 shows the code to generate the Pareto plot and % pcs-squared for the contribution of predictors but this time **numData** matrix is **scaled** between 0 and 1 for each predictor variable.

```
load('winered.mat')
WineRed.quality = categorical(WineRed.quality);
% Extract numeric data
origData = WineRed{:,1:end-1};
MaxVal=max(origData,[], 1);
MinVal=min(origData,[], 1);
for j=1:size(origData,2)
numData(:,j)=(origData(:,j)-MinVal(j))/(MaxVal(j)-MinVal(j));
end
%% Perform PCA to transform numeric data and save to scrs.
[pcs,scrs,~,~,explained]=pca(numData);
%[pcs,scrs,~,~,explained]=pca(numData,'Algorithm','als',...
%'Centered',false,'NumComponents',4);
figure (40)
pareto(explained);
pcssqrd=pcs.^2;
Vars=WineRed.Properties.VariableNames;
figure(41)
pcs100=pcssqrd*100.0;
imagesc(pcs100(:,1:4));
labels={Vars{1:end-1}};
%yticks([1 2 3 4 5 6 7 8 9 10 11]);
yticklabels(labels);
xticks([1 2 3 4]);
xticklabels({'PC#1', 'PC#2', 'PC#3','PC#4'});
colorbar;
```

Figure 1.39 Code for scaling [0→1] the original data and plotting Pareto and image screen plot.

Figure 1.40 (*left*) shows the Pareto plot where more than two principal components are needed to describe the variances in the predictors' data. Figure 1.40 (*right*) shows how the impact and order of predictors are changed from previous two cases. **PC#1** signifies the importance '**citricAcid**' and '**fixedAcidity**' variable. **PC#2** signifies the importance '**alcohol**' variable. **PC#3** signifies the importance '**freeSulfurDioxide**' and '**totalSulfurDioxide**' variable. This is a kind of prioritization of the importance of such predictors in explaining the variance in the original data.

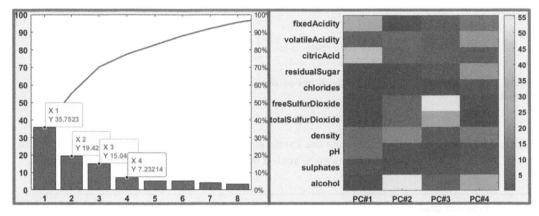

Figure 1.40 The Pareto plot (*left*) and the % pcs-squared predictors' contribution to PCs (*right*), for scaled data.

Figure 1.41 shows the insertion of four name-value pairs to **pca** statement. Please, refer to Tables 1.2 up to 1.4. The code in Figure 1.39 is used here, except for the following pca statement:

```
[pcs,scrs,~,~,explained]=pca(numData,'Algorithm','svd',...
'Centered',false,'Rows','complete','NumComponents',4);
```

Figure 1.41 pca statement where the insertion, of four name-value pairs, is made.

Figure 1.42 (left) shows the Pareto plot where four principal components are needed to describe the variances in the predictors' data. Figure 1.42 (*right*) shows how the impact and order of predictors for each of the four PC#'s.

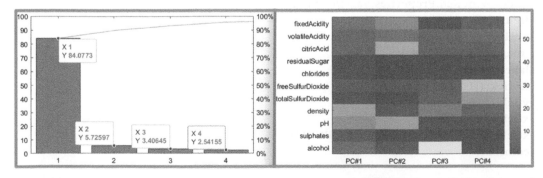

Figure 1.42 The insertion of four name-value pairs to pca statement to generate Pareto and %pcs-squared predictors plot.

Table 1.7 shows a summary of the attempted pca algorithms and whether or not the data is centered. Please, refer to Tables 1.2 up to 1.4 for more information on **pca** name-value pairs.

Table 1.7 Results of red wine using PCA with different algorithms with and without predictor column data centering.

Algorithm	'Centered' Value	PC#'s to Describe 95% Variance	Most Significant Predictor Columns (pcs100)
'als'	true	PC#1 (46%)	PC #1: citricAcid (44%) and fixedAcid (26%)
		PC#2 (25%)	PC#2: alcohol (55%) and density (17%)
		PC#3 (19%)	PC#3: freeSO2 (56%) and totSO2 (25%)
		PC#4 (9.4%)	PC#4: alcohol (25%) and residSugar (20%)
'als'	false	PC#1 (88%)	PC #1: density (26%) and pH (21%)
		PC#2 (6%)	PC#2: citricAcid (36%) and pH (25%)
		PC#3 (3.6%)	PC#3: alcohol (60%) and density (15%)
'eig'	true	PC#1 (35.8%)	PC #1: citricAcid (44%) and fixedAcid (26%)
		PC#2 (19.4%)	PC#2: alcohol (55%) and density (17%)
		PC#3 (15%)	PC#3: freeSO2 (56%) and totSO2 (25%)
		PC#4 (7.2%)	PC#4: alcohol (25%) and residSugar (20%)
		PC#5 (5.3%)	
		PC#6 (5.2%)	
		PC#7 (4.0%)	
		PC#8 (3.5%)	
'eig'	false	PC#1 (84%)	PC #1: density (26%) and pH (20.6%)
		PC#2 (5.7%)	PC#2: citric acid (36%) and pH (25%)
		PC#3 (3.4%)	PC#3: alcohol (60%) and density (15%)
		PC#4 (2.5%)	PC#4: freeSO2 (50%) and totSO2 (23.5%)
'svd'	true	PC#1 (35.8%)	PC #1: citricAcid (44%) and fixedAcid (26%)
		PC#2 (19.4%)	PC#2: alcohol (55%) and density (17%)
		PC#3 (15%)	PC#3: freeSO2 (56%) and totSO2 (25%)
		PC#4 (7.2%)	PC#4: alcohol (25%) and residSugar (20%)
		PC#5 (5.3%)	
		PC#6 (5.2%)	
		PC#7 (4.0%)	
		PC#8 (3.5%)	
'svd'	false	PC#1 (84%)	PC #1: density (26%) and pH (20.6%)
		PC#2 (5.7%)	PC#2: citric acid (36%) and pH (25%)
		PC#3 (3.4%)	PC#3: alcohol (60%) and density (15%)
		PC#4 (2.5%)	PC#4: freeSO2 (50%) and totSO2 (23.5%)

The following points can be concluded:

1) Alternating least squares **'als'** algorithm with false **'Centered'** value gave the lowest number of PC#'s needed to describe the variances in the predictors' data.
2) To a large extent and regardless of the applied algorithm, false **'Centered'** value cases gave the same information on the variances of the predictors' data as well as the number of PC#'s needed to describe such a variance.

3) To a large extent and regardless of the applied algorithm, true **'Centered'** value cases gave the same information on the variances of the predictors' data; nevertheless, the number of PC#'s needed to describe such a variance is the least for alternating least squares **'als'** algorithm case.

4) If we need to describe variance in terms of minimum number of PC#'s needed and in terms of red wine features, one would select the second case (Table 1.7), that is, alternating least squares **'als'** algorithm with false **'Centered'** value.

5) Figure 1.43 shows pcs100 (pcs100=pcssqrd×100.0) values for the case of alternating least squares **'als'** algorithm with false **'Centered'** value. The explained value for **PC#1** is 88%; for **PC#2** 6%; and for **PC#3** 3.6%. See Table 1.7 case #2.

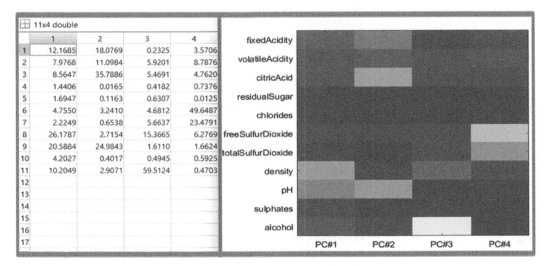

Figure 1.43 pcs100 contribution values (*left*) within each principal component (PC) for red wine features (*right*).

The most influential features are density and pH. Thus, the percentage variance contribution of the density and pH can be estimated by the following formula:

$$\sum_{i=1}^{PC\#}\left(\frac{PC\#\,wt\%}{100\%}\right)_i \times pcs100_{feature}$$

For density:

$$\sum_{i=1}^{PC\#}\left(\frac{PC\#\,wt\%}{100\%}\right)_i \times pcs100_{density} = \frac{88}{100}\times26.2\% + \frac{6}{100}\times2.7\% + \frac{3.6}{100}15.4\% = 23.77\%$$

For pH:

$$\sum_{i=1}^{PC\#}\left(\frac{PC\#\,wt\%}{100\%}\right)_i \times pcs100_{pH} = \frac{88}{100}\times20.6\% + \frac{6}{100}25\% + \frac{3.6}{100}1.6\% = 19.68\%$$

With the 2 out of 11 red wine features, density and pH, one could explain about 23.8%+19.7% = 43.5% of the total variance in predictors' data.

Feature Selection Using Laplacian (fsulaplacian) for Unsupervised Learning

The command

>>[idx,scores] = fsulaplacian(X, Name, Value)

will return the feature **scores**. A large score value indicates that the corresponding feature is important.

Distance metric, specified as the comma-separated pair consisting of '**Distance**' and a character vector, string scalar, or function handle, as described earlier in Table 1.6.

When we use the '**seuclidean**', '**minkowski**', or '**mahalanobis**' distance metric, we can specify the additional name-value pair argument '**Scale**', '**P**', or '**Cov**', respectively, to control the distance metrics.

Example 1.1 'Distance','minkowski','P',3
specifies to use the Minkowski distance metric with an exponent of 3.

Example 1.2 Covariance matrix for the **Mahalanobis** distance metric, specified as the comma-separated pair consisting of '**Cov**' and a positive definite matrix.
This argument is valid only if '**Distance**' is '**mahalanobis**'.

'Cov',eye(4)

Example 1.3 Scale factor for the kernel, specified as the comma-separated pair consisting of '**KernelScale**' and '**auto**' or a positive scalar. The software uses the scale factor to transform distances to similarity measures.

The '**auto**' option is supported only for the '**euclidean**' and '**seuclidean**' distance metrics.

'NumNeighbors',10,'KernelScale','auto'

specifies the number of nearest neighbors as *10* and the kernel scale factor as '*auto*'.

Figure 1.44 shows the code of using Laplacian score **fsulaplacian** function to rank features for unsupervised learning of red wine data.
Figure 1.45 shows **fsulaplacian** feature selection for the case '**Distance**', '**spearman**', where the most five important features are shown. Table 1.8 shows different scenarios of '**Distance**' value, using **fsulaplacian** feature selection. See Table 1.6 for '**Distance**' description.

The conclusion that can be drawn here is:

The order of features is '**Distance**'-specific; however, there is a common order among five different cases, that is, [3 1 11] which corresponds to {'citricAcid'}{'fixedAcidity'}{'alcohol'} features (Table 1.8). For other remaining cases, the order is shuffled for the first five important features.

```
load('winered.mat')
WineRed.quality = categorical(WineRed.quality);
% Extract numeric data
origData = WineRed{:,1:end-1};
MaxVal=max(origData,[], 1);
MinVal=min(origData,[], 1);
for j=1:size(origData,2)
numData(:,j)=(origData(:,j)-MinVal(j))/(MaxVal(j)-MinVal(j));
end
%[idx,scores] = fsulaplacian(numData);
%[idx,scores] = fsulaplacian(numData,'NumNeighbors',10,'KernelScale','auto');
[idx,scores] = fsulaplacian(numData,'NumNeighbors',12,...
'Distance', 'spearman');
Vars=WineRed.Properties.VariableNames;
labels={Vars{1:end-1}};
bar(scores(idx))
xticklabels(labels(idx));
ylabel('Feature importance score')
%Select the top five most important features. Find the columns of these features in X.
idx(1:5)
top5=labels(idx(1:5))
```

Figure 1.44　Feature selection using **fsulaplacian** function classifying features in order of importance.

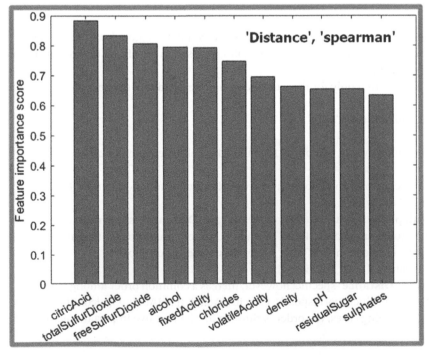

Figure 1.45　Feature selection order of the most important features of red wine using **fsulaplacian**, specifying '**Distance**' as '**spearman**'.

Table 1.8 Feature selection using **fsulaplacian** function for different '**Distance**' values.

Distance	The First Five Important Features
'**euclidean**'	{'citricAcid'}{'fixedAcidity'}{'alcohol'}{'density'}{'freeSulfurDioxide'} 3 1 11 8 6
'**seuclidean**'	{'totalSulfurDiox'}{'fixedAcidity'}{'citricAcid'}{'alcohol'}{'density'} 7 1 3 11 8
'**mahalanobis**'	{'residualSugar'}{'chlorides'}{'totalSulfurDiox'}{'citricAcid'}{'alcohol'} 4 5 7 3 11
'**cityblock**'	{'citricAcid'}{'fixedAcidity'}{'alcohol'}{'density'}{'totalSulfurDiox'} 3 1 11 8 7
'**minkowski**'	{'citricAcid'}{'fixedAcidity'}{'alcohol'}{'density'}{'freeSulfurDioxide'} 3 1 11 8 6
'**chebychev**'	{'citricAcid'}{'fixedAcidity'}{'alcohol'}{'freeSulfurDioxide'}{'density'} 3 1 11 6 8
'**cosine**'	{'citricAcid'}{'alcohol'}{'fixedAcidity'}{'freeSO2'}{'totalSO2'} 3 11 1 6 7
'**correlation**'	{'citricAcid'}{'alcohol'}{'fixedAcidity'}{'freeSO2'}{'totalSO2'} 3 11 1 6 7
'**hamming**'	{'citricAcid'} {'fixedAcidity'} {'alcohol'} {'pH'} {'density'} 3 1 11 9 8
'**jaccard**'	{'fixedAcidity'} {'citricAcid'} {'alcohol'} {'density'} {'pH'} 1 3 11 8 9
'**spearman**'	{'citricAcid'}{'totalSO2'} {'freeSO2'} {'alcohol'} {'fixedAcidity'} 3 7 6 11 1

CHW 1.1   The Iris Flower Features Data

Consider the Iris flower (**load fisheriris;**) MATLAB built-in database. Number of Instances: 150 (50 in each of three classes) Number of Attributes: 4 numeric, predictive attributes and the class Attribute Information:

1) sepal length in cm
2) sepal width in cm
3) petal length in cm
4) petal width in cm
5) class:
   - Iris Setosa
   - Iris Versicolour
   - Iris Virginica

Reference: https://archive.ics.uci.edu/ml/machine-learning-databases/iris/iris.data
Carry out the following tasks:
Scale the data between zero and one for each predictor column.

1) Using classical multidimensional scaling, in the first step, calculate pairwise distance among observations using the function **pdist**. In the second step, use the dissimilarity vector, **dist**, as an input to the function **cmdscale**, to find **eigen**, the Eigenvalues. In the third step, use the pareto function to visualize relative magnitudes of **eigen** vector in descending order. How many reconstructed coordinates in **p**-dimensional space can be used to describe at least 95% of variance of the original predictor variables?

2) Using Principal Component Analysis (PCA), carry out dimensionality reduction. Obtain **pcs**, p×p matrix of principal components, **scrs**, n×p matrix containing the transformed data using the linear coordinate transformation matrix, **pcs**, and **pexp**, p×1 column vector of the percentage of variance, explained by each principal component. Construct Pareto plot for **pexp** and make an image screen for **pcs100=pcssqrd*100.0**. Describe variance in terms of minimum number of PC#'s needed and in terms of Iris flower features.

3) Use k-means clustering with replications (perform the analysis multiple times with different starting centroids) property to get optimum clustering, which will minimize the sum of distances between the observations and their centroids.

4) Find the optimum number of GMM clusters, utilizing the AIC over varying numbers of clusters or components. Notice that AIC stands for Akaike's Information Criterion. According to Akaike's theory, the most accurate model has the smallest or minimum AIC.

5) Use silhouette plot for observed data, grouped by clusters, to find the best clustering schemes in which most of the observations have high silhouette value. Use **'kmeans'**, **'linkage'**, or **'gmdistribution'** in **evalclusters** function statement.

6) Using the **linkage** function to estimate the linkage distances, find **Z**, a tree structure. Make a dendrogram plot of **Z**, the tree cluster. Use **cluster** function to assign observations into groups, according to the linkage distances **Z**. Using **parallelcoords** function, create a parallel coordinates plot of the numeric matrix of predictors, **X**, to visualize the clusters. Use **pdist** to compute the pairwise distance of the original data, **X**. Find the cophenetic correlation coefficient which quantifies how accurately the **Z** tree cluster represents the distances between observations.

7) Using feature selection of Laplacian (**fsulaplacian**) function, find the feature scores for the best highest two features. Use different distance metric, specified as the comma-separated pair consisting of '**Distance**' and a character vector, string scalar, or function handle, as described in Table 1.6.

CHW 1.2   The Ionosphere Data Features

Consider the ionosphere (**load ionosphere;**) MATLAB built-in database. Relevant Information: This radar data was collected by a system in Goose Bay, Labrador. This system consists of a phased array of 16 high-frequency antennas with a total transmitted power on the order of 6.4 kilowatts. The targets were free electrons in the ionosphere. "Good" radar returns are those showing evidence of some type of structure in the ionosphere. "Bad" returns are those that do not; their signals pass through the ionosphere. Received signals were processed using an autocorrelation function whose arguments are the time of a pulse and the pulse number. There were 17 pulse numbers for the Goose Bay system. Instances in this database are described by 2 attributes per pulse number, corresponding to the complex values returned by the function resulting from the complex electromagnetic signal. Number of Instances: 351 Attributes Information

- All 34 predictor attributes are continuous.
- The 35th attribute is either "good" or "bad" according to the definition summarized above. This is a binary classification task.

Reference: https://archive.ics.uci.edu/ml/machine-learning-databases/ionosphere/ionosphere.data
Carry out the following tasks:
Scale the data between zero and one for each predictor column, except the second column. In fact, exclude column #2 as all entries are zero.

1) Using classical multidimensional scaling, in the first step, calculate pairwise distance among observations using the function **pdist**. In the second step, use the dissimilarity vector, **dist**, as an input to the function **cmdscale**, to find **eigen**, the Eigenvalues. In the third step, use the pareto function to visualize relative magnitudes of **eigen** vector in descending order.How many reconstructed coordinates in **p**-dimensional space can be used to describe at least 95% of variance of the original predictor variables?

2) Using Principal Component Analysis (PCA), carry out dimensionality reduction. Obtain **pcs**, p×p matrix of principal components, **scrs**, n×p matrix containing the transformed data using the linear coordinate transformation matrix, **pcs**, and **pexp**, p×1 column vector of the percentage of variance, explained by each principal component. Construct Pareto plot for **pexp** and make an image screen for **pcs100=pcssqrd*100.0**. Describe variance in terms of minimum number of PC#'s needed and in terms of ionosphere features.

3) Use k-means clustering with replications (perform the analysis multiple times with different starting centroids) property to get optimum clustering, which will minimize the sum of distances between the observations and their centroids.

4) Find the optimum number of GMM clusters, utilizing the AIC over varying numbers of clusters or components. You may have to increase the maximum number of iterations.

5) Use silhouette plot for observed data, grouped by clusters, to find the best clustering schemes in which most of the observations have high silhouette value. Use **'kmeans'**, **'linkage'**, or **'gmdistribution'** in **evalclusters** function statement.

6) Using the **linkage** function to estimate the linkage distances, find **Z**, a tree structure. Make a dendrogram plot of **Z**, the tree cluster. Use **cluster** function to assign observations into groups, according to the linkage distances **Z**. Using **parallelcoords** function, create a parallel coordinates plot of the numeric matrix of predictors, **X**, to visualize the clusters. Use **pdist** to compute the pairwise distance of the original data, **X**. Find the cophenetic correlation coefficient which quantifies how accurately the **Z** tree cluster represents the distances between observations.

7) Using feature selection of Laplacian (**fsulaplacian**) function, find the feature scores for the best highest two features. Use different distance metric, specified as the comma-separated pair consisting of **'Distance'** and a character vector, string scalar, or function handle, as described in Table 1.6.

CHW 1.3 The Small Car Data

Consider the small car database. We will use MATLAB built-in small car data (**load carsmall;**) as the source of input datasets. Select four columns, representing predictor variables which are given as X:

>>X = [Weight,Horsepower,Acceleration,Displacement];

Carry out the following tasks:
Scale the data between zero and one for each predictor column. Remove rows having any NaN value.

1) Using classical multidimensional scaling, in the first step, calculate pairwise distance among observations using the function **pdist**. In the second step, use the dissimilarity vector, **dist**, as an input to the function **cmdscale**, to find **eigen**, the Eigenvalues. In the third step, use the pareto function to visualize relative magnitudes of **eigen** vector in descending order.How many reconstructed coordinates in **p**-dimensional space can be used to describe at least 95% of variance of the original predictor variables?

2) Using Principal Component Analysis (PCA), carry out dimensionality reduction. Obtain **pcs**, p×p matrix of principal components, **scrs**, n×p matrix containing the transformed data using the linear coordinate transformation matrix, **pcs**, and **pexp**, p×1 column vector of the percentage of variance, explained by each principal component. Construct Pareto plot for **pexp** and make an image screen for **pcs100=pcssqrd*100.0**. Describe variance in terms of minimum number of PC#'s needed and in terms of small car features.

3) Use k-means clustering with replications (perform the analysis multiple times with different starting centroids) property to get optimum clustering, which will minimize the sum of distances between the observations and their centroids.

4) Find the optimum number of GMM clusters, utilizing the AIC over varying numbers of clusters or components. You may have to increase the maximum number of iterations.

5) Use silhouette plot for observed data, grouped by clusters, to find the best clustering schemes in which most of the observations have high silhouette value. Use **'kmeans'**, **'linkage'**, or **'gmdistribution'** in **evalclusters** function statement.

6) Using the **linkage** function to estimate the linkage distances, find **Z**, a tree structure. Make a dendrogram plot of **Z**, the tree cluster. Use **cluster** function to assign observations into groups, according to the linkage distances **Z**. Using **parallelcoords** function, create a parallel coordinates plot of the numeric matrix of predictors, **X**, to visualize the clusters. Use **pdist** to compute the pairwise distance of the original data, **X**. Find the cophenetic correlation coefficient which quantifies how accurately the **Z** tree cluster represents the distances between observations.

7) Using feature selection of Laplacian (**fsulaplacian**) function, find the feature scores for the best highest two features. Use different distance metric, specified as the comma-separated pair consisting of **'Distance'** and a character vector, string scalar, or function handle, as described in Table 1.6.

CHW 1.4   Seeds Features Data

The wheat seeds dataset (**load('SeedDataSet.mat');**) involves the prediction of species given measurements of seeds from different varieties of wheat. It is a multiclass (3-class) classification problem. The number of observations for each class is balanced. There are 210 observations with 7 input variables and 1 output variable. The variable names are as follows:

1) Area.
2) Perimeter.
3) Compactness.
4) Length of kernel.

5) Width of kernel.
6) Asymmetry coefficient.
7) Length of kernel groove.
8) Class (1, 2, 3).

Reference: https://archive.ics.uci.edu/ml/machine-learning-databases/00236/seeds_dataset.txt-
Carry out the following tasks:
Scale the data between zero and one, for each predictor column. Remove rows having any NaN value.

1) Using classical multidimensional scaling, in the first step, calculate pairwise distance among observations using the function **pdist**. In the second step, use the dissimilarity vector, **dist**, as an input to the function **cmdscale**, to find **eigen**, the Eigenvalues. In the third step, use the pareto function to visualize relative magnitudes of **eigen** vector in descending order. How many reconstructed coordinates in **p**-dimensional space can be used to describe at least 95% of variance of the original predictor variables?

2) Using Principal Component Analysis (PCA), carry out dimensionality reduction. Obtain **pcs**, p×p matrix of principal components, **scrs**, n×p matrix containing the transformed data using the linear coordinate transformation matrix, **pcs**, and **pexp**, p×1 column vector of the percentage of variance, explained by each principal component. Construct Pareto plot for **pexp** and make an image screen for **pcs100=pcssqrd*100.0**. Describe variance in terms of minimum number of PC#'s needed and in terms of seeds features.

3) Use k-means clustering with replications (perform the analysis multiple times with different starting centroids) property to get optimum clustering, which will minimize the sum of distances between the observations and their centroids.

4) Find the optimum number of GMM clusters, utilizing the AIC over varying numbers of clusters or components. You may have to increase the maximum number of iterations.

5) Use silhouette plot for observed data, grouped by clusters, to find the best clustering schemes in which most of the observations have high silhouette value. Use **'kmeans'**, **'linkage'**, or **'gmdistribution'** in **evalclusters** function statement.

6) Using the **linkage** function to estimate the linkage distances, find **Z**, a tree structure. Make a dendrogram plot of **Z**, the tree cluster. Use **cluster** function to assign observations into groups, according to the linkage distances **Z**. Using **parallelcoords** function, create a parallel coordinates plot of the numeric matrix of predictors, **X**, to visualize the clusters. Use **pdist** to compute the pairwise distance of the original data, **X**. Find the cophenetic correlation coefficient which quantifies how accurately the **Z** tree cluster represents the distances between observations.

7) Using feature selection of Laplacian (**fsulaplacian**) function, find the feature scores for the best highest two features. Use different distance metric, specified as the comma-separated pair consisting of '**Distance**' and a character vector, string scalar, or function handle, as described in Table 1.6.

2

ML Supervised Learning

Classification Models

In this chapter, we will handle the case where we have a set of predictor data on the one hand and a response variable on the other hand. However, the response variable will resonate between two or more values (i.e., categorical or discrete) rather than having continuous values. In the latter case, it will be dealt with under regression, not classification, models.

Our approach will consist of the following steps, as shown in Figure 2.1:

1) Data preparation
2) Selection of algorithm
3) Model fitting
4) Model evaluation
5) Model update
6) Make predictions

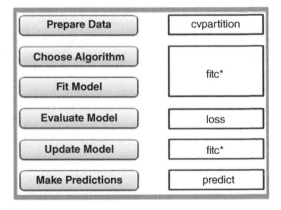

Figure 2.1 Typical MATLAB machine learning workflow.

Fitting Data Using Different Classification Models

We can use the function

fitc*

(replace * by the name of the classifier) to fit a classifier.

Machine and Deep Learning Using MATLAB: Algorithms and Tools for Scientists and Engineers, First Edition. Kamal I. M. Al-Malah.
Companion Website: www.wiley.com/go/al-malah/machinelearningmatlab

For example, the following syntax creates a k-nearest neighbor (k-NN) classification model.

>>**knnModel = fitcknn(tbl,'response')**

Where

knnModel: Variable containing information about the classification model
tbl: Training data, stored in a table or matrix
response: Name of the response variable stored in the input table.

Customizing a Model

Because of the prior knowledge we have about the data or after looking at the classification results, we may want to customize the classifier. We can update and customize the model by setting different options using the fitting functions.

Set the options by providing additional inputs for the option name and the option value.

>> **model = fitc*(tbl,'response','optionName', optionValue)**

How do we know if the model will generalize well to the data that it has never seen?

In machine learning, we can simulate this scenario by dividing the data into two sets that are used for training and testing. We can fit the model using the training data and calculate its accuracy using the test data.

Creating Training and Test Datasets

Suppose we want to divide the data into a training set containing 60% of the observations and a test set containing the remaining 40%. We can do this by indexing to get a subset of the data. However, MATLAB provides specialized function for partitioning the data. It ensures that the response classes are represented in approximately the same proportion in each partition.

We start using the function **cvpartition** on the vector containing the class information (response variable).

>> **c = cvpartition(y,'HoldOut',p)**

Where

c: A variable containing information about the partition.
y: A vector of categories corresponding to each observation.
'HoldOut': An optional property indicating that the partition should divide the observations into a training set and a test (or holdout) set.
p: Fraction of data that is held out of the complete data to form the test set (between 0 and 1).

Getting Indices for Training and Test Datasets

We can use the functions training and test to extract the logical vectors corresponding to the training and test datasets.

Use the function **training** to extract the logical vector corresponding to the training data.

```
>> c = cvpartition(y, 'HoldOut', 0.4);
```

```
>> trainingIdx = training(c);
```

Use the function **test** to extract the logical vector corresponding to the test data.

```
>>testIdx = test(c);
```

Extracting Training and Test Observations

Finally, we divide the data **completeData** into training and test datasets.

```
>>trainingData = completeData(trainingIdx,:);
```

```
>>testData = completeData(testIdx,:);
```

Throughout this chapter, in most of the examples, we will use the following demonstration shown in Figure 2.2 to illustrate the concept of subgrouping based on the categorical response.

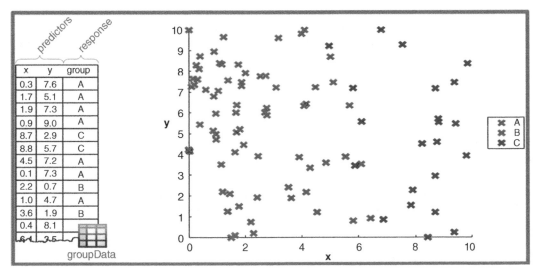

Figure 2.2 The original data contains predictors and response (class) variable. The data analysis will be carried out in light of subgrouping or classification.

After dividing the data into training and test sets, we can *fit a model* to the training data. Suppose we want to create a classification model using the training data.

A model can be considered as a set of rules or mathematical equations which can be used to classify the data. To create a model, we need to apply a learning algorithm to the data.

MATLAB provides fitting functions for several learning algorithms. These function names take the form **fitc***, where * represents the specific type of learning algorithm.

Predicting the Response

Once we train a classification model, we can use it to predict the class of different observations. Suppose we have a new observation which was not part of the original data.

Use the MATLAB function **predict** for predicting the class.

>> yPred = predict(mdl,testData)

To construct a classifier in MATLAB, we can use a **fitc*** function with the training data table as input and specify the name of the response variable in single quotes. For the purpose of this example, we will use a k-NN classifier, which will be described in more detail in a later section.

>> knnModel = fitcknn(tableData,'ResponseVariable');

Evaluating the Classification Model

Evaluating the accuracy of the model can help in choosing the model that generalizes well to the unknown data.

Calculate the Training Error

We can measure the training error (resubstitution loss) using the function **resubLoss**. The **resubLoss** function, the resubstitution loss is calculated as the mean of the loss function (e.g., classification error, mean squared error) evaluated on the training data. The resubstitution loss is a useful metric to evaluate the performance of the model on the training data, and it can be used as an indicator of whether the model is overfitting or underfitting the data. However, it is important to note that relying solely on the resubstitution loss to evaluate a machine learning model can be misleading. The model may perform well on the training data but poorly on new, unseen data, a phenomenon known as overfitting. Therefore, it is recommended to use additional evaluation techniques, such as cross-validation, hold-out, or resubstitution validation, to assess the model's performance on new data.

>> trainErr = resubLoss(mdl)

Where **trainErr** is the training classification error, **mdl** is the classification model variable. The function uses the training data stored in the model variable. The training error as related to misclassification is shown in Figure 2.3. The plot shows the predictions for the training data. On the plot, inaccurate predictions are where there exists a discrepancy between predicted and actual value, shown as two dissimilar colors.

NOTE

---

resubLoss(mdl) returns the classification loss by resubstitution, or the in-sample classification loss, for the trained classification model mdl using the training data stored in **mdl.X** and the corresponding class labels stored in **mdl.Y**.

The interpretation of resubstitution loss depends on the loss function ('**LossFun**') and weighting scheme (mdl.W). In general, better classifiers yield smaller classification loss values. The default '**LossFun**' value varies depending on the model object mdl.

Notice that we do not need to give the **resubLoss** function the training set, since this is already known to the model.

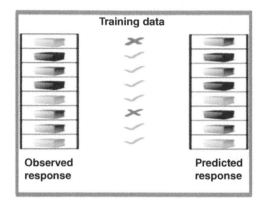

Figure 2.3   The training error as a result of misclassification of the observed response variable.

Calculate the Test Error

Use the function **loss** to calculate the test or validation error. The **loss** function thus measures the predictive inaccuracy of the model on the test data.

>> testErr = loss (mdl, observed_Xdata, observed_Ydata)

Where **testErr** is the test classification error, **mdl** the classification model variable, **observed_ Xdata** is a matrix containing the observed predictors from the test data, and **observed_Ydata** is a vector containing the observed class from the test data. The test error as related to misclassification is shown in Figure 2.4. The plot shows the predictions for the test data. On the plot, inaccurate predictions are where there exists a discrepancy between predicted and actual value, shown as two dissimilar colors.

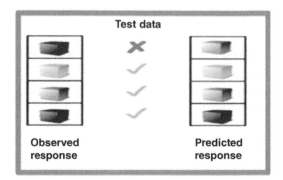

Figure 2.4   The test error as a result of misclassification of the observed response variable.

Quiz (T/F)
Given a model named **mdl** and training data named **trainData** with a trained response named **ytrain**, the following two expressions are equivalent (i.e., err1 == err2)

>> **err1 = resubLoss (mdl);**

>> **err2 = loss (mdl, trainData, ytrain);**

Confusion Matrix

We can find out the distribution of all predicted responses and how they compare to their true classes by calculating the confusion matrix.

The confusion matrix is given by:

>> **[cm,grp] = confusionmat(yObserved,yPred);**

The confusion matrix, **cm**, and the group names, **grp**, are returned. The function heatmap can then be used to visualize the confusion matrix.

>>**heatmap(grp,grp,cm)**

KNN Model for All Categorical or All Numeric Data Type

The following line of code will fit a *k*-nearest neighbor (*k*-NN) model to the training data.

>> **knnModel = fitcknn(trainData,'responseVarName')**

Using **fitcknn** as a classification method has the following features:

Performance: The fit time is fast. The prediction time is fast and proportional with (Data Size)^2. The memory overhead is small.

Some common properties are given in Table 2.1.

Table 2.1 Properties of KNN model.

Property	Description/Value
'NumNeighbors'	Number of neighbors used for classification.
'Distance'	The metric used for calculating distances between neighbors. See Table 1.6 for different values. Examples are: **'cityblock'** | **'chebychev'** | **'correlation'** | **'cosine'** | **'euclidean' (default)**.
'BreakTies'	Tie-breaking algorithm: **'smallest' (default)** | **'nearest'** | **'random'**. By default, ties occur when multiple classes have the same number of nearest points among the K nearest neighbors.
'DistanceWeight'	Weighting factor given to different neighbors, which takes on values, like: **'equal'** No weighting **'inverse'** Weight is 1/distance **'squaredinverse'** Weight is 1/distance^2
'NSMethod'	Nearest neighbor search method. Examples are: **'exhaustive'** for any distance metric of exhaustive searcher (Table 1.6). **'kdtree'** for **'cityblock'**, **'chebychev'**, **'euclidean'**, or **'minkowski'**.

NOTE

We cannot use any cross-validation name-value argument together with the '**Optimize Hyperparameters**' name-value argument. We can modify the cross-validation for '**OptimizeHyperparameters**' only by using the '**HyperparameterOptimizationOptions**' name-value argument. See Figure 2.12.

Moreover, the k-nearest neighbors (k-NN) classification model requires predictors that *are all numeric or all categorical*. Many data sets contain predictors that are numeric and categorical. If we have mixed predictors, we cannot use all of our data with a k-NN model.

KNN Model: Heart Disease Numeric Data

The table **heartData** contains several features of different subjects and whether or not they have heart disease, which is saved in the **HeartDisease** variable. It has 11 numerical columns and 11 categorical columns, as well.

We will divide the data in **heartData** such that 70% of it is used for training and the remainder is used for testing. We will create two tables called **hdTrain** and **hdTest** to store each part. If the you double-click on (i.e., edit) **cvpt**, below, you will notice that it contains '**Trainsize**' with *299* and '**Testsize**' with *128*, totaling *427* data points.

Figure 2.5 shows the complete code for **KNN** classification model of heart disease data with model error assessment for both training and test phase.

```
%% Import data
heartData = readtable('heartDiseaseData.xlsx');
heartData.HeartDisease = categorical(heartData.HeartDisease);
%% Split the data into training and test sets
% Data is already scaled
numData = heartData{:,1:11};
cvpt=cvpartition(heartData.HeartDisease,'HoldOut',0.30);
hdTrain=numData(training(cvpt),:);
hdTest=numData(test(cvpt),:);
ytrain = heartData.HeartDisease(training(cvpt));
ytest = heartData.HeartDisease(test(cvpt));
knnModel = fitcknn(hdTrain,ytrain);
knnModel.NumNeighbors=3;
knnModel.DistanceWeight='squaredinverse';
%knnModel.DistanceWeight='equal';
%knnModel.Cost =[0 1;2 0];
yPred = predict(knnModel,hdTest);
% Calculate the predictive loss on the training data.
trainErr=resubLoss (knnModel);
disp(['Training Error: ',num2str(trainErr)])
```

Figure 2.5 The code for classification of heart disease data via splitting the data into training and testing subsets and for calculating the error of misjudgments of data classification.

```
% Calculate the predictive loss on the test data.
testErr = loss(knnModel,hdTest,ytest);
disp(['Test Error: ',num2str(testErr)])
%The confusion matrix, cm, and the classes, cl, are returned. The function heatmap
can %be used to visualize the confusion matrix
[cm,cl] = confusionmat(ytest,yPred);
%Distribute classification results over all permutated values of true and false.
heatmap(cl,cl,cm)
%Calculate the percentage of patients who were predicted not to have heart disease
%(false) but actually did have it (true) out of all of the patients. Name this value
%falseNeg. The percentage of misclassified heart disease cases can be calculated by
%determining the number of misclassified heart disease cases, then dividing that by
%the total number of observations in hdTest (i.e., height of hdTest).
misClass = cm(cl=='true',cl=='false');
falseNeg=100*misClass/height(hdTest);
disp(['% False Negatives: ',num2str(falseNeg),'%'])
```

Figure 2.5   (Cont'd)

Here are the final results for calculating the error of data classification misjudgment.

```
Training Error: 0 (=resubLoss (knnModel));
Test Error: 0.35698 (= loss(knnModel,hdTest,ytest);)
Percentage of False Negatives: 12.5% (i.e., equivalent to 16 cases below)
```

Notice that

```
>> testErr = loss(knnModel,hdTest,ytest);%testErr=0.357
>> testErr2 = loss(knnModel,hdTest,yPred);%testErr2=0.0
>> testErr3 = loss(knnModel,hdTrain,ytrain);%testErr3=0.0
```

Figure 2.6 shows the heatmap of the confusion matrix between predicted (**yPred**) and observed (**ytest**) data. The total number is *128* response datapoints reserved for testing the model predictability and the prediction results were distributed as follows

28 cases were identified false by both observed and predicted
30 cases were identified false by observed but true by predicted
16 cases were identified true by observed but false by predicted $((16/128) \times 100\% = 12.5\%)$
54 cases were identified true by both observed and predicted

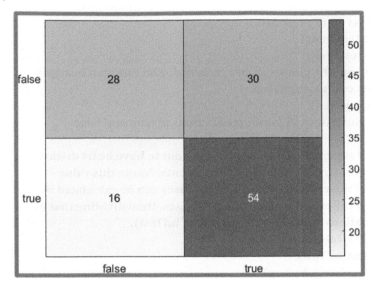

Figure 2.6 The heatmap of confusion matrix showing the matching and mismatching between predicted (**yPred**), column-wise, and observed y (**ytest**) data, row-wise.

NOTE

Different values of training, test, and % of false negatives might be reported from one to another run; do not panic. This is expected as we have stochastic not deterministic models.

Viewing the Fitting Model Properties

To see a description of the created class model, type in its name at the prompt and press enter.

>> knnModel

```
knnModel =
  ClassificationKNN
     ResponseName: 'Y'
  CategoricalPredictors: []
        ClassNames: [false    true]
     ScoreTransform: 'none'
     NumObservations: 299
          Distance: 'euclidean'
       NumNeighbors: 1
  Properties, Methods
```

To see **knnModel**'s different properties, click on '**Properties**' hyperlink shown in the last line, or double click on its name from **Workspace** browser and you will be able to see different features of the class model, as shown in Figure 2.7.

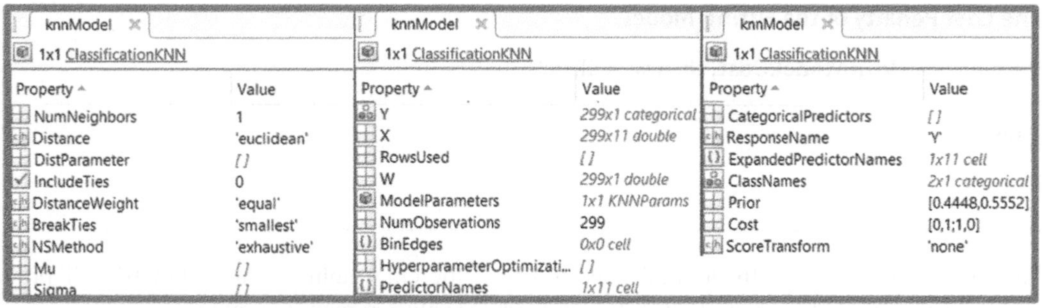

Figure 2.7   Features of the classification model **knnModel**.

We can specify property values when we create the classifier.

>> knnMdl = fitcknn(tableData,'Response',

'PropertyName',PropertyValue);

The Fitting Model: Number of Neighbors and Weighting Factor

We can modify number of neighbors for the created classifier as follows.

knnModel = fitcknn(hdTrain,ytrain, 'NumNeighbors', 3);

If we change the number of lovely neighbors from 1 to 3, then we have

Training Error: 0 for 1 neighbor	**Training Error**: 0.117 for 3 neighbors
Test Error: 0.352 for 1 neighbor	**Test Error**: 0.4224 for 3 neighbors
% False Negatives: 16.4% for 1 neighbor	**% False Negatives**: 18.7% for 3 neighbors

If we change the number of lovely neighbors from 1 to 3 with a weighting factor set to '**squaredinverse**', then we have:

Training Error: 0 for 3 neighbors with a weighting factor
Test Error: 0.29 for 3 neighbors with a weighting factor
% False Negatives: 14.8% for 3 neighbors with a weighting factor

---

NOTE

With **knnModel.DistanceWeight='squaredinverse'**; the classification model gives better accuracies (or predictions) for both training and testing step. The learning lesson here is simply that the user ought to try different tune-up model parameters, like number of neighbors and distance weight factor, such that to minimize both the training and test error for the given classification model.

The Cost Penalty of the Fitting Model

If we type >>**knnModel.Cost**, then we will get

```
ans =
   0   1
   1   0
```

The cost matrix is a part of the loss calculation. Thus, different weights in the cost matrix will produce different loss values even though the misclassified values remain the same.

Should we consider the model's misclassification rate for individual classes? Keep in mind that misclassifying sick people as healthy can have a higher cost than misclassifying healthy people as sick. The first type of misclassification will be weighted as 2 and the second type will be weighted by 1.

If we modify the cost indices as

>> **knnModel.Cost =[0 1;2 0];**

Reevaluate the predictive loss on the test data, starting from:

testErr = loss(knnModel,hdTest,ytest);

```
Training Error: 0
Test Error: 0.54
% False Negatives: 17.97%
So, testErr=0.54 is higher than the previous value testErr=0.35.
```

KNN Model: Mushroom Edibility Data

This data set includes several samples of mushrooms. The mushrooms are classified as either **'edible'** or **'poisonous'**. You can imagine if you misclassify an edible mushroom, the impact would not be very severe. However, if you misclassify a poisonous mushroom, the result could be fatal.

The table mushroom contains several features of mushroom's attributes and the last column, the corresponding response, **Class**, which has two attributes: Either **poisonous** or **edible**. Figure 2.8 shows the complete code for fitting KNN model and calculating both training and test error.

```
%Read data from Mushroom.csv file
Mushroom=readtable('Mushroom.csv','Format', 'auto');
Mushroom.Class = categorical(Mushroom.Class);
% Convert RingNumber column into categorical. Columns are all categorical
Mushroom.RingNumber = categorical(Mushroom.RingNumber);
%Split the data into training and test sets
```

Figure 2.8 The code for classification of mushroom data via splitting the data into training and testing subsets and for calculating the error of misjudgments of data classification.

```
pt = cvpartition(Mushroom.Class,'HoldOut',0.5);
% Create the training and test tables
mushTrain = Mushroom(training(pt),:);
mushTest = Mushroom(test(pt),:);
trainclass = Mushroom.Class(training(pt));
testclass = Mushroom.Class(test(pt));
% Create a k-NN classifier model
knnModel = fitcknn(mushTrain,trainclass);
predictclass = predict(knnModel,mushTest);
% Predict the loss on the test set and the re-substitution loss
errTrain=resubLoss(knnModel);
errTest=loss(knnModel,mushTest,testclass);
disp(['Training Error: ',num2str(errTrain)])
disp(['Test Error: ',num2str(errTest)])
% Calculate the percentage of poisonous mushrooms that were classified as edible
[cm,cl] = confusionmat(testclass, predictclass);
heatmap(cl,cl,cm);
misClass = cm(cl=='poisonous',cl=='edible');
errPoison=100*misClass/height(mushTest);
disp(['Percentage of misclassified poisons: ',num2str(errPoison),'%'])
```

Figure 2.8   (Cont'd)

Here are the final results for calculating the error of data classification misjudgment.

```
Training Error: 0
Test Error: 0
Percentage of misclassified poisons: 0%
```

Figure 2.9 (*left*) shows the heatmap of the confusion matrix between predicted (**yPred**) and observed (**ytest**) y data. The total number is *4062* response datapoints reserved for testing the model predictability and the prediction results were distributed as follows:

2104 cases were identified edible by both observed and predicted
0 cases were identified edible by observed but poisonous by predicted
0 cases were identified poisonous by observed but edible by predicted
1958 cases were identified poisonous by both observed and predicted

The workspace variables are shown on the *right* of Figure 2.9.

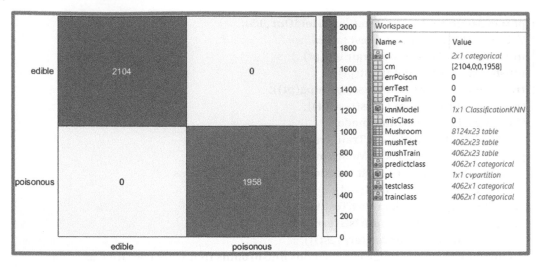

Figure 2.9 The heatmap of confusion matrix showing the matching and mismatching between predicted (**predictclass**), column-wise, and observed (**testclass**) data, row-wise.

```
load('winered.mat')
WineRed.quality = categorical(WineRed.quality);
% Extract numeric data
origData = WineRed{:,1:end-1};
MaxVal=max(origData,[], 1);
MinVal=min(origData,[], 1);
for j=1:size(origData,2)
numData(:,j)=(origData(:,j)-MinVal(j))/(MaxVal(j)-MinVal(j));
end
pt = cvpartition(WineRed.quality,'HoldOut',0.30);
% Create the training and test tables
RWTrain = numData(training(pt),:);
RWTest = numData(test(pt),:);
ytrain = WineRed.quality(training(pt));
ytest = WineRed.quality(test(pt));
%% Create a k-NN classifier model
knnModel = fitcknn(RWTrain,ytrain);
predictquality = predict(knnModel,RWTest);
errTrain=resubLoss(knnModel);
errTest=loss(knnModel,RWTest,ytest);
disp(['Training Error: ',num2str(errTrain)])
disp(['Test Error: ',num2str(errTest)])
[cm,cl] = confusionmat(ytest, predictquality);
figure(11);
heatmap(cl,cl,cm);
trueclass=0.0;
```

Figure 2.10 Code for analyzing red wine data and fitting the data to a classifier while showing the model goodness.

```
for k=1:height(cm)
trueclass = trueclass+sum(cm(k,k));
end
TC=100*trueclass/height(ytest);
disp(['Percentage of true classification: ',num2str(TC),'%'])
```

Figure 2.10 (Cont'd)

KNN Model: Red Wine Data

Let us recall the red wine example but this time we have to predict the response, that is, the wine quality. Figure 2.10 shows the complete code for reading and scaling raw data, carrying out classification model fitting, and presenting the results of the goodness of the classification model. The training and testing error are shown below. Figure 2.11 shows the matrix for observed versus predicted red wine quality.

```
Training Error: 0
Test Error: 0.38475
Percentage of true classification: 61.5866%
```

Figure 2.11 The heatmap of confusion matrix of red wine data showing the matching and mismatching between predicted (**predictquality**), column-wise, and observed (**ytest**) data, row-wise.

Figure 2.11 shows the heatmap of confusion matrix indicating the matching and mismatching between predicted (**predictquality**), column-wise, and observed (**ytest**) data, row-wise.

Figure 2.12 shows the modification made to KNN model statement; the rest are the same as those coded in Figure 2.10. Such a modification will enable auto hyper parameter optimization where

```
%% Create a k-NN classifier model
rng(1)
knnModel = fitcknn(RWTrain,ytrain,'OptimizeHyperparameters','auto',...
    'HyperparameterOptimizationOptions',...
    struct('AcquisitionFunctionName','expected-improvement-plus'))
```

Figure 2.12 Modification of KNN model statement to enable auto hyper parameter optimization.

the estimated minimum objective will try to reach minimum observed function, as shown in Figure 2.13 (left).

The results are:

Best estimated feasible point (according to models):
NumNeighbors: 1
Distance: mahalanobis
Estimated objective function value = 0.40251
Estimated function evaluation time = 0.096223
Training Error: 0
Test Error: 0.39339
Percentage of true classification: 60.5428%

Figure 2.13 (*left*) shows that 30 iterations are needed to complete the optimization process where the best observed feasible point is attained at '**NumNeighbors**' =1 and '**Distance**' = '**cosine**' with an observed objective function value = 0.3875 and an estimated objective function value = 0.41079. On the other hand, the best estimated feasible point (according to models) is attained at '**NumNeighbors**' =1 and '**Distance**' = '**mahalanobis**' with an estimated objective function value = 0.40251.

Notice that the auto hyper parameter optimization probes all different permutations of '**Distance**' values, as shown in Figure 2.13 (*right*).

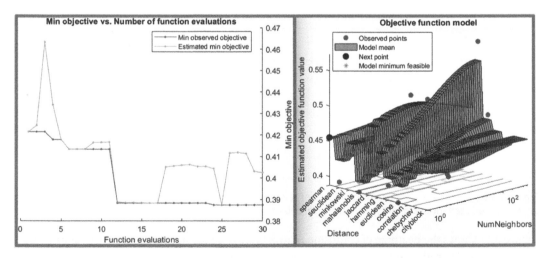

Figure 2.13 Iterations needed to complete the optimization process (*left*) and the convergence of best estimated feasible to the best observed feasible point (*right*).

Figure 2.14 (*left*) shows the heatmap of confusion matrix indicating the matching and mismatching between predicted (**predictquality**), column-wise, and observed (**ytest**) data, row-wise and the workspace variables (*right*).

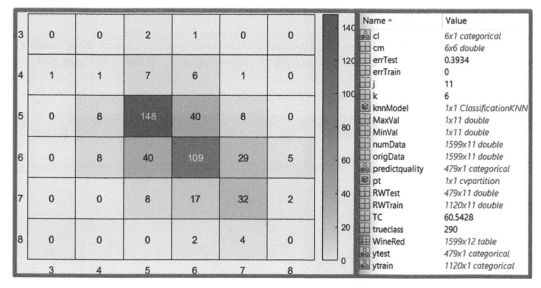

Figure 2.14 (*left*) The heatmap of confusion matrix of red wine data showing the matching and mismatching between predicted (**predictquality**), column-wise, and observed (**ytest**) data, row-wise and workspace variables (*right*).

Using MATLAB Classification Learner

To open the Classification Learner app, we can either use MATLAB Toolstrip: On the **Apps** tab, under "**Machine Learning and Deep Learning**" category, click on "**Classification Learner**" app icon, or use MATLAB command prompt:

Let us use the Classification Learner by executing the command **classificationLearner** in the Command Window of MATLAB. We will use the "**WineRed**" table for both the predictor variables and response variable

Figure 2.15 shows the initial code needed to prepare the raw data to be utilized in the **classificationLearner** environment. We scaled the first 11 columns, representing predictor variables and the 12th column represents the categorical quality column; a drinker-defined rank value between 3 and 8.

At this stage, we are supposed to have **WineRed** table with normalized column values between the first and the 11th column. There are 1599 rows. At the command, type in:

```
load('winered.mat')
WineRed.quality = categorical(WineRed.quality);
% Extract numeric data
origData = WineRed{:,1:end-1};
MaxVal=max(origData,[], 1);
MinVal=min(origData,[], 1);
for j=1:size(origData,2)
WineRed{:,j}=(origData(:,j)-MinVal(j))/(MaxVal(j)-MinVal(j));
end
```

Figure 2.15 Scaling the predictor variables and converting the 12th column to a categorical column.

>>classificationLearner

Figure 2.16 shows the first pop-up window of the classification learner toolbox. Click on **"New Session"** button and select **"From Workspace"** option from the drop-down menu, as shown in Figure 2.17.

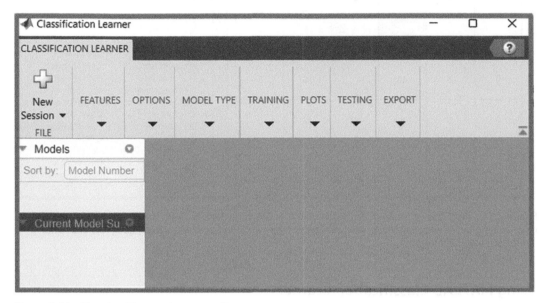

Figure 2.16   The classification learner toolbox.

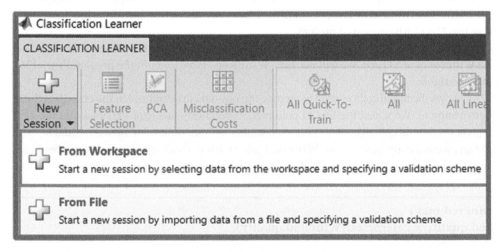

Figure 2.17   Source of data will be borrowed from MATLAB Workspace.

After selection, the data set window will pop-up and we have to populate the required input predictor columns and response column as shown in Figure 2.18. The validation method can be selected as one out of three. We will select the middle one as we have already tried this in previous examples and is recommended for large size data. Click on **"Start Session"** button and the classifier learner main window will pop-up as in Figure 2.19. We will start from top left and move along the top toolbar to the right, briefly showing what each item means or does.

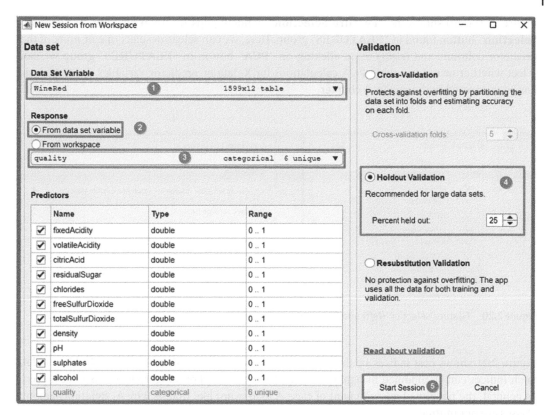

Figure 2.18 Defining the data sets for both the predictor and response variables as well as selection of the validation method.

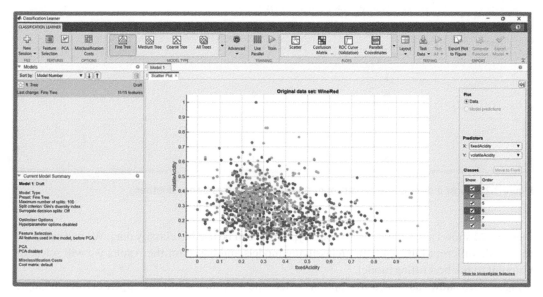

Figure 2.19 The main window of classification learner with numerous tune-up features for classification models and later for presentation of results.

Figure 2.20 (*left*) shows the "**Feature Selection**" window pops up upon clicking on "**Feature Selection**" button found in "**FEATURES**" group. Here, we can select/deselect one or more of the predictor columns and (*right*) upon clicking on "**PCA**" button in "**FEATURES**" group we can select whether or not to enable PCA calculations for X-data or predictor variables. PCA analysis has been covered in the previous chapter.

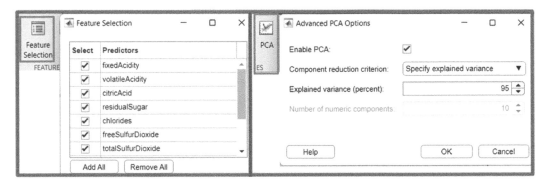

Figure 2.20   Feature selection (*left*) and PCA enabling (*right*).

Figure 2.21 shows cost indices or the penalty of misjudging a classification made by the model. Such indices can be modified in light of real situation how fatal or dangerous to misjudge a class. In general, the penalty of misjudgment of class A by B is not like that of B by A. Relations among classes are not mutual.

		Predicted Class					
		3	4	5	6	7	8
True Class	3	0	1	1	1	1	1
	4	1	0	1	1	1	1
	5	1	1	0	1	1	1
	6	1	1	1	0	1	1
	7	1	1	1	1	0	1
	8	1	1	1	1	1	0

Figure 2.21   Cost indices or the penalty of misjudgment of classes by one another.

Figure 2.22 shows a list of potential model types that can be used to classify data. We will select all to include different kinds of fitting/classifying models and based on the results we will select or pinpoint to the best fitting model.

Figure 2.23 shows advanced tree options where the user may select the "**Split criterion**" to be used in splitting the total data into classes or groups.

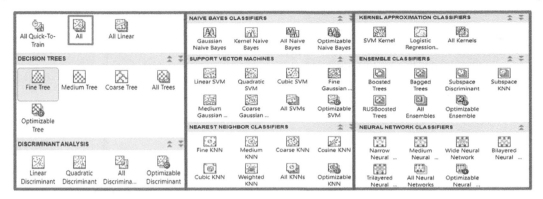

Figure 2.22 A list of fitting/classifying models that can be used to fit the given data.

NOTE

The "**Surrogate decision splits**" option is kept by default *off*. If we have data with missing values, then the "**Surrogate decision splits**" option ought to be enabled to improve the accuracy of prediction.

In regression, using the mean-squared error (MSE), a split is used to minimize the MSE of predictions compared to the training data, upon response classification and one of three measures are used: The Gini's diversity index (default), the Twoing rule, and the maximum deviance reduction. This has to deal with impurity and node error involved in estimates of predictor importance for classification tree. A decision tree splits nodes based on either impurity or node error. Impurity means one of several things, depending on one's choice of the split criterion name-value pair argument:

1) The Gini's Diversity Index ('gdi'): The Gini index of a node is: $1 - \sum_i p^2(i)$

   where **the sum is over the classes *i*** at the node, and ***p(i)*** is the observed fraction of classes, with class *i*, which reach the node. A node with just one class (a pure node) has Gini index 0; otherwise, the Gini index is positive. So, the Gini index is a measure of node impurity.

2) The Deviance ('deviance'): With *p(i)* defined the same as for the Gini index, the deviance of a node is: $-\sum_i p(i) log_2 p(i)$
   A pure node has 0 deviance; otherwise, the deviance is positive.

3) The Twoing Rule ('twoing'): Twoing is not a purity measure of a node, but is a different measure for deciding how to split a node. Let ***L(i)*** denote the fraction of members of class *i* in the left child node after a split, and ***R(i)*** denote the fraction of members of class *i* in the right child node after a split. Choose the split criterion to maximize

$$P(L)P(R) = \left[\sum_i |L(i) - R(i)| \right]^2$$

where ***P(L)*** and ***P(R)*** are the fractions of observations that split to the left and right, respectively. If the expression is large, the split will make each child node purer. Similarly, if the expression is small, the split will make each child node similar to each other, and therefore similar to the parent node; therefore, the split did not increase node purity.

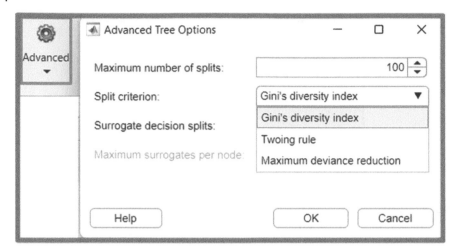

Figure 2.23 The advanced tree options can be tuned up using different maximum number of splits and split criteria.

Next to "**Advanced Tree Options**", there exists "**Use Parallel**" button which can be clicked on to train in parallel.

NOTE

"**Use Parallel**" is disabled for optimizable presets. Under each category of model types (see Figure 2.22), there exists an optimizable item.

Let us click on "**Train**" button after selecting "**All**" from the "**MODEL TYPE**" group (see Figure 2.22). Wait for a while as we have selected to examine all model types. Figure 2.24 shows the training results of the trained models and their corresponding validation (training) accuracy. We can see that the model with "**Ensemble**" category with "**KNN**" subspace got the highest rank in terms of validation accuracy.

NOTE

The best model may vary or change from one to another classification learner session. Such models are stochastic in nature.

Notice that you can sort models based on model number, validation (training) accuracy, test accuracy, and on total cost, as shown in Figure 2.25.

After sorting the examined models in light of a selected criterion (as shown in Figure 2.25), you can see the results pertaining to a specific model. Highlight the model from the models' list, go to "**PLOTS**" group (see Figure 2.19), click on the drop-down menu, select the required plot, for example the confusion matrix for training (validation). Figure 2.26 shows the validation confusion matrix for the best model in terms of accuracy. Notice that the validation accuracy is calculated by summing the on-diagonal entries and dividing the result by the sum of all entries. Thus, we have:

Validation accuracy $= (272/399) \times 100\% = 68.2\%$.

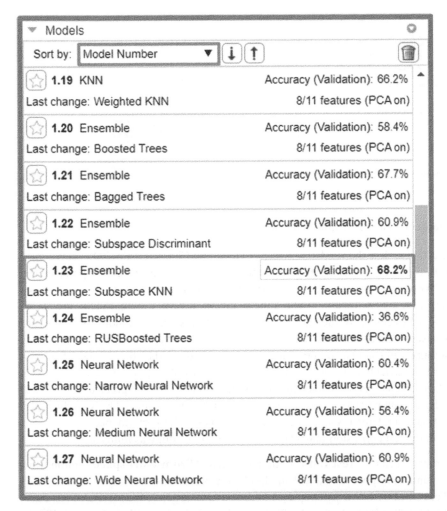

Figure 2.24   The validation (training) accuracy of each trained model and the model with the highest rank, the **bolded** font type, is for ensemble model type with subspace KNN.

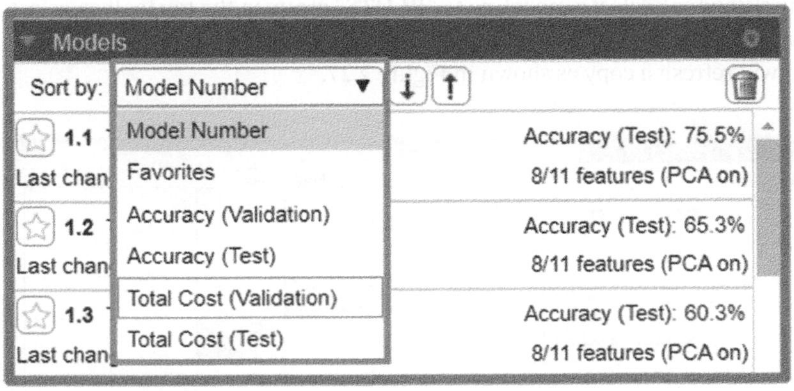

Figure 2.25   Examined models can be sorted out based on different criteria.

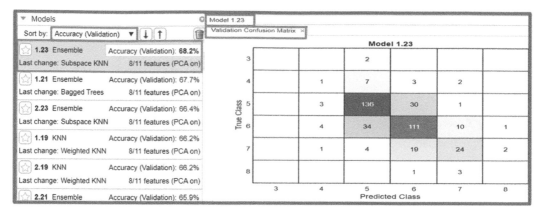

Figure 2.26 The validation confusion matrix for the best model.

At this stage, we can also view the validation Receiver Operating Characteristic (ROC) curve from "PLOTS" group. ROC is, more or less, similar to the confusion matrix. For a single class problem, the function takes a matrix of boolean values indicating class membership and a matrix of outputs values in the range [0,1]. The ROC is a metric used to check the quality of classifiers. For each class of a classifier, roc applies threshold values across the interval [0,1] to outputs. For each threshold, two values are calculated, the True Positive Ratio (TPR) and the False Positive Ratio (FPR). For a particular class i, TPR is the number of outputs whose actual and predicted class is class i, divided by the number of outputs whose predicted class is class i. On the other hand, FPR is the number of outputs whose actual class is not class i, but predicted class is class i, divided by the number of outputs whose predicted class is not class i. The area under the curve (AUC) is indicated in ROC plot. The maximum AUC is 1, which corresponds to a perfect classifier. Larger AUC values indicate better classifier performance.

On the other hand, the test confusion matrix for all models can be done as follows:

1) From the top toolbar, click on **"Test Data"** button and select **"From Workspace"** option from the drop-down menu, to import test data from MATLAB workspace.
2) The **"Import Test Data"** window will pop-up where you have to select the MATLAB test dataset (**WineRed** table). The import wizard will show a layout similar to that shown in Figure 2.18. Click on **"Import"** button to proceed.
3) Click on **"Test All"** button and select **"Test All"** option from the list to test all trained models.
4) From the model menu (left pane), select a model, go to **"PLOTS"** group in the top toolbar, select **"TEST RESULTS"** subcategory, and click on **"Confusion Matrix (Test)"** icon.
5) Classification learner will refresh a copy as shown in Figure 2.27.

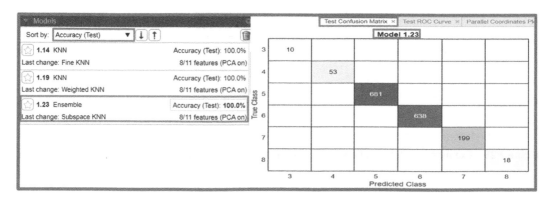

Figure 2.27 The test confusion matrix for the best model. The test accuracy is 100%.

The last step is to present the results to other users, using "**Export**" group found on the right side of the top toolbar. In addition to presenting the results in the platform itself, the user may wish to export a given plot to an external figure so that it can be printed or saved.

NOTE

Do not forget to save the file anywhere within MATALB defined paths, or add its location to MATLAB path via "Set Path button found in "**ENVIRONMENT**" group of MATLAB Toolstrip.

The first choice in "**Export**" group is to click on "**Export Plot to Figure**" button to carry out the export step and a new figure will show up holding the currently shown figure in the classification learner platform itself.

The second choice is to generate an m-file holding the name **trainClassifier** and will be saved as **trainClassifier.m**. The m-file appears first as untitled but once you save it, it will default to **trainClassifier.m** file.

This m-file function accepts the **trainingData** as input argument. The **trainingData** is a table containing the same predictor and response columns as those imported into the classification learner app. On the other hand, the first output of the m-file will be: **trainedClassifier**, a struct containing the trained classifier. The struct contains various fields with information about the trained classifier. For example, **trainedClassifier.predictFcn** is a function to make predictions on new data. The second output argument is **validationAccuracy** which is a double containing the accuracy as a percentage. In the classification learner app, the Models pane displays this overall accuracy score for each model. The **trainClassifier** m-file starts with a function statement holding its name as shown below:

>>**function [trainedClassifier, validationAccuracy] = trainClassifier(trainingData)**

To make use of this generated m-file, we have to call it while providing an input argument in the form of a table which contains the same predictor and response columns. For example, the following statement shows how to use the m-file with **WineRed** table.

>> **[mytrainedcls myvalidaccuracy] =trainClassifier (WineRed);**

If we run the above statement, we will notice that MATLAB will create two variables in its workspace. The first is **mytrainedcls** which is 1×1 struct and **myvalidaccuracy** which has a double value of 0.6767. Of course, the value depends upon the selected model that you highlight in classification learner app under Models pane. In addition, you can search for the following text portion in the created m-file.

```
↓
subspaceDimension = max(1, min(6, width(trainingPredictors) - 1));
classificationEnsemble = fitcensemble(...
    trainingPredictors, ...
```

```
    trainingResponse, ...
    'Method', 'Subspace', ...
    'NumLearningCycles', 30, ...
    'Learners', 'knn', ...
    'NPredToSample', subspaceDimension, ...
    'ClassNames', categorical({'3'; '4'; '5'; '6'; '7'; '8'}));
↓
```

Notice that **fitcensemble** can be replaced by another **fitc*** model, depending on the selected model, like **fitcecoc**, **fitcdiscr**, **fitcnet**, and so on.

The third and last choice to do is to export the model. Click on "**Export Model**" button and the user has three formats by which to export the model.

1. The Export Model: Export the currently selected model in the Models pane to the workspace to make predictions with new data. The training data are included in the created classification model. For example, the created model is named as **trainedModel** which is 1×1 struct. It includes seven fields: **predictFcn**, **RequiredVariables**, **ClassificationEnsemble**, **PCACenters**, **PCACoefficients**, **About**, and **HowToPredict**.

We can make use of this created classification model to predict the response value for a given set of predictor columns. The columns have to be titled and sorted in a way similar to what we used in the training session using the classification learner app. Here is an example on how to make use of **trainedModel** struct, as shown in Figure 2.28.

```
Xdata=WineRed(:,1:11);
ydata=WineRed{:,12};
yfit = trainedModel.predictFcn(Xdata);
[cm,cl] = confusionmat(ydata, yfit);
heatmap(cl,cl,cm);
trueclass=0.0;
for k=1:height(cm)
trueclass = trueclass+sum(cm(k,k));
end
TC=100*trueclass/height(yfit);
disp(['Percentage of true classification: ',num2str(TC),' %'])
```

Figure 2.28 Re-use of the **trainedModel** deployed by **classificationLearner** app.

The result for true classification, **TC** is:

Percentage of true classification: 100 %

Figure 2.29 (*left*) shows the heat map of the confusion matrix of the reused **trainedModel** deployed by **classificationLearner** app and the workspace variables (*right*).

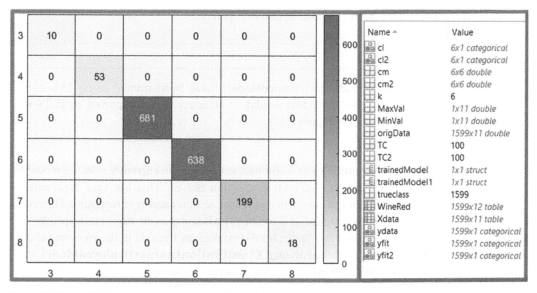

Figure 2.29 *(left)* The confusion matrix for the reused **trainedModel** deployed by **classificationLearner** app and the workspace variables *(right)*.

2. Export Compact Model: Export the currently selected model in the Models pane without its training data to the workspace to make predictions with new data. Again, the created model is named as **trainedModel1** which is 1×1 struct. It also includes the same seven fields.

To show how to make use of the created classifier, we will use the same example as the previous case but we need to change the name from **trainedModel** to **trainedModel1** which is 1×1 struct, as shown in Figure 2.30.

```
Xdata=WineRed(:,1:11);
ydata=WineRed{:,12};
yfit2 = trainedModel1.predictFcn(Xdata);
[cm2,cl2] = confusionmat(ydata, yfit2);
heatmap(cl2,cl2,cm2);
trueclass=0.0;
for k=1:height(cm2)
trueclass = trueclass+sum(cm2(k,k));
end
TC2=100*trueclass/height(yfit2);
disp(['Percentage of true classification: ',num2str(TC2),' %'])
```

Figure 2.30   Reuse of the **trainedModel1** deployed by **classificationLearner** app.

The result for true classification, **TC2** is:

Percentage of true classification: 100 %

and the generated figure is exactly the same as that of Figure 2.29.

NOTE

Both created models **trainedModel** and **trainedModel1** have been trained using 25% held out portion (see Figure 2.18). Also, the test validation accuracy is also reported as 100% in classification learner platform (see Figure 2.27).

---

3. Export Model for Deployment: Export a compact version of the currently selected model in the Models pane for deployment to MATLAB Production Server. This last case represents a situation where you want to run or test the created model on another server or computer. However, it requires either MATLAB itself installed or have MATLAB component runtime (https://www. mathworks.com/products/compiler/mcr/index.html) installed on the side of the receiving client. Notice that MATLAB will create a folder named **ClassificationLearnerDeployedModel** with two sub-folders named **for_redistribution** and **for_testing**. There exists readme.txt file in each subfolder on how to run/test the created classifier on the remote server and what the prerequisites are prior to running the package. I guess this is beyond the scope of this book and the learning material will be dedicated to directly using MATLAB platform as a comprehensive tool for machine learning-based toolboxes and running user-defined codes and functions.

Binary Decision Tree Model for Multiclass Classification of All Data Types

In this section, we will learn how to fit and customize decision trees which do not make any assumptions about the data.

To construct a classification tree in MATLAB, use the **fitctree** function with a table as input and specify the name of the response variable in single quotes.

>> **treeModel = fitctree(tableData, 'ResponseVariable');**

Moreover, we can change the level of branching of the classification tree using the prune (clip or trim) function.

>> **prunedTreeModel = prune(treeModel, 'Level', integer);**

Where integer is a numeric scalar from 0 (no pruning) to the largest pruning level of this tree max(tree.PruneList). **prune** returns the tree pruned to this level.

Using **fitctree** as a classification method has the following features:

Performance: The fit time is $\propto$ (Data Size). The prediction time is fast. The memory overhead is small.

Table 2.2 shows important common properties of decision tree model, **fitctree**, in addition to a description for each property and its possible values, as quoted from MATLAB help.

Table 2.2 Important common properties of decision tree model.

Property	Description/Value
'AlgorithmForCategorical' See MATLAB built-in help on **"Splitting Categorical Predictors in Classification Trees"** for more information on such algorithms. See NOTES below.	Algorithm to find the best split on a categorical predictor with C categories for data and $K \geq 3$ classes, specified as the comma-separated pair consisting of **'AlgorithmForCategorical'** and one of the following values: **'Exact'**: It considers all $2^{C-1} - 1$ combinations. **'PullLeft'**: The pull left by purity algorithm. It considers moving each category to the left branch as it achieves the minimum impurity for the K classes among the remaining categories. **'PCA'**: The principal component-based partitioning algorithm. It computes a score for each category using the inner product between the first principal component of a weighted covariance matrix and the vector of class probabilities for that category. **'OVAbyClass'**: One-versus-all by class algorithm. For the first class, it considers moving each category to the left branch in order, recording the impurity criterion at each move. It repeats for the remaining classes. From this sequence, it chooses the split that has the minimum impurity.
'MaxNumCategories'	Default is *10*. Passing a small value can lead to loss of accuracy and passing a large value can increase computation time and memory overload.
'MaxNumSplits':	Maximum number of splits allowed in the decision tree. $n - 1$ where n is the training sample size.
'MergeLeaves'	**'on'** (default) | **'off'**. If **MergeLeaves** is 'on', then **fitctree**: • Merges leaves that originate from the same parent node, and that yields a sum of risk values greater or equal to the risk associated with the parent node. • Estimates the optimal sequence of pruned subtrees, but does not prune the classification tree.
'MinLeafSize':	Minimum number of observations in each leaf node. Default is *1*. Each leaf has at least **MinLeafSize** observations per tree leaf. If we supply both **MinParentSize** and **MinLeafSize**, **fitctree** will use the setting that gives larger leaves: MinParentSize = max(MinParentSize,2×MinLeafSize).
'MinParentSize'	*10* (default) | positive integer value. Minimum number of branch node observations. Each branch node in the tree has at least **MinParentSize** observations
'PredictorSelection'	Algorithm used to select the best split predictor. **'allsplits'** (default) | **'curvature'** | **'interaction-curvature'**
'SplitCriterion':	Formula used to determine optimal splits at each level. **'gdi'** (Gini's diversity index, the default), **'twoing'** | **'deviance'**

NOTE

When fitting the tree, **fitctree** considers **NaN**, ' ' (empty character vector), " " (empty string), <missing>, and <undefined> values in Y to be missing values. **fitctree** does not use observations with missing values for Y in the fit. For numeric Y, consider fitting a regression tree using **fitrtree**, instead. Data types of Y can be: single \| double \| categorical \| logical \| char \| string \| cell.

(Continued)

NOTE (Continued)

   Predictor data, specified as a numeric matrix. Each row of X corresponds to one observation, and each column corresponds to one predictor variable. **fitctree** considers NaN values in X as missing values. **fitctree** does not use observations with all missing values for X in the fit. **fitctree** uses observations with some missing values for X to find splits on variables for which these observations have valid values. Data types of X can be as those of Y.

   If we do not specify an algorithm value for '**AlgorithmForCategorical**', the software selects the optimal algorithm for each split using the known number of classes and levels of a categorical predictor. If the predictor has at most **MaxNumCategories** levels, the software splits categorical predictors using the exact search algorithm. Otherwise, the software chooses a heuristic search algorithm based on the number of classes and levels. The default **MaxNumCategories** level is *10*. Depending on the platform, the software cannot perform an exact search on categorical predictors with more than 32 or 64 levels.

   We cannot use any cross-validation name-value argument together with the '**OptimizeHyperparameters**' name-value argument. We can modify the cross-validation for '**OptimizeHyperparameters**' only by using the '**HyperparameterOptimizationOptions**' name-value argument.

Quiz

How many times can a classification tree make a split based on each predictor variable?
a) 0 or 1   b) 1   c) 0, 1, or 2   d) 2   e) Any # of times (up to the # of observations)

Quiz

Which of the following properties does not belong to a fitted binary classification decision tree, **fitctree**?
a) MaxNumSplits   b) MinLeafSize   c) NumNeighbors   d) SplitCriterion

---

We can visualize the decision tree with the view function.
>>**view(treeModel,'mode','graph');**

Classification Tree Model: Heart Disease Numeric Data Types

Figure 2.31 shows the complete code for importing numeric heart disease predictor data, implementing **fitctree**, and presenting the classification results and their accuracy.

```
% Import data
heartData = readtable('heartDiseaseData.xlsx');
heartData.HeartDisease = categorical(heartData.HeartDisease);
numData = heartData{:,1:11};
cvpt=cvpartition(heartData.HeartDisease,'HoldOut',0.30);
hdTrain=numData(training(cvpt),:);
```

Figure 2.31 The code for fitting numeric heart disease data, using **fitctree** as classifier, and presenting the results.

```
hdTest=numData(test(cvpt),:);
ytrain = heartData.HeartDisease(training(cvpt));
ytest = heartData.HeartDisease(test(cvpt));
treeModel = fitctree(hdTrain,ytrain);
%Calculate the loss errTrain and errTest
errTrain = resubLoss(treeModel);
errTest = loss(treeModel,hdTest,ytest);
disp(['Training Error for treeModel: ',num2str(errTrain)])
disp(['Test Error for treeModel: ',num2str(errTest)])
prunedModel=prune(treeModel,"Level",3);
errTrain2 = resubLoss(prunedModel);
errTest2 = loss(prunedModel,hdTest,ytest);
disp(['Training Error for prunedModel: ',num2str(errTrain2)])
disp(['Test Error for prunedModel: ',num2str(errTest2)])
view(treeModel,'mode','graph');
view(prunedModel,'mode','graph');
```

Figure 2.31 (Cont'd)

Results copied from Command Window:

Training Error for treeModel: 0.090301
Test Error for treeModel: 0.37205
Training Error for prunedModel: 0.14381
Test Error for prunedModel: 0.26337

Figure 2.32 shows a view of the branched tree class for both the original trees model (*left*) and its pruned version (*right*). Notice that from the Pruning level drop-down/up arrow, you can control the degree of pruning level for each view. The maximum number of branching level is 10 and such levels are shown on the left side of Figure 2.32.

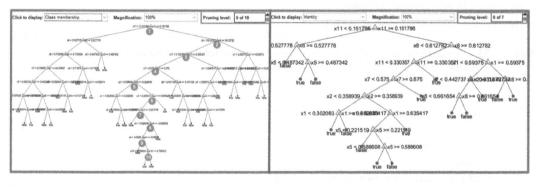

Figure 2.32   View of the treed-class with extended (*left*) and pruned level of branching (*right*).

Classification Tree Model: Heart Disease All Predictor Data Types

Figure 2.33 shows the code of handling *all types of heart disease predictor data*, enabling auto hyper parameters optimization of **fitctree** model, and presenting the results.

```
%% Import data
heartData = readtable('heartDiseaseData.xlsx');
heartData.HeartDisease = categorical(heartData.HeartDisease);
% Split the data into training and test sets. Data is already scaled
PrdData = heartData(:,1:21);
Y= heartData.HeartDisease;
cvpt=cvpartition(heartData.HeartDisease,'HoldOut',0.30);
hdTrain=PrdData(training(cvpt),:);
hdTest=PrdData(test(cvpt),:);
ytrain = heartData.HeartDisease(training(cvpt));
ytest = heartData.HeartDisease(test(cvpt));
%treeModel = fitctree(hdTrain,ytrain);
rng('default')
treeModel = fitctree(PrdData,Y,...
    'OptimizeHyperparameters','auto','Surrogate','on',...
    'HyperparameterOptimizationOptions',struct('Holdout',0.3,...
    'AcquisitionFunctionName','expected-improvement-plus'))
%% Calculate the loss errTrain and errTest
errTrain = resubLoss(treeModel);
errTest = loss(treeModel,hdTest,ytest);
disp(['Training Error for treeModel: ',num2str(errTrain)])
disp(['Test Error for treeModel: ',num2str(errTest)])
prunedModel=prune(treeModel,"Level",1);
errTrain2 = resubLoss(prunedModel);
errTest2 = loss(prunedModel,hdTest,ytest);
disp(['Training Error for prunedModel: ',num2str(errTrain2)])
disp(['Test Error for prunedModel: ',num2str(errTest2)])
view(treeModel,'mode','graph');
view(prunedModel,'mode','graph');
```

Figure 2.33 The code for fitting all types of heart disease predictor data, using auto hyper parameters-optimized **fitctree** classifier, and presenting the results.

The results are copied from Command Window:

```
Best observed feasible point:
   MinLeafSize = 22
Observed objective function value = 0.25781
Estimated objective function value = 0.26468
Function evaluation time = 0.031519
Best estimated feasible point (according to models):
   MinLeafSize = 24
```

> **Estimated objective function value = 0.25674**
> Estimated function evaluation time = 0.032056
> Training Error for treeModel: 0.1897
> Test Error for treeModel: 0.1798
> Training Error for prunedModel: 0.21077
> Test Error for prunedModel: 0.18802

Figure 2.34 (*left*) shows that 30 iterations are needed to converge the solution where the estimated minimum objective tries to reach minimum observed objective function and (*right*) the estimated objective function versus the minimum leaf size.

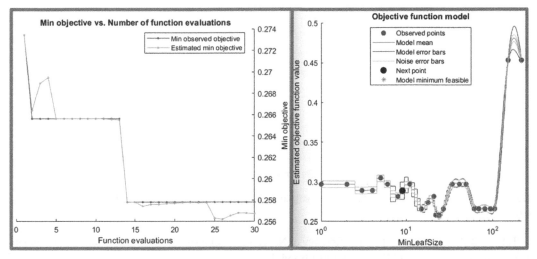

Figure 2.34 The estimated minimum objective and minimum observed objective function with iterations (*left*) and the estimated objective function versus the minimum leaf size (*right*).

Figure 2.35 shows a view of the branched tree class for both the original trees model (*left*) and its pruned version (*right*). Notice that from Pruning level drop-down/up arrow, we can control the degree of pruning level for each view. The maximum # of branching level is 3.

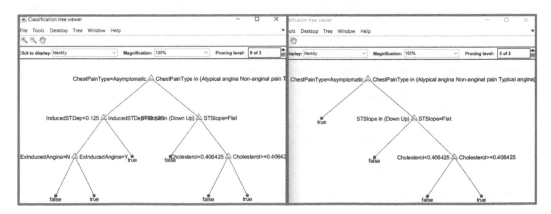

Figure 2.35 The view for the treed-class with extended (*left*) and pruned level of branching (*right*).

Naïve Bayes Classification Model for All Data Types

K-NN and decision trees do not make any assumptions about the distribution of the underlying data. If we assume that the data comes from a certain underlying distribution, we can treat the data as a statistical sample. This can reduce the influence of the outliers on our model. A Naïve Bayes classifier assumes the independence of the predictors within each class, as shown in Figure 2.36. This classifier is a good choice for relatively simple problems.

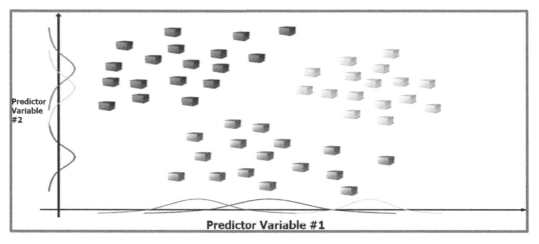

Figure 2.36   The independence of predictors within each class as projected by Naïve Bayes classification approach.

Table 2.3 shows computational efficiency of Naïve Bayes classification model under both modes: Gaussian (unimodal) and Kernel (bi/multimodal).

NOTE

Naïve Bayes is a good choice when there is a significant amount of missing data.

Table 2.4 shows some important common properties of Naïve Bayes classification model, quoted from MATLAB help.

Table 2.3   Computational Performance of Naïve Bayes (**fitcnb**)
model under Gaussian (unimodal) and Kernel (bi/multimodal) modes.

Property	Performance
Fit Time	Normal Distribution: Fast
	Kernel Distribution: Slow
Prediction Time	Normal Distribution: Fast
	Kernel Distribution: Slow
Memory Overhead	Normal Distribution: Small
	Kernel Distribution: Moderate to Large

Table 2.4   Important common properties of Naïve Bayes classification model.

Property	Description/Value
'CrossVal'	Cross-validation flag, specified as the comma-separated pair consisting of '**Crossval**' and '**on**' or '**off**' (default).
	If we specify '**on**', then the software implements tenfold cross-validation.
'DistributionNames'	Data distributions.
	'**kernel**': Kernel smoothing density estimate.
	'**mn**': Multinomial distribution. If '**mn**' is specified, then all predictor features are components of a multinomial distribution. Therefore, we cannot include '**mn**' as an element of a string array or a cell array of character vectors.
	'**mvmn**': Multivariate multinomial distribution.
	'**normal**': Normal (Gaussian) distribution.
'Kernel'	Kernel smoother type. A kernel smoother is a statistical technique to estimate a real valued function as the weighted average of neighboring observed data in a way that closer points are given heavier weights. The level of smoothness is set by a single parameter.
	'**box**': Box (uniform)
	'**epanechnikov**': Epanechnikov
	'**normal**': Gaussian
	'**triangle**': Triangular
'OptimizeHyperparameters'	Parameters to optimize.
	'**none**' (default) | '**auto**' | '**all**' | string array or cell array of eligible parameter names | vector of **optimizableVariable** objects.

To construct a Naïve Bayes classifier in MATLAB, use the **fitcnb** function.

>> **nbModel = fitcnb(*tableData*, '*ResponseVariable*');**

We can change the distribution using the '**DistributionNames**' property.

>> **nbModel = fitcnb(*trainingData*, '*ResponseVariableName*',... 'DistributionNames','dist Name');**

NOTE

By default, the numeric predictors are modeled as normal (Gaussian) distributions, with estimated means and standard deviations. However, the assumption of normality may not always be appropriate.

Fitting Heart Disease Numeric Data to Naïve Bayes Model

The table **heartDiseaseData** contains several features of different subjects and whether or not they have heart disease, which is saved in the **HeartDisease** variable. Figure 2.37 shows the complete code for importing numeric predictor data, fitting the data to Naïve Bayes classifier, using the **fitcnb** function, and presenting the classification model goodness results.

```
%% Import data
heartData = readtable('heartDiseaseData.xlsx');
heartData.HeartDisease = categorical(heartData.HeartDisease);
%Split the data into training and test sets
% Data standardization;
numData = zscore(heartData{:,1:11});
cvpt=cvpartition(heartData.HeartDisease,'HoldOut',0.30);
hdTrain=numData(training(cvpt),:);
hdTest=numData(test(cvpt),:);
ytrain = heartData.HeartDisease(training(cvpt));
ytest = heartData.HeartDisease(test(cvpt));
% By default, distribution is assumed Gaussian.
NBGsMdl = fitcnb(hdTrain,ytrain);
%% Calculate the loss errTrain and errTest
errTrain = resubLoss(NBGsMdl);
errTest = loss(NBGsMdl,hdTest,ytest);
disp(['Training Error for Gauss Dist. Naïve Bayes Model: ',num2str(errTrain)])
disp(['Test Error for Gauss Dist. Naïve Bayes: ',num2str(errTest)])
Ypred = resubPredict(NBGsMdl);
figure (38);
ConfusionMat1 = confusionchart(ytrain,Ypred);
NBKrnlMdl =fitcnb(hdTrain,ytrain,"DistributionNames","kernel");
%Calculate the loss errTrain2 and errTest2
errTrain2 = resubLoss(NBKrnlMdl);
errTest2 = loss(NBKrnlMdl,hdTest,ytest);
disp(['Training Error for Kernel Dist. Naïve Bayes Model: ',num2str(errTrain2)])
disp(['Test Error for Kernel Dist. Naïve Bayes Model: ',num2str(errTest2)])
Ypred = resubPredict(NBKrnlMdl);
figure (39);
ConfusionMat2 = confusionchart(ytrain,Ypred);
```

Figure 2.37 Naïve-Bayes classification model fit to heart disease data.

Copied from Command Windows, results of training and test errors for both modes are:

```
Training Error for Gauss Dist. Naïve Bayes Model: 0.31104
Test Error for Gauss Dist. Naïve Bayes: 0.21925
Training Error for Kernel Dist. Naïve Bayes Model: 0.24415
Test Error for Kernel Dist. Naïve Bayes Model: 0.22702
```

Figure 2.38 (*left*) shows the confusion matrix plot of **ytrain** (true) vs. **Ypred** for Gaussian distribution assumption and, on the *right*, that for kernel distribution. Kernel distribution trained better than Gaussian because the former has one more degree of freedom in terms of model predictability.

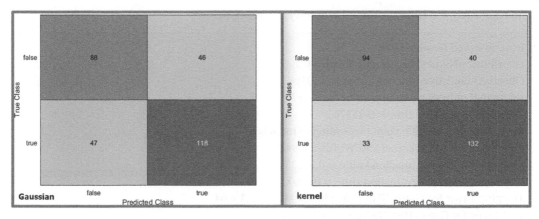

Figure 2.38   The confusion matrix for the trained model under both modes: Gaussian (*left*) and Kernel (*right*).

Fitting Heart Disease All Data Types to Naïve Bayes Model

On the other hand, one may select the entire collection of predictor data, including numeric and non-numeric, as well. Figure 2.39 shows the code for utilizing all predictor data while only standardizing the numeric type. At the same time the hyper parameters optimization mode is set to '**auto**'.

```
%% Import data
heartData = readtable('heartDiseaseData.xlsx');
heartData.HeartDisease = categorical(heartData.HeartDisease);
PrdData = heartData(:,1:21);
% Standardize all numeric predictor columns.
PrdData{:,1:11} = zscore(heartData{:,1:11});
Y= heartData.HeartDisease;
cvpt=cvpartition(heartData.HeartDisease,'HoldOut',0.30);
hdTrain=PrdData(training(cvpt),:);
hdTest=PrdData(test(cvpt),:);
ytrain = heartData.HeartDisease(training(cvpt));
ytest = heartData.HeartDisease(test(cvpt));
rng default;
% Normal data distribution and a "normal" kernel smoothing function.
NBGsMdl = fitcnb(hdTrain,ytrain, 'OptimizeHyperparameters', 'auto', 'kernel',
'normal',...
   'HyperparameterOptimizationOptions',struct('AcquisitionFunctionName',...
   'expected-improvement-plus'))
%Calculate the loss errTrain and errTest
errTrain = resubLoss(NBGsMdl);
errTest = loss(NBGsMdl,hdTest,ytest);
disp(['Training Error for Gauss Dist. Naïve Bayes Model: ',num2str(errTrain)])
disp(['Test Error for Gauss Dist. Naïve Bayes: ',num2str(errTest)])
Ypred = resubPredict(NBGsMdl);
figure (40);
ConfusionMat = confusionchart(ytrain,Ypred);
```

Figure 2.39   Fitting heart disease all predictor data types to Naïve Bayes Model while setting hyper parameters optimization to '**auto**'.

The results (copied from Command Window) are shown below:

Best observed feasible point:
DistributionNames: kernel and Width =0.7965
Observed objective function value = 0.19398
Estimated objective function value = 0.2013
Function evaluation time = 0.25376
Best estimated feasible point (according to models):
DistributionNames: kernel and Width = 0.84299
Estimated objective function value = 0.20133
Estimated function evaluation time = 0.25075
Training Error for Gauss Dist. Naïve Bayes Model: 0.19064
Test Error for Gauss Dist. Naïve Bayes: 0.21841

Figure 2.40 (*left*) shows that 30 iterations are needed to converge the solution where the estimated minimum objective tries to reach minimum observed objective function and (*right*) the estimated objective function versus the distribution type and its width value.

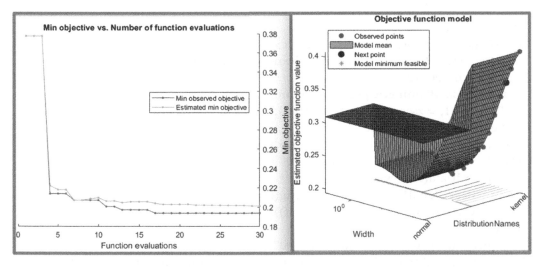

Figure 2.40 Minimum objective versus # of iterations needed to converge the solution (*left*) and optimization of distribution type and its width (*right*).

Figure 2.41 (*left*) shows the confusion matrix plot of **ytrain** (true) vs. **Ypred** for Gaussian distribution assumption with a normal kernel smoothing function and, on the *right*, the workspace variables.

Quiz
Which of the following properties is not associated with multiclass Naïve Bayes model?
a) DistributionNames   b) Kernel   c) MinLeafSize   d) Width

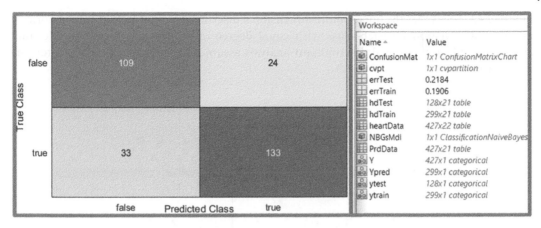

Figure 2.41 (*left*) The confusion matrix plot of **ytrain** (true) vs. **Ypred** for Gaussian distribution assumption with a "normal" kernel smoothing function and (right) workspace variables.

Discriminant Analysis (DA) Classifier for Numeric Predictors Only

Similar to Naïve Bayes, discriminant analysis works by assuming that the observations in each prediction class can be modeled with a normal probability distribution. However, there is no assumption of independence in each predictor. Hence, a multivariate normal distribution is fitted to each class. Figure 2.42 shows that each prediction class is modeled by a normal probability distribution (i.e., $\mu = 0$ and $\sigma = 1$) and the dividing curve is drawn such that the probability value is the same as given by both distributions.

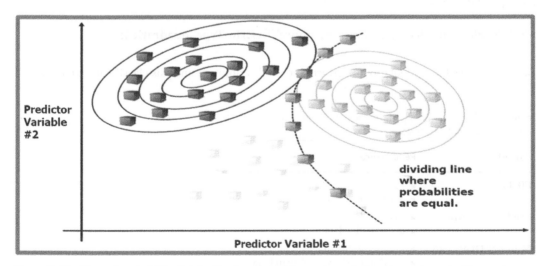

Figure 2.42 Normal probability distribution is assumed for each prediction class data and with equal probability dividing curve.

Figure 2.43 shows that the linear multi-dimensional discriminant analysis assumes a normal distribution function for each class with equal degree of scatter; on the other side, the quadratic multi-dimensional discriminant analysis assumes different degrees of scatter for each class.

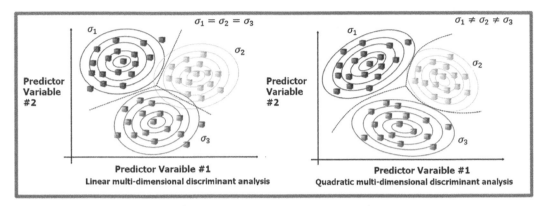

Figure 2.43 Linear- versus quadratic- multidimensional discriminant analysis approach.

By default, the covariance for each response class is assumed to be the same. This results in linear boundaries between classes.

>> **daModel = fitcdiscr(dataTrain,'response');**

Removing the assumption of equal covariances results in a quadratic boundary between classes. Use the '**DiscrimType**' option to do this:

>> **daModel = fitcdiscr(dataTrain,'response','DiscrimType','quadratic');**

Table 2.5 shows the computational performance of discriminant analysis classification model.

Table 2.5 Computational performance of discriminant analysis classification model.

Property	Performance
Fit Time	• Fast • $\propto$ size of the data
Prediction Time	• Fast • $\propto$ size of the data
Memory Overhead	• Linear DA: Small • Quadratic DA: Moderate to large • $\propto$ number of predictors

Table 2.6 shows important common properties of discriminant analysis classification model, quoted from MATLAB help.

Table 2.6 Important common properties of discriminant analysis classification model.

Property	Description/Value
'Delta'	Delta: Linear coefficient threshold for including predictors in a linear boundary. (Default 0) | a nonnegative scalar value. Set Delta to a higher value to eliminate more predictors. Delta must be 0 for QDA models.
'DiscrimType'	**'linear'** (default) **'diaglinear'** | **'pseudolinear'** | **'quadratic'** | **'diagquadratic'** | **'pseudoquadratic'** The first three arguments are of type regularized linear discriminant analysis (LDA) and the last three are quadratic discriminant analysis (QDA). With LDA, all classes have the same, diagonal covariance matrix. On the other side, QDA discriminants tolerate classes with variable covariance matrices.
'FillCoeffs'	Coefficients property flag. **'on'** | **'off'** Setting the flag to **'on'** populates the **Coeffs** property in the classifier object. This can be computationally intensive, especially when cross-validating. The default is **'on'**, unless we specify a cross-validation name-value pair, in which case the flag is set to **'off'** by default.
'Gamma'	Regularization factor to use when estimating the covariance matrix for LDA. A scalar value in the interval [0,1]. Gamma provides finer control over the covariance matrix structure than DiscrimType. If gamma is set to 1, then **fitcdiscr** will restrict the covariance matrix to be diagonal.
'OptimizeHyperparameters'	Parameters to optimize. **'none'** (default) | **'auto'** | **'all'** | string array or cell array of eligible parameter names | vector of **optimizableVariable** objects.
'SaveMemory'	Flag to save covariance matrix. **'off'** (default) | **'on'** If we specify **'on'**, then **fitcdiscr** does not store the full covariance matrix, but instead stores enough information to compute the matrix.
"ScoreTransform"	Score transformation, specified as a character vector, string scalar, or function handle. **"none"** (default) | **"doublelogit"** | **"invlogit"** | "ismax" | **"logit"** | **"sign"** | **"symmetric"** | **"symmetricmax"** | **"symmetriclogit"**.

NOTE

Linear discriminant analysis works well for "wide" data (more predictors than observations).

The **Delta** parameter in **LDA** represents a threshold for including predictors in the linear boundary. Only predictors with a linear coefficient (also known as a "loading") greater than or equal to **Delta** will be included in the boundary. Setting **Delta** to a higher value will eliminate more predictors from the boundary, potentially leading to a simpler and more interpretable model. The default value for **Delta** is *0*, which means that all predictors are included in the boundary.

NOTE (Continued)

Gamma parameter is a scalar value in the interval [0,1] that provides finer control over the covariance matrix structure than other parameters such as the **DiscrimType**. When **Gamma** is set to a value less than 1, it introduces a form of shrinkage regularization that "shrinks" the estimated covariance matrix towards a diagonal matrix. This helps to stabilize the estimate and can improve classification performance.

When **Gamma** is set to 1, **fitcdiscr** will restrict the covariance matrix to be diagonal. This can be useful in situations where the predictor variables are thought to be uncorrelated with each other, or when the number of predictor variables is much larger than the number of observations. In these cases, a diagonal covariance matrix may provide a simpler and more interpretable model. However, setting **Gamma** to 1 should be done with caution, as it may result in a loss of information if the predictor variables are actually correlated.

Quiz

(T/F) A Gaussian distribution for each class of each predictor is assumed for discriminant analysis.

Quiz

(T/F) A quadratic discriminant classifier will always give better results than a linear one. However, the calculations generally take more time.

Discriminant Analysis (DA): Heart Disease Numeric Predictors

Figure 2.44 shows the complete code for importing data, fitting the data to DA classifier, using the **fitcdiscr** function with auto hyper parameters optimization, and presenting the classification model goodness results.

```
%% Import data
heartData = readtable('heartDiseaseData.xlsx');
heartData.HeartDisease = categorical(heartData.HeartDisease);
% Split the data into training and test sets
% Predictors Data is standardized as Gaussian.
numData = zscore(heartData{:,1:11});
cvpt=cvpartition(heartData.HeartDisease,'HoldOut',0.30);
hdTrain=numData(training(cvpt),:);
hdTest=numData(test(cvpt),:);
```

Figure 2.44 Linear- versus quadratic- multidimensional discriminant analysis approach for heart disease data classification.

```
ytrain = heartData.HeartDisease(training(cvpt));
ytest = heartData.HeartDisease(test(cvpt));
%daModel=fitcdiscr(hdTrain,ytrain,'DiscrimType','linear');
rng(1)
daModel = fitcdiscr(hdTrain,ytrain,'OptimizeHyperparameters','auto',...
   'HyperparameterOptimizationOptions',...
   struct('AcquisitionFunctionName','expected-improvement-plus'))
%% Calculate the loss errTrain and errTest
errTrain = resubLoss(daModel);
errTest = loss(daModel,hdTest,ytest);
disp(['Training Error for Auto HP Optim Discriminant Analysis Model:
',num2str(errTrain)])
disp(['Test Error for Auto HP Optim Discriminant Analysis: ',num2str(errTest)])
```

Figure 2.44   (Cont'd)

The results, copied from Command Window, are shown below for both standardized and non-standardized predictors data:

With predictors data standardization:

Best observed feasible point:
Delta = 1.1079e-05   Gamma = 0.0012698
Observed objective function value = 0.24415
Estimated objective function value = 0.24541
Function evaluation time = 0.042119
Best estimated feasible point (according to models):
Delta = 1.9332e-05   Gamma = 8.1669e-05
Estimated objective function value = 0.24497
Estimated function evaluation time = 0.047509
Training Error for Auto HP Optim Discriminant Analysis Model: 0.23077
Test Error for Auto HP Optim Discriminant Analysis: 0.30486

Without predictors data standardization:

Best observed feasible point:
Delta = 6.3462e-06   Gamma = 0.0013377
Observed objective function value = 0.26087
Estimated objective function value = 0.26088
Function evaluation time = 0.047192
Best estimated feasible point (according to models):
Delta = 1.2156e-06   Gamma = 0.00012874
Estimated objective function value = 0.26078
Estimated function evaluation time = 0.046974
Training Error for Auto HP Optim Discriminant Analysis Model: 0.25084
Test Error for Auto HP Optim Discriminant Analysis: 0.26597

Figure 2.45 shows the minimum objective function versus the number of function evaluations (*left*) and the estimated objective function value as a function of gamma and delta parameters (*right*).

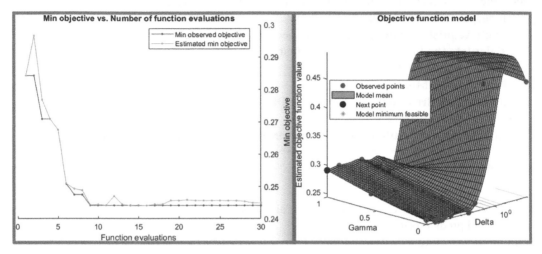

Figure 2.45   The minimum objective function versus the number of function evaluations (*left*) and the estimated objective function value as a function of gamma and delta parameters (*right*).

Support Vector Machine (SVM) Classification Model for All Data Types

fitcsvm function trains or cross-validates a support vector machine (SVM) model for one-class and two-class (binary) classification on a low-dimensional or moderate-dimensional predictor data set. **fitcsvm** supports mapping the predictor data using kernel functions, and supports sequential minimal optimization (SMO), iterative single data algorithm (ISDA), or L1 soft-margin minimization via quadratic programming for objective-function minimization.

NOTES

To train a linear SVM model for binary classification on a high-dimensional data set, that is, a data set that includes many predictor variables, use **fitclinear** instead. So, **fitclinear** is used for predictor data with many features (or columns) yet the response variable is a binary class, at most.

  For multiclass learning with combined binary SVM models, use error-correcting output codes (ECOC). For more details, see **fitcecoc** (next section).

  To train an SVM regression model, see **fitrsvm** for low-dimensional and moderate-dimensional predictor data sets, or **fitrlinear** for high-dimensional data sets.

Suppose we want to classify the data into two classes as shown in Figure 2.46. Which of the three linear boundaries do you think is better at classifying the data?

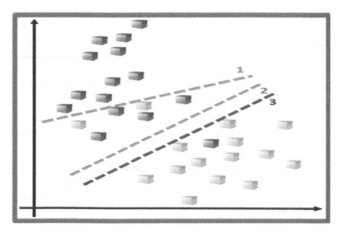

Figure 2.46 Drawing the best border line between two groups of data.

- Boundary 1 misclassifies several data points.
- Boundaries 2 and 3 misclassify fewer data points but would you consider boundary 2 better than boundary 3?

A Support Vector Machine (SVM) algorithm classifies data by finding the "best" hyperplane that separates all data points. In this case, an SVM approach will calculate the boundary that is close to boundary 2. Figure 2.47 shows that SVM is used for a multi-binary variable separation, where the original set of data can be transformed into two separable sets of data. SVM attempts to maximize the margin between the boundary and the nearest neighbor observation (i.e., support vectors).

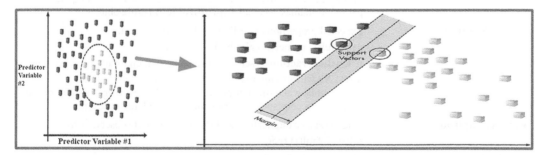

Figure 2.47 SVM data transformation into separable sets of data.

Properties of SVM Model

Table 2.7 shows important common properties of SVM model, quoted from MALTAB help.

Table 2.7 Important properties of SVM model.

Property	Description/Value
'BoxConstraint'	1 (default)	positive scalar. For one-class learning, the software always sets the box constraint to 1.
'CacheSize'	Cache size of reserved memory. *1000* (default)	**maximal**	*positive scalar*. If CacheSize is 'maximal', then the software reserves enough memory to hold the entire n-by-n Gram matrix.
'ClipAlphas'	Flag to clip alpha coefficients. **true** (default)	**false**. If set to **true**, then at each iteration, if αj is near 0 or near Cj, then MATLAB sets αj to *0* or to *Cj*, respectively, where *Cj* is the box constraint of observation j = 1, ..., n.
'KernelFunction'	Kernel function used to compute the elements of the Gram matrix. **linear**	**gaussian**	**rbf**	**polynomial**	function name 'gaussian' or 'rbf': Gaussian or Radial Basis Function (RBF) kernel, *default* for one-class learning. 'linear': Linear kernel, *default* for two-class learning. 'polynomial': Polynomial kernel. Use 'PolynomialOrder', q to specify a polynomial kernel of order q.
'KernelScale'	*1* (default)	**auto**	*positive scalar*. If 'auto' is set, then **fitcsvm** will select an appropriate scale factor using a heuristic procedure.
'PolynomialOrder'	Polynomial kernel function order. *3* (default)	*positive integer*.
'KernelOffset'	Nonnegative scalar, ≥0. The software adds KernelOffset to each element of the Gram matrix.
'OptimizeHyperparameters'	Parameters to optimize. 'none' (default)	'auto'	'all'	string array or cell array of eligible. parameter names	vector of **optimizableVariable** objects.
'OutlierFraction'	Expected proportion of outliers in the training data. *0* (default)	numeric scalar in the interval [0,1] If 'OutlierFraction' is set *outlierfraction*>0, then for a 2-class learning, MATLAB will attempt to remove 100*$outlierfraction$% of the observations when the optimization algorithm converges. The removed observations correspond to gradients that are large in magnitude.
'RemoveDuplicates'	Flag to replace duplicate observations with single observations. **false** (default)	**true**.
'Standardize'	Flag to standardize the predictor data. **false** (default)	**true**.

In addition, the SVM solver optimization routine is specified as the comma-separated pair consisting of 'Solver' and a value in Table 2.8, quoted from MATLAB help.

The default value will be 'ISDA' if you set 'OutlierFraction' to a positive value for two-class learning, and will be 'SMO' otherwise.

Table 2.8 SVM solver optimization routines.

Property	Value/Description
'Solver'	**'ISDA'**: The Iterative Single Data Algorithm (**ISDA**) is a simple and efficient algorithm for solving the **SVM** optimization problem. **ISDA** works by iteratively selecting a single data point and updating the model parameters based on that point. The advantage of **ISDA** is its simplicity and efficiency, especially for large datasets. However, **ISDA** may not always find the global optimum and may be sensitive to the choice of learning rate.
	'L1QP': The **L1QP** solver uses **quadprog** (from the Optimization Toolbox) to solve the optimization problem associated with L1 soft-margin SVM. **L1QP** is a quadratic programming algorithm that can handle linear and nonlinear constraints. The advantage of **L1QP** is that it is guaranteed to find the global optimum for the L1 soft-margin SVM problem. This option requires an Optimization Toolbox™ license. For more details, see *Quadratic Programming Definition* (Optimization Toolbox).
	'SMO': Sequential Minimal Optimization (**SMO**) is a popular algorithm for solving the **SVM** optimization problem. **SMO** works by iteratively selecting a pair of data points and updating the model parameters based on those points. The advantage of **SMO** is its efficiency and scalability, especially for large datasets. **SMO** can also handle nonlinear SVM problems by using the kernel trick. However, **SMO** may not always find the global optimum and may be sensitive to the choice of kernel function.

Quiz

An SVM requires how many groups in the response?
   a) 1   b) 1 or 2   c) ≥ 2 d) ≤ 4

SVM Classification Model: Heart Disease Numeric Data Types

Figure 2.48 shows the complete code for importing data, fitting the numeric predictor data to SVM classifier, using the **fitcsvm** function with auto hyper parameters optimization and presenting the classification model goodness results.

```
%% Import data
heartData = readtable('heartDiseaseData.xlsx');
heartData.HeartDisease = categorical(heartData.HeartDisease);
%% Split the data into training and test sets
% Predictor data is standardized.
numData = zscore(heartData{:,1:11});
cvpt=cvpartition(heartData.HeartDisease,'HoldOut',0.30);
hdTrain=numData(training(cvpt),:);
hdTest=numData(test(cvpt),:);
ytrain = heartData.HeartDisease(training(cvpt));
ytest = heartData.HeartDisease(test(cvpt));
```

Figure 2.48   The code for importing data, fitting the numeric predictor data to SVM classifier, using the **fitcsvm** function with auto hyper parameters optimization.

```
rng default;
svmModel = fitcsvm(hdTrain,ytrain,'OptimizeHyperparameters','auto', ...
  'HyperparameterOptimizationOptions',struct('AcquisitionFunctionName', ...
  'expected-improvement-plus'))
%Calculate the loss errTrain and errTest
errTrain = resubLoss(svmModel);
errTest = loss(svmModel,hdTest,ytest);
disp(['Training Error for Auto Hyper Param Optim SVM Model: ',num2str(errTrain)])
disp(['Test Error for Auto Hyper Param Optim SVM Model: ',num2str(errTest)])
Ypred = resubPredict(svmModel);
figure (50);
ConfusionMat = confusionchart(ytrain,Ypred);
```

Figure 2.48 (Cont'd)

The results are copied from Command Window:

Best observed feasible point:
BoxConstraint = 0.0010153   **KernelScale** = 0.0010751
Observed objective function value = 0.27425
Estimated objective function value = 0.27832
Function evaluation time = 6.7578
Best estimated feasible point (according to models):
BoxConstraint = 920.85   **KernelScale** = 3.2626
Estimated objective function value = 0.2768
Estimated function evaluation time = 1.2511
Training Error for Auto Hyper Param Optim SVM Model: 0.25418
Test Error for Auto Hyper Param Optim SVM Model: 0.25784

Figure 2.49 (*left*) shows the minimum objective function versus the number of function evaluations or iterations and (*right*) the estimated objective function value versus **KernScale** and **BoxConstarint** varied parameters. The **BoxConstraint** is a positive scalar value that specifies the upper bound on the sum of the slack variables, which represent the degree of violation of the margin constraint. The advantage of using **BoxConstraint** in an SVM model is that it provides a way to control the trade-off between maximizing the margin and minimizing the classification error. By adjusting the value of **BoxConstraint**, we can tune the model to emphasize either a larger margin or a smaller classification error. In addition, the **BoxConstraint** can help to prevent overfitting of the SVM model. Overfitting occurs when the model is too complex and captures noise or irrelevant features in the training data, which leads to poor generalization performance on new data. By controlling the value of **BoxConstraint**, we can regularize the model and prevent it from overfitting.

On the other hand, the **KernScale** parameter in an SVM model is used to adjust the kernel function that maps input data to a high-dimensional feature space where a linear classifier is used to find the decision boundary. By scaling the input data before applying the kernel function, the **KernScale** parameter can adjust the sensitivity of the kernel function to the input data, which affects the smoothness and curvature of the decision boundary. Tuning the **KernScale** parameter can help find a balance between overfitting and underfitting, leading to better generalization performance on new data. Additionally, using **KernScale** can make the model more robust to outliers and noise in the input data. Overall, the **KernScale** parameter is a useful parameter in SVM modeling that can improve the performance and robustness of the model by adjusting the sensitivity of the kernel function to the input data.

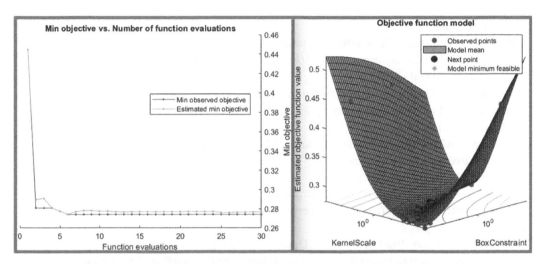

Figure 2.49 (*left*) shows the minimum objective function versus iterations and (*right*) the estimated objective function value versus **KernScale** and **BoxConstarint** varied parameters.

Figure 2.50 (*left*) shows the confusion matrix plot of **ytrain** (true) vs. **Ypred** for SVM model with "**linear**" kernel, *default* for two-class learning (see Table 2.7) and, on the *right*, the workspace variables. From the workspace, click on **svmModel** 1×1 SVM classifier to explore further properties.

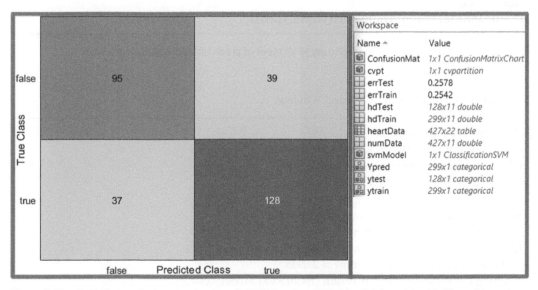

Figure 2.50 (*left*) The confusion matrix plot of **ytrain** (true) vs. **Ypred** for **SVM** model with "**linear**" kernel function and (right) workspace variables.

SVM Classification Model: Heart Disease All Data Types

Figure 2.51 shows that we include both numeric and categorical predictor columns to represent the predictor features. Again, SVM classifier, **fitcsvm** function with auto hyper parameters optimization, is used. The classification model goodness results are shown at the end.

```
%% Import data
heartData = readtable('heartDiseaseData.xlsx');
heartData.HeartDisease = categorical(heartData.HeartDisease);
PrdData = heartData(:,1:21);
% Standardize all numeric predictor columns.
PrdData{:,1:11} = zscore(heartData{:,1:11});
Y= heartData.HeartDisease;
cvpt=cvpartition(heartData.HeartDisease,'HoldOut',0.30);
hdTrain=PrdData(training(cvpt),:);
hdTest=PrdData(test(cvpt),:);
ytrain = heartData.HeartDisease(training(cvpt));
ytest = heartData.HeartDisease(test(cvpt));
rng default;
svmModel = fitcsvm(hdTrain,ytrain,'OptimizeHyperparameters','auto', ...
    'HyperparameterOptimizationOptions',struct('AcquisitionFunctionName', ...
    'expected-improvement-plus'))
errTrain = resubLoss(svmModel);
errTest = loss(svmModel,hdTest,ytest);
disp(['Training Error for Auto Hyper Param Optim SVM Model: ',num2str(errTrain)])
disp(['Test Error for Auto Hyper Param Optim SVM Model: ',num2str(errTest)])
Ypred = resubPredict(svmModel);
figure (52);
ConfusionMat = confusionchart(ytrain,Ypred);
```

Figure 2.51 SVM classification model fitting to all types of predictor data and fitting results.

The results are copied from Command Window:

```
Best observed feasible point:
BoxConstraint = 0.0010646 KernelScale = 0.0050739
Observed objective function value = 0.23746
Estimated objective function value = 0.23754
Function evaluation time = 2.7744
Best estimated feasible point (according to models):
BoxConstraint = 0.029754 KernelScale = 0.022311
Estimated objective function value = 0.23727
Estimated function evaluation time = 2.0935
Training Error for Auto Hyper Param Optim SVM Model: 0.1806
Test Error for Auto Hyper Param Optim SVM Model: 0.21112
```

Figure 2.52 (*left*) shows the minimum objective function versus the number of function evaluations or iterations and (*right*) the estimated objective function value versus **KernScale** and **BoxConstarint** varied parameters.

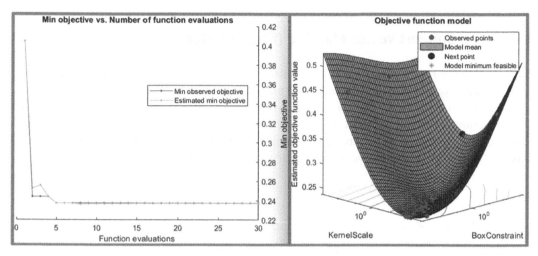

Figure 2.52   (*left*) shows the minimum objective function versus iterations and (*right*) the estimated objective function value versus **KernScale** and **BoxConstarint** varied parameters.

Figure 2.53 (*left*) shows the confusion matrix plot of **ytrain** (true) vs. **Ypred** for SVM model with **"linear"** kernel function, *default* for two-class learning (see Table 2.7) and, on the *right*, the workspace variables. From the workspace, click on **svmModel** 1×1 SVM classifier to explore further properties.

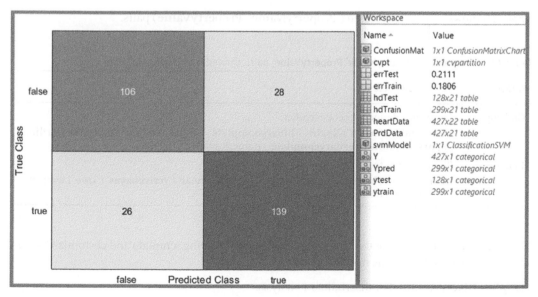

Figure 2.53   (*left*) The confusion matrix plot of **ytrain** (true) vs. **Ypred** for **SVM** model with "**linear**" kernel function and (*right*) workspace variables.

Quiz
(T/F) You should only use SVMs with data that is separated by a linear hyperplane.

Multiclass Support Vector Machine (fitcecoc) Model

Suppose that the risk of heart attack is discretized into multiple categories instead of the binary categorization we had seen so far. Can we still use the support vector machine for classification?

The underlying calculations for classification with support vector machines are binary by nature. We can perform multiclass SVM classification by creating an error-correcting output codes (ECOC) classifier. The function **fitcecoc** creates a multiclass SVM, which reduces the model into multiple, binary classifiers using the one-versus-one design, as demonstrated in Figure 2.54.

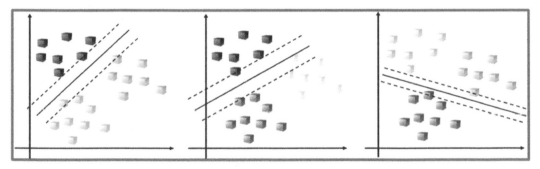

Figure 2.54   Transformation of multiple categories data into multiple, binary data classifiers.

Table 2.9 shows some **ecoc**-related ('**PropertyName**', **PropertyValue**) pairs.

Table 2.9   Some **ecoc** ('**PropertyName**', **PropertyValue**) pairs. Quoted from MATLAB help.

Property	Description/Value
'Coding'	The coding design for ecoc classifier. 'onevsone' (default) | 'allpairs' | 'binarycomplete' | 'denserandom' | 'onevsall' | 'ordinal' | 'sparserandom' | 'ternarycomplete' | numeric matrix.
'Learners'	Binary learner templates. 'svm' (default) | 'discriminant' | 'kernel' | 'knn' | 'linear' | 'naivebayes' | 'tree' | template object | cell vector of template objects

For each '**Learners**' type, the user may refer to the corresponding template and customize its pertinent property value pairs, as mapped below:

templateDiscriminant: For discriminant analysis.
templateKernel: For kernel classification.
templateKNN: For k-nearest neighbors.

templateLinear: For linear classification.
templateNaiveBayes: For Naïve Bayes.
templateSVM: For SVM.
templateTree: For classification trees.

Two example templates are exemplified below:

Creating a multiclass classifier using SVM template learner:

1. Create a template for a binary classifier: Create a template for a binary SVM using the function **templateSVM**.

>>**svmtemplate = templateSVM('PropertyName', PropertyValue)**

Where

templatesvm:	An SVM binary classifier
'PropertyName':	Property value name, e.g., **'KernelFunction'**.
PropertyValue:	Property value, e.g., **'Polynomial'**. You can provide multiple property name-value pairs.

Please, refer to Table 2.7 for other different **'PropertyName', PropertyValue** pairs.

2. Create multiclass SVM classifier: Use the function **fitcecoc** to create a multiclass SVM classifier.

>>**ecocModel = fitcecoc(dataTrain,'y','Learners',svmtemplate)**

Where

ecocModel:	ECOC classifier.
dataTrain:	Training data.
'y':	Response variable name.
'Learners':	Property name for specifying the binary classifier.
svmtemplate:	Binary classifier using **templateSVM** function.

Creating a multiclass classifier using linear template learner:

1. Create a linear classification model to high-dimensional data for multiclass problems using the function **templateLinear.**

>>**lnrtemplate = templateLinear('PropertyName', PropertyValue)**

Where

templateLinear:	linear binary classification model
'PropertyName':	Property value name, e.g., **'KernelFunction'**.
PropertyValue:	Property value, e.g., **'Polynomial'**. You can provide multiple property name-value pairs.

Please, refer to Table 2.9 below for other different **'PropertyName', PropertyValue** pairs.

2. Create multiclass SVM classifier: Use the function **fitcecoc** to create a multiclass SVM classifier.

>>**ecocModel = fitcecoc(dataTrain,'y','Learners', lnrtemplate)**

Where

ecocModel:	ECOC classifier.
dataTrain:	Training data.
'y':	Response variable name.
'Learners':	Property name for specifying the binary classifier.
lnrtemplate:	Binary classifier using **templateLinear** function.

Table 2.10 shows some important Name-Value pair arguments to be used in **templateLinear()**, quoted from MATLAB help.

Table 2.10 Name-Value Pair arguments to be used in **templateLinear()**.

Property	Description/Value
'Lambda'	**Regularization term strength** 'auto' (default) | *nonnegative scalar* | *vector of nonnegative values.* For **'auto'**: Lambda = 1/n.
'Learner'	Linear classification model type **'svm'** (default) | **'logistic'** **'svm'**: Support vector machine **'logistic'**: Logistic regression
'NumCheckConvergence'	Number of passes through entire data set to process before next convergence check *5* (default) | positive integer.
'ObservationsIn'	Predictor data observation dimension. **'rows'** (default) | **'columns'**
'PassLimit'	Maximal number of passes through the data. *10* (default) | positive integer. When MATLAB completes one pass through the data, it will mean all observations are processed. When MATLAB passes through the data **PassLimit** times, it will terminate optimization.

Table 2.10 (Continued)

Property	Description/Value
'Regularization'	Complexity penalty type '**lasso**' | '**ridge**' MATLAB composes the objective function for minimization from the sum of the average loss function and the regularization term. '**lasso**': Lasso (L1) penalty: $\lambda \sum_{j=1}^{p} \lvert \beta_j \rvert$. p number of parameters or coefficients. '**ridge**': Ridge (L2) penalty: $\dfrac{\lambda}{2} \sum_{j=1}^{p} \beta_j^2$. See Ch. 4 on Lasso and Ridge factor.
'Solver'	Objective function minimization technique. '**sgd**' | '**asgd**' | '**dual**' | '**bfgs**' | '**lbfgs**' | '**sparsa**' | *string array* | *cell array of character vectors*. '**sgd**': Stochastic gradient descent (SGD). '**asgd**': Average stochastic gradient descent (ASGD). '**dual**': Dual SGD for SVM. For '**ridge**' and Learner '**svm**' only. '**bfgs**': Broyden-Fletcher-Goldfarb-Shanno quasi-Newton algorithm (BFGS). '**lbfgs**': Limited-memory BFGS (LBFGS). For '**ridge**' only. '**sparsa**': Sparse Reconstruction Separable Approximation (SpaRSA). For '**lasso**'.

Multiclass Support Vector Machines Model: Red Wine Data

Notice that the red wine data contains six levels: 3, 4, 5, 6, 7, 8 whereas in the previous heart disease interactions, it was categorized as true or false.

Figure 2.55 shows the complete code for using the multiclass SVM classifier, **fitcecoc** function, while calling, one at a time, one of the two indicated templates: **templateSVM** or **templateLinear** to carry out the multiple, one-versus-one and/or one-versus-all binary classification. In addition, the model goodness is shown at the end.

```
load('winered.mat')
WineRed.quality = categorical(WineRed.quality);
origData = WineRed{:,1:end-1};
MaxVal=max(origData,[], 1);
MinVal=min(origData,[], 1);
for j=1:size(origData,2)
numData(:,j)=(origData(:,j)-MinVal(j))/(MaxVal(j)-MinVal(j));
end
pt = cvpartition(WineRed.quality,'HoldOut',0.30);
RWTrain = numData(training(pt),:);
```

Figure 2.55 **fitcecoc** function calls one of the two indicated templates: **templateSVM** or **templateLinear** to carry out the multiple, one-versus-one and/or one-versus-all binary classification. The model goodness is shown at the end.

```
RWTest = numData(test(pt),:);
ytrain = WineRed.quality(training(pt));
ytest = WineRed.quality(test(pt));
svmtemplate = templateSVM('Standardize',true,'KernelFunction','gaussian');
ecocModel = fitcecoc(RWTrain,ytrain,'Learners', svmtemplate,...
'OptimizeHyperparameters','auto', ...
   'HyperparameterOptimizationOptions',struct('AcquisitionFunctionName', ...
   'expected-improvement-plus'));
predictquality = predict(ecocModel,RWTest);
errTrain=resubLoss(ecocModel);
errTest=loss(ecocModel,RWTest,ytest);
disp(['Training Error for SVM-ecocModel: ',num2str(errTrain)])
disp(['Test Error for SVM-ecocModel: ',num2str(errTest)])
[cm,cl] = confusionmat(ytest, predictquality);
figure (56);
heatmap(cl,cl,cm);
trueclass=0.0;
for k=1:height(cm)
trueclass = trueclass+sum(cm(k,k));
end
TC=100*trueclass/height(ytest);
disp(['Percentage of true classification for SVM-ecocModel: ',num2str(TC),'%'])
%Using templateLinear function
lnrtemplate = templateLinear('Lambda', 'auto','Regularization','ridge',...
'Learner','svm');
rng default;
ecocModel2= fitcecoc(RWTrain,ytrain,'Learners',lnrtemplate,...
'OptimizeHyperparameters','auto',...
'HyperparameterOptimizationOptions',struct('AcquisitionFunctionName',...
'expected-improvement-plus'));
%errTrain2=resubLoss(ecocModel2);
errTest2=loss(ecocModel2,RWTest,ytest);
%disp(['Training Error for Linear-ecocModel2: ',num2str(errTrain2)])
disp(['Test Error for Linear-ecocModel2: ',num2str(errTest2)])
predictquality2 = predict(ecocModel2,RWTest);
[cm2,cl2] = confusionmat(ytest, predictquality2);
figure (57);
heatmap(cl2,cl2,cm2);
trueclass2=0.0;
for k2=1:height(cm2)
trueclass2 = trueclass2+sum(cm(k2,k2));
end
TC2=100*trueclass2/height(ytest);
disp(['Percentage of true classification for Linear-ecocModel2: ',num2str(TC2),'%'])
```

Figure 2.55 (Cont'd)

The results are copied from Command Window:

Best observed feasible point:
Coding = onevsone   BoxConstraint = 97.776   KernelScale = 1.1431
Observed objective function value = 0.36161
Estimated objective function value = 0.36266
Function evaluation time = 0.66221
Best estimated feasible point (according to models):
Coding = onevsone   BoxConstraint = 89.408   KernelScale =1.1003
Estimated objective function value = 0.36202
Estimated function evaluation time = 0.65139
Training Error for SVM-ecocModel: 0
Test Error for SVM-ecocModel: 0.36024
Percentage of true classification for SVM-ecocModel: 64.0919%
****

Best observed feasible point:
Coding = onevsone   Lambda = 7.2189e-5   Learner = logistic
Observed objective function value = 0.40804
Estimated objective function value = 0.40888
Function evaluation time = 0.87205
Best estimated feasible point (according to models):
Coding = onevsone   Lambda = 6.4769e-5   Learner = logistic
Estimated objective function value = 0.40883
Estimated function evaluation time = 0.87511
Test Error for Linear-ecocModel2: 0.38929
Percentage of true classification for Linear-ecocModel2: 64.0919%

NOTE

Although we selected the learner type as "**svm**" in **templateLinear**, nevertheless, the auto hyper parameters optimization algorithm found that the best learner is "**logistic**". In addition, "**Standardization**" is not an option in the **templateLinear** case.

Figure 2.56 shows the heatmap of confusion matrix indicating the matching and mismatching between predicted, column-wise, and observed data, row-wise, using **templateSVM** (*left*) and **templateLinear** (*right*).

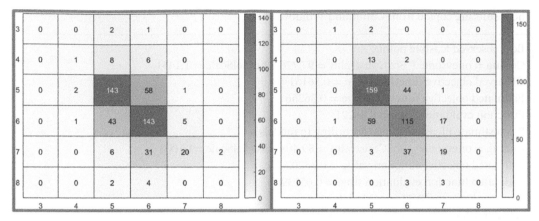

Figure 2.56 The heatmap of confusion matrix of red wine showing the matching and mismatching between predicted, column-wise, and observed data, row-wise, using **templateSVM** (*left*) and **templateLinear** (*right*).

Binary Linear Classifier (fitclinear) to High-Dimensional Data

fitclinear trains linear classification models for two-class (binary) learning with high-dimensional, full or sparse predictor data. Available linear classification models include regularized support vector machines (SVM) and logistic regression models. **fitclinear** minimizes the objective function using techniques that reduce computing time (e.g., stochastic gradient descent).

>>lnrMdl = fitclinear(dataTrain,'y', 'PropertyName', PropertyValue);

Please, refer to Table 2.10 for different '**PropertyName**', **PropertyValue** pairs.
    For auto hyper parameters optimization, one may use:

```
>>rng default;
>>lnrMdl = fitclinear(X,Y,'ObservationsIn','rows','Solver','sparsa',...
    'OptimizeHyperparameters','auto','HyperparameterOptimizationOptions',...
    struct('AcquisitionFunctionName','expected-improvement-plus'));
Or,
>>lnrMdl = fitclinear(X',Y,'ObservationsIn','columns','Solver','sparsa',...
    'OptimizeHyperparameters','auto','HyperparameterOptimizationOptions',...
    struct('AcquisitionFunctionName','expected-improvement-plus'));
```

NOTE

For both classifiers: **fitcecoc** and **fitclinear**, if we orient predictor matrix so that observations correspond to columns and specify '**ObservationsIn**','**columns**', then we might experience a significant reduction in optimization execution time.

Figure 2.57 shows fitting heart disease data to **fitclinear** classification model.
Test Error for Auto Hyper Param Optim LNR Model: $0.17976 = (12+11)/(12+11+46+59)$.

```
heartData = readtable('heartDiseaseData.xlsx');
heartData.HeartDisease = categorical(heartData.HeartDisease);
PrdData = heartData(:,1:21);
% Standardize all numeric predictor columns.
PrdData{:,1:11} = zscore(heartData{:,1:11});
Y= heartData.HeartDisease;
cvpt=cvpartition(heartData.HeartDisease,'HoldOut',0.30);
hdTrain=PrdData(training(cvpt),:);
hdTest=PrdData(test(cvpt),:);
ytrain = heartData.HeartDisease(training(cvpt));
ytest = heartData.HeartDisease(test(cvpt));
rng default;
lnrMdl = fitclinear(hdTrain,ytrain,'ObservationsIn','rows','Solver','sparsa',...
   'OptimizeHyperparameters','auto','HyperparameterOptimizationOptions',...
   struct('AcquisitionFunctionName','expected-improvement-plus'));
%errTrain = resubLoss(lnrMdl);
errTest = loss(lnrMdl,hdTest,ytest);
%disp(['Training Error for Auto Hyper Param Optim LNR Model:
%',num2str(errTrain)])
disp(['Test Error for Auto Hyper Param Optim LNR Model: ',num2str(errTest)])
Ypred = predict(lnrMdl,hdTest);
figure (58);
ConfusionMat = confusionchart(ytest,Ypred);
```

Figure 2.57 Fitting heart disease data to **fitclinear** classification model.

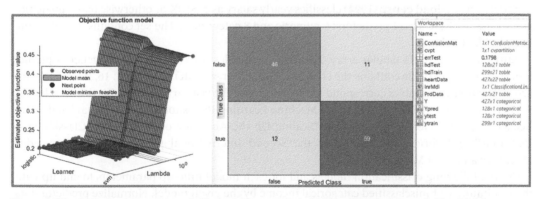

Figure 2.58 Optimized parameters (*left*), confusion matrix (*middle*) and workspace variables (*right*).

Figure 2.58 shows the optimized parameters (*left*), the confusion matrix (*middle*), and the workspace variables (*right*). **Lambda** is a regularization parameter that controls the strength of regularization in the model. Regularization is a technique used to prevent overfitting, which occurs when the model becomes too complex and performs well on the training data but poorly on test data. A higher value of **Lambda** results in stronger regularization, which can help prevent overfitting. On the other hand, a lower value of **Lambda** results in weaker regularization, which can allow the model to fit the training data more closely but may lead to overfitting. On the other hand, the choice of **Learner** depends on the specific problem being solved and the type of data being used.

For example, '**svm**' is a good choice for binary classification problems, while '**ridge**' is a good choice for regression problems with many features.

CHW 2.1 Mushroom Edibility Data

This data set (>>**Mushroom=readtable('Mushroom.csv','Format', 'auto');**) includes several samples of mushrooms. The mushrooms are classified as either edible or poisonous. The table mushroom contains several features of a mushroom's attributes and the corresponding response, edibility. Because all of the predictor variables are categorical, we can specify a multivariate multinomial distribution.

Reference: https://datahub.io/machine-learning/mushroom

Use the following classification models and for each model tune-up parameters to end up with 0% percentage of misclassified poisons by the given model. Show the confusion matrix for each case.

1) KNN (**fitcknn**)
2) Binary Decision Tree (**fitctree**)
3) Naïve Bayes (**fitcnb**)
4) Support Vector Machine (**fitcsvm**)
5) Multi Support Vector Machine (**fitcecoc**)
6) Binary Linear Classifier (**fitclinear**)

Why does not the Discriminant Analysis (DA) Classifier, **fitcdiscr**, work with mushroom edibility data? Also, use MATLAB classification learner toolbox to find the best method with the highest validation and testing accuracy.

Suggest the best model(s) with the lowest model training and test error.

CHW 2.2 1994 Adult Census Income Data

This data set (>>**load census1994**) classifies yearly salary as > \$50K or otherwise based on demographic data. There are 14 predictors: 6 numeric and 8 categorical. The set contains the following two MATLAB tables:

adultdata: 32,561×15 tabular array. The first 14 variables correspond to heterogeneous predictors and the last variable contains the class labels (>50K or <=50K). **adulttest**: 16,281×15 tabular array. The first 14 variables correspond to predictors and the last variable contains the class labels.

This data set was donated to the UCI Machine Learning Repository by: Ron Kohavi and Barry Becker in 1994. Reference: Kohavi, R. "Scaling Up the Accuracy of Naïve-Bayes Classifiers: A Decision-Tree Hybrid." Proceedings of the Second International Conference on Knowledge Discovery and Data Mining, 1996.

Use the following classification models and for each model tune-up parameters to end up with the minimum % of misclassified categorical income by the given model. Normalize predictor data if necessary or recommended for a specific method. Show the confusion matrix for each case.

1) Binary Decision Tree (fitctree)
2) Naïve Bayes (fitcnb)
3) Support Vector Machine (fitcsvm)
4) Multi Support Vector Machine (fitcecoc)
5) Binary Linear Classifier (**fitclinear**)

Notice that the CPU elapsed time in step #3 is seven hours long and in step #4 four hours long to complete the task running on a laptop, having Intel(R) Core(TM) i5-10210U CPU @ 1.60GHz 2.11 GHz processor and 8.00 GB (7.84 GB usable) installed RAM.

Moreover, use MATLAB classification learner toolbox to find the best method with the highest validation and testing accuracy.

Suggest the best model(s) with the lowest model training and test error.

CHW 2.3   White Wine Classification

This data set red wine data can be accessed via

>>**load('winewhite.mat');**

Or,

>>**readtable('winequality-white.csv','Format', 'auto');**

contains several features of white wine and the corresponding quality. The wine quality dataset involves predicting the quality of white wines on a scale given chemical measures of each wine.

Reference: (https://archive.ics.uci.edu/ml/machine-learning-databases/wine-quality/winequality-white.csv).

It is a multiclass classification problem and the number of observations for each class is not balanced. There are 4898 observations with 11 input variables and one output variable. The variable names are as follows:

1) fixed acidity
2) volatile acidity
3) citric acid
4) residual sugar
5) chlorides
6) free sulfur dioxide
7) total sulfur dioxide
8) density
9) pH
10) sulphates
11) alcohol
12) quality (score between 3 and 9).

Create a classifier to predict the quality for the data. Calculate the misclassification loss. Find both training and testing error as explained in the chapter.

Use the following classification models and for each model tune-up parameters to end up with 0% percentage or minimum percent of misclassified wine quality by the given model. Show the confusion matrix for each case.

1) KNN (**fitcknn**)
2) Binary Decision Tree (**fitctree**)
3) Naïve Bayes (**fitcnb**)
4) Discriminant Analysis (DA) Classifier, **fitcdiscr**
5) Multi Support Vector Machine (**fitcecoc**)

Why does not **fitcsvm** work for white wine data?
Why does not **fitclinear** work for white wine data?

Also, use MATLAB classification learner toolbox to find the best method with the highest validation and testing accuracy.

Suggest the best model(s) with the lowest model training and test error.

CHW 2.4   Cardiac Arrhythmia Data

Quoted from: https://www.mayoclinic.org/diseases-conditions/heart-arrhythmia/symptoms-causes/syc-20350668. "A heart arrhythmia (uh-RITH-me-uh) is an irregular heartbeat. Heart rhythm problems (heart arrhythmias) occur when the electrical signals that coordinate the heart's beats don't work properly. The faulty signaling causes the heart to beat too fast (tachycardia), too slow (bradycardia) or irregularly."

Cardiac arrhythmia data from the UCI machine learning repository.

http://archive.ics.uci.edu/ml/datasets/Arrhythmia.

See **VarNames** for names of 279 input variables.

Y is the class attribute ranging from 1 to 16.

---

1 = No arrhythmia
2–15 = Various classes of arrhythmia
16 = Unclassified cases

Nominal attributes are coded as numeric variables with values 0 and 1.

Carry out multi support vector machine (**fitcecoc**) classification. To optimize the performance, you may attempt to change both '**Coding**' and '**Learners**' default values to some other valid values. See Table 2.9 in this regard. Show the confusion matrix.

CHW 2.5   Breast Cancer Diagnosis

Many features are computed from digitized images of breast masses which describe characteristics of the cell nuclei. The diagnosis of the mass is classified as either benign (**B**) or malignant (**M**).

Ref: https://datahub.io/machine-learning/breast-cancer

Ref: https://archive.ics.uci.edu/ml/machine-learning-databases/breast-cancer-wisconsin

Use the following classification models and for each model tune-up parameters to end up with 0% percentage of misclassified breast cancer, predicted benign (**B**) although the observed case is malignant (**M**). Show the confusion matrix for each case.

1) KNN (**fitcknn**)
2) Binary Decision Tree (**fitctree**)
3) Naïve Bayes (**fitcnb**)
4) Discriminant Analysis (DA) Classifier, **fitcdiscr**
5) Support Vector Machine (**fitcsvm**)
6) Multi Support Vector Machine (**fitcecoc**)
7) Binary Linear Classifier (**fitclinear**)

Also, use MATLAB classification learner toolbox to find the best method with the highest validation and testing accuracy.

Suggest the best model(s) with the lowest model training and test error.

3

Methods of Improving ML Predictive Models

Accuracy and Robustness of Predictive Models

Multiple classification algorithms are available for training a predictive model on a given dataset. To identify the model that exhibits superior generalizability, a portion of the data can be set aside for validation. Prior to finalizing the model for production, it is worth exploring further refinements. Is it possible to enhance confidence in loss estimates by altering the division between training and validation data? Additionally, can we streamline the model by reducing the number of predictor variables?

Keep in mind that a simpler model is easier to interpret, more computationally efficient, and less prone to overfitting. So far, we have selected a method/classifier to fit the data using a training set, calculated the model loss, validated the model using a test set, and evaluated the validation accuracy, as shown in Figure 3.1.

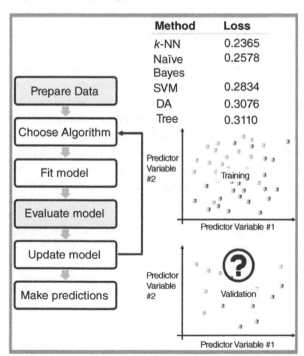

Figure 3.1 Partition of data into training and testing to tune-up the fitting model accuracy.

Machine and Deep Learning Using MATLAB: Algorithms and Tools for Scientists and Engineers, First Edition. Kamal I. M. Al-Malah.
© 2024 Kamal I. M. Al-Malah. Published 2024 by John Wiley & Sons, Inc.
Companion Website: www.wiley.com/go/al-malah/machinelearningmatlab

Figure 3.2 shows more methods to be adopted in this chapter to better refine the predictability of the classifying model without touching the extremities of the reliability, i.e., neither over-fitting nor oversimplification case. The model ought to be robust and accurate.

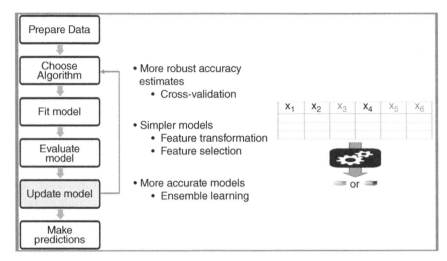

Figure 3.2   Methods for further refinement of model predictability in terms of robustness and accuracy.

Evaluating a Model: Cross-Validation

One way to assess the effectiveness of a classification model is by splitting our data into training and test sets. By training the model with the training data and then evaluating its performance using the test data, we can calculate the loss. When comparing various learning algorithms or methods, we can compute the loss for each method and select the one with the lowest value. However, it's important to note that this loss calculation is specific to a particular test dataset. While a learning algorithm may perform well on that specific test data, it may not generalize effectively to other datasets.

We can reduce the effect of a specific test data on model evaluation using cross-validation. The general idea of a cross-validation is to repeat the above process by creating different training and test data, fit the model to each training data, and calculate the loss using the corresponding test data.

The model variable contains the information that mathematically represents a model and the data, which includes the training and the test data. The information about the partitions or the folds needed for cross-validation should also be stored in the model variable. The model with a cross-validation scheme can be represented as shown in Figure 3.3

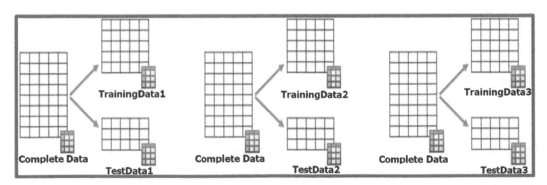

Figure 3.3   Using different partition schemes (KFolds) which will be part of the classifier itself.

Cross-Validation Tune-up Parameters

To create a model with cross-validation, provide one of the following options in the fitting function.

>> **mdl = fitcknn(data,'responseVarName','optionName','optionValue')**

The **'optionName','optionValue'** pairs are shown in Table 3.1, quoted from MATLAB help.

NOTE

For a given classification model, to create a cross-validated model, we can specify only one of these four name-value arguments: **CVPartition**, **Holdout**, **KFold**, or **Leaveout**.

Table 3.1 cvpartition options to use for cross-validation purposes.

cvpartition Option NameOption Value	Description
c = cvpartition(n,'KFold',k)	Returns a cvpartition object **c** that defines a random non-stratified partition for **k-fold** cross-validation on **n** observations. The partition randomly divides the observations into **k** disjoint subsamples, or folds, each of which has approximately the same number of observations. k (scalar).
c = cvpartition(n,'Holdout',p)	Creates a random non-stratified partition for **holdout** validation on **n** observations. This partition divides the observations into a training set and a test, or holdout, set. **p** is scalar from 0 to 1.
c = cvpartition(group,'KFold',k)	Creates a random partition for stratified **k-fold** cross-validation. Each fold, has approximately the same number of observations and contains almost the same class proportions as in group. k (scalar).
c = cvpartition(group,'Holdout',p)	Randomly partitions observations into a training set and a test, or holdout, set with stratification, using the class information in group. Both the training and test sets have approximately the same class proportions as in group. **p** is scalar from 0 to 1.
c = cvpartition(n,'Leaveout')	Creates a random partition for leave-one-out cross-validation on n observations. **Leave-one-out** is a special case of '**KFold**' in which the number of folds equals the number of observations.
c = cvpartition(n,'Resubstitution')	Creates an object c that does not partition the data. Both the training set and the test set contain all of the original n observations.
Options to add to the model statement	
'CrossVal'	'**off**': (default) '**on**': tenfold cross-validation
'Holdout'	Holdout with the given fraction reserved for validation. scalar from 0 to 1.
'KFold'	k-fold cross-validation. k (scalar).
'Leaveout'	'**off**': (default) '**on**': Leave-one-out cross-validation

If we already have a partition created using the **cvpartition** function, we can also provide that to the fitting function.

>> part = cvpartition(y,'KFold',k);

>> mdl = fitcknn(data,'responseVarName','CVPartition',part);

Calculating the Model Loss

To evaluate a cross-validated model, use the **kfoldLoss** function.

>> kfoldLoss(mdl)

NOTE

In general, **kfoldLoss(mdl)** returns the classification loss obtained by the cross-validated classification model **mdl**. For every fold, **kfoldLoss** function computes the classification loss for validation-fold observations using a classifier trained on training-fold observations. **MdL.X** and **mdL.Y** contain both sets of predictors and response data. It computes the loss (e.g., mean squared error, cross-entropy) of a model using k-fold cross-validation. Cross-validation is a technique used to estimate the performance of a model on unseen data by dividing the dataset into **k** equally sized folds and performing multiple iterations, where each fold is used as a validation set once while the remaining folds are used for training. It takes a trained model and the training dataset as input. It internally performs the necessary splitting of the data into folds and trains and evaluates the model on each fold. It returns the average loss across all the folds, providing an estimate of the model's performance on unseen data.

Quiz #1

(T/F): A smaller value of k will generally give a more reliable evaluation.

>> part = cvpartition(y,'KFold', k);

Quiz #2

(T/F): A larger value of k will generally take more computational effort.

>> part = cvpartition(y,'KFold',k);

Partition with K-Fold: Heart Disease Data Classification

The table **heartData** contains several features of different subjects and whether or not they have heart disease, which is saved in the **HeartDisease** variable. It has 11 numerical and 11 categorical columns, as well. The table contains several features of different subjects and whether or not they have heart disease. It includes information such as the age, cholesterol, and heart rate of 427 patients.

Using the data stored in the variable **heartData**, we will create the following models using kfold cross-validation and calculate their loss values, as shown in Table 3.2.

Table 3.2 Partition K-Fold creation and the corresponding loss.

Classifier	Model Variable Name	Loss Variable Name
k-Nearest Neighbor	**LowKFknn**	**lossLowKFknn**
Discriminant Analysis	**LowKFDa**	**lossLowKFDa**
k-Nearest Neighbor	**HiKFknn**	**lossHiKFknn**
Discriminant Analysis	**HiKFDa**	**lossHiKFDa**

Figure 3.4 shows the code for creating a sevenfold cross-validation subsets or partitions of data, fitting the data to both KNN and DA classifiers, and presenting the model loss.

```
heartData = readtable('heartDiseaseData.xlsx','Format', 'auto');
heartData.HeartDisease = categorical(heartData.HeartDisease);
myy= heartData.HeartDisease;
numData = zscore(heartData{:,1:11});
% Create the cvpartition variable
n=size(myy,1);
c = cvpartition(n, "KFold",2);
pt = cvpartition(myy, "KFold",8);
LowKFknn =fitcknn(numData,myy,'CVPartition',c);
HiKFknn=fitcknn(numData,myy,'CVPartition',pt);
LowKFDa=fitcdiscr(numData,myy,'CVPartition',c);
HiKFDa=fitcdiscr(numData,myy,"KFold",9);
lossLowKFknn= kfoldLoss(LowKFknn);
lossHiKFknn =kfoldLoss(HiKFknn);
lossLowKFDa= kfoldLoss(LowKFDa);
lossHiKFDa=kfoldLoss(HiKFDa);
modelNames = {'LowKFknn', 'HiKFknn','LowKFDa','HiKFDa'};
results=table([lossLowKFknn; lossHiKFknn; lossLowKFDa;lossHiKFDa],...
'RowNames',modelNames,'VariableNames',{'kFoldLoss'});
disp('K-fold cross-validated results')
disp(results)
```

Figure 3.4 K-fold cross-validation approach to better improve the prediction of the given classifier.

NOTE

With **c = cvpartition(n, 'Resubstitution')**, both **lossNoKFknn** and **lossNoKFDa** give **NaN** values. The k-fold loss varies from one run to another even for the same size of k-fold and for the same model. The reason for this is simply that the partition mechanism randomly assigns data sets within each partition fold or within holdout used for validation purposes.

In general, one has to select the proper partition size for a given model, which interprets as k-fold or holdout value.

The model loss for 12-fold cross-validation-based classifiers are shown in Figure 3.5 for both classifiers: KNN and DA with and without K-Fold option. There is a slight improvement in model predictability under K-Fold option.

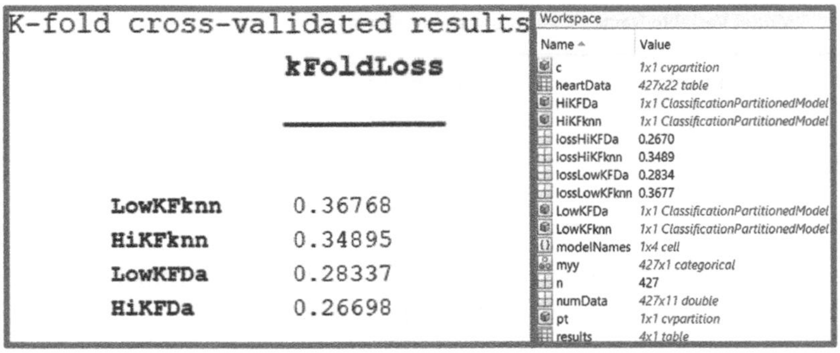

Figure 3.5 Model loss with and without K-Fold option.

Quiz #3

Given a dataset with a small number of observations, which kind of validation method will generally give you the most reliable model?

a) Holdout validation   b) k-fold validation   c) Leave-one-out validation

---

Reducing Predictors: Feature Transformation and Selection

Machine learning tasks frequently encompass data with numerous predictors in high-dimensional spaces, reaching hundreds or even thousands. For instance:

- Facial recognition tasks may entail images composed of thousands of pixels, each of which serves as a predictor.
- Weather prediction tasks may involve examining temperature and humidity measurements across thousands of different locations.

Learning algorithms typically demand substantial computational resources, and reducing the number of predictors can yield notable advantages in terms of computation time and memory usage. Moreover, the reduction of predictors leads to simpler models that can be generalized more effectively and are simpler to interpret. In this and next section, focus will be made on two common ways to reduce the number of predictors in a model, as shown in Figure 3.6.

- **Feature Transformation:** Transform the coordinate space of the observed variables.
- **Feature Selection:** Choose a subset of the observed variables.

Feature Transformation

Principal Component Analysis (PCA) transforms an n-dimensional feature space into a new n-dimensional space of orthogonal components. The components are ordered by the variation explained in the data.

PCA can therefore be used for dimensionality reduction by discarding the components beyond a chosen threshold of explained variance. The principal components by themselves have no physical meaning. However, the coefficients of the linear transformation indicate the contribution of each variable in the principal component. This has been explained earlier in Chapter 1.

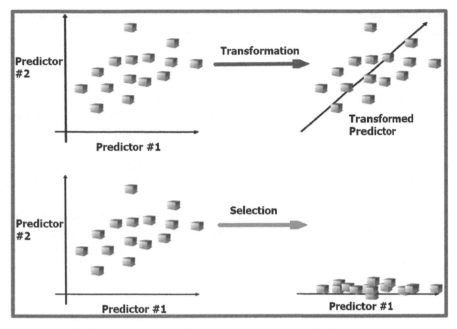

Figure 3.6   Predictors' reduction by feature transformation and selection.

Parallel Coordinates Plot

PCA can be performed independent of the response variable. However, when the data has a response variable that has multiple categories (heart disease true-false), a parallel coordinates plot of the principal component scores can be useful.

In Figure 3.7, observe that the observations from one group (false) have high values of the first principal component and the observations from the second group (true) have low values. The x-coordinate represents the principal components #1 up to #11.

>> **parallelcoords(scrs,'Group',y,'Quantile',0.25)**

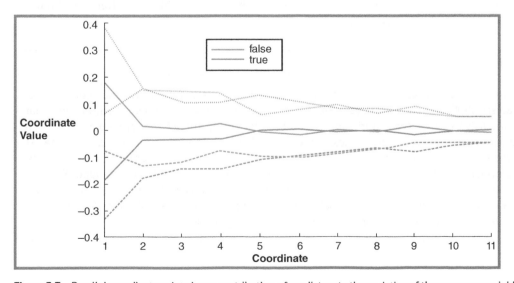

Figure 3.7   Parallel coordinates plot shows contribution of predictors to the variation of the response variable, y.

Parallel coordinates plot has been explained in detail in Chapter 1 | Observations and Clusters Visualization section.

Factor Analysis

Multivariate data frequently consists of numerous measured variables, and there are instances where these variables exhibit overlap, indicating that certain groups of variables may be inter-dependent. Take, for instance, a decathlon where each athlete participates in 12 events. Among these events, some can be categorized as speed-related, while others can be categorized as strength-related, and so on. Consequently, the scores of a competitor across the 12 events can be seen as primarily reliant on a smaller subset of approximately three or four types of athletic abilities.

Factor analysis is a statistical technique used for multivariate dimensionality reduction. It is a method used to identify underlying factors or latent variables that explain the patterns of correlations within a set of observed variables. In multivariate analysis, researchers often encounter situations where a large number of variables are measured, and it can be challenging to interpret and analyze the data due to the complexity and redundancy of information. Factor analysis aims to simplify this complexity by identifying a smaller number of unobserved variables, called factors, that account for the covariation among the observed variables. The underlying assumption in factor analysis is that the observed variables are influenced by a smaller number of latent factors. These factors are not directly observable but are inferred based on the patterns of correlations among the observed variables. The goal of factor analysis is to determine the number of factors and understand their nature and relationship with the observed variables. The factor analysis process involves extracting factors from the data and estimating their relationships with the observed variables. There are different methods for factor extraction, such as principal component analysis (PCA), maximum likelihood estimation, and principal axis factoring. Once the factors are extracted, they can be rotated to achieve a simpler and more interpretable factor structure. Common rotation methods include: **varimax, oblimin**, and **promax**. Factor analysis can be used for various purposes, including data reduction, variable selection, identifying latent constructs, and exploring the underlying structure of a dataset. It is widely employed in fields such as psychology, social sciences, market research, and finance, where researchers aim to uncover hidden patterns and reduce the dimensionality of complex datasets.

NOTE

There is a good deal of overlap in terminology and goals between Principal Components Analysis (**PCA**) and Factor Analysis (**FA**). Much of the literature on the two methods does not distinguish between them, and some algorithms for fitting the **FA** model involve **PCA**. Both are dimension-reduction techniques, in the sense that they can be used to replace a large set of observed variables with a smaller set of new variables. They also often give similar results. However, the two methods are different in their goals and in their underlying models. Roughly speaking, one should use **PCA** when there is a need to summarize or approximate the predictor data using fewer dimensions (to visualize it, for example), and one should use **FA** when there is a need for a model to explore the correlations among predictor data.

---

factoran function provides several useful outputs that can help us understand and interpret the results of the factor analysis. Here are some of the key outputs:

Factor Loadings: The factor loadings indicate the relationship between the observed variables and the underlying factors. These loadings represent the strength and direction of the association between each variable and each factor. They can be interpreted as correlation coefficients or regression coefficients. The factor loadings help identify which variables are most strongly related to each factor and contribute the most to its interpretation.

Variance Explained: **factoran** function provides information about the proportion of variance explained by each factor. It calculates the proportion of total variance accounted for by each factor and provides insights into how much information is captured by the extracted factors.

Rotation Matrix: The $m \times m$ factor loadings rotation matrix represents the orthogonal transformation applied to the original factor loadings to obtain the rotated factor loadings. The matrix contains the coefficients of the linear combinations of the original factor loadings that produce the rotated factor loadings.

stats Structure: refers to a structure containing information related to the null hypothesis (H0) that the number of common factors is m. The **stats** structure provides various statistical measures and diagnostics that can be helpful for assessing the quality and fit of the factor analysis model. The **stats** structure typically includes the following fields:

1) **stats.dfe**: The error degrees of freedom associated with the chi-square statistic. It represents the number of independent pieces of information available for estimating the parameters in the factor analysis model.
2) **stats.p**: The p-value corresponding to the chi-square statistic. It indicates the statistical significance of the chi-square test, which assesses the goodness-of-fit of the factor analysis model.

Factor Scores: Factor scores are the estimated scores for each observation on each extracted factor. These scores represent the position of each observation along each factor dimension. Factor scores can be useful for further analyses or for creating composite scores that summarize the information from multiple variables.

In general, the factor analysis can be put in the form:

```
>>[Loadings,specVar,T,stats,F] = factoran(X,m)
```

Where

Loadings: The factor loadings matrix for the data matrix X with **m** common factors.
specVar: The maximum likelihood estimates of the specific variances.
T: The m×m factor loadings rotation matrix.
stats: The structure containing information relating to the null hypothesis H_0 that the number of common factors is **m**.
F: Predictions of the common factors (factor scores).
m: Number of common factors.

Table 3.3 shows some properties of PropertyName-Value Arguments which can be used with **factoran** statement, quoted from MATLAB help.

Table 3.3 PropertyName-Value pairs which can be used with **factoran** statement.

Property	Description/Value
'Nobs'	Number of observations used to estimate X. positive integer. Used only when **'Xtype'** is **'covariance'**.
'Rotate'	Method used to rotate factor loadings and scores. **'varimax'** (**default**) | **'none'** | **'quartimax'** | **'equamax'** | **'parsimax'** | **'orthomax'** | **'promax'** | **'procrustes'** | **'pattern'** | **function handle**
'Scores'	Method for predicting factor scores **'wls'** or the equivalent 'Bartlett' (**default**) | **'regression'** or the equivalent 'Thomson'. **'wls'**: Weighted least-squares estimate treating F as fixed. **'regression'**: Minimum mean squared error prediction that is equivalent to a ridge regression
'Xtype'	Input data type. **'data'** (**default**) | **'covariance'**

NOTE

The factor analysis approach is to reduce the observed variables into **m** common variables. The lowest value is 1. To see **stats** properties, just click on **stats** variable in workspace window. The null hypothesis, that such **m** common variables are representative for the original, will be accepted if and only if both **stats.dfe**=1 and **stats.p**≈1.0.

chisq: Approximate chi-squared statistic for the null hypothesis.
dfe: Error degrees of freedom = $((d-m)^2 - (d+m))/2$.
p: Right-tail significance level for the null hypothesis.

For the specific variance, a value of 1 would indicate that there is no common factor component in that variable, while a specific variance of 0 would indicate that the variable is entirely determined by common factors.
  See MATLAB built-in help on "Perform Factor Analysis on Exam Grades".

If we start with **m=1** (see the code below), then we have:

stats.dfe = *44* and **stats.p** = *6e-141≈0.0*.

The errors degrees of freedom, **dfe**, is 44 which is >> 1 and the p-value≈0.0 <<1. This means that the null hypothesis of a single common factor is rejected, so we need to refit the model.

  However, for heart disease data (see below), the number of predictors could not be reduced based on the factor analysis approach. The method fails to compute the significance (p-value) as some unique variances are zero. One can say that the 11 variables are highly independent; in other words, there is no correlation (or common factors as the language of factor analysis says) among the 11 variables.

  On the other hand, we can see from pareto plot of **pexp** that it requires at least 9 out of 11 variables to explain 95% of the total variation in predictors' data.

Feature Transformation and Factor Analysis: Heart Disease Data

The table **heartData** contains several features of different subjects and whether or not they have heart disease, which is saved in the **HeartDisease** variable. It has 11 numerical and 11 categorical columns, as well.

Figure 3.8 shows the code to read heat disease data, define predictor and response variables, transforming X data from 11 to 8 dimensions, fitting to Naïve-Bayes model with a cross-validation, presenting the first three principal components, carrying out factor analysis, creating a tenfold cross-validated Naïve-Bayes model using the kernel distribution with the transformed data, and calculating the loss of the latter model.

One can see that using only the first 8 out of 11 principal components by a process called feature transformation, the model loss, **pcaLoss=0.2787**, for tenfold cross-validation-based classifier was practically the same as the loss, **AllDataLoss=0.2763**, of the same classifier but using the entire data, without reducing the number of variables of predictors. Moreover, at **m=2**, **stats.dfe = 34** and **stats.p = 8e-60≈0.0**. Based on factor analysis approach the null hypothesis, that we have in hand two common factors, is rejected. As explained in the previous page, the heart disease predictor data (columns or variables) are apparently independent of one another. Again, if we increment **m** by *1* (i.e., **m**=*3*) then the solution will not converge and the null hypothesis will again be rejected.

```
heartData = readtable('heartDiseaseData.xlsx','Format', 'auto');
heartData.HeartDisease = categorical(heartData.HeartDisease);
vars = heartData.Properties.VariableNames(1:11);
%% Split data into predictors (numData) and response (resp)
numData = zscore(heartData{:,1:11});
resp = heartData.HeartDisease;
%% Perform PCA and fit model on the reduced data
[pcs,scrs,~,~,pexp]=pca(numData);
figure(9);
pareto(pexp);
temp=scrs(:,1:9);
mdl=fitcnb(temp,resp,"DistributionNames","kernel","KFold",10);
pcaLoss=kfoldLoss(mdl);
mdl2 = fitcnb(numData,resp,'Distribution','kernel','KFold',10);
AllDataLoss=kfoldLoss(mdl2);
%% Visualize components in a heatmap
figure(10);
imagesc(abs(pcs(:,1:3)));
yticklabels(vars);
colorbar
%% Factor Analysis
[Loadings1,specVar1,T1,stats1,F1]= factoran(numData,2, 'Scores','wls', 'Rotate', 'varimax' );
[Loadings2,specVar2,T2,stats2,F2]=factoran(temp,5, 'Scores','regression', 'Rotate', 'none' );
%[Loadings1,specVar1,T1,stats1,F1]= factoran(numData,2, 'Scores','regression',
%'Rotate', 'varimax' ); %Same results as above
%[Loadings2,specVar2,T2,stats2,F2]=factoran(temp,5, 'Scores','wls', 'Rotate', 'none' );
```

Figure 3.8 Heart-disease data fitted to Naïve-Bayes tenfold cross-validation model with and without feature transformation using principal component analysis, **pca**.

Figure 3.9 shows the Pareto plot (*left*) and the first three PCs (*right*). Figure 3.10 shows all workspace variables. Neither **p** nor **dfe** value approaches one. In addition, the specific variances are roughly closer to one than to zero, indicating that there is no common factor component in such variables to use a two-factor model. However, it can be seen that for the transformed variables (i.e., **temp** a subset of **scrs**) which count nine transformed variables, we can describe the predictor data by five principal components having the highest specific variance values, as can be seen from both **specVar2** values and **stat2** parameters. Such five transformed variables are standalone and do not participate in common factors. This can be explained by looking at Table 3.4.

From the estimated loadings coefficients, which represent the factor loadings matrix **loadings2** for the data matrix **temp** with 5 common factors, we can see that the first unrotated factor places approximately one weight on the fifth variable only, while the second unrotated factor places one weight on the second variable only, and so on. That proves the previous conclusion that such variables are completely independent or standalone and no overlap exists among them. Or, put it this way: Common factors are uncommon among the original predictors. With **pca** transformation, we have 5 common independent factors.

Table 3.4 Estimated loadings (weights) of the nine transformed principal components to the five common factors.

Variables	Common Factor #1	Common Factor #2	Common Factor #3	Common Factor #4	Common Factor #5
PC#1	0	0	0	0	0
PC#2	−6.63e-17	0.60616	1.67e-13	1.91e-15	9.93e-15
PC#3	1.48e-15	−1.07e-14	−5.90e-16	1.71e-14	0.5859
PC#4	−4.07e-16	6.16e-16	−1.92e-16	4.06e-18	2.90e-16
PC#5	0.6642	−2.40e-16	−1.06e-15	−2.77e-16	−1.03e-15
PC#6	1.39e-15	−1.67e-13	0.6059	4.88e-15	4.93e-16
PC#7	3.14e-16	−1.97e-15	−5.12e-15	0.5925	−1.67e-14
PC#8	−3.11e-16	7.41e-18	9.64e-16	2.41e-16	3.29e-16
PC#9	−3.15e-16	1.62e-16	4.88e-16	1.06e-16	−1.95e-16

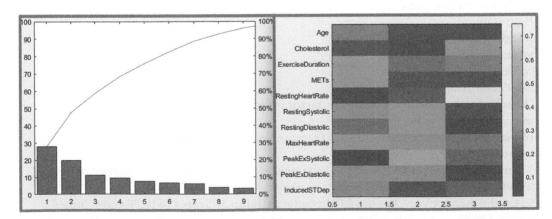

Figure 3.9 Pareto plot (*left*) and first three PCs (*right*).

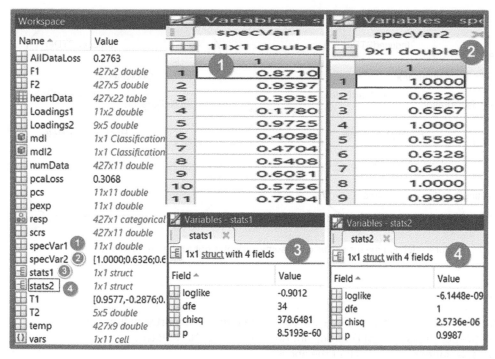

Figure 3.10 Workspace general variables, **stat1** struct, **specVar1**, **stat2** struct, and **specVar2** variable.

Quiz #4
Which of the following functions performs feature transformation? a) biplot   b) pareto   c) pca

Feature Selection

The data often contains predictors which do not have any relationship with the response. These predictors should not be included in a model. For example, the heart health data may include a patient-id. This id does not have any relationship with the risk of heart disease, but a model will treat it like all the other predictors.

The data can also contain highly correlated predictors so that only one of them needs to be included in the model.

Feature selection techniques choose a subset of predictors to be included in the model, as shown, for example, in Figure 3.11.

NOTE
Some classifiers, such as decision trees, have their own built-in methods of feature selection.   Check the *methods* associated with the decision tree model, **fitctree**. One of the methods, **predictorImportance**, can be used to identify the predictor variables that are important for creating an accurate model. The output of **predictorImportance** is a vector of the importance values for each predictor.

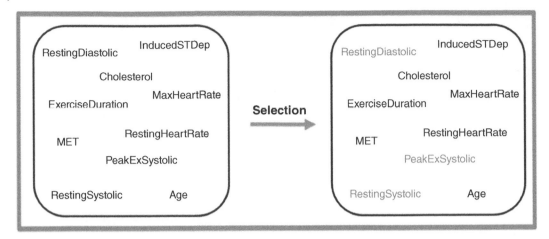

Figure 3.11 Feature selection means exclusion of some variables out of the predictors' list.

Quiz #5
Which classification technique allows the use of the **predictorImportance** function? a) Classification Trees b) Discriminant Analysis c) Naïve Bayes d) Nearest Neighbor e) All of the above

Feature Selection Using predictorImportance Function: Health Disease Data

The table **heartData** contains several features of different subjects and whether or not they have heart disease, which is saved in the **HeartDisease** variable. It has 11 numerical and 11 categorical columns, as well.

Using the **predictorImportance** function, we calculate the importance (**p**) values for each of the predictors, which are saved in the row vector, **p**. The goal of this exercise is to compare the model built using all the predictors to the model built with only the important predictors. Predictor variables that correspond to values of p that are greater than 10% of the maximum value in p are selected. The tenfold cross-validated model with all the predictors is named **mdlAll** and the calculated loss value is saved in a variable named **lossAll**. On the other hand, the cross-validated model containing only the selected predictors is saved as **mdlSel** with the corresponding loss as **lossSel**. Figure 3.12 shows the entire code for what is explained above.

```
heartData = readtable('heartDiseaseData.xlsx','Format', 'auto');
heartData.HeartDisease = categorical(heartData.HeartDisease);
mdlFull = fitctree(heartData,"HeartDisease");
p = predictorImportance(mdlFull);% p-val for each predictor column
figure (13);
bar(p)
%Compare full and reduced models.
```

Figure 3.12 The code for testing heart disease data using a cross-validated **fitctree** model with both a full utilization of all predictors and with a reduced number of predictors.

```
pv=p>0.1*max(p);% Setting the threshold for keeping the predictor column
mdlAll=fitctree(heartData,"HeartDisease","KFold",10);% All predictors' data.
lossAll=kfoldLoss(mdlAll); %Loss for KFold cross-validation-based model for all.
mdlSel=fitctree(heartData(:,[pv true]),"HeartDisease","KFold",10); % For reduced
lossSel=kfoldLoss(mdlSel);% %Loss for KFold cross-validated model for reduced
SelPred_Resp=heartData(:,[pv true]);%Selected columns and response column %included.
SelPred_NoResp=heartData(:,[pv false]);%Selected predictors columns without
%response column
ImpVariables= SelPred_NoResp.Properties.VariableNames
```

Figure 3.12   (Cont'd)

The results are copied from Command Window. Notice that out of *21* predictor variables, only *8* variables are selected as important in explaining the total variance of the model. Thus, **fitctree** model has an advantage over other classification models, because we can benefit from **predictorImportance** function to prioritize the importance of predictor columns.

  In addition, the loss difference between **lossAll**=*0.2693* (all variables) and **lossSel**=*0.2951* (only eight variables) is practically negligible.

ImpVariables = 1×8 cell array

{'Age'}{'Cholesterol'}{'ExerciseDuration'}{'METs'}{'MaxHeartRate'}{'InducedSTDep'}   {'ChestPainType'}{'STSlope'}

Figure 3.13 shows the bar plot for p-values of all predictors (*left*) and the generated workspace variables for reference to help users compare results (*right*).

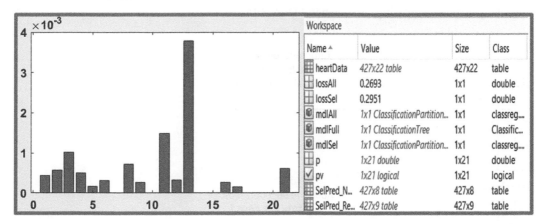

Figure 3.13   The bar plot for p-values of all predictors (*left*) and the generated workspace variables for reference (*right*).

Sequential Feature Selection (SFS): sequentialfs Function with Model Error Handler

A feature selection approach that can be used with any learning algorithm is sequential feature selection. The idea is to incrementally add predictors to the model as long as there is reduction in the prediction error. Sequential feature selection requires an error function that builds a model and calculates the prediction error. The process starts with building the model and calculating the error using one of the predictors, as shown in Figure 3.14.

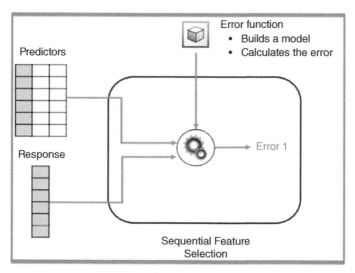

Figure 3.14 The calculated error will be used as a criterion for telling the fitting model goodness.

Add another predictor and calculate the error again. More predictors are added to the model as long as the error is decreasing, as shown in Figure 3.15.

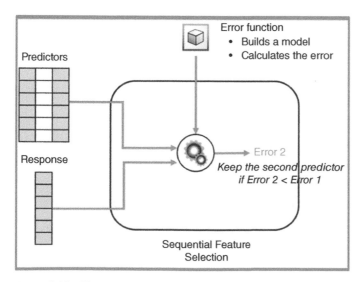

Figure 3.15 The decrease in the latest calculated error is an indicator of model improvement.

Let's start by writing the error function. The error function needs to have the following structure:

Four inputs representing the predictors (matrix) and response (vector) for both training and test. The output is the scalar value representing the prediction error.

Note that you do not have to create the training or test data. The sequential feature selection function internally creates the training and test data and calls this function (saved under **errorFun.m**). Figure 3.16 shows the code for **errorFun.m** file. Save it anywhere but within MATLAB path.

```
function error = errorFun(Xtrain,ytrain,Xtest,ytest)
% Create the model with the learning method of your choice
mdl = fitcknn(Xtrain,ytrain); %It can be changed; you may use any fitc* model.
% Calculate the number of test observations misclassified
ypred = predict(mdl,Xtest);
error = nnz(ypred ~= ytest);
```

Figure 3.16   The code for the m-file **errorFun.m**, where the user may use any **fitc*** model.

You can now use the function **sequentialfs** to perform sequential feature selection. Note that the first input is a handle to the error function. The output is a logical vector indicating the selected predictors. You can provide additional options to specify the cross-validation method. Here, a ten-fold cross-validation is selected.

>> tokeep = sequentialfs(@errorFun,X,y,'cv',10);

Notice that **sequentialfs** will take care of data partition into training and test set.

An alternative to writing the error function in a separate file is to create an anonymous function. To create an anonymous function that calculates the number of misclassifications given training and test data, use the following pattern:

>> af = @(Xtrain,ytrain,Xtest,ytest)...

nnz(ytest ~= predict(...

fitc*(Xtrain,ytrain),Xtest))

Figure 3.17 shows the code where we use k-nearest neighbor selection to determine which variables to keep in the model. Then we create a sevenfold cross-validated k-nearest neighbor model with only those selected features named **mdlSF** and compare it with that where we preserve all predictor columns. Finally, we calculate the model loss under both cases: with SFS, named **lossSF** and without SFS, named **lossNoSF**.

Figure 3.18 shows MATLAB workspace variables. The estimated value of the model error (loss) without sequential feature selection, **lossNoSF**, was found to be 0.3583. On the other hand, the estimated value of the model error (loss), with sequential feature selection, **lossSF**, was found to be 0.3466.

```
%% Import data
heartData = readtable('heartDiseaseData.xlsx','Format', 'auto');
heartData.HeartDisease = categorical(heartData.HeartDisease);
vars = heartData.Properties.VariableNames(1:11);
%% Perform Sequential Feature Selection with k-NN while using anonymous
function.
ferror = @(Xtrain,ytrain,Xtest,ytest)...
nnz(ytest ~=predict(fitcknn(Xtrain,ytrain),Xtest));
Sel=sequentialfs(ferror,heartData{:,1:11},heartData.HeartDisease);
mdlNoSF=fitcknn(heartData(:,1:11), heartData.HeartDisease,"KFold",7);
lossNoSF=kfoldLoss(mdlNoSF);
mdlSF=fitcknn(heartData(:,[Sel false]), heartData.HeartDisease,"KFold",7);
lossSF=kfoldLoss(mdlSF);
```

Figure 3.17 SFS-enabled, k-NN-based classifier using an anonymous function to help calculate the model error (loss).

Name ▲	Value	Size
ferror	@(Xtrain,ytrain,Xtest,ytest)nn...	1x1
heartData	*427x22 table*	427x22
lossNoSF	0.3583	1x1
lossSF	0.3466	1x1
mdlNoSF	*1x1 ClassificationPartitioned...*	1x1
mdlSF	*1x1 ClassificationPartitioned...*	1x1
Sel	*1x11 logical*	1x11
vars	*1x11 cell*	1x11

Figure 3.18 The MATLAB workspace variables showing the model loss for both cases.

Figure 3.19 shows an alternative code for that shown in Figure 3.17, where it utilizes a function handle pointing @ **errorFun**, defined in Figure 3.16.

```
heartData = readtable('heartDiseaseData.xlsx','Format', 'auto');
heartData.HeartDisease = categorical(heartData.HeartDisease);
vars = heartData.Properties.VariableNames(1:11);
mdlNoSF=fitcknn(heartData(:,1:11), heartData.HeartDisease,"KFold",7);
lossNoSF=kfoldLoss(mdlNoSF);
%Perform Sequential Feature Selection with k-NN
%Sel = sequentialfs(@errorFun, heartData{:,1:11},heartData.HeartDisease,'cv',7);
Sel = sequentialfs(@errorFun, heartData{:,1:11},heartData.HeartDisease);
mdlSF=fitcknn(heartData(:,[Sel false]), heartData.HeartDisease,"KFold",7);
lossSF=kfoldLoss(mdlSF);
```

Figure 3.19 SFS-enabled, k-NN-based classifier, using a function handle pointing @ **errorFun** to help estimate the model loss.

Figure 3.20 shows MATLAB workspace variables. The estimated value of the model error (loss), without SFS, **lossNoSF**, was found to be 0.3607 and that with SFS to be 0.3513.

Workspace		
Name ▲	Value	Size
▦ heartData	*427x22 table*	427x22
▦ lossNoSF	0.3607	1x1
▦ lossSF	0.3513	1x1
◉ mdlNoSF	*1x1 ClassificationPartitioned...*	1x1
◉ mdlSF	*1x1 ClassificationPartitioned...*	1x1
✓ Sel	*1x11 logical*	1x11
{} vars	*1x11 cell*	1x11

Figure 3.20   The MATLAB workspace variables where both model errors are shown.

NOTE

As shown in both Figure 3.18 and Figure 3.20, if we click on **mdlSF** icon found in the workspace, we will notice that the 11 predictors were substantially reduced to one predictor, that is, **'InducedSTDep'**. That simply means the response Y can be expressed as a function of X, where X is reduced to '**InducedSTDep**' predictor variable, without jeopardizing the model reliability.

Accommodating Categorical Data: Creating Dummy Variables

Some algorithms and functions (e.g., SFS) require predictors in the form of a numeric matrix. If our data contains categorical predictors, how can we include these predictors in the model?

One option is to assign a number to each category. However, this results in imposing a false numerical structure on the observations. For example, if you assign the numbers 1 through 4 to four categories in a predictor, it implies that the distance between the categories 1 and 4 is longer than the distance between the categories 3 and 4.

A better approach is to create new dummy predictors for each category. Each dummy predictor can have only two values: 0 or 1. For any given observation, only one of the dummy predictors can have the value equal to 1.

We can create a matrix of dummy variables using the function **dummyvar**.

>>d = dummyvar(c)

Concatenate this matrix with a matrix containing the numeric predictors.

Figure 3.21 shows 8×1 observations labeled with three categories (i.e., colors) where we transform them into 8×3 matrix and assign 1 for the corresponding color and 0 elsewhere.

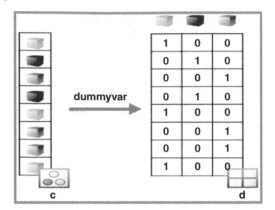

Figure 3.21 The **dummyvar** function transforms the original observations labeled with their categories into dummy variables such that 1 is assigned for the corresponding color and 0 elsewhere.

Use the **dummyvar** function to convert a categorical vector to a numeric matrix.

>> **numericMatrix = dummyvar(catVector)**

Quiz #6
If you have an *n*-element categorical vector with *k* categories, the **dummyvar** function will create a matrix of what size? a) 1×k   b) 1×k×n   c) 1×n   d) k×n   e) n×k   f) k×k

Quiz #7
(T/F): You can either perform feature transformation or feature selection, but not both.

Feature Selection with Categorical Heart Disease Data

The table **heartData** contains several features of different subjects and whether or not they have heart disease, which is saved in the **HeartDisease** variable. It has 11 numerical and 11 categorical columns, as well. Figure 3.22 shows the m-file for **cat2dum** function specifically made for heart disease data to show how to create dummy variables out of the 11 categorical predictor variables {12:21}.

```
function [Xdum]=cat2dum(XDat)
Xdum=[];
for j=12:size(XDat,2)-1
dumvar=dummyvar(XDat{:,j});
[Xdum]=[Xdum dumvar];
end
```

Figure 3.22 The m-file for **cat2dum** function converting each categorical column to dummy columns, depending on how many subcategories are found in each column.

Figure 3.23 shows the code to carry out feature selection (FS) on Naïve Bayes classifier for the heart disease data. FS is computationally intensive, and may take a long time to run.

```
%% Load the data
heartData = readtable('heartDiseaseData.xlsx','Format', 'auto');
heartData.HeartDisease = categorical(heartData.HeartDisease);
numData = heartData{:,1:11};
heartData = convertvars(heartData,12:21,"categorical");
% Extract the response variable
HD = heartData.HeartDisease;
%Convert categorical variables to numeric dummy variables
dumcat=cat2dum(heartData);
heartData2=[numData dumcat];
% Make a partition for evaluation
rng(1234)
part = cvpartition(HD,'KFold',10);
%Fit a  Naïve Bayes model to the full data
dists = [repmat({'kernel'},1,11),repmat({'mvmn'},1,10)];
mFull = fitcnb(heartData,'HeartDisease','Distribution',dists,'CVPartition',part);
%% Perform sequential feature selection
disp([num2str(kfoldLoss(mFull)),' = loss with all data'])
rng(1234)
fmodel = @(X,y) fitcnb(X,y,'Distribution','kernel');
ferror = @(Xtrain,ytrain,Xtest,ytest) nnz(predict(fmodel(Xtrain,ytrain),Xtest) ~=
ytest);
tokeep = sequentialfs(ferror,heartData2,HD,'cv',part,...
   'options',statset('Display','iter'));
% Fit a model with just the kept variables.
mPart = fitcnb(heartData2(:,tokeep),HD,'Distribution','kernel','CVPartition',part);
disp([num2str(kfoldLoss(mPart)),' = loss with selected data'])
```

Figure 3.23 Testing the Naïve Bayes classifier without and with inclusion of categorical dummy variables. The SFS is implemented in the latter case.

The results, copied from Command Window, are:

```
0.21311 = loss with all data
Start forward sequential feature selection:
Initial columns included: none
Columns that cannot be included: none
Step 1, added column 14, criterion value 0.234192
Step 2, added column 8, criterion value 0.222482
Step 3, added column 11, criterion value 0.213115
Step 4, added column 17, criterion value 0.208431
Step 5, added column 25, criterion value 0.203747
```

Step 6, added column 2, criterion value 0.199063
Step 7, added column 12, criterion value 0.194379
Step 8, added column 31, criterion value 0.185012
Step 9, added column 1, criterion value 0.18267
Final columns included: 1 2 8 11 12 14 17 25 31
0.18267 = loss with selected data

Figure 3.24 shows the m-file code for **cattbl2mat** function which converts a categorical predictor column into a number of dummy predictors columns depending on the number of subcategories present in the original predictor column.

```
function [data,vars]=cattbl2mat(data)
% Makes a matrix from a table, with categ. variables replaced by
% numeric dummy variables
vars=string(data.Properties.VariableNames);
idxCat=varfun(@iscategorical,data,"OutputFormat","uniform");
for k=find(idxCat)
    % get list of categories
    c=categories(data.(vars(k)));
    % replace variable with matrix of dummy variables
    data=convertvars(data,vars(k),@dummyvar);
    % split dummy variable and make new variable names (by appending
    % the category value to the categorical variable name).
    varnames=vars(k) + "_" + replace(c, " ", "_");
    data=splitvars(data, vars(k),"NewVariableNames",varnames);
end
vars=string(data.Properties.VariableNames);
```

Figure 3.24 The m-file code for **cattbl2mat** function for creation of dummy out of categorical predictor columns.

Figure 3.25 shows the analogous code for that of Figure 3.23 but this time utilizing **cattbl2mat** instead of **cat2dum** function.

NOTE

Before you run Figure 3.25 code, do not forget to clear the workspace variables. You can click on "**Clear Workspace**" button found in "**VARIABLE**" group under "**HOME**" tab.

```
heartData = readtable('heartDiseaseData.xlsx','Format', 'auto');
heartData.HeartDisease = categorical(heartData.HeartDisease);
numData = heartData{:,1:11};
```

Figure 3.25 Testing the Naïve Bayes classifier without and with inclusion of categorical dummy variables. The SFS is implemented in the latter case.

```
heartData = convertvars(heartData,12:21,"categorical");
% Extract the response variable
HD = heartData.HeartDisease;
%Convert categorical variables to numeric dummy variables
[heartData2 Varnames]=cattbl2mat(heartData(:,1:end-1));
heartData3 = table2array(heartData2);
% Make a partition for evaluation
rng(1234)
part = cvpartition(HD,'KFold',10);
%% Fit a  Naïve Bayes model to the full data
dists = [repmat({'kernel'},1,11),repmat({'mvmn'},1,10)];
mFull = fitcnb(heartData,'HeartDisease','Distribution',dists,'CVPartition',part);
%% Perform sequential feature selection
disp([num2str(kfoldLoss(mFull)),' = loss with all data'])
rng(1234)
fmodel = @(X,y) fitcnb(X,y,'Distribution','kernel');
ferror = @(Xtrain,ytrain,Xtest,ytest) nnz(predict(fmodel(Xtrain,ytrain),Xtest) ~=
ytest);
tokeep = sequentialfs(ferror,heartData3,HD,'cv',part,...
  'options',statset('Display','iter'));
% Which variables are in the final model?
Varnames(tokeep)
mPart = fitcnb(heartData3(:,tokeep),HD,'Distribution','kernel','CVPartition',part);
disp([num2str(kfoldLoss(mPart)),' = loss with selected data'])
```

Figure 3.25   (Cont'd)

Results are copied from Command Window where only 9 out of 35 columns are the winners in the final international championship:

```
0.21311 = loss with all data
Start forward sequential feature selection:
Initial columns included: none
Columns that cannot be included: none
Step 1, added column 14, criterion value 0.234192
Step 2, added column 8, criterion value 0.222482
Step 3, added column 11, criterion value 0.213115
Step 4, added column 17, criterion value 0.208431
Step 5, added column 25, criterion value 0.203747
Step 6, added column 2, criterion value 0.199063
Step 7, added column 12, criterion value 0.194379
Step 8, added column 31, criterion value 0.185012
Step 9, added column 1, criterion value 0.18267
Final columns included: 1 2 8 11 12 14 17 25 31
ans = 1×9 string array
  Columns 1 through 6
```

```
   "Age"   "Cholesterol"   "MaxHeartRate"   "InducedSTDep"   "Sex_F"   "ChestPainType_
Asy..."
   Columns 7 through 9
   "ChestPainType_Typ..."   "NitratesUsed_N"   "ExInducedAngina_N"
0.18267 = loss with selected data
```

Figure 3.26 shows the workspace variables for Figure 3.23 code (*left*) and for Figure 3.25 (*right*). One can see from Figure 3.26 workspace variables that the new predictor data, **heartData2** (427×35) matrix, combines both the numeric **numData** (427×11) and dummy numeric variables, **dumcat** (427×24).

Workspace				Workspace		
Name ▲	Value	Size		Name ▲	Value	Size
dists	*1x21 cell*	1x21		ans	*1x9 string*	1x9
dumcat	*427x24 double*	427x24		dists	*1x21 cell*	1x21
ferror	*@(Xtrain,ytrain,Xtest,ytest)nnz(predict(fmodel...*	1x1		ferror	*@(Xtrain,ytrain,Xtest,ytest)nnz(predict(fmodel...*	1x1
fmodel	*@(X,y)fitcnb(X,y,'Distribution','kernel')*	1x1		fmodel	*@(X,y)fitcnb(X,y,'Distribution','kernel')*	1x1
HD	*427x1 categorical*	427x1		HD	*427x1 categorical*	427x1
heartData	*427x22 table*	427x22		heartData	*427x22 table*	427x22
heartData2	*427x35 double*	427x35		heartData2	*427x35 table*	427x35
mFull	*1x1 ClassificationPartitionedModel*	1x1		heartData3	*427x35 double*	427x35
mPart	*1x1 ClassificationPartitionedModel*	1x1		mFull	*1x1 ClassificationPartitionedModel*	1x1
numData	*427x11 double*	427x11		mPart	*1x1 ClassificationPartitionedModel*	1x1
part	*1x1 cvpartition*	1x1		numData	*427x11 double*	427x11
tokeep	*1x35 logical*	1x35		part	*1x1 cvpartition*	1x1
				tokeep	*1x35 logical*	1x35
				Varnames	*1x35 string*	1x35

Figure 3.26   The workspace variables for Figure 3.23 code (*left*) and that of Figure 3.25 (*right*).

Quiz #8

(T/F) Suppose feature selection is performed on two dummy variables that were created from a binomial variable. Either both variables will be selected or neither variable will be selected. Selecting just one of the two dummy variables is not likely to happen.

Ensemble Learning

Classification trees, being weak learners, exhibit a strong sensitivity to the training data they are provided. As a result, even minor variations in the training data can lead to entirely distinct trees and subsequently, disparate predictions. However, this limitation can be transformed into an advantage by generating multiple trees, forming what is known as a forest. This allows for the application of new observations to all the trees, enabling a comparison of the resulting predictions.

fitensemble can boost or bag decision tree learners or discriminant analysis classifiers. The function can also train random subspace ensembles of KNN or discriminant analysis classifiers.

NOTE

For simpler interfaces that fit classification and regression ensembles, instead use **fitcensemble** and **fitrensemble**, respectively. Also, both **fitcensemble** and **fitrensemble** provide options for Bayesian optimization.

Use the **fitensemble** function to create ensembles for weak learners.

>>**mdl = fitensemble(data,responseVarName,Method,N,Learner)**

Table 3.5 shows the description of **fitensemble** statement.

Table 3.5   Description of **fitensemble** components.

mdl	Ensemble learning model variable.
data	Table containing the predictors and response values.
responseVarName	Response variable name.
Method	Ensemble learning method (see Table 3.6).
N	Number of ensemble learning cycles.
Learner	Weak learning method (see Table 3.7).

Table 3.6 shows different **fitensemble** aggregation methods, based on the type of fitting model.

Table 3.6   Selection of **fitensemble** aggregation method. Quoted from MATLAB help.

Fitting Model Type	Ensemble Aggregation Method
Classification with two classes	• 'AdaBoostM1' • 'LogitBoost' • 'GentleBoost' • 'RobustBoost' (requires Optimization Toolbox™) • 'LPBoost' (requires Optimization Toolbox) • 'TotalBoost' (requires Optimization Toolbox) • 'RUSBoost' • 'Subspace' • 'Bag'
Classification with three or more classes:	• 'AdaBoostM2' • 'LPBoost' (requires Optimization Toolbox) • 'TotalBoost' (requires Optimization Toolbox) • 'RUSBoost' • 'Subspace' • 'Bag'
Regression	• 'LSBoost' • 'Bag'

NOTE

If we specify '**Method**','**Bag**', then we need to specify the problem type using the Type name-value pair argument, because we can specify '**Bag**' for either classification or regression problem. Choose '**Type**','**Classification**' or '**Type**','**Regression**'.

Table 3.7 shows the weak learners to use in the ensemble, specified as a weak-learner name, weak-learner template object, or cell array of weak-learner template objects.

Table 3.7 Weak learners to use with the ensemble aggregation method.

Weak Learner	Weak-Learner Name	Template Object Creation Function	Method Settings
Discriminant analysis	**'Discriminant'**	templateDiscriminant	Recommended for **'Subspace'**
k nearest neighbors	**'KNN'**	templateKNN	For **'Subspace'** only
Decision tree	**'Tree'**	templateTree	All methods except **'Subspace'**

>>t = **templateDiscriminant('DiscrimType','linear')**
'DiscrimType': Discriminant types are:
'linear' (default) | **'quadratic'** | **'diaglinear'** | **'diagquadratic'** | **'pseudolinear'** | **'pseudoquadratic'**.
>>t = **templateKNN('NumNeighbors',5,'Standardize',1)**

For other property types and values see Chapter 2 | KNN Model Properties section.

>> t = **templateTree(Name,Value);**

For tree property types and values see Chapter 2 | Decision Tree Model section.
Table 3.8 shows aggregation methods that can be used with **fitcensemble** function statement.

Table 3.8 **fitcensemble** Ensemble aggregation method, specified as the comma-separated pair consisting of **'Method'** and one of the following values. Quoted from MATLAB help.

Value	Method Description	Classification Problem Support
'Bag'	Bootstrap aggregation (bagging with random predictor selections at each split (random forest) by default.	Binary and multiclass
'Subspace'	Random subspace	Binary and multiclass
'AdaBoostM1'	Adaptive boosting	Binary only
'AdaBoostM2'	Adaptive boosting	Multiclass only
'GentleBoost'	Gentle adaptive boosting	Binary only
'LogitBoost'	Adaptive logistic regression	Binary only
'LPBoost'	Linear programming boosting. Requires Optimization Toolbox™	Binary and multiclass

Table 3.8 (Continued)

Value	Method Description	Classification Problem Support
'RobustBoost'	Robust boosting. Requires Optimization Toolbox	Binary only
'RUSBoost'	Random under-sampling boosting	Binary and multiclass
'TotalBoost'	Totally corrective boosting. Requires Optimization Toolbox	Binary and multiclass

---

NOTE

The defaults are:

'**LogitBoost**' for binary problems and '**AdaBoostM2**' for multiclass problems if '**Learners**' includes only tree learners

'**AdaBoostM1**' for binary problems and '**AdaBoostM2**' for multiclass problems if '**Learners**' includes both tree and discriminant analysis learners

'**Subspace**' if '**Learners**' does not include tree learners

---

fitrensemble Function:
Use the following model statements for **fitrensemble** function with the understanding that we have a regression case.

```
>>enMdl=fitrensemble(Xnum, Ynum,"KFold",N);% N positive integer.
>> kfLoss=kfoldLoss(enMdl); %Cross-validated regression error
```

NOTE

In general, **kfoldLoss(CVMdl)** returns the regression loss obtained by the cross-validated kernel regression model **CVMdl**. For every fold, **kfoldLoss** function computes the regression loss for observations in the validation fold, using a model trained on observations in the training fold.

Alternatively, one may use the automated hyper parameters optimization algorithm. For example, this is called Optimizable Ensemble in both classification learner and regression learner app.

```
enMdl=fitrensemble(Xnum,Ynum,'OptimizeHyperparameters','auto')
ResubLoss=resubLoss(enMdl);
Or,
rng('default')
t = templateTree('Reproducible',true);
enMdl = fitrensemble(Xnum,Ynum,'OptimizeHyperparameters','auto','Learners',t, ...
'HyperparameterOptimizationOptions',struct('AcquisitionFunctionName','expected-i
mprovement-plus'))
ResubLoss=resubLoss(enMdl);
```

NOTE

In general, **L= resubLoss(Mdl)** returns the regression loss by resubstitution, or the in-sample regression loss, for the trained regression model **Mdl** using the training data stored in **Mdl.X** and the corresponding responses stored in **Mdl.Y**. The interpretation of **L** depends on the loss function ('**LossFun**') and weighting scheme (**Mdl.W**). In general, better models yield smaller loss values. The default '**LossFun**' value is '**mse**' (mean squared error).

Quiz #9

(T/F) Ensemble learning can only be applied to classification trees.

Quiz #10

What is generally a benefit of using an ensemble?
a) Faster model creation time   b) Better predictive model performance
c) Less memory usage   d) Less data requirement

Creating Ensembles: Heart Disease Data

The table **heartData** contains several features of different subjects and whether or not they have heart disease, which is saved in the **HeartDisease** variable. Figure 3.27 shows the code for creating an ensemble of 50 sevenfold cross-validated classification trees using the bag method. The model is named **baggedTrees**. When we use the bag method, we need to supply the '**Type**' property with a value of either '**Classification**' or '**Regression**'. Moreover, the loss is calculated and saved to a variable named **lossBT**.

NOTE

The number of bags and the KFold number are parameters that can be adjusted to minimize the estimated loss of the model.

```
load('HeartDisease.mat')
% Build an ensemble of bagged trees
baggedTrees=fitensemble(HeartDisease,'HeartDisease','Bag',50,'Tree', ...
'Type','Classification','KFold',7);
lossBT=kfoldLoss(baggedTrees);
```

Figure 3.27 Creating an ensemble of 50 sevenfold cross-validated classification trees using the bag method.

Figure 3.28 shows the workspace variables, including the model loss.

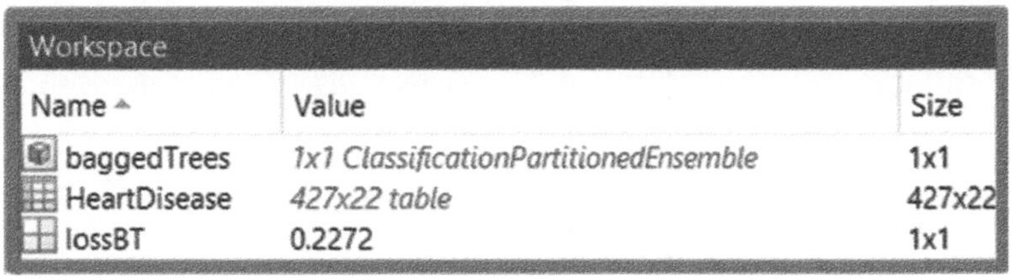

Name ▲	Value	Size
baggedTrees	*1x1 ClassificationPartitionedEnsemble*	1x1
HeartDisease	*427x22 table*	427x22
lossBT	0.2272	1x1

Figure 3.28 The workspace variables where it shows the model loss for the ensemble method based on 50-bag and sevenfold cross-validated classification trees learning method.

Ensemble Learning: Wine Quality Classification

The table **winered** contains several features of red wine and the corresponding quality, which is saved in quality variable. Figure 3.29 shows the code for creating an ensemble with 40 learners using sevenfold cross-validated k-Nearest Neighbor models where $k=3$. We use **'Subspace'** as the ensemble creation method. We named the ensemble model **mdlEn**. The loss is calculated and saved to a variable named **lossEn**. Moreover, the data are fitted using **'Bag'** ensemble method with 100 learners and tree decision weak learner. The resubstitution MSE for the model is saved as **BTmse**.

```
load('winered.mat')
WineRed.quality = categorical(WineRed.quality);
origData = WineRed{:,1:end-1};
MaxVal=max(origData,[], 1);
MinVal=min(origData,[], 1);
for j=1:size(origData,2)
numData(:,j)=(origData(:,j)-MinVal(j))/(MaxVal(j)-MinVal(j));
end
%%Create an ensemble of KNN classifiers
X=numData;
Y= WineRed.quality;
knnTemplate = templateKNN('NumNeighbors',3','DistanceWeight','squaredinverse');
mdlEn = fitensemble(X,Y,'Subspace',40,knnTemplate,'KFold',7);
lossEn=kfoldLoss(mdlEn);
t = templateTree('Surrogate','On');
mdlEn2 = fitensemble(X,Y, 'Bag',100,t, 'Type','Classification');
%Estimate the re-substitution MSE
BTmse = resubLoss(mdlEn2);
```

Figure 3.29 Creating two ensembles based on **'KNN'** and **'Tree'** weak learner.

Figure 3.30 shows workspace variables and the ensemble model loss value of 0.525. The resubstitution MSE for the second model is zero.

Workspace		
Name ⏶	**Value**	**Size**
BTmse	0	1x1
j	11	1x1
knnTemplate	*1x1 FitTemplate*	1x1
lossEn	0.5247	1x1
MaxVal	*1x11 double*	1x11
mdlEn	*1x1 ClassificationPartitionedEnsemble*	1x1
mdlEn2	*1x1 ClassificationBaggedEnsemble*	1x1
MinVal	*1x11 double*	1x11
numData	*1599x11 double*	1599x11
origData	*1599x11 double*	1599x11
t	*1x1 FitTemplate*	1x1
WineRed	*1599x12 table*	1599x12
X	*1599x11 double*	1599x11
Y	*1599x1 categorical*	1599x1

Figure 3.30 The workspace variables showing the ensemble model loss and resubstitution MSE.

Improving fitcensemble Predictive Model: Abalone Age Prediction

The table abalone has several predictor variables: One is categorical, and the remainders are numeric and a response, named **Rings**, which corresponds to the age of the abalone.

Figure 3.31 shows the code for creating a tree ensemble of learners for classification, named **en1**, **en2**, **e3**, and **en4**. For **en1**, we train an ensemble of boosted classification trees by using **fitcensemble**. To speed up the training process, we can use the 'NumBins' name-value pair argument. If we specify the 'NumBins' value as a positive integer scalar, then the algorithm bins every numeric predictor into a specified number of equiprobable bins, and then grows trees on the bin indices instead of the original data. The algorithm *does not bin* categorical predictors. Notice that the process upon using 'NumBins' name-value pair is about three times faster than using the original data. The elapsed time, however, can vary depending on the operating system. To estimate the elapse time, we bracket the command by tic and toc variables. For **en2**, we cross-validate an ensemble of 150 boosted classification trees using 5-fold cross-validation. Using a tree template, we set the maximum number of splits to 16, and the learning rate to 0.5. For **en3**, we can find hyperparameters that minimize fivefold cross-validation loss by using automatic hyperparameter optimization. For reproducibility, we set the random seed and use the 'expected-improvement-plus' acquisition function. Also, for reproducibility of random forest algorithm, we specify the 'Reproducible' name-value pair argument as true for tree learners. For **en4**, we have a simplified version of auto hyper parameters optimization.

```
load('Abalone.mat', 'abalone'); % load abalone;
%% Fit a cross-validated ensemble
dv = dummyvar(abalone.SEX);
pv= [abalone{:,2:end-1} dv];
X=pv;
Y=categorical(abalone.Rings);
% tic and toc are used to bracket the elapsing time for finding the ensemble
tic
```

Figure 3.31 Four different ensembles of boosted classification trees using **fitcensemble**.

```
en=fitcensemble(X,Y,'NumBins',50);
toc
lossen=resubLoss(en);
tic
t = templateTree('MaxNumSplits',16);
en2 = fitcensemble(X,Y,'NumLearningCycles',150,...
'Learners',t,'KFold',5,'LearnRate',0.5);
lossen2=kfoldLoss(en2);
toc
tic
rng('default')
t = templateTree('Reproducible',true);
en3= fitcensemble(X,Y,'OptimizeHyperparameters','auto','Learners',t, ...
   'HyperparameterOptimizationOptions',...
struct('AcquisitionFunctionName','expected-improvement-plus'))
lossen3=resubLoss(en3);
toc
tic
en4 = fitcensemble(X,Y,'OptimizeHyperparameters','auto')
lossen4=resubLoss(en4);
toc
```

Figure 3.31   (Cont'd)

Results are copied from Command window. Such values respectively represent the elapsing time needed for creating the classification ensemble models **en**, **en2**, **en3**, and **en4**.

```
Elapsed (CPU) time is 2.955601 seconds.
Elapsed (CPU) time is 15.848934 seconds.
Elapsed (CPU) time is 920.268294 seconds.
Elapsed (CPU) time is 646.516412 seconds.
```

Figure 3.32 shows the results in the form of tabulating different hyper parameters pertaining to the used method and the elapsed and function evaluation time.

```
| Iter | Eval    | Objective | Objective | BestSoFar  | BestSoFar |     Method | NumLearningC-| LearnRate | MinLeafSize
|      | result  |           | runtime   | (observed) | (estim.)  |            | ycles        |           |
|================================================================================================================
|    1 | Best    |    0.74   |   23.18   |    0.74    |   0.74    | AdaBoostM2 |      228     | 0.002784  |          3
|    2 | Accept  |      1    |  1.0382   |    0.74    | 0.76071   |  RUSBoost  |       13     | 0.013633  |         60
|   29 | Accept  | 0.83505   |  25.586   |  0.72085   | 0.72152   | AdaBoostM2 |      349     | 0.96522   |       1984
|   30 | Accept  | 0.72827   |  1.9022   |  0.72085   | 0.72151   | AdaBoostM2 |       18     | 0.97212   |         31

Optimization completed.
MaxObjectiveEvaluations of 30 reached.           Observed objective function value = 0.72085
Total function evaluations: 30                   Estimated objective function value = 0.72168
Total elapsed time: 911.4449 seconds             Function evaluation time = 48.795
Total objective function evaluation time: 902.0416
                                                 Best estimated feasible point (according to models):
                                                    Method     NumLearningCycles    LearnRate    MinLeafSize
Best observed feasible point:                     _____    _____    _____    _____
   Method     NumLearningCycles    LearnRate    MinLeafSize
 _____    _____    _____    _____   AdaBoostM2         493            0.97531          3

 AdaBoostM2         487            0.94717          4          Estimated objective function value = 0.72151
                                                               Estimated function evaluation time = 50.2982
```

Figure 3.32 Optimization of hyper model parameters to converge the estimated to observed function value.

Figure 3.33 shows the workspace variables, including the loss of each created classification ensemble model.

Name ▲	Value	Size
abalone	*4177x9 table*	*4177x9*
dv	*4177x3 double*	*4177x3*
en	*1x1 ClassificationEnsemble*	*1x1*
en2	*1x1 ClassificationPartitionedEnsemble*	*1x1*
en3	*1x1 ClassificationEnsemble*	*1x1*
en4	*1x1 ClassificationEnsemble*	*1x1*
lossen	0.7141	*1x1*
lossen2	0.7240	*1x1*
lossen3	0.7086	*1x1*
lossen4	0.7086	*1x1*
pv	*4177x10 double*	*4177x10*
t	*1x1 FitTemplate*	*1x1*
X	*4177x10 double*	*4177x10*
Y	*4177x1 categorical*	*4177x1*

Figure 3.33 Workspace variables, including model loss for each ensemble model.

Improving fitctree Predictive Model with Feature Selection (FS): Credit Ratings Data

The table **creditR** has several predictor variables and a response named **class**.

Figure 3.34 shows the code for creating a reduced data set with three predictor variables, using the feature selection based on the output of **predictorImportance** function which computes estimates of predictor importance for tree by summing changes in the risk due to splits on every predictor and dividing the sum by the number of branch nodes. Zero represents the smallest possible importance. Then, a sevenfold cross-validated tree model named **tmdl** and the model loss, **tLoss**, will be calculated.

```
load('creditR.mat')
creditR.class = categorical(creditR.class);
%%Model the data with three or fewer predictors
[CR2 Varnames]=cattbl2mat(creditR(:,1:end-1));
X=CR2;
Y=creditR.class;
p = predictorImportance(fitctree(X,Y));
% View predictor importance on a bar plot
figure(35)
bar(p)
[~,idxOrder] = sort(p, 'descend');
Sel=[idxOrder(1:3)];
tmdl = fitctree(X(:,Sel),Y,'KFold',7);
tLoss=kfoldLoss(tmdl);
```

Figure 3.34 The fitting of binary decision tree for multiclass classification model to credit ratings data with the feature selection of reducing predictor variables.

Figure 3.35 shows workspace variables, including **tmdl** model and its loss, **tLoss**. Notice that the number of predictor variables was reduced from 61 down to 3 variables. These three important predictor variables are found to be: **'credit_amount'**, **'duration'**, and **'checking_status_no_checking'**. The properties of the reduced model, via feature selection, can be accessed via clicking on **"tmdl"** icon found in the work space. The selected three important predictor variables can be reached via clicking on **"PredictorNames"** icon. The bar plot for p-values of all 61 predictors is not shown here as it is similar to that shown in Figure 3.13.

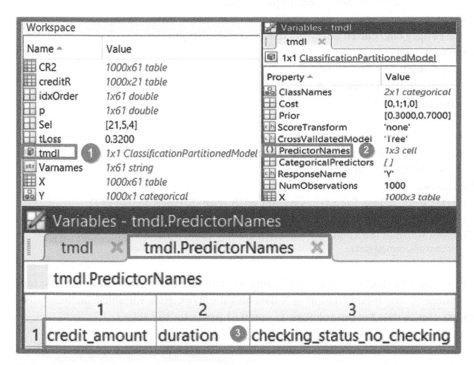

Figure 3.35 Workspace variables showing the remaining three important predictor variables.

Improving fitctree Predictive Model with Feature Transformation (FT): Credit Ratings Data

If we run the show with the original **X** data, then MATLAB tells us that columns of X are linearly dependent to within machine precision and it only used the first 48 components to compute TSQUARED. Consequently, I used the covariance of **X** to see the strongly correlated X-columns which then were removed from the original **X** predictor data.

Figure 3.36 shows the code where we converted a categorical to a matrix followed by conversion into numeric array, testing the covariance of **X** data to eliminate highly correlated columns (with a correlation coefficient > 0.95), carrying out principal component analysis, PCA, selecting the sevenfold cross-validated classification trees to fit the first three transformed principal components, **scrs(:,3)**, and finally estimating the model loss.

```
load('creditR.mat')
creditR.class = categorical(creditR.class);
%% Converting a categorical into a table matrix
[CR2 Varnames]=cattbl2mat(creditR(:,1:end-1));
% Converting a table into numeric matrix
CR3 = table2array(CR2);
X=CR3;
Y=creditR.class;
%Testing X predictor columns for linear dependence (or correlation).
Xcov=cov(X);
% Removing strongly correlated columns to avoid redundancy.
remove=any(Xcov>0.95);
Xclean=X(:,~remove);
% Feature Transformation with PCA.
[pcs,scrs,~,~,pexp] = pca(Xclean);
figure(36)
pareto(pexp)
% Create reduced model using the first three transformed principal components.
t = fitctree(scrs(:,1:3),Y,'KFold',7);
tLoss = kfoldLoss(t);
disp([num2str(tLoss),' = loss for a trees model with transformed features'])
```

Figure 3.36 Predictor variables reduction via FT followed by a sevenfold cross-validated classification trees model to fit the transformed data.

Figure 3.37 (*left*) shows the Pareto plot for explaining the percentile contribution of the first ten principal components to the total variation in X data and the workspace variables, including the model loss (*right*). Notice that **Xclean** is left by 31 out of 61 predictor variables.

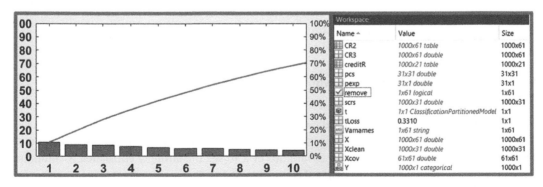

Figure 3.37 The Pareto plot for the first ten principal components (*left*) and workspace variables (*right*).

Using MATLAB Regression Learner

The MATLAB regression learner is a platform to train regression models to predict data using supervised machine learning. Using this app, we can explore data, select features, specify validation schemes, train models, and assess results. We can perform automated training to search for the best regression model type, including linear regression models, regression trees, Gaussian process regression models, support vector machines, ensembles of regression trees, and neural network regression models.

Perform supervised machine learning by supplying a known set of observations of input data (predictors) and known responses. Use the observations to train a model that generates predicted responses for new input data. To use the model with new data, or to learn about programmatic regression, we can export the model to the workspace or generate MATLAB® code to recreate the trained model.

We will use MATLAB built-in big car data as the source of input datasets. Figure 3.38 shows the initial code needed to prepare the raw data to be utilized in the **regressionLearner** environment. We have selected four columns, representing predictor variables and the column called **MPG** as the **M**ileage **P**er **G**allon economic indicator. Both table and numeric matrix versions are created for training and test reasons.

```
load carbig
X = [Weight,Horsepower,Acceleration,Displacement];
Y=MPG;
XY=[X Y];
temp=rmmissing(XY);
Xnum=temp(:,1:end-1);
Ynum=temp(:,end);
[r,c] = size(Xnum) ;
P = 0.70 ;
idx = randperm(r);
XTrain = Xnum(idx(1:round(P*r)),:) ;
XTest = Xnum(idx(round(P*r)+1:end),:) ;
YTrain = Ynum(idx(1:round(P*r)),:);
YTest = Ynum(idx(round(P*r)+1:end),:) ;
XnumTbl=array2table(Xnum,'VariableNames',{'Weight','HP','Accel', 'Disp'});
```

Figure 3.38   The big car data uploaded to MATLAB workspace.

To Open the Regression Learner App, we can either use MATLAB Toolstrip: On the **Apps** tab, under "**Machine Learning and Deep Learning**" category, click on "**Regression Learner**" app icon, or use MATLAB command prompt:

>>regressionLearner

Let us use the Regression Learner by executing the command **regressionLearner** in the Command Window of MATLAB. The first pop-up window of the regression learner app will be similar to that shown in Figure 2.16 for classification learner. Click on "**New Session**" button and select "**From Workspace**" option from the drop-down menu.

After selection, the data set window will pop-up and we have to populate the required input predictor columns and response column as shown in Figure 3.39. The validation method can be selected as one out of three. We will select the top one as we have already tried this in previous examples and also to avoid overfitting scenario. Click on "**Start Session**" button and the regression learner main window will pop-up.

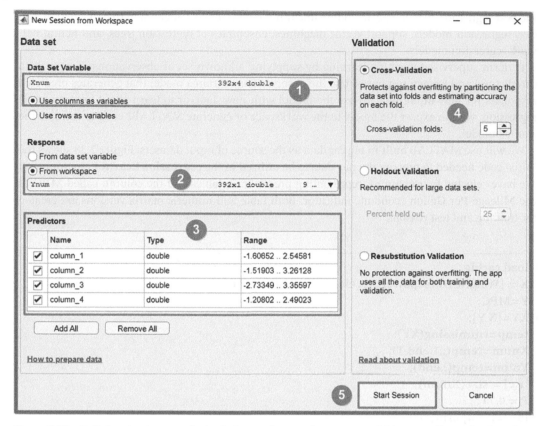

Figure 3.39 Defining the data sets for both the predictor and response variables as well as selection of the validation method.

It is worth mentioning here that features of regression learner are essentially the same as those of the classification learner, shown earlier in Chapter 2 | Using MATLAB Classification Learner section. Please, refer to figures 2.17 through 2.27 for further details.

Under "**MODEL TYPE**" tab, we will select "**All**" to include all different kinds of fitting/classifying models and based on the results we will select or pinpoint to the best fitting model.

Let us click on "**Train**" button after selecting "**All**" from the "**MODEL TYPE**" group. Wait for a while as we have selected to examine all model types. Figure 3.40 shows the training results of the trained models and their corresponding validation (training) accuracy. We can see that the model with "**Gaussian Process Regression, Exponential GPR**" category got the highest rank (lowest RMSE) in terms of validation accuracy.

For the sake of extra learning, I attempted to use "**Gaussian Process Regression, Optimizable**" category. Under an optimizable case of any model that entertains such a feature, the regression learner app will attempt to optimize one or more parameters such that to minimize the model error being expressed in terms of mean squared errors (**MSE**), mean absolute percent error (**MAPE**), and root mean of squared errors (**RMSE**). Figure 3.41 shows such a case where there exists a parameter called sigma which varies between 0.0001 and 78.05 for this case study and the best value of sigma was found to be 65.3745 such that **MSE** is minimum. The situation where we use what is called hyper parameter optimization is dealt with in previous and subsequent chapters for applicable models. In later chapters, the user can realize that after finding the optimum parameter(s) for model fitting, he/she may then reuse the model at such best fixed values, which minimize the error of the fitting model under concern.

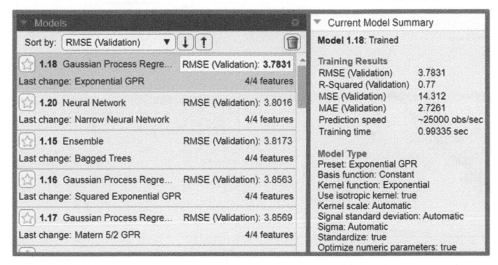

Figure 3.40 Training results using "**Regression Learner**" app with Exponential GPR model as the best.

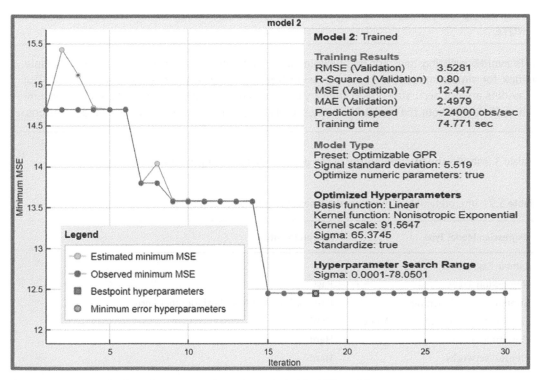

Figure 3.41 Optimization of parameter sigma for optimizable GPR model.

Figure 3.42 (*left*) shows the optimizable GPR predicted versus observed **MPG** value and the residual ($Y_{observed} - Y_{model}$) for each response datapoint (*right*).

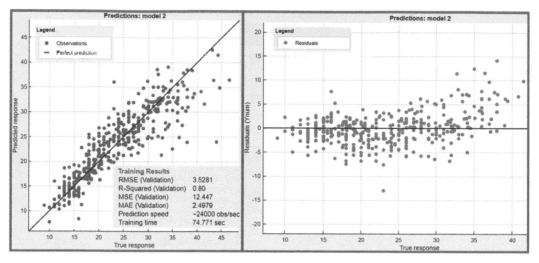

Figure 3.42 The optimizable GPR model goodness in terms of predicted versus observed value (*left*) and also the residual versus true value (*right*).

NOTE

To avoid overfitting, look for a less flexible model that provides sufficient accuracy. For example, look for simple models, such as regression trees that are fast and easy to interpret. If your models are not accurate enough, then try other models with higher flexibility, such as ensembles. To learn about the model flexibility, see MATLAB help: Choose Regression Model Options.

Table 3.9 shows the interpretability of the general classes of regression model types.

Table 3.9 Interpretability of the general classes of regression model types.

Regression Model Type	Interpretability
Linear Regression Models	Easy
Regression Trees	Easy
Support Vector Machines	Easy for linear SVMs. Hard for other kernels.
Gaussian Process Regression Model	Hard
Ensembles of Trees	Hard
Neural Networks	Hard

At this stage, we can proceed to the final step; however, it is preferred to carry out the test step for double check. Be sure to select the optimizable GPR (shown in Figure 3.42) as it has the lowest **RMSE**. From the Toolstrip, click on "**Test Data**" button found in "**TESTING**" group and fill in the required test data sets as shown in Figure 3.43. Click on "**Import**" button to proceed.

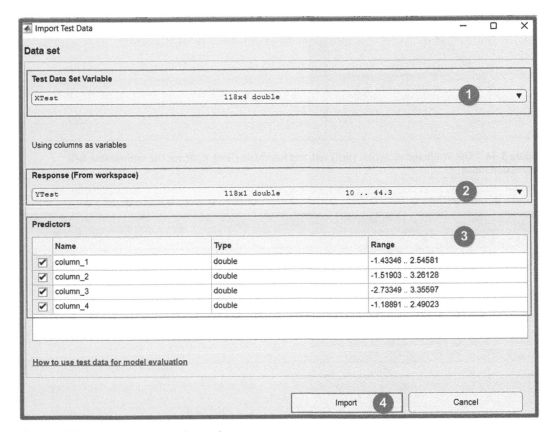

Figure 3.43   Importing data sets for testing step.

From the Toolstrip, click on "**Test All**" button found in "**TESTING**" group and select "**Test Selected**" item from the sub-menu to test only the currently selected model, that is, the optimizable GPR. MATLAB will proceed and carry out the test step. You can then view the results by selecting any plot you wish from the "**PLOTS**" group. For example, we select from "**TEST RESULTS**" menu the Predicted vs. Actual (Test) and Residuals (Test) as shown in Figure 3.44. The test results show that the model behaves more precisely than in the validation step.

Click on "**Generate Function**" button found in "**EXPORT**" group, and a temporary m-file will show up in MATLAB editor where you can modify its name and save it to the same name as an m-file function, for example, named **bigcarGPR.m**, as shown in Figure 3.45.

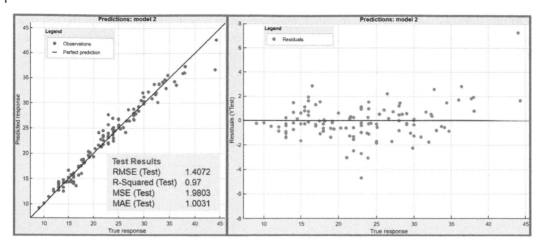

Figure 3.44 The Predicted vs. Actual (Test), *left*, and Residuals (Test), *right*, for the optimizable GPR regression model.

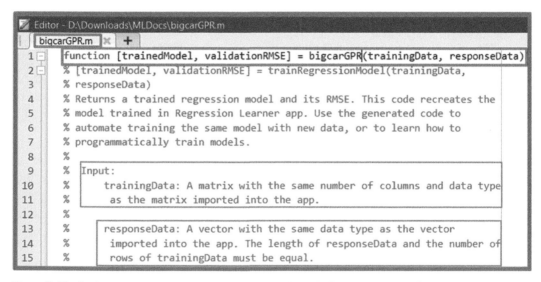

Figure 3.45 Saving the generated hyper parameters optimized GPR regression model as an m-file.

NOTE

Do not forget to save the file anywhere within MATALB defined paths, or add its location to MATLAB path via **"Set Path"** button found in **"ENVIRONMENT"** group of MATLAB Toolstrip.

To make use of this generated function, at the command prompt, enter:

>> [optimGPR vRMSE]=bigcarGPR(Xnum,Ynum);
optimGPR struct is created as well as **vRMSE** variable which has **RMSE** of the validated model itself.

Enter at the prompt,
>> vRMSE
vRMSE = 3.4629

To make predictions on a new array, **X**, we use: yfit = optimGPR.predictFcn(X).

>> Yfit = optimGPR.predictFcn(XTest);

Yfit will be a column vector (118×1) with a size equal to that of **XTest** (i.e., size(XTest,1)).

Alternatively, one may enter four numerical values representing [Weight, Horsepower, Acceleration, Displacement]:

>>Xsample= Xnum(1,:);
Xsample= [3504, 130, 12, 307];
>>Ysample = optimGPR.predictFcn(Xsample)
Ysample = 17.6237

This represents the predicted value, **MPG**. The observed value is:

>>Yobs=Ynum(1)

Yobs= 18.0

$$PRE1(\%) = \frac{\left|Y_{obs} - Y_{pred}\right|}{Y_{obs}} \times 100\% = \frac{\left|18.0 - 17.6237\right|}{18.0} \times 100\% = 2.1\% < 10\%$$

>>Xsample2= Xnum(314,:);

Xsample2=[3003, 90, 20.1, 151];

>> Ysample2 = optimGPR.predictFcn(Xsample2)

Ysample2 = 24.5064

>>Yobs2=Ynum(314)

Yobs2= 24.3

$$PRE2\left(\%\right) = \frac{\left|Y_{obs2} - Y_{pred}\right|}{Y_{obs2}} \times 100\% = \frac{\left|24.3 - 24.5\right|}{24.3} \times 100\% = 0.82\% \ll 10\%$$

Click on "**Export Model**" button found in "**EXPORT**" group, and select "**Export Model**", MATLAB will pop-up the user to enter a name for the exported model in order to be saved in MATLAB Workspace. We can save the workspace variables as *.mat file and load it later. Moreover, MATLAB command shows the following message:

Variables have been created in the base workspace. Structure 'bigcarGPR1' exported from Regression Learner. To make predictions on a new predictor column matrix, X:

yfit = bigcarGPR1.predictFcn(X).

So, we can make use of workspace variable, the structure '**bigcarGPR1**', in a similar fashion as we did ago with the created m-file function.

>> **Xsample2=[3003, 90, 20.1, 151];**
>> **Ysample2 = bigcarGPR1.predictFcn(Xsample2)**
Ysample2 = 24.5064.

Click on "**Export Model**" button found in "**EXPORT**" group, and select "**Export Compact Model**", MATLAB will pop-up the user to enter a name for the exported model in order to be saved in MATLAB Workspace. We can save the workspace variables as *.mat file and load it later. Moreover, MATLAB command shows a message similar to the previous message in the first case. The difference between the previous case and this case is simply whether or not to keep the training data set as part of the created structure.

Finally, the last choice, that is, "**Export Model for Deployment**", it deploys a compact version of the currently selected model for deployment to MATLAB Production server, which requires a PC running an installed version of MATLAB itself or running MATLAB component runtime. This last option is explained in Chapter 2 | Using MATLAB Classification Learner section.

Finally, Figure 3.46 shows workspace variables, including those of exported regression models.

Workspace						
Name ▲	Value	Model_Year	406x1 double	XnumTbl	392x4 table
Acceleration	406x1 double	MPG	406x1 double	Xsample	[3504,130,12,307]
bigcarGPR1	1x1 struct	optimGPR	1x1 struct	Xsample2	[3003,90,20.1000,151]
bigcarGPRc	1x1 struct	org	406x7 char	XTest	118x4 double
c	4	Origin	406x7 char	XTrain	274x4 double
cyl4	406x5 char	P	0.7000	XY	406x5 double
Cylinders	406x1 double	r	392	Y	406x1 double
Displacement	406x1 double	temp	392x5 double	Ynum	392x1 double
Horsepower	406x1 double	vRMSE	3.4629	Ysample	17.6237
idx	1x392 double	Weight	406x1 double	Ysample2	24.5064
Mfg	406x13 char	when	406x5 char	YTest	118x1 double
Model	406x36 char	X	406x4 double	YTrain	274x1 double
		Xnum	392x4 double		

Figure 3.46 Workspace variables as related to regression learner app.

NOTE

It is worth mentioning here that I tried both options with and without standardization of predictor data and the model predictability turns out to be the same.

If we select **XnumTbl** table instead of **Xnum** numeric array as the source for predictor data, explained earlier in Figure 3.39 and if we proceed with a table argument, then all generated functions and explored models will accept only a table as an input argument. The table has a head or name for each column, unlike the numeric matrix. Moreover, if we attempt to make use of explored models or m-file functions, the column names must match those of the original table used in validation phase.

Feature Selection and Feature Transformation Using Regression Learner App

In Regression Learner, use the response plot to try to identify predictors that are useful for predicting the response. To visualize the relation between different predictors and the response, under **X**-axis, select different variables in the **X** list, as shown in Figure 3.47. Notice that to see X predictors as names, create a new regression state and select **XnumTbl** table instead of **Xnum** array as the source for predictor data, as explained in Figure 3.39.

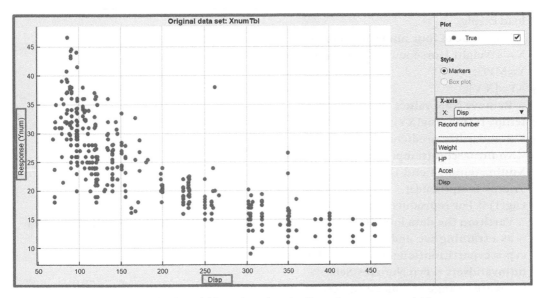

Figure 3.47   One can select X-axis variable and see how it affects the response variable.

Before you train a regression model, the response plot shows the training data. If you have trained a regression model, then the response plot also shows the model predictions. Observe which variables are associated most clearly with the response. When you plot the **carbig** data set, the predictor Horsepower shows a clear negative association with the response. Look for features that do not seem to have any association with the response and use "**Feature Selection**" to remove (or deselect) those features from the set of used predictors, as shown in Figure 3.48.

Figure 3.48   Using "**Feature Selection**" the user may select/deselect a predictor variable from/out of the list.

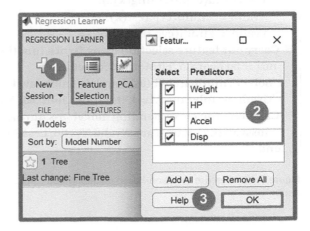

Feature Selection Using Neighborhood Component Analysis (NCA) for Regression: Big Car Data

The big car data will be used here. The code to use NCA regression with feature selection enabled is shown in Figure 3.49. The code is augmented by comments throughout.

```
load carbig
% X is made of four numeric predictors
X = [Weight,Horsepower,Acceleration,Displacement];
Y=MPG;
XY=[X Y];
% Remove NaN values
temp=rmmissing(XY);
% Standardize predictors
%Xnum=zscore(temp(:,1:end-1));
Xnum=temp(:,1:end-1);
Ynum=temp(:,end);
rng(1) % For reproducibility
% Partition the data into five folds. For each fold, cvpartition assigns 4/5th of the data
% as a training set, and 1/5th of the data as a test set.
cvp = cvpartition(length(Ynum),'kfold',5);
numvalidsets = cvp.NumTestSets;
% Create a range of lambda values.
n = length(Ynum);
lambdavals = linspace(0,50,20)*std(Ynum)/n;
% Initialize a matrix of zeros for loss values.
lossvals = zeros(length(lambdavals),numvalidsets);
% Permutate i from 1 upto length of lambdavals row vector.
for i = 1:length(lambdavals)
   for k = 1:numvalidsets
      X = Xnum(cvp.training(k),:);
      Y = Ynum(cvp.training(k),:);
      Xvalid = Xnum(cvp.test(k),:);
      Yvalid = Ynum(cvp.test(k),:);
% Carry out feature selection using fsrnca function for a set of lambda values.
      nca = fsrnca(X,Y,'FitMethod','exact', ...
      'Solver','minibatch-lbfgs','Lambda',lambdavals(i), ...
      'GradientTolerance',1e-4,'IterationLimit',60, 'Standardize',true);
% Calculate the corresponding model loss during validation step.
      lossvals(i,k) = loss(nca,Xvalid,Yvalid,'LossFunction','mse');
   end
```

Figure 3.49 Feature selection-based neighborhood component analysis (NCA) regression fit to big car data.

```
end
meanloss = mean(lossvals,2);
figure
plot(lambdavals,meanloss,'ro-')
xlabel('Lambda')
ylabel('Loss (MSE)')
grid on
[~,idx] = min(meanloss);
bestlambda = lambdavals(idx);
bestloss = meanloss(idx);
% Retrain at the best value of lambda.
nca = fsrnca(X,Y,'FitMethod','exact', 'Solver','lbfgs',...
'Lambda',bestlambda, 'Standardize',true);
figure
plot(nca.FeatureWeights,'ro')
xlabel('Feature Index')
ylabel('Feature Weight')
```

Figure 3.49 (Cont'd)

Figure 3.50 (*left*) shows the model **MSE** versus the regularization parameter, **lambda**. The optimum occurs at **lambda**=*0.0524* where **MSE** has the least value of *14.1943*. On the *right*, the feature weight value for each predictor is shown. The most influential predictor on the response **MPG** is predictor #2, **Horsepower**, followed by #4, **Displacement**, and the least important predictor is #3, **Acceleration**. More on model regularization parameter is covered in Chapter 4.

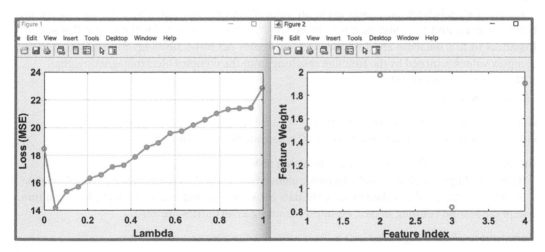

Figure 3.50 The optimum regularization parameter, lambda (*left*) and the feature weight for predictors (*right*).

Finally, the workspace variables are shown in Figure 3.51.

Workspace		Workspace		Workspace	
Name ▲	Value	Name ▲	Value	Name ▲	Value
Acceleration	*406x1 double*	k	5	org	*406x7 char*
bestlambda	0.0524	lambdavals	*1x20 double*	Origin	*406x7 char*
bestloss	14.1943	lossvals	*20x5 double*	temp	*392x5 double*
cvp	*1x1 cvpartition*	meanloss	*20x1 double*	Weight	*406x1 double*
cyl4	*406x5 char*	Mfg	*406x13 char*	when	*406x5 char*
Cylinders	*406x1 double*	Model	*406x36 char*	X	*314x4 double*
Displacement	*406x1 double*	Model_Year	*406x1 double*	Xnum	*392x4 double*
Horsepower	*406x1 double*	MPG	*406x1 double*	Xvalid	*78x4 double*
i	20	n	392	XY	*406x5 double*
idx	2	nca	*1x1 FeatureSele*	Y	*314x1 double*
		numvalidsets	5	Ynum	*392x1 double*
				Yvalid	*78x1 double*

Figure 3.51 The workspace variables showing values of best lambda and best model loss.

CHW 3.1   The Ionosphere Data

Consider the ionosphere (**load ionosphere;**) MATLAB built-in database.
Relevant Information:

This radar data was collected by a system in Goose Bay, Labrador. This system consists of a phased array of 16 high-frequency antennas with a total transmitted power on the order of 6.4 kilowatts. The targets were free electrons in the ionosphere. "Good" radar returns are those showing evidence of some type of structure in the ionosphere. "Bad" returns are those that do not; their signals pass through the ionosphere.

Received signals were processed using an autocorrelation function whose arguments are the time of a pulse and the pulse number. There were 17 pulse numbers for the Goose Bay system. Instances in this database are described by 2 attributes per pulse number, corresponding to the complex values returned by the function resulting from the complex electromagnetic signal.

Number of Instances: 351
Attributes Information

- All 34 predictor attributes are continuous.
- The 35th attribute is either "good" or "bad" according to the definition

summarized above. This is a binary classification task.

Reference: https://archive.ics.uci.edu/ml/machine-learning-databases/ionosphere/ionosphere.data

Standardize the predictor data or scale the data between zero and one for each predictor column, except the second column. In fact, exclude column #2 as all entries are zero.

Carry out the following tasks:

1) For a given classification model, create a cross-validated model by specifying only one of these four name-value arguments: CVPartition, Holdout, KFold, or Leaveout. Show the effect of low and high k-fold or holdout size on the kfold loss of the model.

2) Use **PCA** feature transformation and factor analysis to see if it is possible to reduce the number of predictor variables without sacrificing the model rigor. In other words, find the kfoldLoss for the model both utilizing all predictor data and reduced data. Find the maximum number of common factors to describe model variance.

3) Using the **predictorImportance** function, calculate the importance (**p**) values for each of the predictors. Reduce some predictors by selecting a threshold. Calculate the model loss for the case with all predictors versus the case with selected predictors via applying the feature selection. Enforce k-fold validation for the examined method. Show the most important selected predictors as an outcome of feature selection elimination process.

4) Using the function **sequentialfs** to perform sequential feature selection, which incrementally adds a predictor to the model as long as there is reduction in the prediction error, to find the number of predictor variables needed to minimize the prediction to the lowest possible value. You should either refer to an m-file that defines the error function or use a handle to the error function. Provide additional options to specify the cross-validation method. Calculate the model loss under both cases: with **SFS** and without **SFS**. Name the remaining selected predictor variables.

5) Create an ensemble with 50 learners using different ensemble aggregation methods.
   You may use different weak learners with the ensemble aggregation method.
   Calculate the loss of the ensemble model or resubstitution MSE for the model.

6) Find hyperparameters that minimize **fitcensemble** method using automatic hyperparameter optimization. For reproducibility, set the random seed and use the 'expected-improvement-plus' acquisition function. Also, for reproducibility of random forest algorithm, specify the '**Reproducible**' name-value pair argument as *true* for tree learners.

7) Using MATLAB classification learner app, find the best classification method, characterized by the highest validation accuracy. Specify both the predictor and response data and train all available models to see the best performing classification model. Show the confusion matrix plot for the best model and its **ROC** curve. At the end, generate an m-file function which can be used to return a trained classifier and its accuracy. The created m-file code recreates the classification model trained in classification learner app.

CHW 3.2   Sonar Dataset

The Sonar Dataset involves the prediction of whether or not an object is a mine or a rock given the strength of sonar returns at different angles.

   Reference: https://archive.ics.uci.edu/ml/machine-learning-databases/undocumented/connectionist-bench/sonar/sonar.all-data

   It is a binary (2-class) classification problem. The number of observations for each class is not balanced. There are 208 observations with 60 input variables and 1 output variable. The variable names are as follows:

Sonar returns at different angles
VN1 up to VN60
Class (M for mine and R for rock)

The baseline performance of predicting the most prevalent class is a classification accuracy of approximately 53%. Top results achieve a classification accuracy of approximately 88%.

   Standardize the predictor data or scale the data between zero and one for each predictor column.

Carry out the following tasks:

1) For a given classification model, create a cross-validated model by specifying only one of these four name-value arguments: CVPartition, Holdout, KFold, or Leaveout. Show the effect of low and high k-fold or holdout size on the kfold loss of the model.

2) Use **PCA** feature transformation and factor analysis to see if it is possible to reduce the number of predictor variables without sacrificing the model rigor. In other words, find the kfoldLoss for the model utilizing all predictor data and reduced data. Find the maximum number of common factors to describe model variance.

3) Using the **predictorImportance** function, calculate the importance (**p**) values for each of the predictors. Reduce some predictors by selecting a threshold. Calculate the model loss for the case with all predictors versus the case with selected predictors via applying the feature selection. Enforce k-fold validation for the examined method. Show the most important selected predictors as an outcome of feature selection elimination process.

4) Using the function **sequentialfs** to perform sequential feature selection, which incrementally adds a predictor to the model as long as there is reduction in the prediction error, to find the number of predictor variables needed to minimize the prediction to the lowest possible value. You should either refer to an m-file that defines the error function or use a handle to the error function. Provide additional options to specify the cross-validation method. Calculate the model loss under both cases: with **SFS** and without **SFS**. Name the remaining selected predictor variables.

5) Create an ensemble with 50 learners using different ensemble aggregation methods.
   You may use different weak learners with the ensemble aggregation method.
   Calculate the loss of the ensemble model or resubstitution MSE for the model.

6) Find hyperparameters that minimize **fitcensemble** method using automatic hyperparameter optimization. For reproducibility, set the random seed and use the 'expected-improvement-plus' acquisition function. Also, for reproducibility of random forest algorithm, specify the 'Reproducible' name-value pair argument as *true* for tree learners.

7) Using MATLAB classification learner app, find the best classification method, characterized by the highest validation accuracy. Specify both the predictor and response data and train all available model to see the best performing classification model. Show the confusion matrix plot for the best model and its **ROC** curve. At the end, generate an m-file function which can be used to return a trained classifier and its accuracy. The created m-file code recreates the classification model trained in classification learner app.

CHW 3.3   White Wine Classification

This data set red wine data can be accessed via

>>load('winewhite.mat');

Or,

>>readtable('winequality-white.csv','Format', 'auto');

contains several features of white wine and the corresponding quality. The wine quality dataset involves predicting the quality of white wines on a scale given chemical measures of each wine.
   Reference: (https://archive.ics.uci.edu/ml/machine-learning-databases/wine-quality/winequality-white.csv).

It is a multiclass classification problem and the number of observations for each class is not balanced. There are 4898 observations with 11 input variables and 1 output variable. The variable names are as follows:

1) fixed acidity
2) volatile acidity
3) citric acid
4) residual sugar
5) chlorides
6) free sulfur dioxide
7) total sulfur dioxide
8) density
9) pH
10) sulphates
11) alcohol
12) quality (score between 3 and 9).

Standardize the predictor data or scale the data between zero and one for each predictor column.

Carry out the following tasks:

1) For a given classification model, create a cross-validated model by specifying only one of these four name-value arguments: CVPartition, Holdout, KFold, or Leaveout. Show the effect of low and high k-fold or holdout size on the kfold loss of the model.
2) Use **PCA** feature transformation and factor analysis to see if it is possible to reduce the number of predictor variables without sacrificing the model rigor. In other words, find the kfoldLoss for the model utilizing all predictor data and reduced data. Find the maximum number of common factors to describe model variance.
3) Using the **predictorImportance** function, calculate the importance (**p**) values for each of the predictors. Reduce some predictors by selecting a threshold. Calculate the model loss for the case with all predictors versus the case with selected predictors via applying the feature selection. Enforce k-fold validation for the examined method. Show the most important selected predictors as an outcome of feature selection elimination process.
4) Using the function **sequentialfs** to perform sequential feature selection, which incrementally adds a predictor to the model as long as there is reduction in the prediction error, to find the number of predictor variables needed to minimize the prediction to the lowest possible value. You should either refer to an m-file that defines the error function or use a handle to the error function. Provide additional options to specify the cross-validation method. Calculate the model loss under both cases: with **SFS** and without **SFS**. Name the remaining selected predictor variables.
5) Create an ensemble with 50 learners using different ensemble aggregation methods.
   You may use different weak learners with the ensemble aggregation method.
   Calculate the loss of the ensemble model or resubstitution MSE for the model.
6) Find hyperparameters that minimize **fitcensemble** method using automatic hyperparameter optimization. For reproducibility, set the random seed and use the 'expected-improvement-plus' acquisition function. Also, for reproducibility of random forest algorithm, specify the '**Reproducible**' name-value pair argument as *true* for tree learners.
7) Using MATLAB classification learner app, find the best classification method, characterized by the highest validation accuracy. Specify both the predictor and response data and train all available model to see the best performing classification model. Show the Predicted vs. Actual plot

for the best model and its **ROC** curve. At the end, generate an m-file function which can be used to return a trained classifier and its accuracy. The created m-file code recreates the classification model trained in classification learner app.

CHW 3.4   Small Car Data (Regression Case)

The small car data (**load carsmall**) can be found in MATLAB built-in database.

1) Using the function **sequentialfs** to perform sequential feature selection, which incrementally adds a predictor to the model as long as there is a reduction in the prediction error, to find the number of predictor variables needed to minimize the prediction to the lowest possible value. You should either refer to an m-file that defines the error function or use a handle to the error function. Use **fitr*** instead of **fitc*** both in error function statement and in the regression model statement, as well. Provide additional options to specify the cross-validation method. Calculate the model loss under both cases: with **SFS** and without **SFS**. Name the remaining selected predictor variable(s).

2) Find hyperparameters that minimize **fitrensemble** method using automatic hyperparameter optimization. For reproducibility, set the random seed and use the '**expected-improvement-plus**' acquisition function. Also, for reproducibility of random forest algorithm, specify the '**Reproducible**' name-value pair argument as *true* for tree learners. Calculate the resubstitution loss for the regression model.

3) Using MATLAB **regression** learner app, find the best regression method, characterized by the highest validation accuracy. Specify both the predictor and response data and train all available model to see the best performing classification model. Show the Validation Predicted vs. Actual and Validation Residuals plot. If the optimizable case of the given model is found to be the best, report optimized parameters and minimum MSE model.

   At the end, generate an m-file function which can be used to return a trained classifier and its accuracy. The created m-file code recreates the classification model trained in classification learner app.

4) Apply feature selection using neighborhood component analysis (**NCA**) for Regression. Show a plot for the model **MSE** versus the regularization parameter, **lambda**. Find the optimum **lambda**, where **MSE** is minimum. Show another plot for the feature weight value for each predictor. Prioritize the predictors in light of their weights on the response **MPG**. Compare the results of small car data with those of big car data. Draw your conclusions.

4

Methods of ML Linear Regression

Introduction

Fuel efficiency is an integral aspect of a motor vehicle's technical specifications, and it holds significant weight for numerous individuals when they are making a car purchase. Various technical specifications, including the vehicle's size and engine capacity, impact its fuel economy. Therefore, it would be advantageous for an automobile manufacturer to forecast fuel efficiency based on other specifications. The objective is to comprehend the correlation between these factors, indicating the need for constructing a model with a transparent formula that can be easily interpreted. Figure 4.1 shows Y, the fuel economy, as a function of multi-variate X so that the machine learner comes up with a formula for $Y=f(X)$.

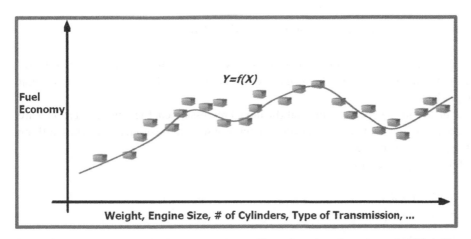

Figure 4.1 Expressing the fuel economy of a car, Y, as a function of pertinent variables, X.

Since the fuel economy is a continuous function, the prediction model will be a regression model, unlike the discretized classification model, with the various technical specifications as the predictor variables and fuel economy as the response, as shown in Figure 4.2.

Machine and Deep Learning Using MATLAB: Algorithms and Tools for Scientists and Engineers, First Edition.
Kamal I. M. Al-Malah.
© 2024 Kamal I. M. Al-Malah. Published 2024 by John Wiley & Sons, Inc.
Companion Website: www.wiley.com/go/al-malah/machinelearningmatlab

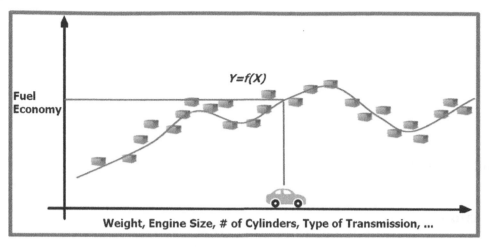

Figure 4.2 Expressing the fuel consumption or price, **Y**, as a continuous function of **X**.

Linear Regression Models

Linear regression is a parametric regression technique in which the response is modeled as some known formula given in terms of the predictor variables. It is a simple, yet a powerful regression technique. Suppose we want to fit a linear regression model to the fuel economy data using the following formula:

$$Fuel\,economy = a + b \times weight + c \times horse\,power + d \times axle\,ratio$$

In this model, the *fuel economy* is considered as a response whereas the *weight*, *horse power*, and *axle ratio* are the predictor variables. The fitting process finds the coefficients a, b, c, and d that minimize the prediction error.

Notice that a linear model simply means that the model regression coefficients: a, b, c, and d are each raised to the power one (or to the power zero for being absent in the model) and it has nothing to deal with the predictor variable X.

Examples of linear regression models are:

$$Y = a + b \times X + c \times X^2$$
$$Y = a + b \times X + c \times \ln(X)$$
$$Y = a + b \times X + c \times e^{X^2}$$
$$Y = a + \frac{b}{X} + c \times X^{0.5}$$

Examples of nonlinear regression models are:

$$Y = a + b \times \ln(cX)$$
$$Y = a + b \times e^{cX^2}$$
$$Y = a + \frac{b}{X + c}$$
$$Y = a + \frac{b}{X} + X^c$$

Quiz #1

Two of the following forms are nonlinear regression models:

A. $Y = a + bX + cX^2$

B. $Y = ax_1 + bx_2 + cx_3$

C. $Y = ax_1x_2 + bx_2x_3 + cx_1x_3$

D. $Y = a + b\sin(x_1) + c\sin(x_2)$

E. $Y = a + b\sin(cx_1)$

F. $Y = a + bx_1 + x_1^c$

Fitting Linear Regression Models Using fitlm Function

Use the function **fitlm** to fit a linear regression model.

>> mdl = fitlm(data,modelspec);

Details are shown in Table 4.1.

Table 4.1 Explanation of **fitlm** regression model statement.

mdl	A regression model variable containing the coefficients and other information about the model.
data	A table containing the data used to fit the regression model. See below for details.
modelspec	Specification of the regression model. See below for details.

How to Organize the Data?

In a regression model, the relationship between the predictors and the response can be described by the following formula. The first input to fitlm is a table containing the predictors and the response, as shown in Figure 4.3.

Figure 4.3 Formula for the regression model (*left*) and the data table format (*right*).

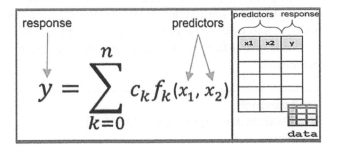

How to Specify fitlm Regression Model?

When modeling a linear regression, we can apply different functions to the predictive variables.

As the second input to **fitlm**, we can use one of the predefined models. The most common predefined models are shown in Table 4.2.

Table 4.2 Predefined models to be used with **fitlm** function.

Model Name	Meaning
'linear'	Intercept and linear terms for each predictor.
'interaction'	Intercept, linear terms, and all products of pairs of distinct predictors (no squared terms).
'quadratic'	Intercept, linear terms, interactions, and squared terms.

We can use a more specific model by providing a formula in Wilkinson–Rogers notation: responseVar ~ terms, as shown in Table 4.3, quoted from MATLAB offline help.

Table 4.3 Wilkinson–Rogers notation for specific models.

Operator	Meaning	Example
+	Include this term.	'y ~ x1 + x2' includes the intercept term, x_1, and x_2: $y = c_0 + c_1 x_1 + c_2 x_2$
-	Exclude this term.	'y ~ x1 + x2 – 1' excludes the intercept term: $y = c_1 x_1 + c_2 x_2$
*	Include product and all lower-order terms.	'y ~ x1*x2' includes the intercept term, x1, x2, and x1*x2: $y = c_0 + c_1 x_1 + c_2 x_2 + c_3 x_1 x_2$
:	Includes the product term only.	'y ~ x1:x2' includes only x1*x2: $y = c_1 x_1 x_2$
^	Include power and all lower-order terms.	'y ~ (x1^2)+(x2^2)' includes the intercept, x1, x2, x1^2, and x2^2: $y = c_0 + c_1 x_1 + c_2 x_2 + c_3 x_1^2 + c_4 x_2^2$

The Linear Model (Intercept and Linear Terms for Each Predictor): Small Car Data

Consider a small car data where we would like to express the miles per gallon, MPG, as a function of relevant car specifications. Let us select the car weight, the horsepower, and the acceleration value as predictors. Some rows do contain not a number, NaN, value. The initial treatment is to remove the entire row if it contains Nan value. Figure 4.4 shows the code for removing a row containing a Nan value, random partition of both the predictor and response matrix, implementing the linear model fitting to training data, showing the model statistical results, and the model coefficients.

```
load carsmall
X = [Weight,Horsepower,Acceleration];
Y=MPG;
XY=[X Y];
temp=rmmissing(XY);
Xnum=temp(:,1:end-1);
Ynum=temp(:,end);
[r,c] = size(Xnum) ;
```

Figure 4.4 Preparing raw data by removing Nan values, random partition into training and test, using **fitlm**, and showing fitting results.

```
P = 0.70 ;
idx = randperm(r)  ;
XTrain = Xnum(idx(1:round(P*r)),:) ;
XTest = Xnum(idx(round(P*r)+1:end),:) ;
YTrain = Ynum(idx(1:round(P*r)),:);
YTest = Ynum(idx(round(P*r)+1:end),:) ;
mdl=fitlm(XTrain,YTrain);
%mdl=fitlm(XTrain,YTrain,'quadratic', 'RobustOpts',"on");
%mdl=fitlm(XTrain,YTrain, 'y ~ x1:x1 + x1:x2+ x3', 'RobustOpts',"on");
Model_Coeffs=mdl.Coefficients
AdjR2=mdl.Rsquared
figure(7)
plot(YTrain, mdl.Residuals{:,1}, 'x')
hold on;
xval=min(YTrain):0.1:max(YTrain);
yval=0.0;
plot(xval,yval, '+')
hold off;
%Evaluate the model at test values.
yPred=predict(mdl,XTest);
Resid=YTest-yPred;
figure (8)
plot (YTest,Resid, '*')
hold on;
xval=min(YTest):0.1:max(YTest);
yval=0.0;
plot(xval,yval, 'o')
hold off;
```

Figure 4.4 (Cont'd)

Figure 4.5 (*left*) shows the workspace variables and (*right*) the regression model properties.

Workspace							1x1 LinearModel			
Name ^	Value	Size	Name ^	Value	Size	Property ^	Value	Property ^	Value
Acceleration	100x1 double	100x1	Resid	28x1 double	28x1	Residuals	65x4 table	NumEstimatedCoefficients	4
ans	1x1 struct	1x1	temp	93x4 double	93x4	Fitted	65x1 double	Coefficients	4x4 table
c	3	1x1	Weight	100x1 double	100x1	Diagnostics	65x7 table	Rsquared	1x1 struct
Cylinders	100x1 double	100x1	X	100x3 double	100x3	MSE	13.4610	ModelCriterion	1x1 struct
Displacement	100x1 double	100x1	Xnum	93x3 double	93x3	Robust	[]	VariableInfo	4x4 table
Horsepower	100x1 double	100x1	XTest	28x3 double	28x3	RMSE	3.6689	NumVariables	4
idx	1x93 double	1x93	XTrain	65x3 double	65x3	Formula	1x1 LinearFormula	VariableNames	4x1 cell
mdl	1x1 LinearModel	1x1	xval	1x311 double	1x311	LogLikelihood	-174.6603	NumPredictors	3
Mfg	100x13 char	100x13	XY	100x4 double	100x4	DFE	61	PredictorNames	3x1 cell
Model	100x33 char	100x33	Y	100x1 double	100x1	SST	3.8617e+03	ResponseName	'y'
Model_Year	100x1 double	100x1	Ynum	93x1 double	93x1	SSE	821.1227	NumObservations	65
MPG	100x1 double	100x1	yPred	28x1 double	28x1	SSR	3.0405e+03	Steps	[]
Origin	100x7 char	100x7	YTest	28x1 double	28x1	CoefficientCovariance	4x4 double	ObservationInfo	65x4 table
P	0.7000	1x1	YTrain	65x1 double	65x1	CoefficientNames	1x4 cell	Variables	65x4 table
r	93	1x1	yval	0	1x1	NumCoefficients	4	ObservationNames	0x0 cell

Figure 4.5 The workspace variables (*left*) and the linear regression model properties (*right*).

Figure 4.6 (*left*) shows the linear model regressed parameters and their statistical significance, as well. The adjusted R^2 (correlation coefficient) of the model is shown on the *right*.

>> mdl.Coefficients				>> mdl.Rsquared
	Estimate	SE	tStat	pValue
(Intercept)	51.274	4.5869	11.178	2.12e-16
x1	-0.0061496	0.001269	-4.8461	9.0151e-06
x2	-0.060188	0.027462	-2.1917	0.032226
x3	-0.16397	0.23582	-0.69532	0.4895

ans =

struct with fields:

Ordinary: 0.7799
Adjusted: 0.7691

Figure 4.6 The regressed parameters and their statistical significance (left) and the adjusted R^2 parameter (right).

Figure 4.7 (*left*) shows the residuals (YTrain-Ymdl) versus the trained response variable, **YTrain**, and the residuals (YTest-Ymdl) versus the tested response variable, **YTest** (*right*). The horizontal line in both figures represents the zero-residual value for a perfect model.

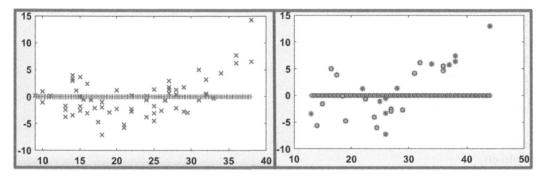

Figure 4.7 Plot of residuals ($Y_{observed}-Y_{mdl}$) for both trained and test sets.

Polynomial Model (Intercept, Linear Terms, Interactions, and Squared Terms): Small Car Data
Figure 4.8 shows the same code as that of Figure 4.4, except for defining the **mdl** statement where we added an extra '**modelspec**' equal to '**quadratic**'.

%% The same code is exactly as that of Figure 4.4, except for the following statement
mdl=fitlm(XTrain,YTrain,'quadratic');

Figure 4.8 The mdl statement contains an extra '**modelspec**' with a value of '**quadratic**'.

Figure 4.9 shows the model regression coefficients (*left*) and the adjusted R^2, or correlation coefficient (*right*). Notice that the p-value gets more significant for each coefficient with a slight increase of R^2.

NOTE

A smaller p-value (typically below a predetermined threshold, such as 0.05) suggests stronger evidence against the null hypothesis, which states that the coefficient is equal to zero (i.e., no relationship). If the p-value is small, it indicates that the observed coefficient is unlikely to have occurred due to random chance alone, and there is evidence to support a relationship between the predictor and response variables. However, the importance or magnitude of a coefficient is determined by its value itself, not just the p-value. The coefficient's size indicates the direction and strength of the relationship between the predictor variable and the response variable. A larger coefficient value signifies a more substantial impact of the predictor variable on the response variable, regardless of the p-value. Therefore, while a smaller p-value suggests stronger statistical evidence, it does not necessarily imply a larger or more important coefficient in terms of practical (physical) significance.

>> mdl.Coefficients	**Estimate**	**SE**	**tStat**	**pValue**	>> mdl.Rsquared
					ans =
(Intercept)	23.566	31.942	0.73779	0.46378	
x1	-0.0023744	0.01785	-0.13302	0.89466	**struct** with fields:
x2	-0.14148	0.35378	-0.39992	0.69077	
x3	3.6014	3.123	1.1532	0.25382	Ordinary: 0.8012
x1:x2	6.0793e-05	7.3315e-05	0.8292	0.41058	Adjusted: 0.7687
x1:x3	0.00017715	0.0008494	0.20855	0.83557	
x2:x3	-0.0091402	0.019419	-0.47069	0.63973	
x1^2	-1.9726e-06	2.352e-06	-0.83867	0.40528	
x2^2	-5.3078e-05	0.00094945	-0.055903	0.95562	
x3^2	-0.11453	0.08805	-1.3008	0.19876	

Figure 4.9 The regressed parameters and their statistical significance (left) and the adjusted R^2 parameter (right).

Figure 4.10 (*left*) shows the residuals (YTrain-Ymdl) versus the trained response variable, **YTrain**, and the residuals (YTest-Ymdl) versus the tested response variable, **YTest** (*right*). The horizontal line in both figures represents the zero-residual value for a perfect model.

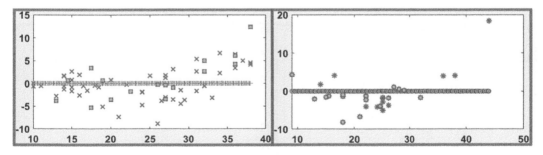

Figure 4.10 Plot of residuals ($Y_{observed}-Y_{mdl}$) for both trained and test sets.

Robust Polynomial Model (Intercept, Linear Terms, Interactions, and Squared Terms): Small Car Data

Figure 4.11 shows the same code as that of Figure 4.4, except for defining the **mdl** statement where we added an extra '**modelspec**' equal to '**quadratic**'. However, we can reduce the influence of the outliers on the model by setting the '**RobustOpts**' property to '**on**'.

%% **The same code is exactly as that of Figure 4.4, except for the following statement**
mdl=fitlm(XTrain,YTrain,'quadratic', 'RobustOpts',"on");

Figure 4.11   The mdl statement contains an extra '**RobustOpts**' property with a value of '**on**'.

Figure 4.12 shows the model regression coefficients (*left*) and the adjusted R^2, or correlation coefficient (*right*).

>> mdl.Coefficients	Estimate	SE	tStat	pValue	>> mdl.Rsquared
(Intercept)	73.955	23.8	3.1074	0.0029853	Ordinary: 0.8439
x1	0.012862	0.015591	0.82498	0.41295	Adjusted: 0.8184
x2	-0.68615	0.32873	-2.0873	0.041507	
x3	-2.5124	2.4803	-1.0129	0.31552	
x1:x2	-1.0304e-05	6.1433e-05	-0.16772	0.86742	
x1:x3	-0.0010447	0.0006832	-1.5292	0.13196	
x2:x3	0.018371	0.015556	1.1809	0.24271	
x1^2	5.7408e-08	2.0273e-06	0.028317	0.97751	
x2^2	0.001289	0.00088304	1.4597	0.15007	
x3^2	0.1131	0.06714	1.6846	0.097734	

Figure 4.12   The regressed parameters and their statistical significance (left) and the adjusted R^2 parameter (right).

Figure 4.13 (*left*) shows the residuals (YTrain-Ymdl) versus the trained response variable, **YTrain**, and the residuals (YTest-Ymdl) versus the tested response variable, **YTest** (*right*). The horizontal line in both figures represents the zero-residual value for a perfect model.

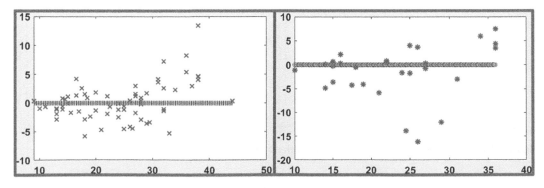

Figure 4.13   Plot of residuals ($Y_{observed}$-Y_{mdl}) for both trained and test sets.

Notice that the partition between training and test set is randomly assigned; hence, the run from one case to another will change as far as the data distribution is concerned.

NOTE

To find the best regression model, it is preferred that the user includes all data and try different forms of the regression model and based on SSE (or MSE) or the adjusted R^2 correlation coefficient, one may then select the best of the best among the rest. Notice that if we check any regression model (mdl) properties, we will find out that it has such statistical parameters that tell the model goodness or fitness, like, total sum of squares (SST), sum of squared error (SSE), sum of squares of the regression (SSR), the mean squared errors $\left(MSE = \dfrac{SSE}{n-p} \right)$, where n is the number of observations and p number of model parameters, and finally the root mean of squared errors (RMSE = $\sqrt{MSE}$).

Let us give an example of how to specifically select a model format and test its model goodness. For the sake of learning, Table 4.4 shows specific forms which were examined using the code shown in Figure 4.14

Table 4.4 Examples of specific model expression used with **fitlm** function.

#	mdl Format	The Model Expression	Adj-R2 (RMSE)
1	'y ~ x1:x1 + x1:x2 + x3'	$Y = a + bx_3 + cx_1x_2 + dx_1^2$	0.6983 (4.3372)
2	'y ~ x1:x1 + x1:x2 + x3-1'	$Y = bx_3 + cx_1x_2 + dx_1^2$	0.5628 (6.3737)
3	'y ~ x1^2 + x1:x2 + x3'	$Y = a + bx_1 + cx_3 + dx_1x_2 + ex_1^2$	0.7592 (3.8870)
4	'y ~ x1^2 + x1:x2 + x3-1'	$Y = bx_1 + cx_3 + dx_1x_2 + ex_1^2$	0.6178 (5.2612)

```
load carsmall
X = [Weight,Horsepower,Acceleration];
Y=MPG;
XY=[X Y];
temp=rmmissing(XY);
Xnum=temp(:,1:end-1);
Ynum=temp(:,end);
mdl=fitlm(Xnum,Ynum, 'y ~ x1:x1 + x1:x2+ x3', 'RobustOpts','on');
%mdl=fitlm(Xnum,Ynum, 'y ~ x1:x1 + x1:x2+ x3 -1', 'RobustOpts','on');
%mdl=fitlm(Xnum,Ynum, 'y ~ x1^2 + x1:x2+ x3', 'RobustOpts','on');
%mdl=fitlm(Xnum,Ynum, 'y ~ x1^2 + x1:x2+ x3 -1', 'RobustOpts','on');
Model_Coeffs=mdl.Coefficients
AdjR2=mdl.Rsquared
RMSE=mdl.RMSE
figure(15)
```

Figure 4.14 Different model formats are tested and results are shown in Table 4.4.

```
plot(Ynum, mdl.Residuals{:,1}, 'x')
hold on;
xval=min(Ynum):0.1:max(Ynum);
yval=0.0;
plot(xval,yval, '+')
hold off;
xlabel('Observed response');
ylabel('Residual for each point');
```

Figure 4.14 (Cont'd)

Results Visualization: Big Car Data

The table car data has been split into a training set, **XTrain**, and a test set, **XTest**. The response variable is **MPG**. We will create a multivariate linear model named **mdl** with the training data. Then, use the model to predict the values, **MPG**.

We will use A function named **evaluateFit2** (Appendix A | Figure A.15). This function requires four input arguments: The actual response values, the predicted response results, the type of the regression model as a text, and the name of the response variable as a text. The data is sorted by the actual response. The first plot shows the actual values (in blue) and the predicted values (in red) against the observed data. The second plot shows the actual value on the x-axis against the predicted value on the y-axis. The third plot is a histogram showing the distribution of the errors between the actual and predicted values. The last plot is another histogram showing the per cent relative error (error as a percentage of the actual value). MAPE stands for mean absolute percent error.

Figure 4.15 shows the code of preparing both X and Y data, randomly splitting the data into a training and test set, constructing the linear regression model, predicting **MPG** value for **XTest** data, and using **evaluateFit2** function to compare predicted versus actual **MPG** value.

```
load carbig
X = [Weight,Horsepower,Acceleration,Displacement];
Y=MPG;
XY=[X Y];
temp=rmmissing(XY);
Xnum=temp(:,1:end-1);
Ynum=temp(:,end);
[r,c] = size(Xnum) ;
P = 0.70 ;
idx = randperm(r)  ;
XTrain = Xnum(idx(1:round(P*r)),:) ;
XTest = Xnum(idx(round(P*r)+1:end),:) ;
YTrain = Ynum(idx(1:round(P*r)),:);
YTest = Ynum(idx(round(P*r)+1:end),:) ;
mdl=fitlm(XTrain,YTrain,"linear");
%Evaluate the model at test values.
YPred=predict(mdl,XTest);
%% Compare predicted and actual responses
evaluateFit2(YTest,YPred,'linear', 'MPG')
```

Figure 4.15 Carrying out liner regression fitting and showing the model goodness.

Figure 4.16 shows the four subplots of **evaluateFit2** function, explained earlier.

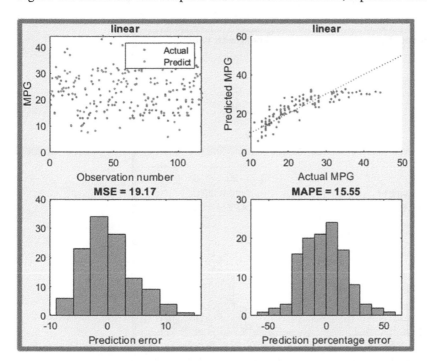

Figure 4.16   **evaluateFit2** function generated subplots for showing the model goodness.

Figure 4.17 shows the workspace variables generated by compiling the code of Figure 4.15.

Name ▲	Value	Size	Min	Max	Name ▲	Value	Size	Min	Max
					P	0.7000	1x1	0.7000	0.7000
Acceleration	406x1 double	406x1	8	24.8000	r	392	1x1	392	392
c	4	1x1	4	4	temp	392x5 double	392x5	8	5140
cyl4	406x5 char	406x5			Weight	406x1 double	406x1	1613	5140
Cylinders	406x1 double	406x1	3	8	when	406x5 char	406x5		
Displacement	406x1 double	406x1	68	455	X	406x4 double	406x4	NaN	NaN
Horsepower	406x1 double	406x1	NaN	NaN	Xnum	392x4 double	392x4	8	5140
idx	1x392 double	1x392	1	392	XTest	118x4 double	118x4	8	5140
mdl	1x1 LinearModel	1x1			XTrain	274x4 double	274x4	8.5000	4997
Mfg	406x13 char	406x13			XY	406x5 double	406x5	NaN	NaN
Model	406x36 char	406x36			Y	406x1 double	406x1	NaN	NaN
Model_Year	406x1 double	406x1	70	82	Ynum	392x1 double	392x1	9	46.6000
MPG	406x1 double	406x1	NaN	NaN	YPred	118x1 double	118x1	5.8504	32.4607
org	406x7 char	406x7			YTest	118x1 double	118x1	10	44.3000
Origin	406x7 char	406x7			YTrain	274x1 double	274x1	9	46.6000

Figure 4.17   The workspace variables created by running the code of Figure 4.15.

Alternatively, we can use the '**cauchy**' value for the '**RobustOpts**' property value. Figure 4.18 shows the same code as that of Figure 4.15, except for the model statement.

```
mdl=fitlm(XTrain,YTrain,"linear","RobustOpts","cauchy");
```

Figure 4.18   Use the code shown in Figure 4.15, except for the above model statement.

Figure 4.19 shows four subplots generated by **evaluateFit2** function, as explained earlier.

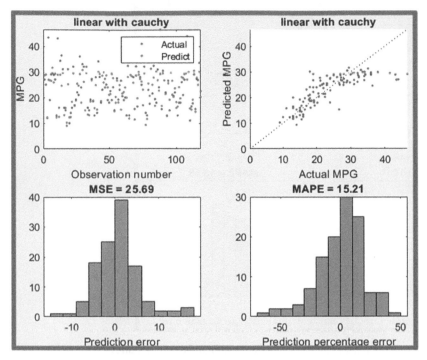

Figure 4.19 **evaluateFit2** function generated subplots for showing the model goodness.

Fitting Linear Regression Models Using fitglm Function

fitglm creates a generalized linear regression model. The advantage over **fitlm** is simply that the user can manipulate the number of maximum iterations needed to converge to a solution. For example, there are cases as in HCW 4.2 where, upon using **fitlm**, the user will face the following warning error:

```
Warning: Iteration limit reached.
In statrobustfit (line 80)
In robustfit (line 114)
In LinearModel/fitter (line 774)
In classreg.regr/FitObject/doFit (line 94)
In LinearModel.fit (line 1035)
In fitlm (line 121)
```

In such a case, **fitglm** has an option where the user may modify the maximum iteration limit.

```
opts = statset('fitglm');
opts.MaxIter = 1000;
opts.Link= 'probit';
mdl=fitglm(Xnum,Ynum, 'linear', 'Distribution',"normal",'Options',opts);
```

Let us give more details on each entry within the function statement.

The term **'linear'** can be replaced by any of the following terms appearing in Table 4.5, quoted from MATLAB offline help:

Table 4.5   x-y relationship for y = f(x) using **fitglm** function.

Value	Model Type
'constant'	Model contains only a constant (intercept) term.
'linear'	Model contains an intercept and linear term for each predictor.
'interactions'	Model contains an intercept, linear term for each predictor, and all products of pairs of distinct predictors (no squared terms).
'purequadratic'	Model contains an intercept term and linear and squared terms for each predictor only (i.e., no multiplication of mixed predictors).
'quadratic'	Model contains an intercept term, linear and squared terms for each predictor, and all products of pairs of distinct predictors.
'poly*ijk*'	Model is a polynomial with all terms up to degree i in the first predictor, degree j in the second predictor, and so on. Specify the maximum degree for each predictor by using numerals 0 through 9. The model contains interaction terms, but the degree of each interaction term does not exceed the maximum value of the specified degrees. For example, **'poly13'** has an intercept and x_1, x_2, x_2^2, x_2^3, $x_1{}^*x_2$, and $x_1{}^*x_2^2$ terms, where x_1 and x_2 are the first and second predictors, respectively.

The value **"normal"** can be replaced by one of the following types of predictor distribution, appearing in Table 4.6, as quoted from MATLAB offline help:

Table 4.6   Different types of data distribution can be assumed with **fitglm** function.

Distribution Name	Description
'normal'	Normal distribution
'binomial'	Binomial distribution
'poisson'	Poisson distribution
'gamma'	Gamma distribution
'inverse gaussian'	Inverse Gaussian distribution

The link option (e.g., **opts.Link = 'probit'**) can be selected out of the following link function types shown in Table 4.7, as quoted from MATLAB offline help. The link function defines the relationship $f(\mu) = X^*b$ between the mean response μ and the linear combination of predictors X^*b.

Table 4.7 The link function types to use as an option with **fitglm** function.

Link Function Name	Link Function	Mean (Inverse) Function
'identity'	$f(\mu) = \mu$	$\mu = Xb$
'log'	$f(\mu) = \log(\mu)$	$\mu = \exp(Xb)$
'logit'	$f(\mu) = \log(\mu/(1{-}\mu))$	$\mu = \exp(Xb) \,/\, (1 + \exp(Xb))$
'probit'	$f(\mu) = \Phi^{-1}(\mu)$, where Φ is the cumulative distribution function of the standard normal distribution.	$\mu = \Phi(Xb)$
'comploglog'	$f(\mu) = \log(-\log(1-\mu))$	$\mu = 1 - \exp(-\exp(Xb))$
'reciprocal'	$f(\mu) = 1/\mu$	$\mu = 1/(Xb)$
p (a number)	$f(\mu) = \mu^p$	$\mu = Xb^{1/p}$
S (a structure)	$f(\mu) = \mathrm{S.Link}(\mu)$	$\mu = \mathrm{S.Inverse}(Xb)$

with three fields. Each field holds a function handle that accepts a vector of inputs and returns a vector of the same size:

- S.Link: The link function
- S.Inverse: The inverse link function
- S.Derivative: The derivative of the link function

NOTE

fitglm reports what is called adjusted generalized R-squared:

$$R^2_{AdjGeneralized} = \frac{1 - \exp^{\left(\frac{2(L_o - L)}{N}\right)}}{1 - \exp^{\left(\frac{2L_o}{N}\right)}}$$

$R^2_{AdjGeneralized}$ is the Nagelkerke adjustment to a formula proposed for logistic regression models. N is the number of observations, L is the loglikelihood of the fitted model, and L_0 is the loglikelihood of a model that includes only a constant term. **L = mdl.LogLikelihood; Lo = mdl.Deviance / 2**.

Nonparametric Regression Models

A parametric regression model assumes a relation that can be specified using a formula. It is easier to interpret. We can measure the change in the response per unit change in the predictor. For example,

$$Y = a + bx_1 + cx_3 + dx_2 x_3 + ex_3^2$$

This implies that

$$\left.\frac{\Delta Y}{\Delta x_1}\right|_{@constant\ x_2, x_3} = b$$

If we have several predictors, choosing a formula that creates a model with minimum prediction error can be quite difficult.

On the other hand, if predicting the response for unknown observations is the primary purpose of the model, we can use *nonparametric* regression methods which do not fit the regression model based on a given formula and can provide more accurate prediction but are difficult to interpret.

In general, we can use the function **fitr*** (replace * by the name of the classifier) to fit other nonparametric models. For example, the support vector machine regression model, **svm**, becomes:

>> svmMdl = fitrsvm(table,'ResponseName')

The decision trees, **fitrtree**, a nonparametric technique, can also be used.

The table data is split into a training set, **dataTrain**, and a test set, **dataTest**. The response is in Y. The loss function, which is the mean squared error (MSE), can be used to evaluate the model.

>> tLoss = loss(svmMdl,XTest,*YTest*);

The predict function can predict responses for new data.

>>predY = predict(linearModel,XTest)

The default **KernelFunction** value is **'linear'**. We can set properties when we train the model.

>> svmMdl = fitrsvm(table,'ResponseName','PropertyName', PropertyValue);

For example, we can let **svmMdl** use a **'polynomial'** kernel function.

>>svmMdl=fitrsvm(dataTrain,"y","KernelFunction","polynomial");

Quiz #2
(Select all that apply.) Which of the following functions creates a nonparametric regression model?
a) fitcnb   b) fitlm   c) fitrsvm   d) fitrtree

---

fitrtree Nonparametric Regression Model: Big Car Data

The table car data has been split into a training set, **XTrain**, and a test set, **XTest**. The response variable is **MPG**. We will create a nonparametric decision tree model named **mdl** with the training data. We calculate the loss of the model and name it **mdlLoss**. We will use the model to predict the test values, **XTest**, and store the predictions in a variable named **YPred**. The code is shown in Figure 4.20.

```
load carbig
X = [Weight,Horsepower,Acceleration,Displacement];
Y=MPG;
XY=[X Y];
temp=rmmissing(XY);
Xnum=temp(:,1:end-1);
Ynum=temp(:,end);
[r,c] = size(Xnum) ;
P = 0.70 ;
idx = randperm(r)  ;
XTrain = Xnum(idx(1:round(P*r)),:) ;
XTest = Xnum(idx(round(P*r)+1:end),:) ;
YTrain = Ynum(idx(1:round(P*r)),:);
YTest = Ynum(idx(round(P*r)+1:end),:) ;
mdl=fitrtree(XTrain,YTrain);
mdlLoss = loss(mdl,XTest,YTest);
disp([num2str(mdlLoss),' = model loss']);
%Evaluate the model at test values.
YPred=predict(mdl,XTest);
%% Compare predicted and actual responses
evaluateFit2(YTest,YPred,'tree', 'MPG')
```

Figure 4.20 Preparing car data for **fitrtree** model, testing and presenting the model goodness.

Figure 4.21 shows the workspace variable with **mdlLoss** value (equal to MSE) of 19.454.

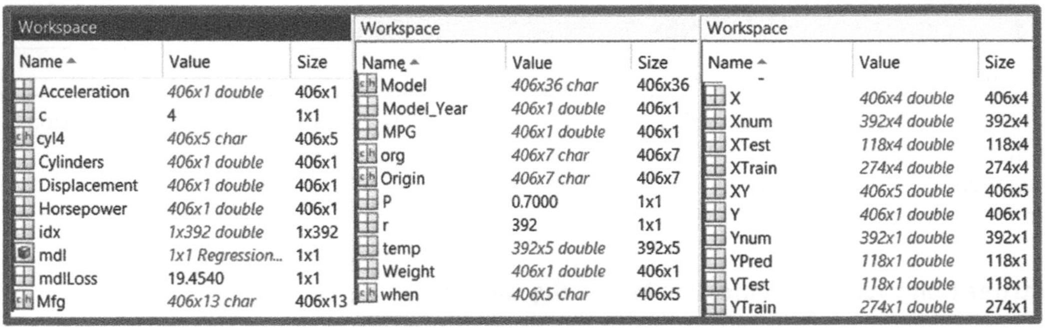

Name ▲	Value	Size	Name ▲	Value	Size	Name ▲	Value	Size
Acceleration	406x1 double	406x1	Model	406x36 char	406x36	X	406x4 double	406x4
c	4	1x1	Model_Year	406x1 double	406x1	Xnum	392x4 double	392x4
cyl4	406x5 char	406x5	MPG	406x1 double	406x1	XTest	118x4 double	118x4
Cylinders	406x1 double	406x1	org	406x7 char	406x7	XTrain	274x4 double	274x4
Displacement	406x1 double	406x1	Origin	406x7 char	406x7	XY	406x5 double	406x5
Horsepower	406x1 double	406x1	P	0.7000	1x1	Y	406x1 double	406x1
idx	1x392 double	1x392	r	392	1x1	Ynum	392x1 double	392x1
mdl	1x1 Regression...	1x1	temp	392x5 double	392x5	YPred	118x1 double	118x1
mdlLoss	19.4540	1x1	Weight	406x1 double	406x1	YTest	118x1 double	118x1
Mfg	406x13 char	406x13	when	406x5 char	406x5	YTrain	274x1 double	274x1

Figure 4.21 The workspace variables with mdlLoss (MSE) value of 19.45.

Figure 4.22 shows the comparison between actual and predicted **MPG** values.

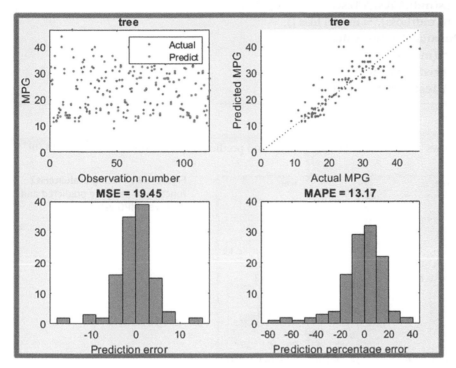

Figure 4.22 Using **evaluateFit2** function to compare predicted and observed MPG values.

Figure 4.23 shows the code for making a refinement for the decision tree model using the prune function which can be used to reduce overfitting and also setting "**MinLeafSize**" property of mdl to *5*.

```
load carbig
X = [Weight,Horsepower,Acceleration,Displacement];
Y=MPG;
XY=[X Y];
temp=rmmissing(XY);
Xnum=temp(:,1:end-1);
Ynum=temp(:,end);
[r,c] = size(Xnum) ;
P = 0.70 ;
idx = randperm(r)  ;
XTrain = Xnum(idx(1:round(P*r)),:) ;
XTest = Xnum(idx(round(P*r)+1:end),:) ;
YTrain = Ynum(idx(1:round(P*r)),:);
YTest = Ynum(idx(round(P*r)+1:end),:) ;
%Set the MinLeafSize property of mdl to 5.
mdl=fitrtree(XTrain,YTrain,"MinLeafSize",5);
%The prune function can be used to reduce overfitting.
```

Figure 4.23 Making a refinement for the decision tree model, **fitrtree**, to have a better fit.

```
mdl = prune(mdl,'Level',10);
mdlLoss = loss(mdl,XTest,YTest);
disp([num2str(mdlLoss),' = model loss']);
%Evaluate the model at test values.
YPred=predict(mdl,XTest);
%% Compare predicted and actual responses
evaluateFit2(YTest,YPred,'tree','MPG')
```

Figure 4.23 (Cont'd)

Figure 4.24 shows a comparison between actual and predicted **MPG** values, MSE, and MAPE.

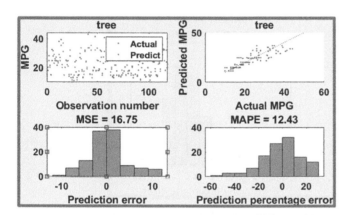

Figure 4.24 Using **evaluateFit2** function to compare predicted and observed **MPG** values.

Support Vector Machine, fitrsvm, Nonparametric Regression Model: Big Car Data

The table car data has been split into a training set, **XTrain**, and a test set, **XTest**. The response variable is **MPG**. We create a nonparametric **SVM** regression model with the training data and name it **mdl**. We use the '**polynomial**' kernel function. We calculate the loss of the model and name it **mdlLoss**. Then, we use the model to predict the test values, **XTest**, and store the predictions in a variable named **YPred**. The code is shown in Figure 4.25.

```
load carbig
X = [Weight,Horsepower,Acceleration,Displacement];
Y=MPG;
XY=[X Y];
temp=rmmissing(XY);
Xnum=temp(:,1:end-1);
Ynum=temp(:,end);
[r,c] = size(Xnum) ;
P = 0.70 ;
idx = randperm(r)  ;
XTrain = Xnum(idx(1:round(P*r)),:) ;
XTest = Xnum(idx(round(P*r)+1:end),:) ;
YTrain = Ynum(idx(1:round(P*r)),:);
YTest = Ynum(idx(round(P*r)+1:end),:) ;
mdl=fitrsvm(XTrain,YTrain,"KernelFunction", "polynomial");
```

Figure 4.25 The nonparametric regression model, **fitrsvm**, used to fit the big car data.

```
mdlLoss = loss(mdl,XTest,YTest);
disp([num2str(mdlLoss),' = model loss']);
YPred=predict(mdl,XTest);
evaluateFit2(YTest,YPred,'SVM', 'MPG')
```

Figure 4.25 (Cont'd)

Figure 4.26 shows a comparison between actual and predicted **MPG** values, MSE, and MAPE.

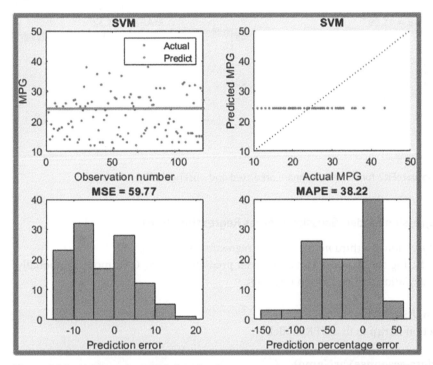

Figure 4.26   Using **evaluateFit2** function to compare predicted and observed **MPG** values.

Let us normalize and center the big car data. This can be done as an option in **fitrsvm** to standardize the variables in **mdl**. Notice that the code used in Figure 4.25 is used here, except for the **mdl** statement which is shown in Figure 4.27. By setting "**Standardize**" to *true*, the predictor data (XTrain) will be standardized, meaning it will be centered and scaled to have zero mean and unit variance. This can be helpful when the predictor variables have different scales or units, as it can improve the performance of the SVM algorithm. On the other hand, the response data (YTrain) is not affected by the "**Standardize**" option. If we want to standardize the response data as well, we would need to perform that separately before calling **fitrsvm** function.

```
mdl=fitrsvm(XTrain,YTrain,"KernelFunction", "polynomial",...
"Standardize",true);
```

Figure 4.27   The nonparametric regression model, **fitrsvm**, used to fit the standardized big car data.

Figure 4.28 shows a comparison between actual and predicted **MPG** values, MSE, and MAPE. Notice that with *standardization* of original data, the model gave better estimates; i.e., lower MSE and MAPE values.

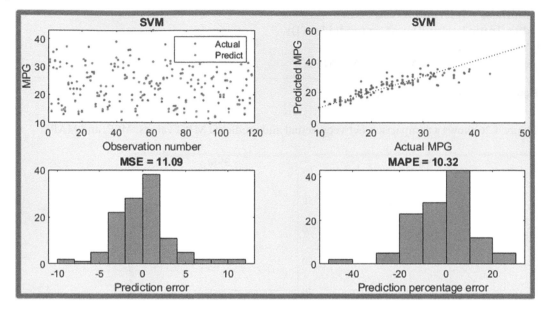

Figure 4.28 Using **evaluateFit2** function to compare predicted and observed **MPG** values.

Nonparametric Regression Model: Gaussian Process Regression (GPR)

Gaussian process regression is a third *nonparametric regression* technique.

In addition to predicting the response value for given predictor values, GPR models optionally return the standard deviation and prediction intervals.

Fitting and Prediction Using GPR

We can use the function **fitrgp** to fit a GPR model.

>> mdl = fitrgp(data,responseVarName)

Details are given in Table 4.8

Table 4.8 **fitrgp** function statement details.

mdl	A GPR model variable.
data	A table containing the predictors and the response values.
responseVarName	Name of the response variable.

We can use the **'KernelFunction'** option to change the kernel to one of the predefined options.

In addition to the predicted response value, the predict function for GPR models optionally returns the standard deviation and prediction intervals for the predicted values.

>> [yPred,yStd,yInt] = predict(mdl,dataNew)

Details of **predict** function statement are shown in Table 4.9.

Table 4.9   **predict** function used to predict the response of new predictor data by a trained model, **mdl**.

yPred	Predicted value.
yStd	Standard deviations for each predicted value.
yInt	A matrix with columns representing the lower and upper limits of the prediction intervals. The default is 95% confidence interval.

GPR Nonparametric Regression Model: Big Car Data

The big car data has been split into a training set, **XTrain**, and a test set, **XTest**. The response variable is **MPG**. We create a nonparametric **GPR** regression model with the training data and name it **mdl**. We fit the **GPR** model using the initial kernel parameter values. We standardize the predictors in the training data. We use the exact fitting and prediction methods. We calculate the loss of the model and name it **MSE**. Then, we use the model to predict the test values, **XTest**, and store the predictions in a variable named **YPred**. The code is shown in Figure 4.29.

```
load carbig
X = [Weight,Horsepower,Acceleration,Displacement];
Y=MPG;
XY=[X Y];
temp=rmmissing(XY);
Xnum=temp(:,1:end-1);
Ynum=temp(:,end);
[r,c] = size(Xnum) ;
P = 0.70 ;
idx = randperm(r)  ;
XTrain = Xnum(idx(1:round(P*r)),:) ;
XTest = Xnum(idx(round(P*r)+1:end),:) ;
YTrain = Ynum(idx(1:round(P*r)),:);
YTest = Ynum(idx(round(P*r)+1:end),:) ;
sigma0 = std(YTrain);
sigmaF0 = sigma0;
d = size(XTrain,2);
sigmaM0 = 10*ones(d,1);
mdl = fitrgp(XTrain,YTrain,'Basis','constant','FitMethod','exact',...
'PredictMethod','exact','KernelFunction','ardsquaredexponential',...
'KernelParameters',[sigmaM0;sigmaF0],'Sigma',sigma0,'Standardize',1);
%Calculate MSE
MSE = loss(mdl,XTest,YTest);
disp(['GPR MSE = ',num2str(MSE)])
%The default for YInt is 95 % Confidence Interval, or alpha=0.05;
%[YPred,~,YInt]=predict(mdl,XTest, 'Alpha',0.05);
[YPred,~,YInt]=predict(mdl,XTest);
evaluateFit2(YTest,YPred,'GPR', 'MPG')
% Plot 95 % Confidence Interval around the predicted values.
figure(31)
plot(YTest, YPred, '*')
```

Figure 4.29   The nonparametric regression model, **fitrgp**, used to fit the big car data and present the model goodness.

```
hold on
plot(YTest,YInt(:,1), 'kx')
plot(YTest,YInt(:,2), 'go')
ezplot('x',[min(YTest) max(YTest)]);
legend('Pred', 'Lower','Upper')
xlabel ('YTest');
ylabel('YPred');
hold off
```

Figure 4.29 (Cont'd)

Figure 4.30 shows a comparison between actual and predicted **MPG** values, **MSE**, and **MAPE**.

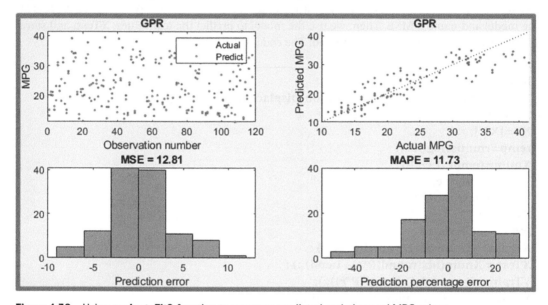

Figure 4.30 Using **evaluateFit2** function to compare predicted and observed **MPG** values.

Figure 4.31 shows **YPred** versus **YTest** as well as the upper and lower 95% confidence interval around the mean (**YPred**) value for each **YTest** datapoint.

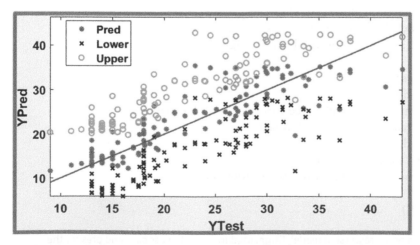

Figure 4.31 **YPred** versus **YTest** as well as the upper and lower 95% confidence interval.

Alternatively, we can find hyperparameters that minimize fivefold cross-validation loss by using automatic hyperparameter optimization. For reproducibility, set the random seed and use the **'expected-improvement-plus'** acquisition function. Figure 4.32 shows only the re-definition of **mdl** statement; the rest of pre-code and post-code are exactly the same as those of Figure 4.29.

```
rng default;
mdl = fitrgp(XTrain,YTrain,'KernelFunction','squaredexponential',...
    'OptimizeHyperparameters','auto','HyperparameterOptimizationOptions',...
    struct('AcquisitionFunctionName','expected-improvement-plus'));
```

Figure 4.32 Automatic hyper parameter optimization of **GPR** model.

The Command Window results of automatic hyper parameter optimization are:

```
Optimization completed.
MaxObjectiveEvaluations of 30 reached.
Total function evaluations: 30
Total elapsed time: 19.9724 seconds
Total objective function evaluation time: 9.0807
Best observed feasible point:
    Sigma = 3.0117
Observed objective function value = 2.8855
Estimated objective function value = 2.8946
Function evaluation time = 0.205
Best estimated feasible point (according to models):
    Sigma = 3.2575
Estimated objective function value = 2.8869
Estimated function evaluation time = 0.2019
```

Figure 4.33 (*left*) shows that 30 function evaluations are needed to reach a steady-state value for the estimated min objective to reach observed minimum objective and the estimated objective function value achieved minimum (best estimated feasible point, according to models) at sigma value of 3.26 (*right*). Sigma is an initial value for the noise standard deviation of the Gaussian process model, equal to std(Y)/sqrt(2) (by default), and it is a positive scalar value.

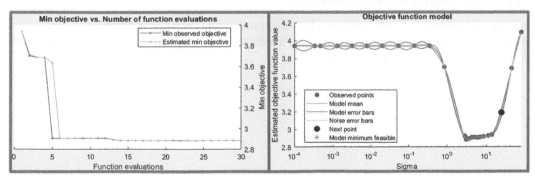

Figure 4.33 The total function evaluations needed to reach minimum (*left*) and the best sigma value for reaching an estimated min objective value (*right*).

Figure 4.34 (*left*) shows a comparison between observed and predicted response values in addition to **MSE** and **MAPE** of **GPR** model and, on the *right*, **YPred** versus **YTest** as well as the upper and lower 95% confidence interval for the point estimate of **YPred** value for each **YTest** datapoint.

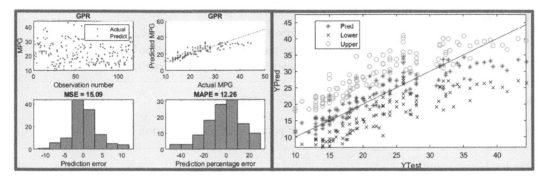

Figure 4.34 Observed vs. predicted response values in addition to **MSE** and **MAPE** of **GPR** model and the upper and lower 95% confidence interval for the point estimate of **YPred**.

Regularized Parametric Linear Regression

Selecting the appropriate parametric regression model can be difficult when dealing with datasets containing numerous predictive variables. Incorporating all of these variables may result in an excessively intricate model that fails to effectively generalize to new data. To address this issue, we will employ methods aimed at eliminating less significant predictive variables or diminishing their impact on the regression model.

In cases where datasets contain a multitude of predictors, it is common for some of these predictors to exhibit correlation. When fitting a linear regression model to such data, the resulting coefficients may display significant variance. This means that the values of the coefficients undergo substantial changes as the training data is altered.

Consider the predictors x1 and x2 that are correlated. Ridge regression and Lasso shrink the regression coefficients. This reduces the variance of the coefficients and can create models with smaller prediction error.

Ridge Linear Regression: The Penalty Term

In linear regression, the coefficients are chosen by minimizing the squared difference between the observed and the predicted response value. This difference is referred to as mean squared error (MSE), as demonstrated in Figure 4.35.

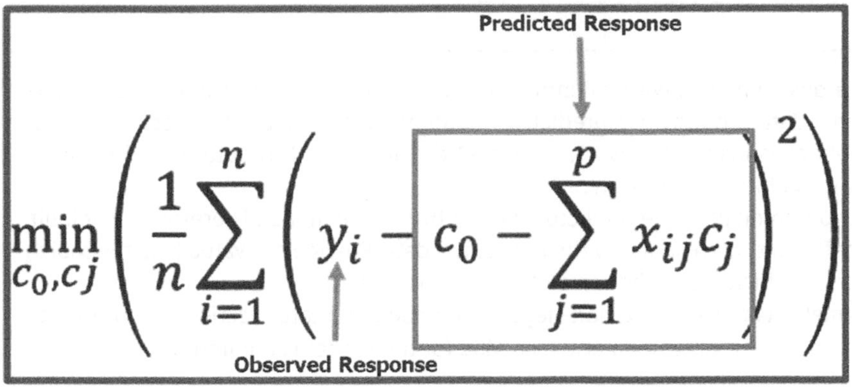

$$\min_{c_0, cj} \left(\frac{1}{n} \sum_{i=1}^{n} \left(y_i - \boxed{c_0 - \sum_{j=1}^{p} x_{ij} c_j} \right)^2 \right)$$

Predicted Response

Observed Response

Figure 4.35 The soul of regression relies on minimization of the difference between reality and in what we believe (i.e., our own modeling or perception).

In ridge regression, a penalty term is added to MSE. This penalty term is controlled by the coefficient values and a tuning parameter, λ. The larger the value of λ, the greater the penalty and, therefore, the more the coefficients are "*shrunk*" toward zero, as shown in Figure 4.36.

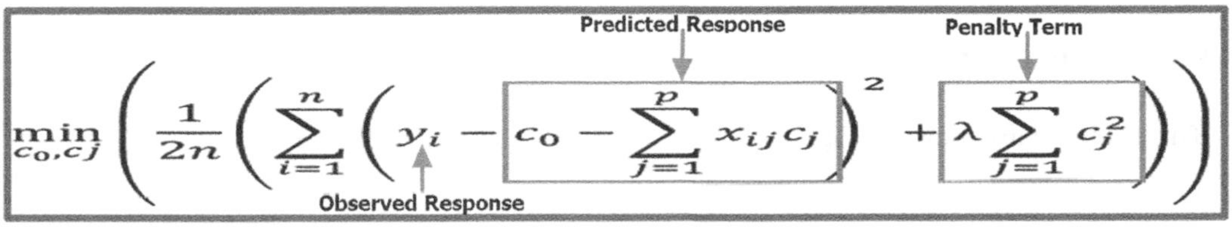

$$\min_{c_0, cj} \left(\frac{1}{2n} \sum_{i=1}^{n} \left(y_i - \boxed{c_0 - \sum_{j=1}^{p} x_{ij} c_j} \right)^2 + \boxed{\lambda \sum_{j=1}^{p} c_j^2} \right)$$

Predicted Response **Penalty Term**

Observed Response

Figure 4.36 The insertion of a tuning parameter, λ, will mitigate heavy dependence of MSE on regression coefficients, c_j.

Fitting Ridge Regression Models

We can use the **ridge** function to fit a ridge regression model to a given data.

>> **b = ridge(y,X,lambda,scaling)**

Table 4.10 gives more details on ridge statement.

Table 4.10   Items appearing in ridge function statement.

b	Ridge regression coefficients whose values depend on the scaling flag.
y	Response values, specified as a vector.
X	Predictor values, specified as a numeric matrix.
lambda	Ridge parameter.
scaling	A {0,1}-valued flag to dictate whether or not to include the intercept in the predicted model parameters. See the below second NOTE.

NOTE

The matrix **X** is a design matrix with columns representing the terms in the regression formula. If the original data contains two predictive variables x1 and x2 but the desired regression model formula contains the terms x1, x2, and x1x2, the matrix X should have 3 columns containing the values of x1, x2, and x1x2.

The ridge function normalizes the predictors before fitting the model. Therefore, by default, the regression coefficients correspond to the *normalized* data. Provide the value 0 as the fourth input to restore the coefficients to the scale of the original data.

The ridge parameter lambda, λ, is a non-negative number. In a later section, we will try to estimate the optimum value of lambda at which **MSE** for the model is minimum.

Predicting Response Using Ridge Regression Models

Given that the regression coefficients were restored to the scale of the original data, we can predict the response by multiplying the matrix containing the predictors by the coefficient vector. Notice that a column of ones is appended at the beginning of the matrix of predictor values to incorporate the intercept in the calculation.

YPred = DataNew * b;

This is pictorially shown in Figure 4.37.

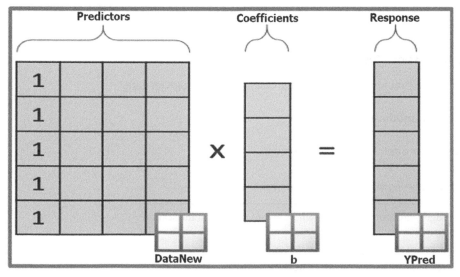

Figure 4.37 Inclusion of a first column of ones to the original **X** data to account for the intercept found in the model.

Determining Ridge Regression Parameter, λ

When modeling the data using the ridge regression, how can we determine the ridge regression parameter, lambda?

>> **b = ridge(y,X,lambda,scaling)**

requires a vector of lambda values as the third input to the function ridge. The output **b** will be a matrix with columns containing the coefficient values for each parameter in the vector lambda.

We can now use each column of the matrix **b** as regression coefficients, predict the response, calculate the mean squared error (MSE), and choose the coefficients with minimum MSE.

The Ridge Regression Model: Big Car Data

The big car data has been split into a training set, **XTrain**, and a test set, **XTest**. The response variable is **MPG**. We perform ridge linear regression with the training data and name the output **b**. Then, we use the model to predict the test values, **XTest**, and store the predictions in a variable named **yhat**. The model loss is **MSE**. The code is shown in Figure 4.38.

NOTE

The **scaling** flag that determines whether the coefficient estimates in **b** are restored to the scale of the original data, specified as either *0* or *1*. If **scaling** is 0, then ridge performs this additional transformation. In this case, **b** contains **p+1** coefficients for each value of **k**, with the first row of **b** corresponding to a constant term in the model. If **scaling** is *1*, then the software omits the additional transformation, and **b** contains **p** coefficients without a constant term coefficient.

Coefficient estimates, returned as a numeric matrix. The rows of **b** correspond to the predictors in **X**, and the columns of **b** correspond to the possible ridge parameter **lambda** values.

If **scaling** is *1*, then **b** will be a **p×m** matrix, where **m** is the number of elements in **lambda**. If **scaling** is *0*, then **b** will be a **(p+1)×m** matrix.

```
load carbig
% X is made of four numeric predictors
X = [Weight,Horsepower,Acceleration,Displacement];
Y=MPG;
XY=[X Y];
temp=rmmissing(XY);
Xnum=temp(:,1:end-1);
Ynum=temp(:,end);
```

Figure 4.38 Implementing the ridge regression model with lambda=5.0 and 0 scaling factor.

```
[r,c] = size(Xnum) ;
P = 0.70 ;
idx = randperm(r) ;
XTrain = Xnum(idx(1:round(P*r)),:) ;
XTest = Xnum(idx(round(P*r)+1:end),:) ;
YTrain = Ynum(idx(1:round(P*r)),:);
YTest = Ynum(idx(round(P*r)+1:end),:) ;
%Find the coefficients of a ridge regression model (with lambda = 5).
lambda = 5;
b = ridge(YTrain,XTrain,lambda,0);
yhat = b(1) + XTest*b(2:end);
scatter(YTest,yhat)
hold on
plot(YTest,YTest)
xlabel('Actual MPG')
ylabel('Predicted MPG')
hold off
%Compare predicted and actual responses
evaluateFit2(YTest,yhat,'Ridge', 'MPG')
```

Figure 4.38 (Cont'd)

Figure 4.39 (*left*) shows predicted (**yhat**) versus observed (**YTest**) response variable and the workspace variables (*right*). **b** is 5×1 matrix; 4 coefficients for 4 predictors plus a constant. **lambda** is scalar thus has one value; hence, # of columns of **b** is 1.

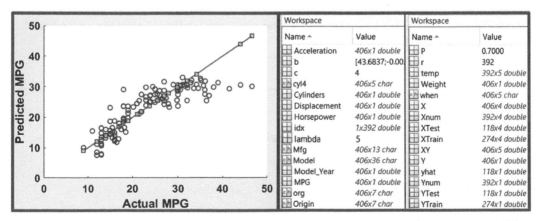

Figure 4.39   (*left*) Predicted (**yhat**) versus observed (**YTest**) response variable and the workspace variables (*right*).

Figure 4.40 shows a comparison between observed and predicted response values in addition to **MSE** and **MAPE** model values.

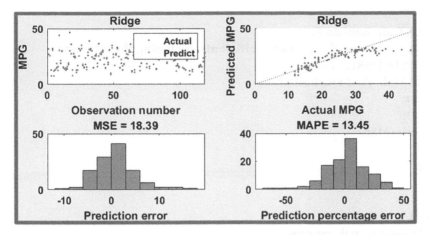

Figure 4.40 Observed vs. predicted response values in addition to **MSE** and **MAPE** of ridge model.

The Ridge Regression Model with Optimum λ: Big Car Data

The big car data has been split into a training set, **XTrain**, and a test set, **XTest**. The response variable is **MPG**. We perform ridge linear regression with the training data and name the output **bR**. Then, we use the model to predict the test values, **XTest**, and store the predictions in a variable named **YPredR**. The model loss is **MSE**. The parameter λ is selected such that **MSE** is minimum. Notice here that **lambda** is a row vector having 111 possible values and thus **bR** will be a matrix of 5×111. This means the ridge function will estimate best five coefficients for each possible **lambda** value. The code is shown in Figure 4.41.

```
load carbig
% X is made of four numeric predictors
X = [Weight,Horsepower,Acceleration,Displacement];
Y=MPG;
XY=[X Y];
temp=rmmissing(XY);
Xnum=temp(:,1:end-1);
Ynum=temp(:,end);
[r,c] = size(Xnum) ;
P = 0.70 ;
idx = randperm(r) ;
XTrain = Xnum(idx(1:round(P*r)),:) ;
XTest = Xnum(idx(round(P*r)+1:end),:) ;
YTrain = Ynum(idx(1:round(P*r)),:);
YTest = Ynum(idx(round(P*r)+1:end),:) ;
%Find the coefficients of a ridge regression model (with parametric lambda values).
```

Figure 4.41 Implementing the ridge regression model with optimum lambda=27.0 and min(MSER)=16.78.

```
lambda = 0:0.5:55;
bR = ridge(YTrain,XTrain,lambda,0);
%Express YPredR as a function of regressed coefficients, including the intercept.
YPredR = bR(1,:) + XTest*bR(2:end,:);
%Calculate and plot MSE. Find smallest MSE.
err = YPredR - YTest;
MSE = mean(err.^2);
[minMSE,idxR] = min(MSE)
clf
plot(lambda,MSE)
xlabel("\lambda")
ylabel("MSE")
title("Ridge model")
%Compare predicted and actual responses.
MPGpred = YPredR(:,idxR);
evaluateFit2(YTest,MPGpred,"Ridge",'MPG')
```

Figure 4.41 (Cont'd)

Figure 4.42 The workspace variables, including the ridge model minimum **MSE**, or optimum λ. $\lambda = 27$ evaluated @ **idxR**=55.

Workspace		Workspace		Workspace	
Name ▲	Value	Name ▲	Value	Name ▲	Value
Acceleration	*406x1 double*	minMSER	16.7788	Weight	*406x1 double*
bR	*5x111 double*	Model	*406x36 char*	when	*406x5 char*
c	4	Model_Year	*406x1 double*	X	*406x4 double*
cyl4	*406x5 char*	MPG	*406x1 double*	Xnum	*392x4 double*
Cylinders	*406x1 double*	MPGpred	*118x1 double*	XTest	*118x4 double*
Displacement	*406x1 double*	MSER	*1x111 double*	XTrain	*274x4 double*
err	*118x111 double*	org	*406x7 char*	XY	*406x5 double*
Horsepower	*406x1 double*	Origin	*406x7 char*	Y	*406x1 double*
idx	*1x392 double*	P	0.7000	Ynum	*392x1 double*
idxR	55	r	392	YPredR	*118x111 double*
lambdaR	*1x111 double*	temp	*392x5 double*	YTest	*118x1 double*
Mfg	*406x13 char*			YTrain	*274x1 double*

Figure 4.42 The workspace variables, including min(**MSE**), or optimum λ for the ridge model.

Figure 4.43 shows a plot of ridge model **MSE** versus λ where min(**MSE**) occurs at λ=27.

Figure 4.43 A plot of ridge model **MSE** versus λ. Optimum **MSE** occurs at $\lambda=27$.

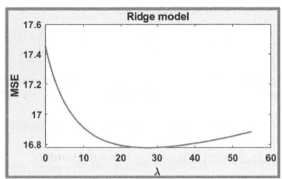

Figure 4.44 shows a comparison between observed and predicted response values in addition to **MSE** and **MAPE** model values.

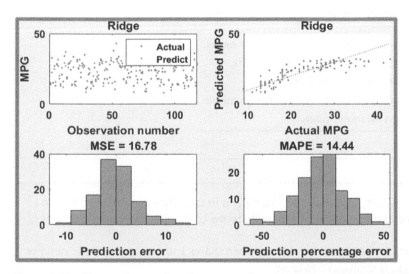

Figure 4.44 Observed vs. predicted response values, in addition to **MSE** and **MAPE** of ridge model.

Quiz #3
(T/F) With ridge regression, you may only model a line.

Regularized Parametric Linear Regression Model: Lasso

Lasso (least absolute shrinkage and selection operator) is a regularized regression method similar to ridge regression. The difference between the two methods is the penalty term. In ridge regression an L^2 norm of the coefficients is used whereas in Lasso an L^1 norm is used, as explained in Figure 4.45.

Ridge Regression	Penalty Term	Lasso	Penalty Term
$\min\limits_{c_0,c_j}\left(\dfrac{1}{2n}\left(\sum\limits_{i=1}^{n}\left(y_i - c_0 - \sum\limits_{j=1}^{p} x_{ij}c_j\right)^2 + \lambda\sum\limits_{j=1}^{p} c_j^2\right)\right)$		$\min\limits_{c_0,c_j}\left(\dfrac{1}{2n}\sum\limits_{i=1}^{n}\left(y_i - c_0 - \sum\limits_{j=1}^{p} x_{ij}c_j\right)^2 + \lambda\sum\limits_{j=1}^{p} |c_j|\right)$	

Figure 4.45 The penalty term has always a positive value but with L^2 norm (*left*) or L^1 norm (*right*) of the coefficients for Ridge and Lasso regression model, respectively.

We can use the function **lasso** to fit a lasso model.

>> **[b,fitInfo] = lasso(X,y,'Lambda',lambda)**

Table 4.11 gives more details on lasso function statement.

Table 4.11 Items of lasso function statement.

Item	Description
b	Lasso coefficients.
fitInfo	A structure containing information about the model.
X	Predictor values, specified as a numeric matrix.
y	Response values, specified as a vector.
'Lambda'	Optional property name for regularization parameter.
lambda	Regularization parameter value.

NOTE

In ridge regression, the penalty term has an L^2 norm and in lasso, the penalty term has an L^1 norm. We can create regression models with penalty terms containing the combination of L^1 and L^2 norms.

Use the property alpha and assign it a value between 0 and 1 to create an elastic net. For alpha equal to 1, the case is equivalent to lasso.

Predicting Response Using Lasso Model

We can predict the response by multiplying the matrix containing the predictors by the coefficient vector. Note that an intercept term is not included in the output coefficients. Instead, it is a field in the output structure **fitInfo**.

>>**YPred = fitInfo.Intercept + XTest*b;**

We can use the lasso function to perform lasso and elastic net regression, using alpha parameter.

>> **lambda = (0:10)/length(y);**

>> **[coef,fitInfo] = lasso(X,y,'Lambda',lambda,"Alpha",0.4);**

Note that lasso uses a different scaling for lambda. Scale by the number of observations to have the same interpretation as for ridge.

Quiz #4

Which property do you set to perform elastic net regression?
a) Alpha   b) KernelFunction   c) Lambda   d) Lasso   e) Ridge

Lasso Regression Model: Big Car Data

The big car data has been split into a training set, **XTrain**, and a test set, **XTest**. The response variable is **MPG**. We perform Lasso regression with the training data, name the coefficients **b**, and name the model fit information **fitInfo**. We use a vector (0:200)/size(XTrain,1) as the lambda values. Name the vector containing the lambda values **lambda**.

Then, we use the model to predict the test values, **XTest**, and store the predictions in a variable named **YPred**. The model loss is **MSE**. The code is shown in Figure 4.46.

```
load carbig
% X is made of four numeric predictors
X = [Weight,Horsepower,Acceleration,Displacement];
Y=MPG;
XY=[X Y];
temp=rmmissing(XY);
Xnum=temp(:,1:end-1);
Ynum=temp(:,end);
[r,c] = size(Xnum) ;
P = 0.70 ;
idx = randperm(r) ;
XTrain = Xnum(idx(1:round(P*r)),:) ;
XTest = Xnum(idx(round(P*r)+1:end),:) ;
YTrain = Ynum(idx(1:round(P*r)),:);
YTest = Ynum(idx(round(P*r)+1:end),:) ;
% Perform lasso regression.
lambda = (0:200)/size(XTrain,1);
[b,fitInfo] = lasso(XTrain,YTrain,'Lambda',lambda);
%[b,fitInfo] = lasso(Xtrain,ytrain,'Lambda',lambda,"Alpha",0.6);
%Make predictions
YPred = fitInfo.Intercept + XTest*b;
figure(47)
scatter(YTest,YPred)
hold on
plot(YTest,YTest)
xlabel('Actual MPG')
ylabel('Predicted MPG')
hold off
%% View the effect of lambda on the coefficients and error
figure(48)
subplot(2,1,1)
```

Figure 4.46 Lasso regression model for big car data and model sensitivity to parameter λ.

```
plot(lambda,b','.')
title('Coefficients')
xlabel('lambda')
subplot(2,1,2)
plot(fitInfo.Lambda,fitInfo.MSE,'.')
title('MSE')
xlabel('lambda')
```

Figure 4.46 (Cont'd)

Figure 4.47 shows the plot of predicted versus actual **MPG** (*left*) and variation of both the model regressed coefficients and associated **MSE** with the parameter λ (*right*). Notice that at low lambda values, the model has the least MSE values; however, one coefficient is more sensitive at low lambda values and they all level off at higher values.

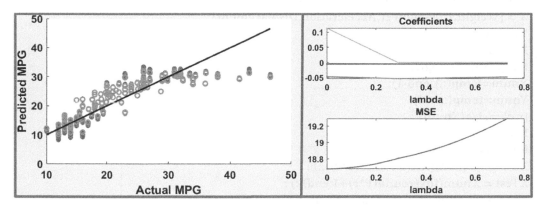

Figure 4.47 Predicted versus Observed **MPG** (*left*) and model parameters sensitivity with λ (*right*).

Quiz #5
Which function performs elastic net regression? a) alpha   b) KernelFunction   c) lambda   d) lasso   e) ridge

Stepwise Parametric Linear Regression

When fitting a linear regression model, there are infinitely many model specifications we can use. How can we choose the model specification which creates the *best* model? This problem becomes worse when we have a large number of predictors.

Stepwise linear regression methods choose a subset of the predictor variables and their polynomial functions to create a compact model. In this chapter, we will learn to use the function **stepwiselm**.

NOTE
This function is used only when the underlying model is *linear regression*. For nonlinear regression and classification problems, use the function **sequentialfs**. See Chapter 3	Sequential Feature Selection (SFS) section for more information on using SFS for classification models.

Fitting Stepwise Linear Regression

Use the function **stepwiselm** to fit a stepwise linear regression model.

>> stepwiseMdl = stepwiselm(data,modelspec)

Table 4.12 gives more details on **stepwiselm** function statement, as quoted from MATLAB offline help.

Table 4.12 Items appearing in **stepwiselm** function statement.

Item	Description
stepwiseMdl	A regression model variable containing the coefficients and other information about the model.
data	A table containing the data used to fit the regression model. See below for details.
modelspec	Starting model for the stepwise regression (see below).

How to Specify stepwiselm Model?

stepwiselm *chooses* the model for us. However, we can provide the following inputs to control the model selection process.

- **modelspec** – The second input to the function specifies the *starting model*. **stepwiselm** starts with this model and adds or removes terms based on certain criteria.
- Commonly used starting values: constant, linear, quadratic (model uses constant, linear, interaction, and quadratic terms).
- "Lower' and 'Upper'. If you want to limit the complexity of the model, use these properties.

Example 1 The following model will definitely contain the intercept and the linear terms but will not contain any terms with a degree of three or more.

>> mdl =

stepwiselm(data,'linear','Lower','linear','Upper','quadratic')

Example 2 The following code is for a model that contains an intercept, linear terms, and all products of pairs of distinct predictors without squared terms.

>> lmMdl = stepwiselm(tableData,'Upper','interactions');

Quiz #6

(T/F) Whenever your predictive model is a regression model, you can use **stepwiselm** instead of **sequentialfs** to perform feature selection.

Quiz #7

Given two predictors, x_1 and x_2, which of the property name-value pairs in *stepwiselm* will create the following model:

$$y = a_0 + a_1 x_1 + a_2 x_2 + a_3 x_1 x_2$$

a) 'Upper', 'constant'   b) 'Upper', 'linear'   c) 'Upper', 'interactions'
d) 'Lower', 'purequadratic'   e) 'Lower', 'quadratic'

Stepwise Linear Regression Model: Big Car Data

The big car data has been split into a training set, **XTrain**, and a test set, **XTest**. The response variable is **MPG**. We perform stepwise linear fitting to create **mdl**. We limit the model to only use a constant and linear terms. Then, we use the model to predict the test values, **XTest**, and store the predictions in a variable named **YPred**. The model loss is shown via **RMSE**. The code is shown in Figure 4.48.

```
load carbig
% X is made of four numeric predictors
X = [Weight,Horsepower,Acceleration,Displacement];
Y=MPG;
XY=[X Y];
temp=rmmissing(XY);
Xnum=temp(:,1:end-1);
Ynum=temp(:,end);
[r,c] = size(Xnum) ;
P = 0.70 ;
idx = randperm(r)  ;
XTrain = Xnum(idx(1:round(P*r)),:) ;
XTest = Xnum(idx(round(P*r)+1:end),:) ;
YTrain = Ynum(idx(1:round(P*r)),:);
YTest = Ynum(idx(round(P*r)+1:end),:) ;
%Perform a stepwise linear fit of fuel economy
mdl = stepwiselm(XTrain,YTrain,'Upper','linear','Verbose',2);
%mdl = stepwiselm(XTrain,YTrain,'Upper','linear','Criterion','aic','Verbose',2);
%mdl = stepwiselm(Xnum,Ynum,'Upper','linear','Criterion','bic','Verbose',2);
YPred = predict(mdl,XTest);
evaluateFit2(YTest,YPred,'Stepwise', 'MPG')
disp(['Root Mean Squared Error: ',num2str(mdl.RMSE)])
```

Figure 4.48   Stepwise linear regression model for big car data with the upper limit as linear.

The results from Command Window are shown but annotated by my own bracketed comments:

---

pValue for adding x1 is 5.69e-70 (*x1 has lowest pValue among all and will be added first*)
pValue for adding x2 is 3.3548e-59 (Good luck for x2; x1 is still 1st Miss World)
pValue for adding x3 is 1.4769e-13 (Good luck for x3; x1 is still 1st Miss World)
pValue for adding x4 is 3.2009e-64 (Good luck for x4; x1 is still 1st Miss World)

1. **Adding x1, FStat = 588.2943, pValue = 5.69407e-70 (x1 is now part of the model)**
   pValue for adding x2 is 1.4527e-05 (x2 has the lowest pValue compared with x3 and x4)
   pValue for adding x3 is 0.0018447 (Good luck for x3; x2 is 2nd Miss World)
   pValue for adding x4 is 0.0022512 (Good luck for x4; x2 is 2nd Miss World)

2. **Adding x2, FStat = 19.50, pValue = 1.453e-05 (x2 will be added in 2nd round, after x1)**
   pValue for adding x3 is 0.84649 (pValue=0.84649>0.05, thus x3 will not be added)
   pValue for adding x4 is 0.31391(pValue=0.31391>0.05, thus x4 will not be added)
   pValue for removing x1 is 1.6092e-16 (pValue=1.609e-16<0.10, thus x1 will be retained)

Root Mean Squared Error: 4.2239

The model starts with a constant, then it adds x1 in the first round, followed by x2 in the second round. Keep in mind that the predictors are four. Thus, x3 and x4 were not included in the model in the final round, because x3 and x4 do not add extra credibility to the model. This is judged by guidelines, shown in the following two tables, for either adding or removing a given predictor.

NOTE

Please, see Tables 4.13 and 4.14 below for the policy of adding a predictor to or removing it from a model. The default p-value (i.e., '**PEnter**' value) for the default criterion ('**PEnter**') is 0.05. It can be modified, however, to be a more or less conservative; for example, using '**PEnter**', *0.075* pair.

Control for the display of information, specified as the comma-separated pair consisting of '**Verbose**' and one of these values:

0: Suppress all display.
1: Display the action taken at each step.
2: Display the evaluation process and the action taken at each step.

The threshold for the criterion to add a term, specified as the comma-separated pair consisting of '**PEnter**' and a scalar value, is described in Table 4.13, as quoted from MATLAB offline help.

Table 4.13 The criterion and its default threshold value to add a column predictor.

Criterion	'PEnter' (default)	Decision
'SSE' (default)	0.05	If the *p*-value of the *F*-statistic is less than **PEnter** (*p*-value to enter), add the term to the model.
'AIC'	0	If the change in the AIC of the model is less than **PEnter**, add the term to the model.
'BIC'	0	If the change in the Bayesian (Schwarz) information criteria, BIC, of the model is less than **PEnter**, add the term to the model.
'Rsquared'	0.1	If the increase in the R-squared value of the model is greater than **PEnter**, add the term to the model.
'AdjRsquared'	0	If the increase in the adjusted R-squared value of the model is greater than **PEnter**, add the term to the model.

On the other hand, Table 4.14 shows the threshold for the criterion to remove a column predictor, specified as the comma-separated pair consisting of '**PRemove**' and a scalar value, as quoted from MATLAB offline help.

Table 4.14 The criterion and its default threshold value to remove a column predictor.

Criterion	'PRemove' (default)	Decision
'**SSE**' (default)	0.10	If the p-value of the F-statistic is greater than **PRemove** (p-value to remove), remove the term from the model.
'**AIC**'	0.01	If the change in the AIC of the model is greater than **PRemove**, remove the term from the model.
'**BIC**'	0.01	If the change in the BIC of the model is greater than **PRemove**, remove the term from the model.
'**Rsquared**'	0.05	If the increase in the R-squared value of the model is less than **PRemove**, remove the term from the model.
'**AdjRsquared**'	−0.05	If the increase in the adjusted R-squared value of the model is less than **PRemove**, remove the term from the model.

NOTE

At each step, the **stepwiselm** function also checks whether a term is redundant (linearly dependent) with other terms in the current model. When any term is linearly dependent with other terms in the current model, the **stepwiselm** function removes the redundant term, regardless of the criterion value.

Figure 4.49 shows a comparison between predicted and observed **MPG** while showing **MSE** and **MAPE** for **YPred** using **XTest** set.

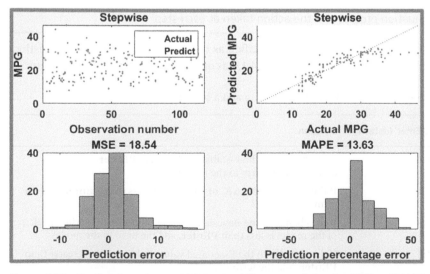

Figure 4.49 Predicted vs. observed **MPG** using **XTest** set and showing **MSE** and **MAPE**.

On the other hand, if we use 'aic' as a 'Criterion' for inclusion or exclusion of a predictor (player), then what player is more important than others will be different from the previous case. 'aic' stands for Akaike information criterion. Akaike's Information Criterion (AIC) provides a measure of model quality obtained by simulating the situation where the model is tested on a different data set. After computing several different models, you can compare them using this criterion.

NOTE

According to Akaike's theory, *the most accurate model has the smallest AIC*. If you use the same data set for both model estimation and validation, the fit always improves as you increase the model order and, therefore, the flexibility of the model structure. The idea of having the smallest AIC extends well to Bayesian (Schwarz) information criterion of having the smallest BIC for the most accurate model.

When applying 'aic' option as a 'Criterion' for **stepwiselm** function, the following statement must replace **mdl** statement in Figurer 4.48.

---

mdl = stepwiselm(XTrain,YTrain,'Upper','linear','Criterion','aic','Verbose',2);

The results from Command Window are shown below but annotated by my own bracketed comments:

Change in AIC for adding x1 is –306.7477 (x1 has lowest AIC among all and will be added first)
Change in AIC for adding x2 is –251.1397 (Good luck for x2; x1 is still 1st Miss World)
Change in AIC for adding x3 is –39.8831 (Good luck for x3; x1 is still 1st Miss World)
Change in AIC for adding x4 is –257.1479 (Good luck for x4; x1 is still 1st Miss World)

1. Adding x1, AIC = 1604.093 (x1 is now part of the model)
Change in AIC for adding x2 is –11.4634 (x2 has the lowest AIC compared with x3 and x4))
Change in AIC for adding x3 is –0.39134 (Good luck for x3; x2 is 2nd Miss World)
Change in AIC for adding x4 is 0.25586 (Good luck for x4; x2 is 2nd Miss World)

2. Adding x2, AIC = 1592.6296 (x2 will be added in 2nd round, after x1)
Change in AIC for adding x3 is 0.1582 (Adding x3 will increase the model AIC by 0.1582, thus x3 will not be added)
Change in AIC for adding x4 is 1.6115 (Adding x4 will increase AIC by 1.6115, thus x4 will not be added)
Change in AIC for removing x1 is 67.0714 (Removing x1 will increase AIC by 67.07, thus x1 will be retained)

NOTE

With selection of either 'aic' or 'bic' option as a 'Criterion' for the **stepwiselm** function, the p-value is not shown although it is calculated by the **stepwiselm** function. The p-values can be reached via exploring **mdl.Coefficients** table where the **Estimate**, **SE**, **tStat**, and **pValue** are shown as columns for each coefficient present in the model, including the intercept.

CHW 4.1   Boston House Price

The Boston House Price Dataset involves the prediction of a house price in thousands of dollars given details of the house and its neighborhood. It is a regression problem. There are 506 observations with 13 input variables and 1 output variable.

The variable names are as follows:

1) **Crime**: per capita crime rate by town.
2) **Zone**: proportion of residential land zoned for lots over 25,000 sq. ft.
3) **Indust**: proportion of nonretail business acres per town.
4) **CHRRiver**: Charles River dummy variable (= 1 if tract bounds river; 0 otherwise).
5) **NOX**: nitric oxides concentration (parts per 10 million).
6) **Room**: average number of rooms per dwelling.
7) **Age**: proportion of owner-occupied units built prior to 1940.
8) **Dist**: weighted distances to five Boston employment centers.
9) **RadHwy**: index of accessibility to radial highways.
10) **Tax**: full-value property-tax rate per $10,000.
11) **PTRatio**: pupil-teacher ratio by town.
12) **BlkRatio**: $1000(Bk - 0.63)^2$ where Bk is the proportion of blacks by town.
13) **LStat**: % lower status of the population.
14) **MedVal**: Median value of owner-occupied homes in $1000s.

The table called housing contains several predictor variables and a response in the last column named **MedVal**.

1) Data Preparation: Prior to any regression implementation, carry out treatment steps need to transform the original table into numeric table and also remove **NaN** values if any. Split the numeric, NaN-free table into a training set, **XTrain**, and a test set, **XTest**. So does the case with the response variable, split into **YTrain** and **YTest**, respectively.
2) Fit a parametric linear regression model to the training set using all predictor variables to predict median house values. Name it **mdl**. Evaluate the fit by showing residuals for **YTest** versus **YPred**. You can also use **evaluateFit2** function (Appendix A | Figure A15) to plot residuals and show statistical parameters related to the model goodness. In addition to **fitlm** function, try **fitglm** function.
3) Fit more than a nonparametric linear regression model to the training set with response **YTrain**. Use automatic hyperparameter optimization to tune-up the regression model. You can use **evaluateFit2** function to plot residuals and show statistical parameters related to the model goodness.
4) Use the Ridge Regression Model with Optimum λ. Show a plot of MSE vs. λ. You can use **evaluateFit2** function to plot residuals and show statistical parameters related to the model goodness.
5) Use Lasso regression model and the model sensitivity to parameter λ. Show a plot of **YTest** versus **YPred**. Show how regression coefficients and MSE change with the parameter λ.
6) Use the Stepwise Linear Regression model to fit the housing data. See which predictor variables are included in the linear regression model. Show RMSE for the model. You can use **evaluateFit2** function to plot residuals and show statistical parameters related to the model goodness.

CHW 4.2   The Forest Fires Data

The table **forestfires** contains 12 predictor variables and a response named **area**, stored as the last variable.

Predictor information:

1) X: x-axis spatial coordinate within the Montesinho park map: 1 to 9
2) Y: y-axis spatial coordinate within the Montesinho park map: 2 to 9
3) month: Month of the year: "jan" to "dec"
4) day: Day of the week: "mon" to "sun"
5) FFMC: FFMC index from the FWI system: 18.7 to 96.20
6) DMC: DMC index from the FWI system: 1.1 to 291.3
7) DC: DC index from the FWI system: 7.9 to 860.6
8) ISI: ISI index from the FWI system: 0.0 to 56.10
9) temp: Temperature in Celsius degrees: 2.2 to 33.30
10) RH: Relative humidity in %: 15.0 to 100
11) wind: Wind speed in km/h: 0.40 to 9.40
12) rain: Outside rain in mm/m2: 0.0 to 6.4
13) area: The burned area of the forest (in ha): 0.00 to 1090.84

1) Data Preparation: Prior to any regression implementation, carry out treatment steps need to transform the original table into numeric table and also remove **NaN** values if any. Moreover, exclude or remove both month and day column as both will add extra superfluous predictor variables, especially when converting into numerical dummy variables. The "**month**" column will produce 12 dummy variables and the "**day**" column will produce 7 dummy variables, totaling up to 19 extra variables. Split the numeric, NaN-free table into a training set, **XTrain**, and a test set, **XTest**. So does the case with the response variable, split into **YTrain** and **YTest**, respectively.

2) Fit a parametric linear regression model to the training set using all predictor variables to predict burned area values. Name it **mdl**. Evaluate the fit by showing residuals for **YTest** versus **YPred**. You can also use **evaluateFit2** function (Appendix A | Figure A.15) to plot residuals and show statistical parameters related to the model goodness. If **fitlm** function does not work, try **fitglm** function.

3) Fit more than a nonparametric linear regression model to the training set with response **YTrain**. Use automatic hyperparameter optimization to tune-up the regression model. You can use **evaluateFit2** function to plot residuals and show statistical parameters related to the model goodness.

4) Use the Ridge Regression Model with Optimum λ. Show a plot of MSE vs. λ. You can use **evaluateFit2** function to plot residuals and show statistical parameters related to the model goodness.

5) Use Lasso regression model and the model sensitivity to parameter λ. Show a plot of **YTest** versus **YPred**. Show how regression coefficients and MSE change with the parameter λ.

6) Use the Stepwise Linear Regression model to fit the forest fire data. See which predictor variables are included in the linear regression model. Show RMSE for the model. You can use **evaluateFit2** function to plot residuals and show statistical parameters related to the model goodness.Based on estimated statistical parameters and generated plots generated, what do you conclude about the pattern of experimental data?

CHW 4.3   The Parkinson's Disease Telemonitoring Data

This dataset is composed of a range of biomedical voice measurements from 42 people with early-stage Parkinson's disease recruited to a six-month trial of a telemonitoring device for remote symptom progression monitoring. The recordings were automatically captured in the patient's homes.

Columns in the table contain subject number, subject age, subject gender, time interval from baseline recruitment date, motor UPDRS, total UPDRS, and 16 biomedical voice measures. Each row corresponds to one of 5875 voice-recordings from these individuals. The main aim of the data is to predict the motor and total UPDRS scores ('motor_UPDRS' and 'total_UPDRS') from the 16 voice measures. The data is in ASCII CSV format. The rows of the CSV file contain an instance corresponding to one voice recording. There are around 200 recordings per patient; the subject number of the patient is identified in the first column.

Thus, the table data contains several predictor variables and a response in an interior column named **total_UPDRS**. The other response variable, called **motor_UPDRS**, will not be included in data analysis.

Attribute Information:

Column(s)	Description
subject	Integer that uniquely identifies each subject.
age	Subject age.
sex	Subject gender '0' – male, '1' – female.
test_time	Time since recruitment into the trial. The integer part is the number of days since recruitment.
motor_UPDRS	Clinician's motor UPDRS score, linearly interpolated (**response**).
total_UPDRS	Clinician's total UPDRS score, linearly interpolated (**response**).
Jitter, JitterAbs, Jitter:RAP, JitterPPQ5, and **JitterDDP**	Several measures of variation in fundamental frequency.
Shimmer, ShimmerdB, ShimmerAPQ3, ShimmerAPQ5, ShimmerAPQ11, and **ShimmerDDA**	Several measures of variation in amplitude.
NHR and **HNR**	Two measures of ratio of noise to tonal components in the voice.
RPDE	A nonlinear dynamical complexity measure.
DFA	Signal fractal scaling exponent.
PPE	A nonlinear measure of fundamental frequency variation.

1) Data Preparation: Prior to any regression implementation, carry out treatment steps need to transform the original table into numeric table and also remove **NaN** values if any. Moreover, exclude or remove the first "**subject**" and the fifth "**motor_UPDRS**" column from predictors' list. Split the numeric, NaN-free table into a training set, **XTrain**, and a test set, **XTest**. So does the case with the response variable, split into **YTrain** and **YTest**, respectively.

2) Fit a parametric linear regression model to the training set using all predictor variables to predict total_UPDRS values. Name it **mdl**. Evaluate the fit by showing residuals for **YTest** versus **YPred**. You can also use **evaluateFit2** function (Appendix A | Figure A.15) to plot residuals and show statistical parameters related to the model goodness. If **fitlm** function does not work, especially for higher X order, try **fitglm** function.

3) Fit more than a nonparametric linear regression model to the training set with response **YTrain**. Use automatic hyperparameter optimization to tune-up the regression model. You can use **evaluateFit2** function to plot residuals and show statistical parameters related to the model goodness.

4) Use the Ridge Regression Model with Optimum λ. Show a plot of MSE vs. λ. You can use **evaluateFit2** function to plot residuals and show statistical parameters related to the model goodness.

5) Use Lasso regression model and the model sensitivity to parameter λ. Show a plot of **YTest** versus **YPred**. Show how regression coefficients and MSE change with the parameter λ.

6) Use the Stepwise Linear Regression model to fit the Parkinson's disease data. See which predictor variables are included in the linear regression model. Show RMSE for the model. You can use **evaluateFit2** function to plot residuals and show statistical parameters related to the model goodness.

CHW 4.4   The Car Fuel Economy Data

This dataset is composed of 14 car properties or features representing the predictor data and one response variable, that is, the fuel economy. Some predictor columns are classified as categorical and the rest as numeric type. There are no NaN entries.

Columns 1 through 15

1) {'Car_Truck'}: Categorical
2) {'EngDisp'}: Numeric
3) {'RatedHP'}: Numeric
4) {'Transmission'}: Categorical
5) {'Drive'}: Categorical
6) {'Weight'}: Numeric
7) {'Comp'}: Numeric
8) {'AxleRatio'}: Numeric
9) {'EVSpeedRatio'}: Numeric
10) {'AC'} : Categorical
11) {'PRP'}: Numeric
12) {'FuelType'}: Numeric
13) {'City_Highway'}: Categorical
14) {'Valves_Cyl'}: Numeric
15) {'FuelEcon'}: Numeric

1) Data Preparation: Prior to any regression implementation, carry out treatment steps need to transform the original table into numeric table. Split the numeric, NaN-free table into a training set, **XTrain**, and a test set, **XTest**. So does the case with the response variable, split into **YTrain** and **YTest**, respectively.

2) Fit a parametric linear regression model to the training set using all predictor variables to predict FuelEcon values. Name it **mdl**. Evaluate the fit by showing residuals for **YTest** versus

YPred. You can also use **evaluateFit2** function (Appendix A | Figure A.15) to plot residuals and show statistical parameters related to the model goodness. In addition to **fitlm** function, try **fitglm** function.

3) Fit more than a nonparametric linear regression model to the training set with response **YTrain**. Use automatic hyperparameter optimization to tune-up the regression model. You can use **evaluateFit2** function to plot residuals and show statistical parameters related to the model goodness.

4) Use the Ridge Regression Model with Optimum λ. Show a plot of MSE vs. λ. You can use **evaluateFit2** function to plot residuals and show statistical parameters related to the model goodness.

5) Use Lasso regression model and the model sensitivity to parameter λ. Show a plot of **YTest** versus **YPred**. Show how regression coefficients and MSE change with the parameter λ.

6) Use the Stepwise Linear Regression model to fit the car fuel data. See which predictor variables are included in the linear regression model. Show RMSE for the model. You can use **evaluateFit2** function to plot residuals and show statistical parameters related to the model goodness.

5

Neural Networks

Introduction

The idea of using neural networks is relevant when we do not really know the equation that dictates the relationship between the response variable on one side and the set of predictor variables on the other side. In many case studies, we have so many inputs with interrelations and yet we do not know what the response behavior will be. If we have an idea of the equation, it will be better to use the conventional linear and nonlinear regression methods. Consequently, the neural network regression is a kind of stochastic, as opposed to deterministic, model in the sense that it tries to understand different relationships among the players (variables) via recognizing what is called patterns. To demonstrate the game of neural network, a person is assigned a duty to separate the male baby from female baby chicken. The accuracy in predicting the type of baby chicken class is more than 90%. This assigned person makes use of brain neural networks-based recognition using the God-granted natural intelligence. The neural networks-based artificial intelligence more or less mimics one of the primary brain main duties.

Neural networks have found applications in various fields due to their ability to learn from data and make predictions or decisions. Here are some life examples where neural networks are commonly used:

1) Image Recognition: Neural networks are extensively used in applications like facial recognition, object detection, and image classification. They can be employed in security systems, social media platforms, and autonomous vehicles to identify and interpret visual information.

2) Natural Language Processing: Neural networks have revolutionized language processing tasks such as machine translation, sentiment analysis, speech recognition, and text generation. Virtual assistants like Siri and Alexa utilize neural networks to understand and respond to user commands.

3) Recommender Systems: Popular platforms like Netflix, Amazon, and Spotify employ neural networks to analyze user preferences and provide personalized recommendations. These systems learn from user behavior and patterns to suggest movies, products, or music tailored to individual tastes.

4) Financial Forecasting: Neural networks are utilized in predicting stock market trends, credit risk assessment, fraud detection, and portfolio optimization. They can analyze historical data and market indicators to make predictions and assist in decision-making.

Machine and Deep Learning Using MATLAB: Algorithms and Tools for Scientists and Engineers, First Edition.
Kamal I. M. Al-Malah.
© 2024 Kamal I. M. Al-Malah. Published 2024 by John Wiley & Sons, Inc.
Companion Website: www.wiley.com/go/al-malah/machinelearningmatlab

5) Autonomous Vehicles: Neural networks are crucial in the development of self-driving cars. They are used for object detection, lane recognition, pedestrian detection, and decision-making algorithms, enabling vehicles to perceive and navigate their environment.

6) Medical Diagnosis: Neural networks are applied in medical fields for disease diagnosis, radiology image analysis, and drug discovery. They can assist doctors in identifying patterns or anomalies in medical images and help in early detection and diagnosis.

7) Gaming: Neural networks have been employed in game-playing systems, such as chess. They can learn strategies and make decisions based on past game data, allowing for intelligent and adaptive gameplay.

8) Customer Service Chatbots: Neural networks power chatbots that provide customer support and respond to inquiries. These systems can understand natural language, provide automated responses, and assist users with common queries.

We will cover some of such applications here in this chapter and in subsequent chapters.

Feed-Forward Neural Networks

Feed-forward neural networks are a type of artificial neural networks which are used for modeling classification and regression problems. Neural net pattern recognition solves pattern recognition problem using two-layer feed-forward networks. A two-layer feed-forward network, also known as a shallow neural network, consists of an input layer, a hidden layer, and an output layer. The input layer receives data, the hidden layer applies weights and biases to the inputs using an activation function, and the output layer produces the final output. The hidden layer captures complex patterns in the data, and the network learns by adjusting the weights and biases during training. Although simpler than deeper architectures, two-layer feed-forward networks can still be effective for certain problems, especially when the data is not too complex or computational resources are limited.

NOTE

Although it is highly recommended to standardize the numeric predictors, as standardization makes predictors insensitive to the scales on which they are measured, I tried both cases with and without predictor data standardization and found that neural network regression model using non-standardized data works as perfectly as the model with data standardization. See the running big car data example under "**Neural Network Regression (nftool)**" section.

On the other hand, linear regression models do not require standardization, or scaling of predictor data, because the regression parameters will acquire the physical units of X and Y. For example, if we say $Y = b_o + b_1X$, then b_o will have dimensions of Y and b_1 those of Y/X (i.e., slope).

We will start with a feed-forward, fully connected neural network for regression (Table 5.1). The first fully connected layer of the neural network has a connection from the network input (predictor data), and each subsequent layer has a connection from the previous layer. As the name suggests, all neurons in a fully connected layer connect to all the neurons in the previous layer. Each fully connected layer multiplies the input by a weight matrix and then adds a bias vector. An activation function follows each fully connected layer, excluding the last. The final fully connected layer produces the network's output, namely predicted response values. Keep in mind that for classification problems, the **outputSize** argument of the last fully connected layer of the network is equal to the number of classes of the data set. For regression problems, the output size must be equal to the number of response variables.

Table 5.1 Typical feed-forward, fully connected neural network for regression.

Structure	Description
Input ↓ FC ↓ ReLU ↓ FC ↓ Output	**Input**: This layer corresponds to the predictor data in **Tbl** or **X**. First Fully Connected (**FC**) layer: This layer has *10* outputs by default. We can widen the layer or add more fully connected layers to the network by specifying the **LayerSizes** name-value argument. We can also find the weights and biases for this layer in the **Mdl.LayerWeights{1}** and **Mdl.LayerBiases{1}** properties of **Mdl**, respectively. Rectified Linear Unit (**ReLU**) activation function: **fitrnet** applies this activation function to the first fully connected layer. We can change the activation function by specifying the **Activations** name-value argument. A **ReLU** layer performs a threshold operation to each element of the input, where any value less than zero is set to zero. Final Fully Connected (**FC**) layer: This layer has one output. We can find the weights and biases for this layer in the **Mdl.LayerWeights{end}** and **Mdl. LayerBiases{end}** properties of **Mdl**, respectively. **Output**: This layer corresponds to the predicted response values.

Feed-Forward Neural Network Classification

The Neural Network Pattern Recognition (**nprtool**) app lets us create, visualize, and train two-layer feed-forward networks to solve data classification problems. Using this app, we can:

- Import data from file, the MATLAB® workspace, or use one of the example data sets.
- Split data into training, validation, and test sets.
- Define and train a neural network.
- Evaluate network performance using cross-entropy error and misclassification error.
- Analyze results using visualization plots, such as confusion matrices and receiver operating characteristic curves.
- Generate MATLAB scripts to reproduce results and customize the training process.
- Generate functions suitable for deployment with MATLAB Compiler™ and MATLAB Coder™ tools, and export to Simulink® for use with Simulink Coder.

Figure 5.1 shows a schematic for a feed-forward classification neural network which can be thought of as an arrangement of interconnected nodes (neurons) that map inputs to responses.

Number of neurons in the input layer = Number of predictors
Number of neurons in the output layer = Number of response classes

In theory, this sandwiched (i.e., one-hidden-layer) layer feed-forward neural network can learn any input-output relationship given enough neurons in the hidden layer. However, too many hidden-layer neurons can lead to overfitting the training data, resulting in poor generalization.

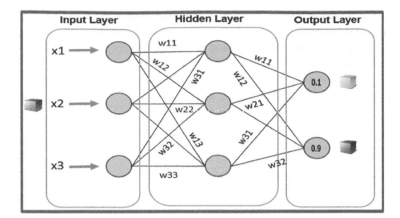

Figure 5.1 A feed-forward classification neural network augmented by a sandwiched hidden layer.

Feed-Forward Neural Network Regression

On the other hand, Figure 5.2 shows a schematic for a feed-forward neural regression network which can be thought of as an arrangement of interconnected nodes (neurons) that map a set of inputs to the corresponding single response.

Number of neurons in the input layer = number of predictors
Number of neurons in the output layer = 1

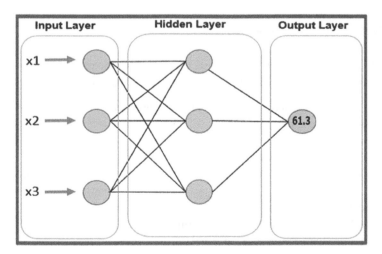

Figure 5.2 A feed-forward regression neural network has only one neuron in the output layer.

Numeric Data: Dummy Variables

To train a feed-forward network, predictors and responses both must be in the form of a numeric matrix. If we have categorical data, we can create *dummy predictors* for each category. Each dummy predictor can have only two values: 0 or 1. For any given observation, only one of the dummy predictors can have the value equal to 1. See Chapter 3 | Accommodating Categorical Data: Creating Dummy Variables section for further information.

The table **heartData** contains several features of different subjects and whether or not they have heart disease, which is saved in the **HeartDisease** variable. It has 11 numerical and 11 categorical columns recorded for 427 patients.

Let us create the workspace variables which will be imported into the Neural Network Pattern Recognition application called **nprtool**. Figure 5.3 shows the code to create predictor variables made of numeric and transformed into numeric categorical columns, in addition to converting the response, **HeartDisease**, variable into numeric two dummy variables equal to the number of classes found in the categorical response variable which are "**yes**" for heart disease and "**no**" for heart-disease-free case.

```
%% Load the data
heartData = readtable('heartDiseaseData.xlsx','Format', 'auto');
heartData.HeartDisease = categorical(heartData.HeartDisease);
numData = heartData{:,1:11};
heartData = convertvars(heartData,12:21,"categorical");
% Extract the response variable
HD = heartData.HeartDisease;
HDDum=dummyvar(HD);
%Convert categorical variables to numeric dummy variables
[heartData2 Varnames]=cattbl2mat(heartData(:,1:end-1));
heartData3 = table2array(hcartData2);
```

Figure 5.3   Preparing the workspace to have variables which can be imported into **nprtool**.

Neural Network Pattern Recognition (nprtool) Application

At the command prompt, key in:

>>**nprtool**

The main window will pop-up as shown in Figure 5.4. Upon clicking on "**Import**" button, Figure 5.4 (left) shows the choices on how to import the data to be dealt with. Select the first option and the "**Import Data from Workspace**" window (*inset*) will pop-up where we have to select **heartData3** dataset which contains 35 predictor columns (or features) with 427 observations. Moreover, select the data as row-wise which means we have 427 observations with 35 features, known as predictors. Click on "**OK**" button to close the wizard.

nprtool is now ready for the training step. Notice that the tool will allow the user to make the desired partition of the original data into training, validation, and test subsets. The layer size, which is by default equal to 10, can be modified, as well.

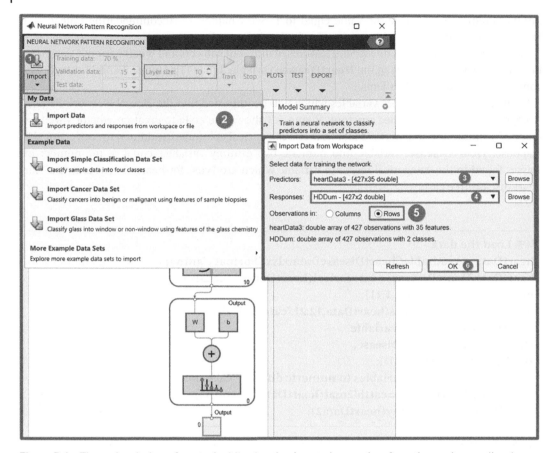

Figure 5.4 The main window of **nprtool** while showing how to import data from the workspace (*inset*).

After clicking on "**Train**" button, shown in the top toolbar, **nprtool** tool will initiate the algorithm called *scaled conjugate gradient* and it will start with zero number of epochs and it will gradually increment the size of epochs until the solution converges.

Training occurs until a maximum number of epochs occurs, the performance goal is met, or any other stopping condition of the function **net.trainFcn** occurs.

Typically, one *epoch* of training is *defined* as a single presentation of all input vectors to the network. The network is then updated according to the results of all those presentations. Some training functions depart from this norm by presenting only one input vector (or sequence) each epoch. An input vector (or sequence) is chosen randomly for each epoch from concurrent input vectors (or sequences).

As can be seen in Figure 5.5, upon converging to a final solution, **nprtool** reports the cross-entropy (measure of performance) error and the error of the classification neural network model for the training, validation, and test data subsets.

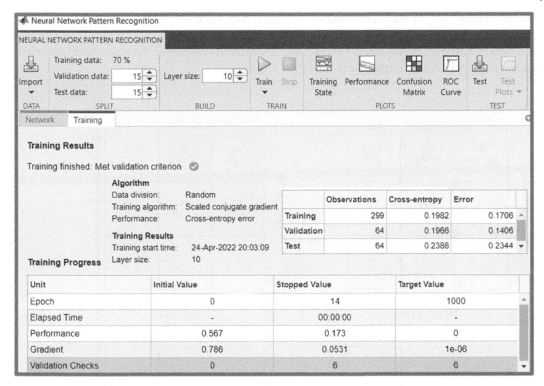

Figure 5.5 Upon training, validation, and testing, **nprtool** reports performance and model error.

The model loss for a classification model is defined as the half mean squared error loss between network predictions and target values for regression tasks. The loss is calculated using the following formula:

$$loss = \frac{1}{2N}\sum_{i=1}^{M}\left(X_i - T_i\right)^2$$

where X_i is the network prediction, T_i is the target value, M is the total number of responses in X (across all observations), and N is the total number of observations in X.

On the other hand, the cross-entropy error is essentially the cross-entropy loss (scalar quantity) between network predictions and target values for single-label and multi-label classification tasks. In MATLAB, the cross-entropy loss for a neural network model can be computed using the **crossentropy** function (part of Deep Learning Toolbox). The **crossentropy** function takes the predicted probabilities (output of the neural network) and the true class labels as inputs and returns the cross-entropy loss. Here is a simple example of using MATLAB **crossentropy** function:

```
predictedProbabilities = [0.3, 0.5, 0.2]; % Example predicted probabilities
trueLabels = [0, 1, 0]; % Example true class labels
loss = crossentropy(predictedProbabilities, trueLabels);
loss =12.046
```

Under "**Plots**" tab in the main toolbar, the user may click on such buttons to create the corresponding plot. For example, Figure 5.6 shows the confusion matrix for training, validation, test, and overall step. Notice that the overall score for correctly predicting the class type of observed data is about 82%. Other steps have at least 77%.

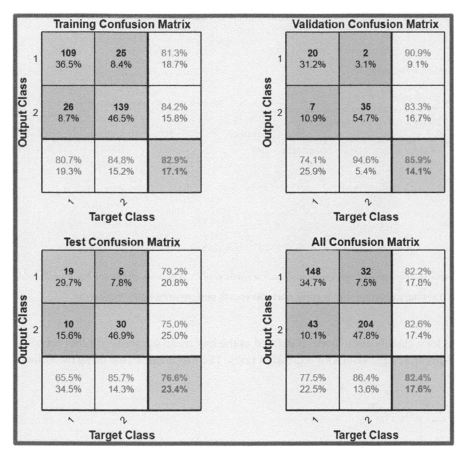

Figure 5.6 The confusion matrix for all phases plus the overall classification predictability.

Besides, there is another plot called receiver operating characteristic (ROC), which is, more or less, similar to the confusion matrix. See Chapter 2 | Using MATLAB Classification Learner section for more information on ROC. The maximum area under the curve (AUC) is 1, which corresponds to a perfect classifier. Larger AUC values indicate better classifier performance.

In MATLAB Toolstrip and under "**TEST**" group, clicking on "**Test**" button will allow the user to test the model over a new test set. This step is really not needed because the previous step divides the entire data set into training (70%), validation (15%), and test (15%) sub-sets. However, the user may attempt to try new data other than the data initially imported into **nprtool** application.

In Neural Network Pattern Recognition (**nprtool**) app and under "**EXPORT**" group, click on "**Generate Code**" drop down menu and select the first choice, that is, "**Generate Simple Training Script**". This will allow the user to save a MATLAB m-code of the entire process of learning and creating the model. The second choice gives more details on four types of errors: overall, training, validation, and test. Both options, however, will give the overall performance and be able to reproduce training workflow.

Figure 5.7 shows the generated code where it allows the user to modify the neural network model parameters, like the size of the hidden layer, the training ratio, validation ratio, and test ratio. For example, I saved the file as **HDDSCG.m.**, which means heart disease data classifier using Scaled Conjugate Gradient algorithm.

To run the code, we need first to have the data available in the MATLAB Workspace. In other words, we have to define what x and y are. To do so, run the code of Figure 5.3, first, as HDDSCG requires definition of both x and t.

```
% This script assumes these two variables are defined:
x = heartData3'; %transposed
t = HDDum'; %transposed
% Choose a Training Function
% 'trainlm' is usually fastest.
% 'trainbr' takes longer but may be better for challenging problems.
% 'trainscg' uses less memory. Suitable in low memory situations.
trainFcn = 'trainscg';  % Scaled conjugate gradient backpropagation.
% Create a Pattern Recognition Network
hiddenLayerSize = 10;
net = patternnet(hiddenLayerSize, trainFcn);
% Setup Division of Data for Training, Validation, Testing
net.divideParam.trainRatio = 70/100;
net.divideParam.valRatio = 15/100;
net.divideParam.testRatio = 15/100;
% Train the Network
[net,tr] = train(net,x,t);
% Test the Network
y = net(x);
e = gsubtract(t,y);
performance = perform(net,t,y);
tind = vec2ind(t);
yind = vec2ind(y);
percentErrors = sum(tind ~= yind)/numel(tind);
% View the Network
view(net)
```

Figure 5.7 The **nprtool** generated **HDDSCG.m** file for creating the neural network classification model.

At the prompt, enter the name of the saved m-file:

>> **HDDSCG**

Figure 5.8 will pop-up where you can see the model-related plots, similar to that shown in Figure 5.6. Click on "**Performance**" button and Figure 5.9 is generated.

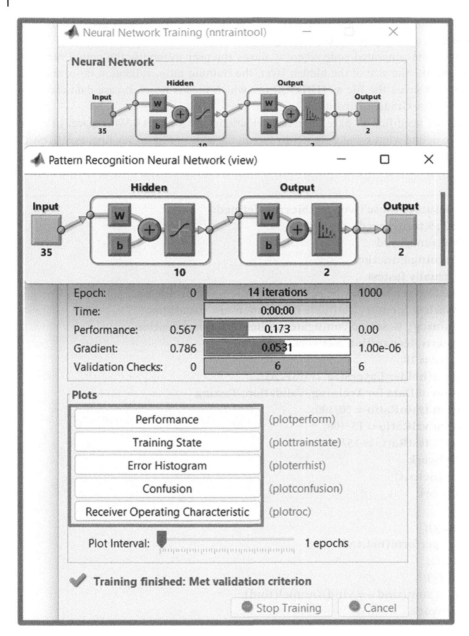

Figure 5.8 HHDSCG-created **nntraintool** where different plots can be generated to show the classification model goodness.

The best validation performance, expressed in cross entropy, of 0.159 is achieved at epoch 9.

Figure 5.9 The neural network model performance measured in cross entropy as *f(epoch)*.

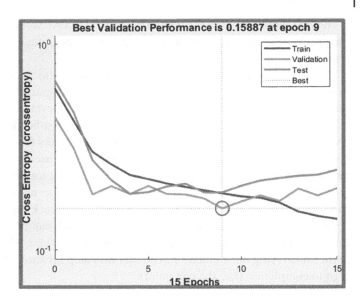

NOTE

If we replace **trainFcn = 'trainscg';** by **trainFcn = 'trainbr'**; in Figure 5.7 code, save the m-code to a new file named **HDDBR**, run the file at the prompt, we will get a model that perform better than HDDSCG. In fact, the overall performance of making a good classification judgment, as given by the confusion matrix, is found to be **94.8%** compared with the **'trainscg'** value of **82.4%**. So, training with **B**ayesian **R**egularization (**'trainbr'**) takes longer time but converges to a more accurate classification judgment.

Figure 5.10 shows the "**Export Model**" button where the user has the choice to select different ways to export the ready-to-go model for predicting the response variable for new data of the same format in terms of number of predictor variables (i.e., 35 predictors). The model will predict the response variable, that is, the heart disease value as either 0 or 1.

Figure 5.10 "**Export Model**" button where it allows the user to save the model and use it later.

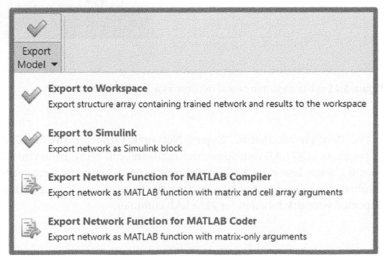

The *first* choice is to save the trained model to MATLAB workspace as a structure variable where the user has the chance to explore the properties of the trained model and summary of its training results.

The *second* choice is to export to Simulink environment. Figure 5.11 shows the exported block to Simulink platform. In brief, Simulink® is a block diagram environment for multidomain simulation and model-based design. It supports system-level design, simulation, automatic code generation, and continuous test and verification of embedded systems. Simulink provides a graphical editor, customizable block libraries, and solvers for modeling and simulating dynamic systems. It is integrated with MATLAB®, enabling the user to incorporate MATLAB algorithms into models and export simulation results to MATLAB for further analysis. This topic is out of the scope of this book and the user may refer to built-in help and relevant literature. The user can test the output y as a function of the input x. The block represents a transfer function which connects *y* to *x*. *x* is a user-defined input signal which can be edited by the user and the monitored *y* will represent the block response to the input load, *x*. The run time has a start and an end point for carrying out simulation. The user may step forward and backward once at a time or let it go to completion (i.e., the end point).

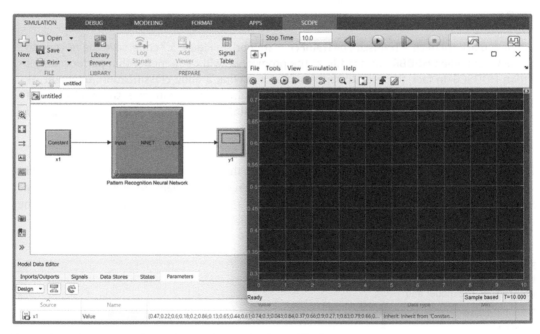

Figure 5.11 The exported neural network as a simulation block.

The *third* choice, that is, "**Export Network Function for MATLAB Compiler**" will export network as MATLAB function with matrix and cell array arguments. It is similar to the last and fourth choice; however, it allows the user to enter tagged or time-spanned input matrices as well as obtaining tagged output response matrix. Figure 5.12 shows a portion of the m-file code for the exported network function for MATLAB compiler.

```
function [Y,Xf,Af] = myNeuralNetworkFunction(X,~,~)
%MYNEURALNETWORKFUNCTION neural network simulation function.
% [Y] = myNeuralNetworkFunction(X,~,~) takes these arguments:
%  X = 1xTS cell, 1 inputs over TS timesteps
%  Each X{1,ts} = 35xQ matrix, input #1 at timestep ts.
% and returns:
%  Y = 1xTS cell of 1 outputs over TS timesteps.
%  Each Y{1,ts} = 2xQ matrix, output #1 at timestep ts.
% where Q is number of samples (or series) and TS is the number of timesteps.
```

Figure 5.12   A portion of the third option m-code file, exported network function for MATLAB compiler.

Figure 5.13 shows a portion of the exported m-file for "**Export Network Function for MATLAB Coder**" choice, shown last in Figure 5.10. Of course, one may change the function name and save it to a file holding the same function name. For example, I saved the function as **HDCNN3.m** file.

NOTE

Do not forget to save the file anywhere within MATLAB defined paths, or add its location to MATLAB path via "**Set Path**" button found in "**ENVIRONMENT**" group of MATLAB Toolstrip.

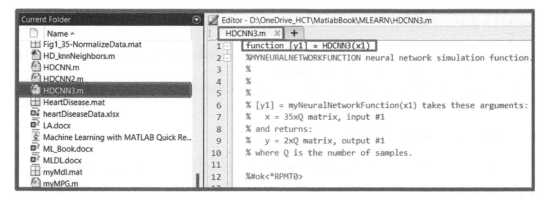

Figure 5.13   A portion of the exported m-file function holding the name of the function specifically made for MATLAB coder.

   To make use of the saved file we can try showing the predicted value for the first four observations as shown below:

>> YP4first4=HDCNN3(heartData3(1:4,:)')

YP4first4 =

0.7366	0.0463	0.0547	0.7603
0.2634	0.9537	0.9453	0.2397

Compare with the observed HDDum values

>> **YO4first4R=HDDum(1:4,:)'**

YO4first4R =

```
1   0   0   1
0   1   1   0
```

Notice that the model predicts well as the non-integer predicted values will be rounded as either *0* or *1*, as shown below:

>> **YP4first4=round(HDCNN3(heartData3(1:4,:)'))**

YP4first4 =

```
1   0   0   1
0   1   1   0
```

Command-Based Feed-Forward Neural Network Classification: Heart Data

1. Initialize a neural network with a given number of hidden-layer neurons.

>> **net = patternnet(3);**

Alternatively, we can specify different numbers of hidden layers and the number of neurons in each layer by providing a vector as input to **fitnet**.

For example, the following command creates a network with three hidden layers with 9, 6, and 4 neurons, respectively.

>> **net = fitnet([9 6 4]);**

We can then view a schematic diagram of the neural network using the view function.

>> **view(net)**

Figure 5.14 shows a view of the created neural network called net.

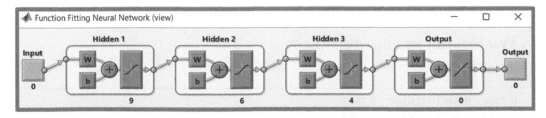

Figure 5.14   View of the architecture of the neural network called **net**.

2. Set the data set ratios and other properties of the network object variable.

>> net.divideParam.trainRatio = pTr/100;

>> net.divideParam.valRatio = pVa/100;

>> net.divideParam.testRatio = pTe/100;

The percentage values: **pTr + pVa + pTe**, = *100*.

3. Train the network. *Remember that each column of the matrix X is an observation.*

>> [net,tr] = train(net,X',targets');

The second output, **tr**, is a structure containing detailed information about the training, test, and validation data.

4. Use the trained network to make predictions on new data.

>> Preds = net(Xnew');

5. See the results by plotting a confusion matrix.

>> plotconfusion(Target, Preds)

NOTE

The output **Preds** is a matrix showing the value of each output neuron for each observation. Typically, each observation is classified according to which output neuron has the largest value.

You can use the index returned by the **max** function to convert the dummy variable output to a numeric equivalent, as demonstrated in Figure 5.15.

>> [~,PredNum] = max(Preds);

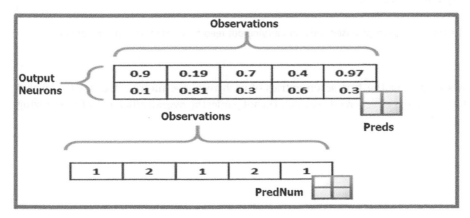

Figure 5.15 Conversion of the dummy variable output to a numeric equivalent, utilizing the index returned by the **max** function.

Figure 5.16 shows the previous example used in **nprtool** but this time, the command prompt is used instead.

```
%% Load the data
heartData = readtable('heartDiseaseData.xlsx','Format', 'auto');
heartData.HeartDisease = categorical(heartData.HeartDisease);
numData = heartData{:,1:11};
heartData = convertvars(heartData,12:21,"categorical");
% Extract the response variable
HD = heartData.HeartDisease;
HDDum=dummyvar(HD);
%Convert categorical variables to numeric dummy variables
[heartData2 Varnames]=cattbl2mat(heartData(:,1:end-1));
heartData3 = table2array(heartData2);
[r,c] = size(heartData3) ;
P = 0.70 ;
idx = randperm(r) ;
XTrain = heartData3(idx(1:round(P*r)),:) ;
XTest = heartData3(idx(round(P*r)+1:end),:) ;
YTrain = HDDum(idx(1:round(P*r)),:);
YTest = IIDDum(idx(round(P*r)+1:end),:) ;
net = patternnet(10);
net.trainFcn = 'trainbr'
net.divideParam.trainRatio = 70/100;
net.divideParam.valRatio = 15/100;
net.divideParam.testRatio = 15/100;
[net,tr] = train(net, heartData3', HDDum');
Preds = net(XTest');
%Figure 5.18 right is generated
plotconfusion(YTest',Preds)
% PredNum, row vector, with either 1 or 2 value, similar to 0 or 1 classification.
[~,PredNum] = max(Preds);
```

Figure 5.16 The command prompt-based code for carrying out feed-forward classification network modeling.

Table 5.2 shows results for the first seven entries of each response matrix. Asterisk (*) labeled values mean the model did not correctly predict. Do not panic; the overall efficiency of recognizing all cases is about 97.7%.

Table 5.2   The first seven entries of each of the response matrices. Asterisk (*) cases are misjudged.

Matrix	1st	2nd	3rd	4th	5th	6th	7th
HD(1:7)'	false	true	true	false	false	false	true
HDDum(1:7,:)'	1	0	0	1	1	1	0
	0	1	1	0	0	0	1
Preds(:,1:7)	0.9992	0.0000	0.0003	0.0133	0.0000	0.9948	0.9805
	0.0008	1.0000	0.9997	0.9867	1.0000	0.0052	0.0195
PredNum(1:7)	1	2	2	2*	2*	1	1*

Figure 5.17 shows the neural network training (**nntraintool**) which will pop up upon executing the above code. The user may click on any of the plot buttons to generate the corresponding plot (left). The workspace variables are shown on the *right* side.

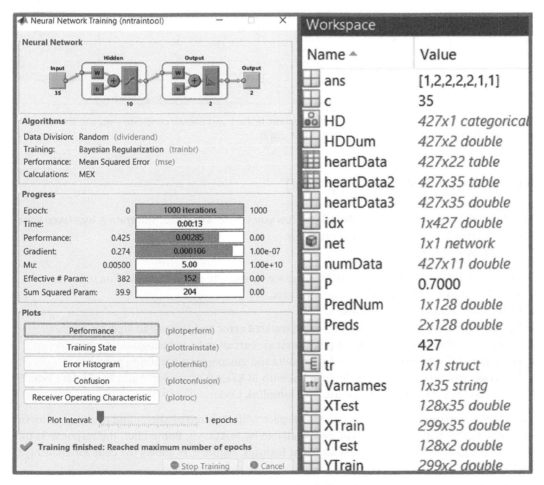

Figure 5.17   The neural network training (**nntraintool**) main window where the user can create plots for the model goodness (*left*) and the workspace variables (*right*).

If, for example, the "**Confusion**" button is clicked on, then Figure 5.18 (*left*) will be the result and the confusion matrix for predicting the response variable for the newly tested matrix data, **XTest**, is shown on the *right*. Notice that the overall score for correctly predicting the class type of observed data is about 96.0%. Other steps have at least 75%.

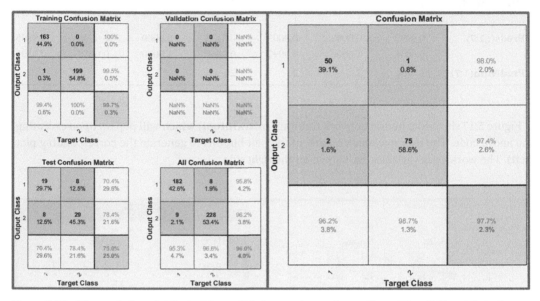

Figure 5.18 The confusion plot for training, validation, testing, and overall condition (*left*) and that of predicting the response of newly tested data, **XTest** (*right*).

Neural Network Regression (nftool)

The **Neural Network Fitting** (**nftool**) app lets you create, visualize, and train a two-layer feed-forward network to solve data fitting problems.

Using this app, we can:

- Import data from file, the MATLAB® workspace, or use one of the example data sets.
- Split data into training, validation, and test sets.
- Define and train a neural network.
- Evaluate network performance using mean squared error and regression analysis.
- Analyze results using visualization plots, such as regression fit or histogram of errors.
- Generate MATLAB scripts to reproduce results and customize the training process.
- Generate functions suitable for deployment with MATLAB Compiler™ and MATLAB Coder™ tools, and export to Simulink® for use with Simulink Coder.

Before we use **nftool**, let us populate the workspace with required predictor and response numeric matrices. Let us make use of big car data available by MATLAB. Remember, the response is the mileage per gallon, **MPG** for a given set of car features. Figure 5.19 shows the code for creating **X** and **Y** data.

```
load carbig
% X is made of four numeric predictors
X = [Weight,Horsepower,Acceleration,Displacement];
Y=MPG;
XY=[X Y];
temp=rmmissing(XY);
%Standardize predictor data
[Xnum, mu,sigma]=zscore(temp(:,1:end-1));
Ynum=temp(:,end);
```

Figure 5.19 Creation of **X** and **Y** data to be used in **nftool** application.

To open the regression learner App, we can either use MATLAB Toolstrip: On the **Apps** tab, under "**Machine Learning and Deep Learning**" category, click on "**Neural Net Fitting**" app icon, or use MATLAB command prompt.

At the command prompt, key in

>>nftool

The main window will pop-up, a portion is shown in Figure 5.20 (*left*). Upon clicking on "**Import**" button, Figure 5.20 (left) also shows the choices on how to import the data to be dealt with. Select the first option and the "**Import Data from Workspace**" window (*right*) will pop-up where we have to select **Xnum** dataset which contains 4 predictor columns (or features) with 392 observations. Hence, select the data as row-wise which means we have 392 observations with 4 features, known as predictors.

Click on "**OK**" button to close the wizard. **nftool** is ready for the training step. Notice that **nftool** will allow the user to make the desired partition of the original data into training, validation, and test subsets and the layer size, which is by default equal to 10, can be modified, as well.

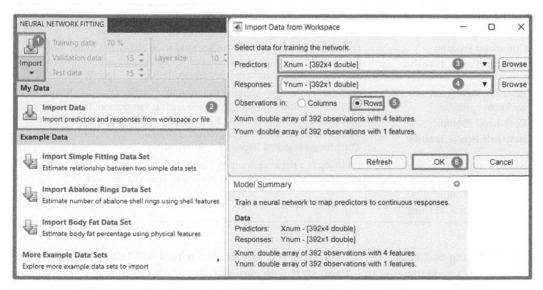

Figure 5.20 A portion of the main window of **nftool** (*left*) while showing how to import data from the workspace (*right*).

If we click on "**Train**" drop-down menu we will notice that **nftool** gives us three choices to select from and are shown in Figure 5.21 where it gives a quick glimpse about the main difference in convergence rate/memory requirement among the three algorithms.

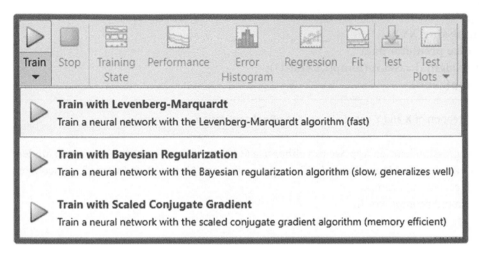

Figure 5.21   Training the regression model with different algorithms.

Table 5.3 gives more information about each algorithm, as quoted from MATLAB offline help. At the command prompt level, the algorithm train calls the function indicated by **net.trainFcn**, using the training parameter values indicated by **net.trainParam**.

Table 5.3   Network training algorithms (functions).

Algorithm	Description
Levenberg-Marquardt backpropagation Command Prompt: **net.trainFcn = 'trainlm'**	**trainlm** is a network training function that updates weight and bias values according to Levenberg-Marquardt optimization. **trainlm** is often the fastest backpropagation algorithm in the toolbox, and is highly recommended as a first-choice supervised algorithm, although it does require more memory than other algorithms.
Bayesian regularization backpropagation Command Prompt: **net.trainFcn = 'trainbr'**	**trainbr** is a network training function that updates the weight and bias values according to Levenberg-Marquardt optimization. It minimizes a combination of squared errors and weights, and then determines the correct combination so as to produce a network that generalizes well. The process is called Bayesian regularization.
Scaled conjugate gradient backpropagation Command Prompt: **net.trainFcn = 'traincg'**	**trainscg** is a network training function that updates weight and bias values according to the scaled conjugate gradient method.

After clicking on "**Train**" button, shown in the Top Toolbar, **nftool** will initiate the algorithm called Levenberg-Marquardt and it will start with zero number of epochs and it will gradually increment the size of epochs until the solution converges. As can be seen in Figure 5.22, upon converging to a final solution, **nftool** reports **MSE** (measure of performance) and **R** of the neural network regression model for the training, validation, and test data subsets.

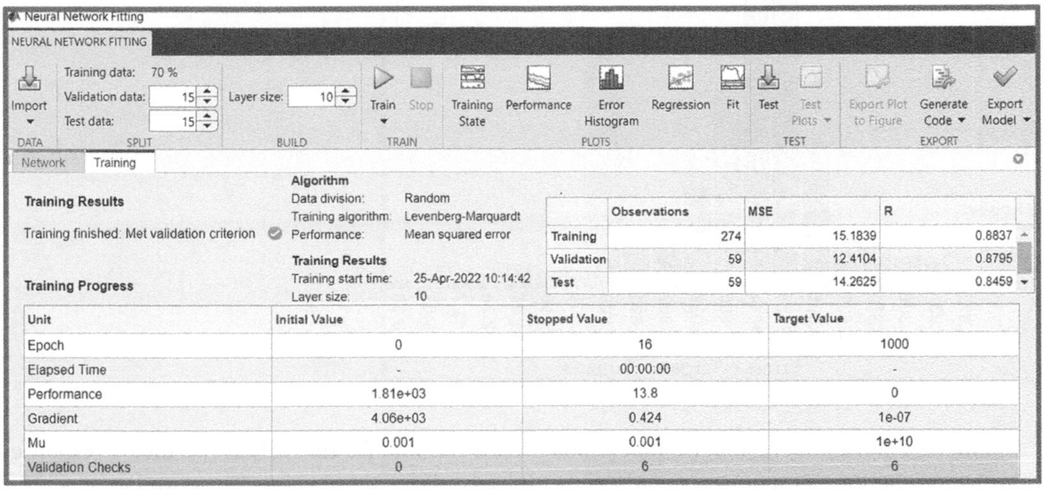

Figure 5.22 The results of the neural network regression model for the training, validation, and test data subsets.

Under "**Plots**" tab in the main toolbar, the user may click on such buttons to create the corresponding plot. For example, Figure 5.23 shows the absolute error $\left(Y_{obs} - Y_{pred}\right)$ histogram with 20 bins for the regressed model. Notice that the absolute error can be negative, zero, or positive, depending on the difference between observed and predicted response value.

The blue bars represent training data; the green ones represent validation data; and the red ones represent testing data. The histogram provides an indication of outliers, which are datapoints where the fit is significantly worse than most of the data. It is a good idea to check the outliers to determine if the data is poor, or if those datapoints are different than the rest of the data set. If the outliers are valid datapoints, but are unlike the rest of the data, then the network is extrapolating for these points. You should collect more data that looks like the outlier points and retrain the network.

NOTE

If you are unhappy with the network performance, you can do one of the following:

1) Train the network again.
2) Increase the number of hidden neurons.
3) Use a larger training data set.

If performance on the training set is good but the test set performance is poor, this could indicate the model is overfitting. Reducing the number of neurons can reduce the overfitting.

You can also evaluate the network performance on an additional test set. To load an additional test data to evaluate the network with, in the "**TEST**" section, click on **Test** button. Then you can proceed following the same procedure as that of the training step. After testing the model, you can explore the **Model Summary** which displays the additional test results. You can also generate plots to analyze the additional test data results.

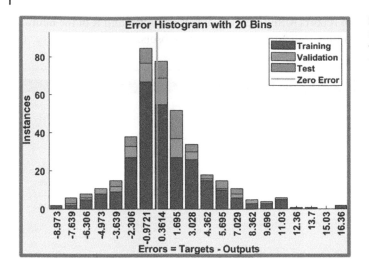

Figure 5.24 shows the regression plot which gives a formula for the output (predicted response value) as a function of input which is the target (observed) response value. The correlation coefficient is also shown for training, validation, testing, and overall.

For a perfect fit, the data should fall along a 45-degree line (i.e., the dashed line), where the network outputs are equal to the responses. For this problem, the fit is reasonably good for all of the data sets. If you require more accurate results, you can retrain the network by clicking **Train** again. Each training will have different initial weights and biases of the network, and can produce an improved network after retraining.

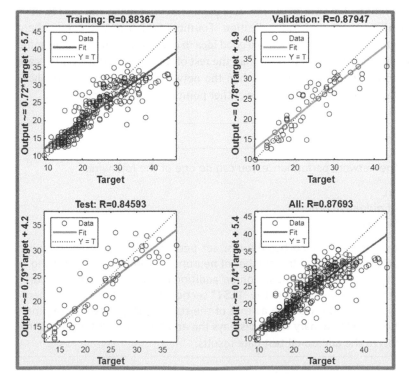

Figure 5.24 The regression model: **Output** = f(**Target**) for all phases of probing the regression model.

In MATLAB Toolstrip and under "**TEST**" group, clicking on "**Test**" button will allow the user to test the model over a new test set. This step is really not needed as the previous step divides the entire data set into training (70%), validation (15%), and test (15%) sub-sets. However, the user may attempt to try new data other than the data initially imported into **nftool** application.

In MATLAB **nftool** app and under "**EXPORT**" group, clicking on "**Generate Code**" button will allow the user to save a MATLAB m-code of the entire process of learning and creating the model. Select the first choice: "**Generate Simple Training Script**" to proceed.

Figure 5.25 shows the generated code where it allows the user to modify the neural network model parameters, like the size of the hidden layer, the training ratio, validation ratio, and test ratio. For example, I saved the file as **MPGNRLM.m**.

```matlab
% Solve an Input-Output Fitting problem with a Neural Network
% This script assumes these variables are defined:
%   Xnum - input data.
%   Ynum - target data.
x = Xnum';
t = Ynum';
% Choose a Training Function
% For a list of all training functions type: help nntrain
% 'trainlm' is usually fastest.
% 'trainbr' takes longer but may be better for challenging problems.
% 'trainscg' uses less memory. Suitable in low memory situations.
trainFcn = 'trainlm';  % Levenberg-Marquardt backpropagation.
% Create a Fitting Network
hiddenLayerSize = 10;
net = fitnet(hiddenLayerSize,trainFcn);
% Setup Division of Data for Training, Validation, Testing
net.divideParam.trainRatio = 70/100;
net.divideParam.valRatio = 15/100;
net.divideParam.testRatio = 15/100;
% Train the Network
[net,tr] = train(net,x,t);
% Test the Network
y = net(x);
e = gsubtract(t,y);
performance = perform(net,t,y)
% View the Network
view(net)
% Plots
% Uncomment these lines to enable various plots.
%figure, plotperform(tr)
%figure, plottrainstate(tr)
%figure, ploterrhist(e)
%figure, plotregression(t,y)
%figure, plotfit(net,x,t)
```

Figure 5.25 The generated simple training script or MATLAB m-code for training, learning, and validation of the network model.

To run the code, we need first to have the data available in MATLAB Workspace. In other words, we must run the code in Figure 5.19 before attempting to use the already saved m-file. Alternatively, we may bring new predictor and response data of the same type and number but, of course, with different observations, and have MATLAB workspace ready in a fashion similar to that after running the code of Figure 5.19.

At the prompt, enter the name of the saved m-file:

>> MPGNRLM

Figure 5.26 will pop-up where we can create the model-related plots, exactly similar to those shown in figures 5.23 and 5.24.

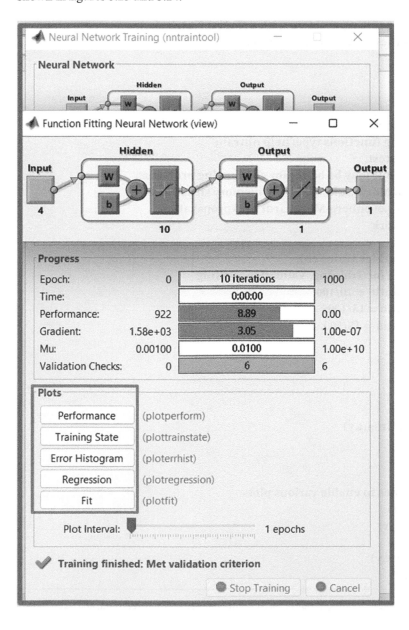

Figure 5.26 **nntraintool** pops up upon running **MPGNRLM** m-file, generated by **nftool**.

Similar to what is shown earlier in Figure 5.10, the user will have the choice to save the regression model in different four formats so that he/she can use it later. The user may refer to the section that exactly follows Figure 5.10 for reviewing such options. Here, we will show how to make use of the exported file using the fourth and last choice in Figure 5.10.

The fourth and last choice is to export the network as MATLAB function with matrix-only arguments. I exported the function and saved the model as **MPGNR.m** file by selecting the last choice shown in Figure 5.10. Figure 5.27 shows a portion of the m-file, showing at the beginning how to use the input argument **X**, a numeric matrix made of **4** rows and **Q** observations. Of course, one may enter a matrix with **4×1** as a single entry and the model will predict the value of **MPG**. The four rows stand for the rows of the *transposed* **X** data. **X'** will have **4** rows and **Q** columns.

```
function [y1] = MPGNR(x1)
%MYNEURALNETWORKFUNCTION neural network simulation function.
% [y1] = myNeuralNetworkFunction(x1) takes these arguments:
%   x = 4xQ matrix, input #1
% and returns:
%   y = 1xQ matrix, output #1
% where Q is the number of samples.
```

Figure 5.27 A portion of **MPGNR.m** file which is saved after running **nftool** for big car data.

To make use of the saved file we can try showing the predicted value for the first observation as shown below:

X1 = [3504 130 12 307];

X1 as it is written is not standardized; we need to standardize such four entries which respectively account for weight, horsepower, acceleration, and displacement value of a big car. Refer to Figure 5.19 where we have:

[Xnum, mu,sigma]=zscore(temp(:,1:end-1));

In addition to transforming temp data in standardized **Xnum** data, we also report both the mean and standard deviation of each column. Thus, **X1** values can be transformed into standardized values by:

$$X1_{stndrd} = \frac{X1 - mu}{sigma}$$

We use the dot division (./) notation which means carrying out division element-wise not division of a matrix by another matrix.

X1stndrd=(X1-mu)./sigma

X1stndrd = 0.6197   0.6633   -1.2836   1.0759

Which are the same as

Xnum1stRow=Xnum(1,:)

Xnum1stRow = 0.6197   0.6633   -1.2836   1.0759

Transpose first **X1stndrd** before using as an argument in the model.

>> MPG1=MPGNR(X1stndrd')

MPG1 = 17.8677

Compare with the observed MPG value

>>MPG1Obs=Ynum(1)

MPG1Obs = 18

$$PRE(\%) = \frac{|Observed - Predicted|}{Observed} \times 100\% = \frac{|18.0 - 17.87|}{18.0} \times 100\% = 0.72\% \ll 10\%$$

Obviously, the model predicts well the **MPG** value with a very reasonable accuracy.

NOTE

If we have a new set of predictor data named **Xnew (n×4 matrix)**, with a sufficiently large number of **n** observations, then we can directly use it as in the following statement:

>> MPGNew=MPGNR(zscore(Xnew)');

However, if we have a small number of observations, then we need to standardize them using the reported values of **mu** and **sigma** for the trained predictor data **Xnum**.

At the end, it is worth-mentioning that I tried the neural network regression model without predictor data standardization and the model is named **MPGSCGNOSTD**. Let us try it on:

>>X1 = [3504 130 12 307];

>>MPG1=MPGSCGNOSTD(X1')

MPG1 = 17.9590

which is even more accurate than MPG1 = 17.8677 (with standardization), given that the observed MPG, MPG1Obs, = 18.

To double check let us consider another datapoint, that is, X2.

>>X2 = [3200 140 14 305];

>> MPG2= MPGSCGNOSTD(X2')

MPG2= 17.9091 which is much closer to the observed value, MPG2Obs, = 18.0

Command-Based Feed-Forward Neural Network Regression: Big Car Data

Let us populate the workspace with required predictor and response numeric matrices. Let us make use of big car data available by MATLAB. Remember, the response is the mileage per gallon, **MPG** for a given set of car features. Figure 5.28 shows the code for creating **X** and **Y** data and using command-based feed-forward neural network regression to fit big car data.

```
load carbig
% X is made of four numeric predictors
X = [Weight,Horsepower,Acceleration,Displacement];
Y=MPG;
XY=[X Y];
temp=rmmissing(XY);
% Standardize the predictor data
Xnum=zscore(temp(:,1:end-1));
Ynum=temp(:,end);
%% Initialize & train the neural network
net = fitnet([8 12]);
net.layers{1}.transferFcn = 'logsig';
net.layers{2}.transferFcn = 'radbas';
cD=Xnum';
eC=Ynum';
[net,tr] = train(net,cD,eC);
%% Predict response & evaluate
% tr struct has three indices for training, validation, and test to partition the X data.
econPred = net(cD(:,tr.testInd));
evaluateFit2(eC(tr.testInd),econPred,'Neural Nets', 'MPG');
```

Figure 5.28 Creation of **X** and **Y** data and using command-based feed-forward neural network regression to fit the big car data.

Figure 5.29 (*left*) shows the neural network training (**nntraintool**) which will pop up upon executing the above code. The user may click on any of the plot buttons to generate the corresponding plot. The workspace variables are shown on the *right* side.

If, for example, the "**Error Histogram**" button is clicked on, Figure 5.30 will be the result where it shows the distribution of (absolute error $= Y_{Obs} - Y_{Pred}$) around zero absolute error for all datapoints lumped in 20 bins.

Figure 5.31 shows the regression plot which gives a formula for the output (predicted response value) as a function of input which is the target (observed) response value. The correlation coefficient is also shown for training, validation, testing, and overall. Again, for a perfect fit, the data should fall along a 45-degree line (i.e., the dashed line), where the network outputs are equal to the responses.

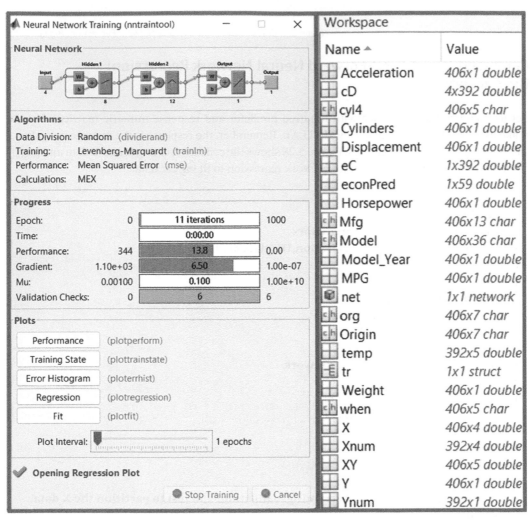

Workspace	
Name ▲	Value
Acceleration	406x1 double
cD	4x392 double
cyl4	406x5 char
Cylinders	406x1 double
Displacement	406x1 double
eC	1x392 double
econPred	1x59 double
Horsepower	406x1 double
Mfg	406x13 char
Model	406x36 char
Model_Year	406x1 double
MPG	406x1 double
net	1x1 network
org	406x7 char
Origin	406x7 char
temp	392x5 double
tr	1x1 struct
Weight	406x1 double
when	406x5 char
X	406x4 double
Xnum	392x4 double
XY	406x5 double
Y	406x1 double
Ynum	392x1 double

Figure 5.29 The neural network training (**nntraintool**) main window where the user can create plots for the model goodness (*left*) and the workspace variables (*right*).

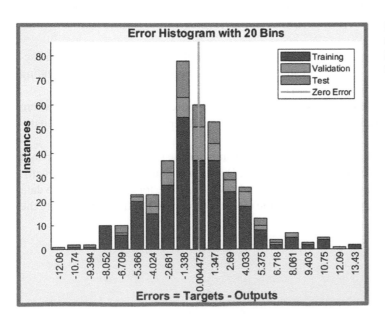

Figure 5.30 The absolute error histogram with 20 bins (bars) for the neural network regressed model.

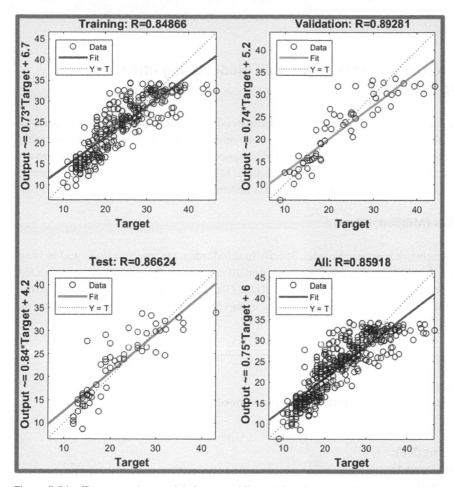

Figure 5.31 The regression model: **Output** = f(**Target**) for all phases of probing the regression model.

Figure 5.32 shows observed vs. predicted **MPG** and the model goodness parameters.

Figure 5.32 Observed vs. predicted **MPG** using feed-forward neural network regression model, showing its MSE and MAPE.

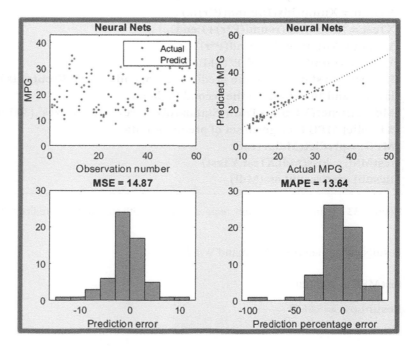

Training the Neural Network Regression Model Using fitrnet Function: Big Car Data

Load the **carbig** data set. We train a neural network regression model by passing the **XTrain** data to the fitrnet function. **Mdl** is a trained neural network regression model. We can use dot notation to access the properties of **Mdl**. We evaluate the performance of the regression model on **XTest** set by computing the test mean squared errors (MSE). Smaller MSE values indicate better performance. To predict the response for a new **Xdata** set of big car features, we use:

>> Yval =predict (Mdlbest, Xdata);

We train a neural network regression model. Specify to *standardize the predictor data,* and to have 30 outputs in the first fully connected (**FC**) layer and 10 outputs in the second fully connected layer. *By default, both layers use a rectified linear unit (ReLU) activation function.* We can change the activation functions for the fully connected layers by using the **Activations** name-value argument. Figure 5.33 shows the code for such steps.

```
load carbig
% X is made of four numeric predictors
X = [Weight,Horsepower,Acceleration,Displacement];
Y=MPG;
XY=[X Y];
temp=rmmissing(XY);
Xnum=temp(:,1:end-1);
Ynum=temp(:,end);
[r,c] = size(Xnum) ;
P = 0.70 ;
idx = randperm(r)  ;
XTrain = Xnum(idx(1:round(P*r)),:) ;
XTest = Xnum(idx(round(P*r)+1:end),:) ;
YTrain = Ynum(idx(1:round(P*r)),:);
YTest = Ynum(idx(round(P*r)+1:end),:) ;
% Specify to standardize the predictor data, and to have 30 outputs in the first FC
layer % and 10 outputs in the second FC layer.
Mdl = fitrnet(XTrain,YTrain,"Standardize",true, "LayerSizes",[30 10]);
%Predict MPG for a given set of predictor data.
MPGP=predict(Mdl,XTest);
testMSE = loss(Mdl,XTest,YTest)
ResubLoss=resubLoss(Mdl)
```

Figure 5.33 Using neural network regression model with standardized predictors and specified layer sizes.

Results are copied from Command Window:

testMSE = 29.1120

ResubLoss = 5.8658

Figure 5.34 shows the generated workspace variables for Figure 5.33 code.

Workspace				Workspace			
Name ▲	Value	Size	Min	Name ▲	Value	Size	Min
Acceleration	*406x1 double*	406x1	8	P	0.7000	1x1	0.7000
c	4	1x1	4	r	392	1x1	392
cyl4	*406x5 char*	406x5		ResubLoss	5.8658	1x1	5.8658
Cylinders	*406x1 double*	406x1	3	temp	*392x5 double*	392x5	8
Displacement	*406x1 double*	406x1	68	testMSE	29.1120	1x1	29.1120
Horsepower	*406x1 double*	406x1	NaN	Weight	*406x1 double*	406x1	1613
idx	*1x392 double*	1x392	1	when	*406x5 char*	406x5	
Mdl	*1x1 RegressionNeuralN...*	1x1		X	*406x4 double*	406x4	NaN
Mfg	*406x13 char*	406x13		Xnum	*392x4 double*	392x4	8
Model	*406x36 char*	406x36		XTest	*118x4 double*	118x4	8.5000
Model_Year	*406x1 double*	406x1	70	XTrain	*274x4 double*	274x4	8
MPG	*406x1 double*	406x1	NaN	XY	*406x5 double*	406x5	NaN
MPGP	*118x1 double*	118x1	0.0512	Y	*406x1 double*	406x1	NaN
org	*406x7 char*	406x7		Ynum	*392x1 double*	392x1	9
Origin	*406x7 char*	406x7		YTest	*118x1 double*	118x1	11

Figure 5.34 Workspace variables generated by executing Figure 5.33 code.

We can access the weights and biases for the FC layers of the trained model by using the **LayerWeights** and **LayerBiases** properties of **Mdl**. The first two elements of each property correspond to the values for the first two **FC** layers, and the third element corresponds to the values for the final **FC** layer for regression.

For example, we can display the weights and biases for the first FC layer:

>> firstFClayerW=Mdl.LayerWeights{1}

```
firstFClayerW = [30×4] matrix
    2.0248  -1.7355   1.5011   1.5368
   -0.1290  -0.8001   0.5087   0.3982
    0.5250   1.4369  -0.0842  -0.8419
   -1.5273  -3.2908   0.6677   3.9806
       ↓       ↓       ↓       ↓ [30×4] matrix
```

>> firstFClayerB = Mdl.LayerBiases{1}

```
firstFClayerB =
    0.4185
   -2.3402
    1.2022
       ↓ [30×1] matrix
```

The final FC layer has one output. The number of layer outputs corresponds to the first dimension of the layer weights and layer biases.

>>FinalFC_layer_W=Mdl.LayerWeights{end}

FinalFC_layer_W =

2.3131 -1.7789 3.0548 -0.3350 0.5285 0.2137 4.9660 -1.1043 4.7042 -0.1684

>> FinalFC_layer_W_Size=size(Mdl.LayerWeights{end})

FinalFC_layer_W_Size =

  1   10

So, the final **FC** layer has one output with ten weight factors.

>>FinalFC_layer_B=Mdl.LayerBiases{end}

FinalFC_layer_B =

  3.1191

So, the final **FC** layer has a single bias value of 3.1191.

>>FinalFC_layer_B_Size=size(Mdl.LayerBiases{end})

FinalFC_layer_B_Size =

  1   1

The final **FC** layer has one output which is one bias factor.
Let us look at the second FC layer.

>> secondFClayer_W_Size=size(Mdl.LayerWeights{2});

secondFClayer_W_Size =

  10   30

NOTE

The size of the weight matrix of the second **FC** layer is 10×30. Let us explain the concept of neuron network **FC** layer's signal transmission mechanism.

The source or input signal will be made of 4 sub-signals; one for each of the 4 predictors. Those 4 different sub-signals will be channeled or pushed into each of receiving 30 neurons found in the first **FC** layer, totaling 4×30=120 possible permutated incomes. We need to assign one weight function for each potential income. This explains why the size of weight factors in the 1st **FC** layer is 30×4 and that of bias is 30×1 (one bias is added to the sum of the multiplication of weight by signal). As explained at the start of this chapter, all neurons in a fully connected layer connect to all the neurons in the previous layer. The single output of each neuron will be transmitted as an input signal to each of the ten neurons found in the second **FC** layer. Thus, each neuron will receive 30 different signals; hence, each is multiplied by a different weight

---

NOTE (Continued)

---

factor and a bias is added to the sum of multiplication. So, we talk about 30 weight factors by 10 nodes. This explains why the size of weight factors in the 2nd **FC** layer is now 10×30 and that of bias is 10×1 (one bias is added to the sum of the multiplication of weight by signal).

Finally, notice that the weigh factor can take on a positive or negative value and by default both layers use a rectified linear unit (**ReLU**) activation function. A **ReLU**-based **FC** layer performs a threshold operation to each element of the input, where any value less than zero is set to zero.

Finding the Optimum Regularization Strength for Neural Network Using Cross-Validation: Big Car Data

We use the same big car data. We will assess the cross-validation loss of neural network models with different regularization strengths, and choose the regularization strength corresponding to the best performing model; i.e., finding the best value of lambda which will minimize the model error. Figure 5.35 shows the code for accomplishing the best regularization strength matching with the best model performance.

To predict the response for a new **Xdata** set of big car features, we use:

`>> Yval =predict (Mdlbest, Xdata);`

```
load carbig
tbl = table(Acceleration,Displacement,Horsepower,Weight,MPG);
cars = rmmissing(tbl);
%Xnum=cars(:,1:4);
%Ynum=cars.MPG;
[r,c] = size(cars) ;
rng("default")
n = size(cars,1);
%cvp = cvpartition(n,"KFold",5);
cvp = cvpartition(n,"Holdout",0.20);
trainingIdx = training(cvp); % Training set indices
carsTrain = cars(trainingIdx,:);
testIdx = test(cvp); % Test set indices
carsTest = cars(testIdx,:);
lambda = (0:0.2:5)*1e-2;
cvloss = zeros(length(lambda),1);
for i = 1:length(lambda)
cvMdl = fitrnet(cars,"MPG","Lambda",lambda(i),"CVPartition",cvp,"Standardize",
true);
cvloss(i) = kfoldLoss(cvMdl);
end
```

Figure 5.35 Finding the good regularization strength for neural network using cross-validation for big car data.

```
% Find best lambda, corresponding to the lowest cross-validation MSE.
plot(lambda,cvloss)
xlabel("Regularization Strength")
ylabel("Cross-Validation Loss")
[~,idx] = min(cvloss);
bestLambda = lambda(idx)
% Train the model at bestlambda
Mdlbest = fitrnet(cars,"MPG","Lambda",bestLambda,"Standardize",true);
testMSE = loss(Mdlbest,carsTest,"MPG")
%Predict MPG value for a given Predictor data.
MPGP=predict(Mdlbest,carsTest);
```

Figure 5.35 (Cont'd)

bestLambda = 0.0040

testMSE = 9.4599

Figure 5.36 shows the cross-validation loss versus the regularization strength, lambda. The minimum loss occurs at the bottom left of the figure where lambda = 0.0045.

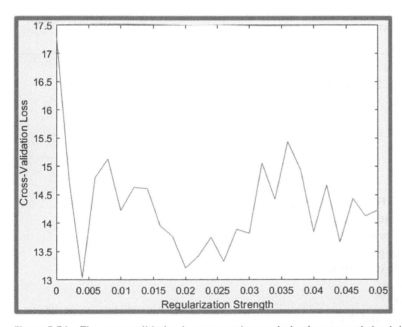

Figure 5.36 The cross-validation loss versus the regularization strength, lambda.

Figure 5.37 shows the workspace variables, including **bestLambda** and minimum **testMSE**.

Name ▲	Value	Size	Min
Acceleration	406x1 double	406x1	8
bestLambda	0.0040	1x1	0.0040
c	5	1x1	5
cars	392x5 table	392x5	
carsTest	78x5 table	78x5	
carsTrain	314x5 table	314x5	
cvloss	26x1 double	26x1	13.0387
cvMdl	1x1 RegressionPartitionedModel	1x1	
cvp	1x1 cvpartition	1x1	
cyl4	406x5 char	406x5	
Cylinders	406x1 double	406x1	3
Displacement	406x1 double	406x1	68
Horsepower	406x1 double	406x1	NaN
i	26	1x1	26
idx	3	1x1	3
lambda	1x26 double	1x26	0

Name ▲	Value	Size	Min
Mdlbest	1x1 RegressionNeuralNetwork	1x1	
Mfg	406x13 char	406x13	
Model	406x36 char	406x36	
Model_Year	406x1 double	406x1	70
MPG	406x1 double	406x1	NaN
MPGP	78x1 double	78x1	8.7288
n	392	1x1	392
org	406x7 char	406x7	
Origin	406x7 char	406x7	
r	392	1x1	392
tbl	406x5 table	406x5	
testIdx	392x1 logical	392x1	
testMSE	9.4599	1x1	9.4599
trainingIdx	392x1 logical	392x1	
Weight	406x1 double	406x1	1613
when	406x5 char	406x5	

Figure 5.37 The workspace variables, including **bestLambda** and minimum **testMSE**.

Custom Hyperparameter Optimization in Neural Network Regression: Big Car Data

Figure 5.38 shows the code to accomplish the automatic hyper parameter optimization for a neural network regression model using big car data. To make use of the model for predicting the MPG values of a new **Xdata** set which represents the big car features, we use:

>> Yval =predict (Mdl, Xdata);

```
load carbig
cars = table(Acceleration,Displacement,Horsepower, Weight,MPG);
rng("default") % For reproducibility of the data partition
c = cvpartition(length(MPG),"Holdout",0.20);
trainingIdx = training(c); % Training set indices
carsTrain = cars(trainingIdx,:);
testIdx = test(c); % Test set indices
carsTest = cars(testIdx,:);
%List the hyperparameters available for this problem of fitting the MPG response.
params = hyperparameters("fitrnet",carsTrain,"MPG");
for ii = 1:length(params)
disp(ii);disp(params(ii));
end
params(1).Range = [1 5];
params(10).Optimize = true;
params(11).Optimize = true;
for ii = 7:11
```

Figure 5.38 Carrying out hyper parameter optimization for the neural network regression model.

```
    params(ii).Range = [1 400];
end
rng("default") % For reproducibility
Mdl = fitrnet(carsTrain,"MPG","OptimizeHyperparameters",params, ...
    "HyperparameterOptimizationOptions", ...
    struct("AcquisitionFunctionName","expected-improvement-plus", ...
    "MaxObjectiveEvaluations",60))
testMSE = loss(Mdl,carsTest,"MPG")
% To predict the response for a given Xdata set
MPGP=predict(Mdl,carsTest);
```

Figure 5.38 (Cont'd)

The result for the model loss is:

testMSE = 17.7807

Figure 5.39 (*left*) shows the hyper parameter optimization (i.e., best lambda value) for a varied layer size of the neural network regression model and (*right*) the workspace variables.

Figure 5.40 shows how both the minimum observed and minimum estimated objective function change with number of iterations until the estimated asymptotically approaches the observed one. Notice that **params** is 11x1 **optimizableVariable** class type, as seen in Figure 5.39.

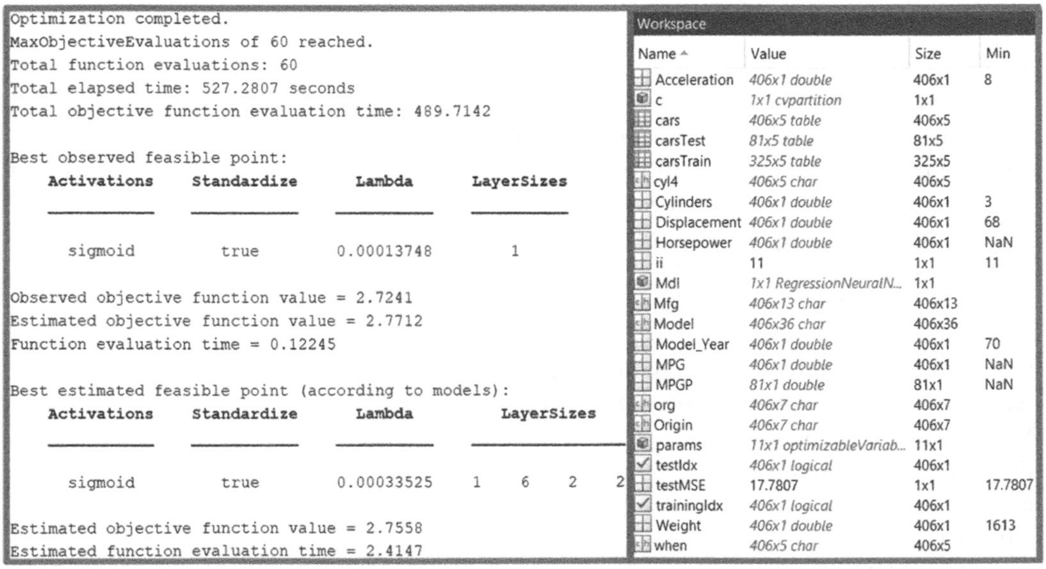

Figure 5.39 (*left*) The hyper parameter optimization (i.e., best lambda value) for a varied layer size of the neural network regression model and the workspace variables (*right*).

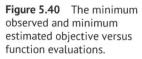

Figure 5.40 The minimum observed and minimum estimated objective versus function evaluations.

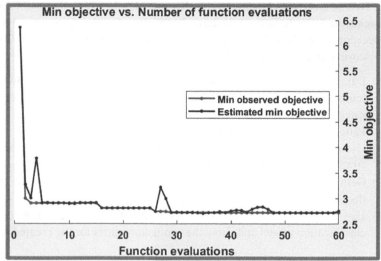

CHW 5.1   Mushroom Edibility Data

This data set (>>Mushroom=readtable('Mushroom.csv','Format', 'auto');) includes several samples of mushrooms. The mushrooms are classified as either edible or poisonous. The table mushroom contains several features of a mushroom's attributes and the corresponding response, edibility. All of the predictor variables are categorical.

1) Data Preparation: You must convert categorical columns into numeric types before attempting to use MATLAB Neural Network Pattern Recognition (**nprtool**) Application.
   HINT: Create a categorical table from Mushroom using **convertvars** function which requires two inputs: Mushroom table and its variable names. Then use **cattbl2mat** and **table2array** function to end up with numeric X and Y data.
2) Use **nprtool** to create a neural network classification model and shows different plots demonstrating model goodness and performance.
3) Generate the simple training script using **nprtool** to reproduce the training process and create the neural network classification model.
4) Use the Command-based code to create from scratch a feed-forward neural network classification model and show the confusion matrix for the created model.

CHW 5.2   1994 Adult Census Income Data

This data set (>>load census1994) classifies yearly salary as > $50K or otherwise based on demographic data. There are 14 predictors: 6 numeric and 8 categorical. The set contains the following two MATLAB tables:

adultdata: 32,561×15 tabular array. The first 14 variables correspond to heterogeneous predictors and the last variable contains the class labels (>50K or <=50K). **adulttest**: 16,281×15 tabular array. The first 14 variables correspond to predictors and the last variable contains the class labels.

This data set was donated to the UCI Machine Learning Repository by: Ron Kohavi and Barry Becker in 1994. Reference: Kohavi, R. "Scaling Up the Accuracy of Naive-Bayes Classifiers: A Decision-Tree Hybrid." Proceedings of the Second International Conference on Knowledge Discovery and Data Mining, 1996.

1) Data Preparation: Combine both adultdata (32,561×15) and adulttest (16,281×15) into adult-temp (48,842×15). Use adulttemp as the source predictor and response data. You must convert categorical columns into numeric types before attempting to use MATLAB Neural Network Pattern Recognition (**nprtool**) Application.
2) Use **nprtool** to create a neural network classification model and shows different plots demonstrating model goodness and performance.
3) Generate the simple training script using **nprtool** to reproduce the training process and create the neural network classification model.
4) Use the Command-based code to create from scratch a feed-forward neural network classification model and show the confusion matrix for the created model.

CHW 5.3   Breast Cancer Diagnosis

Many features are computed from digitized images of breast masses which describe characteristics of the cell nuclei. The diagnosis of the mass is classified as either benign (**B**) or malignant (**M**).

1) Data Preparation: Combine both adultdata (32,561×15) and adulttest (16,281×15) into adult-temp (48,842×15). Use adulttemp as the source predictor and response data. You must convert categorical columns into numeric types before attempting to use MATLAB Neural Network Pattern Recognition (**nprtool**) Application.
2) Use **nftool** to create a neural network classification model and shows different plots demonstrating model goodness and performance.
3) Generate the simple training script using **nftool** to reproduce the training process and create the neural network classification model.
4) Use the Command-based code to create from scratch a feed-forward neural network classification model and show the confusion matrix for the created model.

CHW 5.4   Small Car Data (Regression Case)

The small car data (**load carsmall**) can be found in MATLAB built-in database.

1) Data Preparation: Prepare data to be admitted to nftool platform. Try the tool with and without predictor data standardization. See which case gives a more accurate prediction.
2) Use **nftool** to create a neural network regression model and shows different plots demonstrating model goodness and performance.
3) Generate the simple training script using **nprtool** to reproduce the training process and create the neural network regression model and show plots demonstrating the model goodness and performance.
4) Use the Command-based code to create from scratch a feed-forward neural network regression model and show plots demonstrating the model goodness and performance.

5) Train the neural network regression model using **fitrnet** function. Find the model loss (MSE) for testing and resubstitution loss for training.

6) Find the optimum regularization strength, lambda, for the neural network regression model using **fitrnet** function with cross-validation. Change the range of explored lambda to be able to bracket the optimum value. Train the neural network regression model using the optimum lambda. Find the model loss at the optimum lambda.

7) Hyper-optimize the neural network regression model using **fitrnet** function. Find the model loss at the optimized hyper parameters.

CHW 5.5   Boston House Price

The Boston House Price Dataset involves the prediction of a house price in thousands of dollars given details of the house and its neighborhood. It is a regression problem. There are 506 observations with 13 input variables and 1 output variable. Data available with Wiley book companion.

The variable names are as follows:

1) **Crime**: per capita crime rate by town.
2) **Zone**: proportion of residential land zoned for lots over 25,000 sq. ft.
3) **Indust**: proportion of nonretail business acres per town.
4) **CHRRiver**: Charles River dummy variable (= 1 if tract bounds river; 0 otherwise).
5) **NOX**: nitric oxides concentration (parts per 10 million).
6) **Room**: average number of rooms per dwelling.
7) **Age**: proportion of owner-occupied units built prior to 1940.
8) **Dist**: weighted distances to five Boston employment centers.
9) **RadHwy**: index of accessibility to radial highways.
10) **Tax**: full-value property-tax rate per $10,000.
11) **PTRatio**: pupil-teacher ratio by town.
12) **BlkRatio**: $1000(Bk - 0.63)^2$ where Bk is the proportion of blacks by town.
13) **LStat**: % lower status of the population.
14) **MedVal**: Median value of owner-occupied homes in $1000s.

The table called housing contains several predictor variables and a response in the last column named **MedVal**.

1) Data Preparation: Prior to any regression implementation, carry out treatment steps need to transform the original table into numeric table and also remove **NaN** values if any. Split the numeric, NaN-free table into a training set, **XTrain**, and a test set, **XTest**. So does the case with the response variable, split into **YTrain** and **YTest**, respectively.

2) Use **nftool** to create a neural network regression model and shows different plots demonstrating model goodness and performance.

3) Generate the simple training script using **nprtool** to reproduce the training process and create the neural network regression model and show plots demonstrating the model goodness and performance.

4) Use the Command-based code to create from scratch a feed-forward neural network regression model and show plots demonstrating the model goodness and performance.

5) Train the neural network regression model using **fitrnet** function. Find the model loss (MSE) for testing and resubstitution loss for training.

6) Find the optimum regularization strength, lambda, for the neural network regression model using **fitrnet** function with cross-validation. Change the range of explored lambda to be able to bracket the optimum value. Train the neural network regression model using the optimum lambda. Find the model loss at the optimum lambda.

7) Hyper-optimize the neural network regression model using **fitrnet** function. Find the model loss at the optimized hyper parameters.

6

Pretrained Neural Networks

Transfer Learning

Deep Learning: Image Networks

Deep learning is a subset of machine learning (or artificial intelligence in a broader sense) that teaches computers to do what humans naturally do: Learn from experience. Image networks are intrinsically neural networks. Deep learning employs neural networks to directly learn useful representations of features from data. Neural networks are inspired by biological nervous systems and combine multiple nonlinear processing layers using simple elements operating in parallel. Deep learning models can achieve cutting-edge accuracy in object classification, sometimes outperforming humans.

Deep Learning Toolbox includes simple MATLAB® commands for building and connecting deep neural network layers. Even without prior knowledge of advanced computer vision algorithms or neural networks, examples and pretrained networks make it simple to use MATLAB for deep learning.

Transfer learning is frequently employed in deep learning applications, where a pretrained network serves as a foundation for acquiring knowledge in a new task. The process of fine-tuning the network through transfer learning proves to be considerably faster and simpler compared to training it from the beginning. By utilizing a smaller dataset of training images, we can promptly train the network to grasp a new task. The key benefit of transfer learning lies in the fact that the pretrained network has already acquired a comprehensive set of features that can be effectively applied to various related tasks.

To choose whether to use a pretrained network or create a new deep network, refer to the trade-off scenarios in Table 6.1, as quoted from MATLAB offline help.

Table 6.1 Selection between transfer learning and training from scratch.

Feature	Use a Pretrained Network for Transfer Learning	Create a New Deep Network
Training Data	Hundreds to thousands of labeled data (small)	Thousands to millions of labeled data
Computation	Moderate computation (GPU optional)	Compute intensive (requires GPU for speed)
Training Time	Seconds to minutes	Days to weeks for real problems
Model Accuracy	Good; depends on the pretrained model	High, but can overfit to small data sets

Machine and Deep Learning Using MATLAB: Algorithms and Tools for Scientists and Engineers, First Edition.
Kamal I. M. Al-Malah.
© 2024 John Wiley & Sons, Inc. Published 2024 by John Wiley & Sons, Inc.
Companion Website: www.wiley.com/go/al-malah/machinelearningmatlab

Table 6.2 shows MATLAB deep learning tools, quoted from MATLAB offline help. Some of them will be explained in details later.

Table 6.2 MATLAB deep learning tools and description.

Tool	Description
Deep Network Designer (DND)	Build, visualize, edit, and train deep learning networks.
Experiment Manager	Create deep learning experiments to train networks under multiple initial conditions and compare the results.
Deep Network Quantizer	Reduce the memory requirement of a deep neural network by quantizing weights, biases, and activations of convolution layers to 8-bit scaled integer data types.
Reinforcement Learning Designer (Reinforcement Learning Toolbox)	Design, train, and simulate reinforcement learning agents.
Image Labeler (Computer Vision Toolbox)	Label ground truth data in a collection of images.
Video Labeler (Computer Vision Toolbox)	Label ground truth data in a video, in an image sequence, or from a custom data source reader.
Ground Truth Labeler (Automated Driving Toolbox)	Label ground truth data in multiple videos, image sequences, or lidar point clouds.
Lidar Labeler (Lidar Toolbox)	Label objects in a point cloud or a point cloud sequence. The app reads point cloud data from PLY, PCAP, LAS, LAZ, ROS and PCD files.
Signal Labeler (Signal Processing Toolbox)	Label signals for analysis or for use in machine learning and deep learning applications.

Quoted from MATLAB offline help, Table 6.3 shows pretrained networks that can be used in deep learning for either transfer learning (i.e., re-training) or any can be used as a template to build a network from scratch by modifying some of its given layers, insertion of new layers, or deletion of existing layers. The network depth is defined as the largest number of sequential convolutional or fully connected layers on a path from the input layer to the output layer. The inputs to all networks are RGB 3-D images. However, if modifications are made to the given network, it can be used, for example, for gray-scale 2-D images or even 4-D images.

We can use a pretrained image classification network that has already learned to extract powerful and informative features from natural images to learn a new task. The vast majority of pretrained networks are trained on a subset of the ImageNet database [1], which is used in the ImageNet Large-Scale Visual Recognition Challenge (ILSVRC) [2]. These networks have been trained on over a million images and can classify images into 1000 different object categories, including keyboards, coffee mugs, pencils, and a variety of animals. Transfer learning on a pretrained network is typically much faster and easier than training a network from scratch.

Table 6.3 Pretrained networks that can be used for RGB 3-D image processing.

Network	Depth	Size	Parameters (Millions)	Image Input Size
squeezenet	18	5.2 MB	1.24	227×227
googlenet	22	27 MB	7.0	224×224
inceptionv3	48	89 MB	23.9	299×299
densenet201	201	77 MB	20.0	224×224
mobilenetv2	53	13 MB	3.5	224×224
resnet18	18	44 MB	11.7	224×224
resnet50	50	96 MB	25.6	224×224
resnet101	101	167 MB	44.6	224×224
xception	71	85 MB	22.9	299×299
inceptionresnetv2	164	209 MB	55.9	299×299
shufflenet	50	5.4 MB	1.4	224×224
nasnetmobile	Non-linear sequence of modules	20 MB	5.3	224×224
nasnetlarge	Non-linear sequence of modules	332 MB	88.9	331×331
darknet19	19	78 MB	20.8	256×256
darknet53	53	155 MB	41.6	256×256
efficientnetb0	82	20 MB	5.3	224×224
alexnet	8	227 MB	61.0	227×227
vgg16	16	515 MB	138	224×224
vgg19	19	535 MB	144	224×224

Thus, we can use a previously trained network for the following tasks:

1) **Classification:** Apply pretrained networks directly to classification problems. To classify a new image, use **classify** function.
2) **Feature Extraction:** Use a pretrained network as a feature extractor by using the layer activations as features. We can use these activations as features to train another machine learning model, such as a support vector machine (SVM).
3) **Transfer Learning:** Take layers from a network trained on a large data set and fine-tune on a new data set.
4) **Network Modification:** Modify some of the pretrained network layers and treat the modified version as a new network from scratch.

A neural network consists of different layers connected to one another. It learns from huge volumes of data and uses complex algorithms to train a neural net.

Neural networks can help solve different problems. Examples are:

- Feed-Forward Neural Network: General Regression and Classification problems.
- Convolutional Neural Network: Image classification and object detection.
- Deep Belief Network (DBN): Healthcare sectors for cancer detection.
- Recurrent Neural Network (RNN): Speech recognition, voice recognition, time series prediction, and natural language processing.

In a feed-forward neural network, the decisions are based on the current input signal. It neither retains the past data nor does it predict the future scope and hope. Feed-forward neural networks are thus used in general regression and classification problems. Feed-forward neural networks are covered in previous machine learning chapters.

Convolutional neural networks (CNNs) are specifically suitable for images as inputs. Compared with a fully connected neural network, a convolutional neural network needs a smaller number of parameters to run the show of deep learning, as a result of reduction in number of connections, shared weights, and downsampling. See Chapter 7 on CNN architecture.

A Deep Belief Network (DBN) is a type of artificial neural network that is composed of multiple layers of interconnected nodes, known as neurons. It is considered a generative model, meaning it can generate new data samples that are similar to the training data it has been exposed to.

DBNs are typically structured as a stack of Restricted Boltzmann Machines (RBMs), which are a type of unsupervised learning algorithm. RBMs are used to pretrain each layer of the DBN in a layer-wise manner. Once the RBMs have been pretrained, the entire network is fine-tuned using supervised learning algorithms, such as backpropagation, to adjust the weights between the layers.

DBNs are known for their ability to automatically learn hierarchical representations of data. The lower layers of the network capture low-level features, such as edges or textures, while the higher layers learn more complex and abstract features. This hierarchical representation allows DBNs to effectively model complex patterns and structures in the data.

DBNs have been successfully applied to various tasks, including image and speech recognition, recommender systems, and natural language processing. They have also been used for unsupervised feature learning and as a building block for other deep learning architectures, such as deep convolutional neural networks. This book will not cover DBNs.

On the other hand, a Recurrent Neural Network (RNN) is a type of artificial neural network designed for sequential data processing. Unlike traditional feedforward neural networks, RNNs have connections that allow information to flow in a loop, enabling them to maintain internal memory of past inputs. This memory-like property makes RNNs particularly useful for tasks that involve sequential or time-dependent data.

The key feature of an RNN is its recurrent connection, which allows the output from a previous step to be used as input for the current step. This mechanism enables the network to consider not only the current input but also the context of previous inputs, making it capable of capturing temporal dependencies in the data.

RNNs are commonly used for tasks such as natural language processing, speech recognition, machine translation, and time series analysis. They can process input sequences of variable lengths, making them flexible for handling different types of data. The basic unit of an RNN is called a recurrent neuron or cell, which maintains an internal hidden state that evolves as it processes each input step.

However, traditional RNNs suffer from the "vanishing gradient" problem, which limits their ability to capture long-term dependencies. To address this issue, more advanced variants of RNNs have been developed, such as Long Short-Term Memory (LSTM) and Gated Recurrent Unit (GRU). These variants incorporate specialized mechanisms that help the network retain and propagate information over longer sequences, alleviating the vanishing gradient problem.

There are four types of RNNs:

1) SISO: Single Input Single Output. Examples are machine learning problems
2) SIMO: Single Input Multiple Output. Example is an image caption.

3) MISO: Multiple Input Single Output. Example is sentiment analysis where a phrase or paragraph is to be scrutinized as either a positive or negative sentiment.
4) MIMO: Multiple Input Multiple Output. Example is machine translation where a set of phrases are translated into a set of phrases in different languages.

Quoted from MATLAB offline help, Table 6.4 shows different neural networks sorted by their tasks, shown in the first column. The second column accounts for layer architecture, represented in MATLAB as an array of layers. Architectures are distinguished by the type of layers in the network. For example, CNNs contain convolutional layers and LSTMs contain LSTM layers. The third column accounts for training data format. For some tasks, there are multiple ways we can organize our training data. The last column accounts for their outputs. The output data type reflects what the network predicts. The output type generally depends on the output layer used in the network architecture.

Table 6.4 Types of neural networks and their properties.

Task	Architecture	Training Data Format	Network Output
Image classification	CNN	Image datastore	Categorical
Image regression	CNN	Table	Numeric
Object detection	YOLO	Table	Bounding box
Sequence to label	LSTM	Cell array	Categorical
Sequence to sequence	LSTM	Cell array	Categorical vector

Data Stores in MATLAB

Quoted from MATLAB offline help, Table 6.5 can be used as a reference guide for this chapter and subsequent chapters; it summarizes types of data or data store, their description, and applications as can be used within MATLAB environment.

Table 6.5 Types of data stores, their description, and applications within MATLAB.

Data/Data Store Type	Description	Applications
ImageDatastore*	Datastore of images saved on disk.	Train image classification neural network with images saved on disk, where the images are the **same size**. When the images are different sizes, use an **AugmentedImageDatastore** object.
AugmentedImageDatastore	Datastore that applies random affine geometric transformations, including resizing, rotation, reflection, shear, and translation.	Train image classification neural network with images saved on disk, where the images are different sizes. Train image classification neural network and generate new data using augmentations.

(Continued)

Table 6.5 (Continued)

Data/Data Store Type	Description	Applications
TransformedDatastore	Datastore that transforms batches of data read from an underlying datastore using a custom transformation function.	Train image regression neural network. Train networks with multiple inputs. Transform datastores with outputs not supported by **trainNetwork**. Apply custom transformations to datastore output.
CombinedDatastore	Datastore that reads from two or more underlying datastores.	Train image regression neural network. Train networks with multiple inputs. Combine predictors and responses from different data sources.
PixelLabelImageDatastore (Computer Vision Toolbox)	Datastore that applies identical affine geometric transformations to images and corresponding pixel labels.	Train neural network for semantic segmentation.
RandomPatchExtractionDatastore (Image Processing Toolbox)	Datastore that extracts pairs of random patches from images or pixel label images and optionally applies identical random affine geometric transformations to the pairs.	Train neural network for object detection.
DenoisingImageDatastore (Image Processing Toolbox)	Datastore that applies randomly generated Gaussian noise.	Train neural network for image denoising.
Custom Mini-Batch Datastore	Custom datastore that returns mini-batches of data.	Train neural network using data in a format that other datastores do not support. For details, see Develop Custom Mini-Batch Datastore.
Numeric array	Images specified as numeric array. If we specify images as a numeric array, then we must also specify the responses argument.	Train neural network using data that fits in memory and does not require additional processing like augmentation.
Table	Images specified as a table. If you specify images as a table, then you can also specify which columns contain the responses using the responses argument.	Train neural network using data stored in a table.

NOTE #0

*ImageDatastore objects support *image classification tasks only*. To use image datastores for regression networks, create a *transformed* or *combined* datastore that contains the images and responses using the transform and combine functions, respectively.

Image and Augmented Image Datastores

A datastore is a variable that references a data source. When performing deep learning for image classification, it can help store images in an **imageDatastore**. This stores information like the **filename, format, and classification label**. The files will not be imported into memory until they are needed. We can also preprocess the entire collection of images with an augmented image datastore.

Figure 6.1 shows a sketch of an image datastore where image files are stored and will serve as input to pre-trained network which will be able to predict or label the given images.

Figure 6.1 The image datastore serves as an input argument for further image analysis.

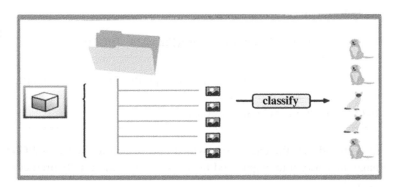

We can create an image datastore with the **imageDatastore** function, read the image files, and show them. Figure 6.2 shows the code to set up the activity.

```
%Specify the location of a directory containing image files.
ImDir=fullfile('D:\OneDrive_HCT\MatlabBook\MLDL\RandomImages');
%Create an imageDatastore, specifying the read function as a handle to imread
function.
JPGDS = imageDatastore(ImDir, ...
  'FileExtensions','.jpg','ReadFcn',@(x) imread(x));
%Read and display the first image in the datastore.
figure (1);
I = read(JPGDS);
% Scale the display range to the pixel values in the image.
imshow(I,[])
%The montage function can be used to display images in a datastore.
figure (2);
montage(JPGDS)
```

Figure 6.2 Creating an image datastore, reading the image files, and showing them.

NOTE #1

Before we run the code below, we must install Deep Learning Toolbox Model for GoogleNet Network. In a similar fashion, any other neural network listed in Table 6.3, we have to install the corresponding deep learning toolbox model for that particular network.

Let us use the **googlenet** function to load the pretrained network GoogleNet and save the result to a variable named **net**.

net=googlenet;

On the other hand, an augmented image datastore can perform simple preprocessing on an entire collection of images. To create this datastore, use the **augmentedImageDatastore** function using the given network's image input size as input. Create an augmented image datastore from **JPGDS** that will resize the images to 224×224. This size is specified by GoogleNet's input layer. Name the new datastore **auds**.

auds=augmentedImageDatastore([224 224],JPGDS);

The classify function will use a network to predict labels for an image.

predictedLabels=classify(net,auds)

Figure 6.3 shows the entire code which covers calling the trained **GoogleNet**, transforming the image datastore into augmented image datastore, and finally predicting the type (or class) of the selected images.

```
ImDir=fullfile('D:\OneDrive_HCT\MatlabBook\MLDL\RandomImages');
%Create an imageDatastore, specifying the read function as a handle to imread
%function.
JPGDS = imageDatastore(ImDir, ...
  'FileExtensions','.jpg','ReadFcn',@(x) imread(x));
%Read and display the first image in the datastore.
figure (1);
I = read(JPGDS);
% Scale the display range to the pixel values in the image.
imshow(I,[])
figure (2);
montage(JPGDS)
net=googlenet;
%inputSize = net.Layers(1).InputSize;
%I = imresize(I,inputSize(1:2));
```

Figure 6.3 Calling the trained GoogleNet, transforming the image datastore into augmented image datastore, and predicting the type (or class) of the selected images.

```
auds = augmentedImageDatastore([224 224],JPGDS);
predictedLabels=classify(net,auds)
% View class names and number of classes
classNames = net.Layers(end).ClassNames;
numClasses = numel(classNames)
% Alternatively, we may use the two commented statements, below.
%classNames2 = net.Layers(end).Classes;
%numClasses2 = numel(classNames2)
```

Figure 6.3 (Cont'd)

Results are copied from Command Window. GoogleNet has 1000 different types of image classes. The predicted labels for the 12 images, shown in Figure 6.4, are listed below. Notice that the predicted labels from GoogleNet are reasonable, but not correct. Pretrained networks are trained on images from many different classes, which may not include classes appropriate to one's data set. In the next section, we will improve these results using *transfer learning*.

```
predictedLabels = 12×1 categorical array
   seashore
   Bernese mountain dog
   bakery
   lakeside
   chiton
   lakeside
   bucket
   studio couch
   Border collie
   vestment
   bucket
   sandbar
numClasses = 1000
```

NOTE #2

Both **AlexNet** and **GoogleNet** network accept 3-d images as arguments to their image input layers. If the original image data store is gray scale (see HCW #1), then we need to convert from 1-d image data store into an rgb-based augmented image data store. Use the following statement to do the job:

3_d_augds = augmentedImageDatastore([227 227],1_d_ImDS,'ColorPreprocessing','gray2rgb');

Figure 6.4   Images stored in the image datastore, which are inspected by **GoogleNet**.

Figure 6.5 shows the workspace variables related to execution of Figure 6.3 code.

Workspace			
Name ▲	Value	Size	Min
auds	*1x1 augmentedImageDatastore*	1x1	
className	*1000x1 cell*	1000x1	
I	*227x227x3 uint8*	227x227x3	0
ImDir	'D:\OneDrive_HCT\MatlabBook\MLDL\RandomIm...	1x44	
JPGDS	*1x1 ImageDatastore*	1x1	
net	*1x1 DAGNetwork*	1x1	
numClasses	1000	1x1	1000
predictedLa...	*12x1 categorical*	12x1	

Figure 6.5   Workspace variables created by executing the code of Figure 6.3

Accessing an Image File

We need to locate the image file. It can be accessed either directly via its full name or by its reference number as part of an augmented image datastore, as shown in Figure 6.6.

```
ImDir=fullfile('D:\OneDrive_HCT\MatlabBook\MLDL\Flowers');
%Create an imageDatastore, specifying the read function as a handle to imread
function.
JPGDS = imageDatastore(ImDir, 'IncludeSubfolders',true,'LabelSource','foldernames',...
'FileExtensions','.jpg','ReadFcn',@(x) imread(x));
% an image can be directly accessed by name
flwrImage1= imread('D:\OneDrive_HCT\MatlabBook\MLDL\Flowers\image_0001.
jpg');
% Access by a reference number in an augmented image datastore.
flwrImage2=imread(JPGDS.Files{241});
```

Figure 6.6 How to access an image file within a regular folder or part of an augmented image datastore.

Alternatively, one can call the GUI image browser and search for the location of the image file as coded in Figure 6.7.

```
[file_name file_path] = uigetfile('*.jpg');
full_file_name = fullfile(file_path, file_name);
flwrImage=imread(full_file_name);
```

Figure 6.7 Using the GUI image browser to locate the image file and save both its name and location.

Figure 6.8 shows **"Select File to Open"** pop-up window where the user can search for the location of an image file. The selected image file name and its location will be saved to two separate workspace variables. The **full_file_name** variable has the file name and its root-folder.

Retraining: Transfer Learning for Image Recognition

[**Transfer learning is the process of retraining a pretrained network to classify a new set of images. The components needed for transfer learning are: pretrained network layers, training data, and algorithm options**].

This section reviews creating each of these inputs to perform transfer learning with **AlexNet**. Figure 6.9 summons the three transfer learning components.

Pretrained networks can classify images into predetermined categories. Instead of starting from scratch, we can modify a pretrained network to fit our problem.

The images in this data set are stored in subfolders. The name of each folder is the corresponding label for the images in that folder. Set the '**IncludeSubfolder**' and '**LabelSource**' options to find and label all the images for the selected datastore.

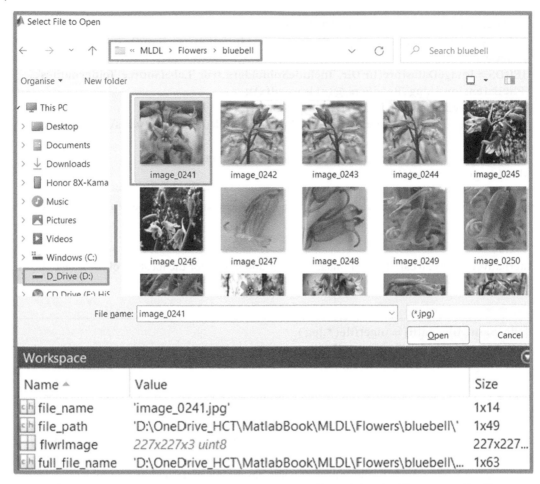

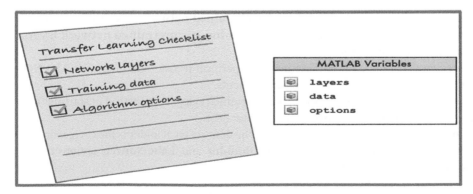

Figure 6.8 (*top*) The image browser for looking up the image file and then the file name and its file location will be saved to two workspace variables (*bottom*).

Figure 6.9 The three components of transfer learning needed for improving image prediction.

>>ds = imageDatastore(location,'IncludeSubfolders',true,'LabelSource',
'foldernames');

Here are the steps needed to proceed with training an image network

Step #1:
Create an image datastore named **imds**. Set the '**IncludeSubfolders**' and '**LabelSource**' options to find images in subfolders labeled with the folder names.

During training, the network learns to associate the training images and labels. The network may have a high training accuracy, but a network is only useful if it can generalize to new images. We should use a separate test data set to check if the network can classify new images not yet seen.

We can split the data set with **splitEachLabel** function.

>>[*ds1,ds2*] = splitEachLabel(*imds,p*)

The proportion **p** (a value from 0 to 1) indicates the proportion of images from each label from **imds** that are contained in **ds1**. The remaining files are assigned to **ds2**.

Step #2:
Create two augmented image datastores that preprocess **trainImgs** and **testImgs** to be sized 227×227, which suits **AlexNet**. Name the corresponding datastores **trainds** and **testds**, respectively. For **GoogleNet**, the size will be 224×224.

Step #3a (AlexNet):
Load the pretrained network **alexnet** and store it in a variable named **net**. The layers are stored in the **Layers** property of a network.

Extract the layers into a variable named **layers**.

We need to modify the final layers from **AlexNet** to suit our data set. We can replace the fully connected and classification layers in layers using standard array indexing.

>>*layers*(*n*) = *newLayer*

When we create a fully connected layer, we should specify the number of output classes.

>>*fc* = fullyConnectedLayer(*numClasses*)

Classification layers do not require any inputs.

>>*classificationLayer*()

Replace the 23rd element of layers with a new fully connected layer that has twelve classes.
Replace the last (25th) layer with a new classification layer.

Step #3b (GoogleNet):
We need to modify the final layers from **GoogleNet** to suit our data set. First, we can create the new layers that will replace the fully connected and classification layers.

When we create a fully connected layer, we should specify the number of output classes.

>>*fc* = fullyConnectedLayer(*numClasses*,"Name","lyname1")

Classification layers only require a name.

>>classificationLayer("Name","lyname2")

We can use **replaceLayer** function to modify **GoogleNet**'s architecture. To replace a layer, we need to know layer's name, which can be found by looking at the output of **net.Layers**.

>>lg = replaceLayer(lg,"NameToReplace",newlayer)

In GoogleNet, the fully connected layer is named "**loss3-classifier**" and the classification layer is named "**output.**"

Step #4:
Algorithm options control how a network is trained. We can create these options with the **trainingOptions** function. Create training options named options. Set the algorithm to '*adam*' and the initial learning rate to *0.0001*.

>>*options* = trainingOptions(*algorithm*,"InitialLearnRate",*rate*)

The first input is the name of the training algorithm to use. A common modification from the default options is to decrease the initial learning rate. There are many other options we can set. To see the possible training algorithms and other options, refer to MATLAB offline help: **trainingOptions**: *Options for training a deep neural network.*

Step #5:
We can predict the labels for new images with the classify function.

>> predictedLabels = classify(net,testds);

Step #6:
We can calculate the testing accuracy by comparing the labels in the test image datastore with the predicted labels from the network. Get the true labels from **testImgs.Labels**. Calculate the percentage accuracy of elements in **testPreds** that are correct.

```
>> n=testds.NumObservations;
>>Accuracy=(nnz(trueLabels == predictedLabels) / n)*100;
```

Here **n** is the number of test images. The **nnz** function counts the number of non-zero elements, or the number of elements where the true label equals the predicted label.

Step #7:
The confusion matrix shows which labels are misclassified.

>> confusionchart(TrueLabels, PredictedLabels);

NOTE #3
Before we run the code below, we must install Deep Learning Toolbox Model for **AlexNet** Network.

Figure 6.10 shows the complete code for the transfer learning for pre-trained **AlexNet**. One portion of an augmented image datastore is used for **AlexNet** training and another portion for **AlexNet** testing. The testing performance is also coded at the end.

```
rng(123);
% Create a datastore
ImDir=fullfile('D:\OneDrive_HCT\MatlabBook\MLDL\Flowers');
imds=imageDatastore(ImDir,'IncludeSubfolders',true,'LabelSource','foldernames');
%Split datastore
[trainImgs,testImgs] = splitEachLabel(imds,0.95);
%Preprocess datastores
trainds=augmentedImageDatastore([227 227],trainImgs);
testds=augmentedImageDatastore([227 227],testImgs);
net = alexnet;
layers = net.Layers;
%Replace final layers of AlexNet series network.
layers(23)=fullyConnectedLayer(12);
layers(25)=classificationLayer();
%Create training options
options=trainingOptions("adam","InitialLearnRate",0.0001);
net = trainNetwork(trainds,layers,options);
% Measure the model performance
predictedLabels = classify(net,testds);
n=testds.NumObservations;
trueLabels=testImgs.Labels;
AlexNaccuracy=(nnz(trueLabels == predictedLabels) / n)*100;
display(['AlexNet Testing Accuracy is: ' num2str(AlexNaccuracy) ' %'])
confusionchart(trueLabels,predictedLabels)
```

Figure 6.10 Creating two augmented image datastores: One for **AlexNet** training and another for testing its accuracy.

The result is copied from Command Window: **AlexNet** Testing Accuracy is: 89.5833%

Figure 6.11 shows **AlexNet** re-training mini-batch accuracy and the elapsing time needed for training.

```
Training on single CPU.
Initializing input data normalization.
|=====================================================================================|
| Epoch   | Iteration | Time Elapsed | Mini-batch | Mini-batch | Base Learning |
|         |           | (hh:mm:ss)   | Accuracy   | Loss       | Rate          |
|=====================================================================================|
|     1 |          1 |    00:00:05 |      7.03% |     5.9779 |    1.0000e-04 |
|     8 |         50 |    00:03:41 |    100.00% |     0.0027 |    1.0000e-04 |
|    15 |        100 |    00:07:17 |    100.00% |     0.0007 |    1.0000e-04 |
|    22 |        150 |    00:11:01 |    100.00% |     0.0011 |    1.0000e-04 |
|    29 |        200 |    00:14:58 |    100.00% | 9.2914e-05 |    1.0000e-04 |
|    30 |        210 |    00:15:47 |    100.00% | 8.9900e-05 |    1.0000e-04 |
|=====================================================================================|
Training finished: Max epochs completed.
```

Figure 6.11 The **AlexNet** re-training results, including training accuracy and training time.

Figure 6.12 shows the confusion matrix for **AlexNet** misclassified labels. Off-diagonal cases are misclassified.

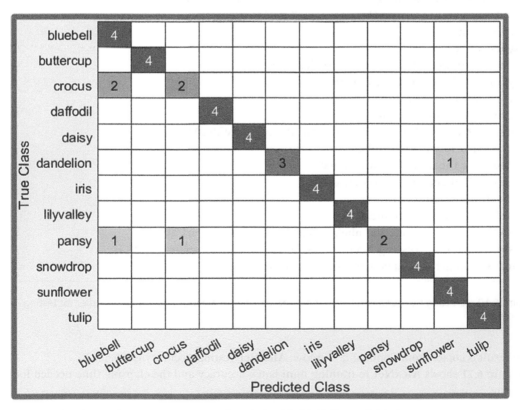

Figure 6.12 The confusion matrix for **AlexNet** misclassified (off-diagonal) labels.

Figure 6.13 shows the generated workspace variables related to **AlexNet** network re-training and testing accuracy.

Workspace		
Name ▲	Value	Size
AlexNaccuracy	89.5833	1x1
fc	*1x1 FullyConnectedLayer*	1x1
ImDir	'D:\OneDrive_HCT\MatlabBook\MLD...	1x39
imds	*1x1 ImageDatastore*	1x1
layers	*25x1 Layer*	25x1
n	48	1x1
net	*1x1 SeriesNetwork*	1x1
options	*1x1 TrainingOptionsADAM*	1x1
out	*1x1 ClassificationOutputLayer*	1x1
predictedLab...	*48x1 categorical*	48x1
testds	*1x1 augmentedImageDatastore*	1x1
testImgs	*1x1 ImageDatastore*	1x1
trainds	*1x1 augmentedImageDatastore*	1x1
trainImgs	*1x1 ImageDatastore*	1x1
trueLabels	*48x1 categorical*	48x1

Figure 6.13   Workspace variables related to **AlexNet** network re-training and testing accuracy.

Figure 6.14 shows the code for re-training **GoogleNet**, which is a type of Directed Acyclic Graphs networks. See Ch. 7 | "Directed Acyclic Graphs Networks" section.

```
rng(123);
% Create a datastore
ImDir=fullfile('D:\OneDrive_HCT\MatlabBook\MLDL\Flowers');
imds=imageDatastore(ImDir,'IncludeSubfolders',true,'LabelSource','foldernames');
%Split datastore
[trainImgs,testImgs] = splitEachLabel(imds,0.95);
%Preprocess datastores
trainds=augmentedImageDatastore([224 224],trainImgs);
testds=augmentedImageDatastore([224 224],testImgs);
net2 = googlenet;
layers2 = net2.Layers;
%Replace final layers
fc=fullyConnectedLayer(12,"Name","new_fc");
out=classificationLayer("Name","new_out");
```

Figure 6.14   Creation of two augmented image datastores for re-training **GoogleNet** DAG network.

```
lgraph = layerGraph(net2);%Creates a layerGraph variable for GoogleNet layers.
% Unlike series AlexNet, with DAG networks, we need to pinpoint to the layer
replacement.
% Else, an error message: "An error using trainNetwork; invalid network" will
%pop-up.
lgraph = replaceLayer(lgraph,"loss3-classifier",fc);
lgraph = replaceLayer(lgraph,"output",out);
options=trainingOptions("adam","InitialLearnRate",0.0001);
net2 = trainNetwork(trainds,lgraph,options);
% Measure the model performance
predictedLabels = classify(net2,testds);
n=testds.NumObservations;
trueLabels=testImgs.Labels;
GNaccuracy=(nnz(trueLabels == predictedLabels) / n)*100;
display(['GoogleNet Testing Accuracy is:' num2str(GNaccuracy) '%'])
confusionchart(trueLabels,predictedLabels)
```

Figure 6.14 (Cont'd)

The result is copied from Command Window: GoogleNet Testing Accuracy is: 91.6667%.

Figure 6.15 shows the accuracy and elapsing time of **GoogleNet** network training. Nevertheless, training takes very longer time than that of **AlexNet** network.

```
Training on single CPU.
Initializing input data normalization.
|==========================================================================================|
|  Epoch  |  Iteration  |  Time Elapsed  |  Mini-batch  |  Mini-batch  |  Base Learning  |
|         |             |   (hh:mm:ss)   |   Accuracy   |     Loss     |      Rate       |
|==========================================================================================|
|      1  |          1  |    00:00:24    |     3.12%    |    4.4660    |   1.0000e-04    |
|      8  |         50  |    00:13:44    |    98.44%    |    0.0382    |   1.0000e-04    |
|     15  |        100  |    00:28:39    |   100.00%    |    0.0007    |   1.0000e-04    |
|     22  |        150  |    00:43:37    |   100.00%    |    0.0009    |   1.0000e-04    |
|     29  |        200  |    00:58:33    |   100.00%    |    0.0007    |   1.0000e-04    |
|     30  |        210  |    03:30:53    |   100.00%    |    0.0003    |   1.0000e-04    |
|==========================================================================================|
Training finished: Max epochs completed.
>>
```

Figure 6.15 **GoogleNet** re-training results, including accuracy and time of training.

Figure 6.16 shows the confusion matrix for **GoogleNet** misclassified (off-diagonal) labels.

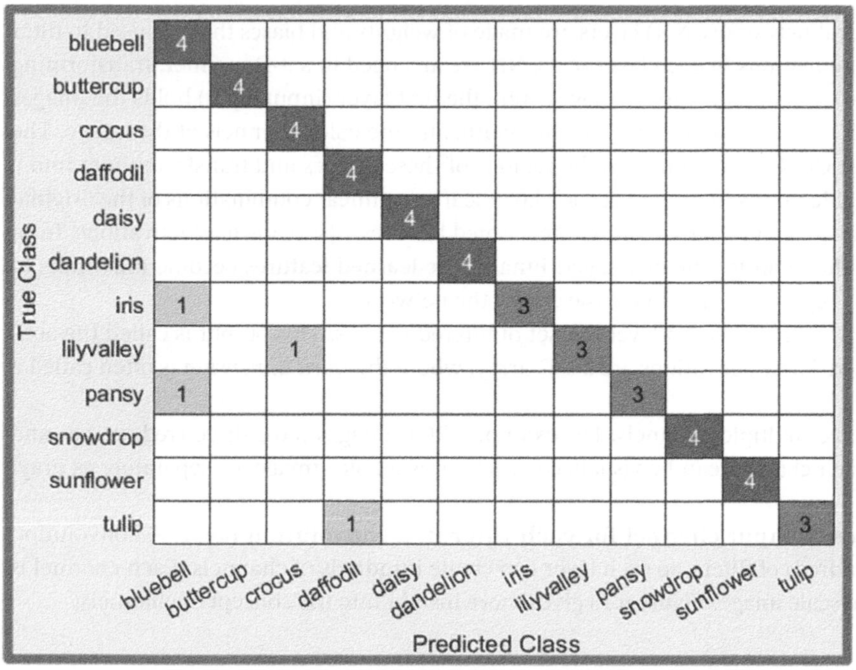

Figure 6.16 The confusion matrix for **GoogleNet** misclassified (off-diagonal) labels.

Figure 6.17 shows workspace variables related to **GoogleNet** network re-training and testing accuracy.

Workspace

Name	Value	Size	Min
fc	*1x1 FullyConnectedLayer*	1x1	
GNaccuracy	91.6667	1x1	91.6667
ImDir	'D:\OneDrive_HCT\MatlabBook\MLD...	1x39	
imds	*1x1 ImageDatastore*	1x1	
layers2	*144x1 Layer*	144x1	
lgraph	*1x1 LayerGraph*	1x1	
n	48	1x1	48
net2	*1x1 DAGNetwork*	1x1	
options	*1x1 TrainingOptionsADAM*	1x1	
out	*1x1 ClassificationOutputLayer*	1x1	
predictedLabels	*48x1 categorical*	48x1	
testds	*1x1 augmentedImageDatastore*	1x1	
testImgs	*1x1 ImageDatastore*	1x1	
trainds	*1x1 augmentedImageDatastore*	1x1	
trainImgs	*1x1 ImageDatastore*	1x1	
trueLabels	*48x1 categorical*	48x1	

Figure 6.17 Workspace variables related to **GoogleNet** network re-training and testing accuracy.

Convolutional Neural Network (CNN) Layers: Channels and Activations

Convolutional neural network (CNN) layers are made of weights and biases that are used to filter an input image. The neurons in each layer of a CNN are arranged in a 3-D manner, transforming a 3-D input to a 3-D output. For a given image input, the first layer (input layer) holds the images as 3-D inputs, with the dimensions being height, width, and the color channels of the image. The neurons in the first CNN layer connect to the regions of these images and transform them into a 3-D output. The hidden units (neurons) in each layer learn nonlinear combinations of the original inputs, which is called feature extraction. These learned features, also known as activations, from one layer become the input for the next layer. Finally, the learned features become the inputs to the classifier or the regression function at the end of the network.

Thus, the output of a convolution layer is a set of filtered images. This output is called the activations of that layer. These activations are a 3-D array, where the *third dimension* is often called a **channel**.

Many images have multiple channels. For example, RGB images have three (red, green, and blue) channels. Each channel can be visualized together as a color image, or separately as grayscale images.

There is only one output channel for each filter in a convolution layer. A convolution layer can have hundreds of filters, so each layer can create hundreds of channels. Each channel is visualized as a grayscale image. Figure 6.18 gives more insight into the concept of channels.

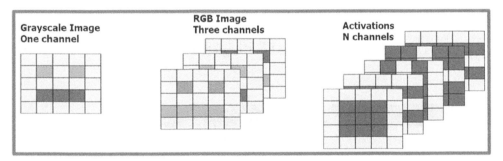

Figure 6.18 The outputs of filters are made of a series of 2-D images but augmented by a third dimension, symbolizing the channel number.

Figure 6.19 shows a grayscale image generated by a filter which will recognize the presence of an edge (color change) throughout the canvas of the image.

Convolution layers process images from left to right. This filter finds edges from dark to light. The **positive activations** are in places where the **dark fur is left of the white fur** (or white wall), as shown in Figure 6.20.

The **negative activations** show the opposite. The **white wall is left of the black fur**, as shown in Figure 6.21.

Figure 6.19 This grayscale image is one channel of the filtered cat image. The filter that produces this image finds an edge (step color change).

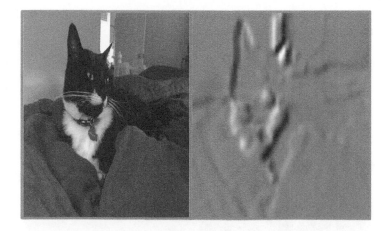

Figure 6.20 Positive activation when there exists a step color change from black (left) to white (right) fur.

Figure 6.21 Negative activation when there exists a step color change from white (left) to black (right) fur.

Gray activations show nothing interesting. This filter did not find any edges on the right half of the image, as shown in Figure 6.22.

Figure 6.22 Zero activations as there are no edges (or step color changes) found on the right side.

Quiz #1

In convolutional neural networks, what is an activation?
A. Weights of a convolution layer   B. Filters of a convolution layer
C. Information passed between the network layers   D. None of the above

---

Quiz #2

If you work with an RGB image stored as a 3-D array, how can you separate the red color?

a) Use logical indexing: redim = im(color=="red")
b) Extract the red layer by indexing into the array: redim = im(:,:,1)

Convolution 2-D Layer Features via Activations

Let us call **AlexNet**, a Convolutional Neural Network (CNN).

```
%load('AlexNet.mat')
net=alexnet;
```

The first convolution layer in **AlexNet** is named '**conv1**'. We can see this by viewing the network's layers, as shown in Figure 6.23. **AlexNet** which is named **net** is a series network, a neural network for deep learning with layers arranged one after the other. It has a single input layer and a single output layer.

Variables - net		Variables - net.Layers		Variables - net.Layers(2, 1)	
net ✕ net.Layers ✕ net.Layers(2, 1) ✕		net ✕ net.Layers ✕ net.Layers(2, 1) ✕		net ✕ net.Layers ✕ net.Layers(2, 1) ✕	
1x1 SeriesNetwork		net.Layers		net.Layers(2, 1)	
Property ▲	Value		1	Property ▲	Value
Layers	25x1 Layer	1 1x1 ImageInputLayer		Name	'conv1'
InputNames	1x1 cell	2 1x1 Convolution2DLayer		FilterSize	[11,11]
OutputNames	1x1 cell	3 1x1 ReLULayer		NumChannels	3

Figure 6.23 **net** has 25 layers (*left*). The first convolution layer (*middle*) and its name (*right*).

We can extract features from an input image using a Convolutional Neural Network (CNN) with the **activations** function.

This function accepts three inputs: The network, the original input image, and the layer to extract features from

>>*features* = activations(*net,img,layer*)

The complete code for loading the network, reading and saving an image file, and calling **activations** function is shown in Figure 6.24.

```
%load('AlexNet.mat')
net=alexnet;
ImDir=fullfile('D:\OneDrive_HCT\MatlabBook\MLDL\Flowers');
%Create an imageDatastore, specifying the read function as a handle to imread
%function.
JPGDS = imageDatastore(ImDir, 'IncludeSubfolders',true,'LabelSource',
'foldernames',...
'FileExtensions',',jpg','ReadFcn',@(x) imread(x));
% an image can be accessed either directly by name or by its reference number in a
datastore.
flwrImage1= imread('D:\OneDrive_HCT\MatlabBook\MLDL\Flowers\bluebell\
image_0241.jpg');
%Or
flwrImage2=imread(JPGDS.Files{1});
%Get activations from first conv layer
actvn=activations(net,flwrImage1,'conv1');
```

Figure 6.24 The code for loading the trained network, reading and saving the image file, and calling **activations** function.

Each convolution layer consists of many 2-D arrays called channels. The '**conv1**' layer has 96 channels, so it is difficult to inspect all channels at once.

The provided function **showActivationsForChannel** (Appendix A.16) will display the activations from a specific channel. Use the image, the features, and the desired channel number as input.

Let us view one of the activations, issuing from the 41st channel of **actvn** (has 96 channels) using **showActivationsForChannel** function. Use **flwrImage1**, **actvn**, and **41** as the input arguments.

```
showActivationsForChannel(flwrImage1,actvn,41)
```

Figure 6.25 shows the original and channel 41st-filtered image of the 1st convolution layer.

Figure 6.25 The original and 41st-channel-filtered image of the 1st convolution layer.

What color do you think this channel is looking for?

Channel 41 activates on colors with high blue values, such as violet.

Let us view the activation from the 11th channel of **actvn** using **showActivationsForChannel** function. Use **flwrImage1**, **actvn**, and **11** as the input arguments.

```
showActivationsForChannel(flwrImage1,actvn,11)
```

Figure 6.26 shows the original and 11th-channel-filtered image. Channel 11 activates on colors with high blue values, such as violet.

Figure 6.26 The original and 11th-channel-filtered image of the 1st convolution layer.

Let us view the activations from the 90th channel of **actvn** using **showActivationsForChannel** function to view the activations from the 90th channel.

```
showActivationsForChannel(flwrImage1,actvn,90)
```

Figure 6.27 shows the original and 90th-channel-filtered image. Channel 90 negatively activates on edges from light to dark (see Figure 6.21).

Figure 6.27 Negative activation (from light left to dark right) for channel 90 filter of the 1st convolution layer.

Most CNNs learn to detect features like color and edges in the first convolution layer. In deeper layers, the network learns more complicated features. The deepest convolution layer of **AlexNet** is '**conv5**'.

Let us extract features from **AlexNet**'s deepest convolution layer and name the output **actvn5**.

```
actvn5=activations(net,flwrImage1,'conv5');
```

Notice that '**conv5**' layer contains 256 channels. Use **showActivationsForChannel** function to view the activations from the 99th channel of the deepest convolution layer.

```
showActivationsForChannel(flwrImage1,actvn5,99)
```

Figure 6.28 shows the original and 99th-channel-filtered image of the 5th convolution layer.

Figure 6.28 The original and channel 99th-channel-filtered image of the 5th convolution layer.

Extraction and Visualization of Activations

We have looked at specific activations at the shallowest and deepest convolution layers, but how does the network learn these features?

We can view the activations to track the evolution of an image through the network. This enables us to see what happens to the image at each layer level.

The activations from the first convolution layer are already displayed. In most of these activations, the flowers are still recognizable. As the image progresses through the network, it will be less like the input image and more like features used to represent the image, as shown in Figure 6.29. In creating networks, we will learn more about how these features are created.

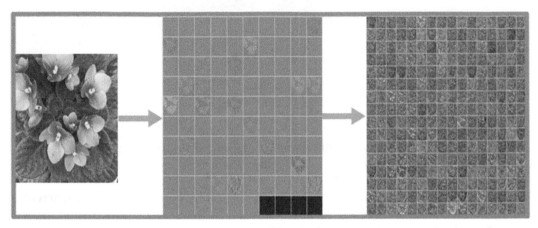

Figure 6.29 The progression from left to right through the network layers is like segmenting the flower image into tiny pieces of segments to pick up tiny features of a flower.

We have used the '**conv5**' layer to extract features. The activations from a layer are the output of that layer. We can learn more about the activations by investigating the layer's properties. These properties can be displayed by accessing any individual layer in a network. We can access the *n*th layer using:

>> layers(*n*)

Leave off the final semicolon to display the result. Figure 6.30 shows the code for loading **AlexNet**, displaying its layers, reading an image file, and showing grayscale activations tiny images.

```
%This code loads AlexNet and displays the layers.
net = alexnet;
layers = net.Layers
%This code loads the daisy image and displays the first convolution layer's activa-
tions.
ImDir=fullfile('D:\OneDrive_HCT\MatlabBook\MLDL\Flowers');
%Create an imageDatastore, specifying the read function as a handle to imread
function.
JPGDS = imageDatastore(ImDir, 'IncludeSubfolders',true,'LabelSource','foldernam
es',...
'FileExtensions','.jpg','ReadFcn',@(x) imread(x));
flwrImage100=imread(JPGDS.Files{100});
actvn = activations(net,flwrImage100,'conv1');
montage(mat2gray(actvn))
```

Figure 6.30 Code for showing **AlexNet** CNN layers and presenting grayscale activations tiny images.

Figure 6.31 shows names of layers of the network, their types, and the type of the output argument of that layer.

```
1   'data'     Image Input                  227×227×3 images with 'zerocenter' normalization
2   'conv1'    Convolution                  96 11×11×3 convolutions with stride [4  4] and padding [0  0  0  0]
3   'relu1'    ReLU                         ReLU
4   'norm1'    Cross Channel Normalization  cross channel normalization with 5 channels per element
5   'pool1'    Max Pooling                  3×3 max pooling with stride [2  2] and padding [0  0  0  0]
6   'conv2'    Grouped Convolution          2 groups of 128 5×5×48 convolutions with stride [1  1] and padding [2  2  2  2]
7   'relu2'    ReLU                         ReLU
8   'norm2'    Cross Channel Normalization  cross channel normalization with 5 channels per element
9   'pool2'    Max Pooling                  3×3 max pooling with stride [2  2] and padding [0  0  0  0]
10  'conv3'    Convolution                  384 3×3×256 convolutions with stride [1  1] and padding [1  1  1  1]
11  'relu3'    ReLU                         ReLU
12  'conv4'    Grouped Convolution          2 groups of 192 3×3×192 convolutions with stride [1  1] and padding [1  1  1  1]
13  'relu4'    ReLU                         ReLU
14  'conv5'    Grouped Convolution          2 groups of 128 3×3×192 convolutions with stride [1  1] and padding [1  1  1  1]
15  'relu5'    ReLU                         ReLU
16  'pool5'    Max Pooling                  3×3 max pooling with stride [2  2] and padding [0  0  0  0]
17  'fc6'      Fully Connected              4096 fully connected layer
18  'relu6'    ReLU                         ReLU
19  'drop6'    Dropout                      50% dropout
20  'fc7'      Fully Connected              4096 fully connected layer
21  'relu7'    ReLU                         ReLU
22  'drop7'    Dropout                      50% dropout
23  'fc8'      Fully Connected              1000 fully connected layer
24  'prob'     Softmax                      softmax
25  'output'   Classification Output        crossentropyex with 'tench' and 999 other classes
```

Figure 6.31   **AlexNet** CNN layers, their types, and the output argument for each layer.

Figure 6.32 shows the 96 grayscale activations tiny images of the 1st convolution layer, '**conv1**'. Each tiny image has an 11×11-pixel size.

Figure 6.32   The 96 grayscale activations tiny images of the 1st convolution layer, '**conv1**'.

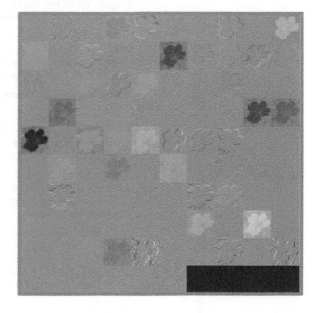

A 2-D (or 2-D Grouped) Convolutional Layer

On one hand, a 2-D convolutional layer applies sliding convolutional filters to 2-D input. The layer convolves the input by moving the filters along the input vertically and horizontally and computing the dot product of the weights and the input, and then adding a bias term.

Grouped convolutional layers are primarily used for channel-wise separable convolution, also known as depth-wise separable convolution. This means that instead of applying a single convolutional filter to the entire input volume, the layer performs separate convolutions on each group of input channels.

For each group, the layer takes the filters and moves them across the input vertically and horizontally. At each position, the layer computes the dot product between the weights of the filter and the corresponding input values. A bias term is then added to the result. These convolutions are performed independently for each group.

If the number of groups in a grouped convolutional layer is equal to the number of input channels, then this layer effectively performs channel-wise convolution. In other words, each input channel is treated as a separate group, and individual convolutions are applied to them. The convolution layer has hyperparameters and learnable parameters. Hyperparameters are set when the network is created and do not change during training. On the other hand, learnable parameters are updated during training.

Table 6.6 shows typical properties of a 2-D (or 2-D grouped) convolution layer, as quoted from MATLAB offline help.

Table 6.6 Typical properties of a 2D (or 2D grouped) convolution layer.

Property	Description
FilterSize	Height and width of the filters, specified as a vector [$h\ w$] of two positive integers, where h is the height and w is the width. **FilterSize** defines the size of the local regions to which the neurons connect in the input.
NumFilters	Number of filters, specified as a positive integer. It corresponds to the number of neurons in the convolutional layer that connect to the same region in the input. This parameter determines the number of channels (feature maps) in the output of the convolutional layer.
NumFiltersPerGroup	Positive integer. The number of output channels is NumFiltersPerGroup × NumGroups.
NumGroups	Number of groups, specified as a positive integer or *'channel-wise'*. If NumGroups is *'channel-wise'*, then the value will be equivalent to setting NumGroups to the number of input channels.
Stride	Step size for traversing the input vertically and horizontally, specified as a vector [$a\ b$] of two positive integers, where a is the vertical step size and b is the horizontal step size. When creating the layer, you can specify **Stride** as a scalar to use the same value for both step sizes.
DilationFactor	Factor for dilated convolution (also known as atrous convolution), specified as a vector [$h\ w$] of two positive integers, where h is the vertical dilation and w is the horizontal dilation. When creating the layer, you can specify DilationFactor as a scalar to use the same value for both horizontal and vertical dilations.

Table 6.6 (Continued)

Property	Description
PaddingSize	Size of padding to apply to input borders, specified as a vector $[t\ b\ l\ r]$ of four nonnegative integers, where t is the padding applied to the top, b is the padding applied to the bottom, l is the padding applied to the left, and r is the padding applied to the right.
PaddingValue	Value to pad data **0 (default) | scalar | 'symmetric-include-edge' | 'symmetric-exclude-edge' | 'replicate'**
NumChannels (NumChannelsPerGroup)	Number of channels for each filter or (per group), specified as **'auto'** or a **positive integer**. The number of channels per group is equal to the number of input channels divided by the number of groups. **'auto' (default) | positive integer.**

Figure 6.33 shows the code for loading **AlexNet**, presenting the grayscale activations tiny images, and showing layer #14, the fifth convolution layer, properties.

```
%This code loads AlexNet and displays the layers.
net = alexnet;
layers = net.Layers
%This code loads the daisy image and displays the first convolution layer's activations.
ImDir=fullfile('D:\OneDrive_HCT\MatlabBook\MLDL\Flowers');
%Create an imageDatastore, specifying the read function as a handle to imread function.
JPGDS = imageDatastore(ImDir, 'IncludeSubfolders',true,'LabelSource', 'foldernames',...
'FileExtensions','.jpg','ReadFcn',@(x) imread(x));
flwrImage150=imread(JPGDS.Files{150});
actvn5 = activations(net,flwrImage150,'conv5');
montage(mat2gray(actvn5))
ly=layers(14)
```

Figure 6.33 Loading **AlexNet** CNN, presenting grayscale activations tiny images and showing properties of the fifth convolution layer, layer #14.

Figure 6.34 (*left*) shows the properties of **AlexNet** fifth, 2-D grouped convolution layer, layer #14, and grayscale activations tiny images (*right*).

The weights of this convolution layer (layer #14) are stored in a 5-D array. This layer has 2 groups made of 128 $3 \times 3 \times 192$ convolutions with stride [1 1] and padding [1 1 1 1]. Referring to Table 6.6, this layer has $2 \times 192 = 384$ input channels and $2 \times 128 = 256$ output channels (feature maps), as shown in the right image of Figure 6.34 where the entire image is sub-divided into $16 \times 16 = 256$ image segments, or 256 different features. For the sake of a better insight, this layer has the following properties:

Number of Groups (NG) = 2
Number of Input Channels Per Group (NICPG) = 192
Number of Output Filters Per Group (NOFPG) = 128
Input Channels (IC) = NG × NICPG = 2 × 192 = 384
Output Channels (OC) = NG × NOFPG = 2 × 128 = 256

```
ly =

    GroupedConvolution2DLayer with properties:

                    Name: 'conv5'

    Hyperparameters
              FilterSize: [3 3]
               NumGroups: 2
     NumChannelsPerGroup: 192
      NumFiltersPerGroup: 128
                  Stride: [1 1]
          DilationFactor: [1 1]
             PaddingMode: 'manual'
             PaddingSize: [1 1 1 1]
            PaddingValue: 0

    Learnable Parameters
                 Weights: [5-D single]
                    Bias: [1×1×128×2 single]
```

Figure 6.34 Layer #14 properties (*left*) and grayscale activations images (*right*).

Let us repeat the same procedure but this time for **GoogleNet** CNN and display the first, convolution 2-D layer, layer #2. The code is shown in Figure 6.35.

```
%This code loads GoogleNet and displays the layers.
net = googlenet;
layers = net.Layers
%This code loads the daisy image and displays the first convolution layer's activations.
ImDir=fullfile('D:\OneDrive_HCT\MatlabBook\MLDL\Flowers');
%Create an imageDatastore, specifying the read function as a handle to imread function.
JPGDS = imageDatastore(ImDir, 'IncludeSubfolders',true,'LabelSource',
'foldernames',...
'FileExtensions',',.jpg','ReadFcn',@(x) imread(x));
flwrImage180=imread(JPGDS.Files{180});
actvn1 = activations(net,flwrImage180,'conv1-7x7_s2');
montage(mat2gray(actvn1))
ly=layers(2)
```

Figure 6.35 Loading **GoogleNet** CNN, presenting grayscale activations tiny images, and showing properties of the 1st convolution 2-D layer, layer #2.

Figure 6.36 (*left*) shows the properties of **GoogleNet** first, convolution 2-D layer, layer #2, and grayscale activations tiny images (*right*). Referring to Table 6.6, this layer has three input channels and 64 output channels (feature maps), as shown in the right image of Figure 6.36 where the entire image is sub-divided into $8 \times 8 = 64$ image segments, or 64 different features.

Number of Groups (NG) = 1

Number of Input Channels Per Group (NICPG) = 3

Number of Output Filters Per Group (NOFPG) = 64

Input Channels (IC) = NG $\times$ NICPG = $1 \times 3 = 3$

Output Channels (OC)= NG $\times$ NOFPG = $1 \times 64 = 64$

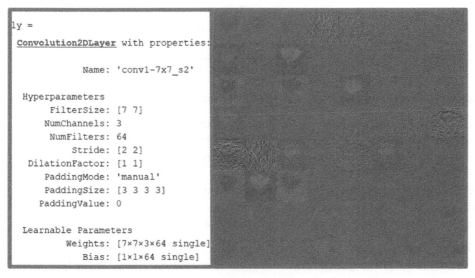

```
ly =

  Convolution2DLayer with properties:

             Name: 'conv1-7x7_s2'

  Hyperparameters
       FilterSize: [7 7]
      NumChannels: 3
       NumFilters: 64
           Stride: [2 2]
   DilationFactor: [1 1]
      PaddingMode: 'manual'
      PaddingSize: [3 3 3 3]
     PaddingValue: 0

  Learnable Parameters
          Weights: [7×7×3×64 single]
             Bias: [1×1×64 single]
```

Figure 6.36   Layer #2 properties (*left*) and grayscale activations tiny images made of 8 × 8 segments (*right*).

NOTE #4

We can use these three functions: **activations, mat2gray,** and **montage** to visualize the output of any layer in the network. Try viewing activations throughout the architecture to see how the violet image progresses through the network. Thus, features extraction using "**activations**" function and exploiting one of ML algorithms may compensate for either transfer learning (i.e., re-training) or training from scratch as will be explained in later chapters.

Features Extraction for Machine Learning

It is difficult to perform deep learning on a computer without a Graphics Processing Unit (GPU) because of the long training time. An alternative is to use pretrained networks for feature extraction. Then we can use traditional machine learning methods to classify these features.

Traditional machine learning is a broad term that encompasses many kinds of algorithms. These algorithms learn from predictor variables. Instead of using a whole raw image as the training data for a model, we need a set of features. Features can be anything, which accurately describe a given data set. For example, features to describe flowers could be the color, the number of petals, and the length of the stem, as shown in Table 6.7.

Table 6.7 Flower features which can be used in flowers classifications.

Color	Petal Count	Stem Length	Class
Yellow	22	275	Sunflower
Yellow	15	260	Sunflower
White	55	60	Daisy
Red	6	15	Tulip
Yellow	18	300	Sunflower
White	34	55	Daisy

Image Features in Pretrained Convolutional Neural Networks (CNNs)

CNNs learn to extract useful features while learning how to classify image data. As shown earlier, the early layers read an input image and extract features. Then, fully connected layers use these features to classify the image. The layers, including the fully connected layers, can be used as a standalone network for feature extraction.

With deep learning, we can use the activations function to extract features. These features can be used as the predictor variables for machine learning.

The rest of the machine learning workflow is very similar to deep learning:

- Extract training features
- Train the model
- Extract test features
- Predict test features
- Evaluate the model

Classification with Machine Learning

Given a set **X** of **n** points and a distance function. Set **X** represents a collection of points in a multi-dimensional space. Each point has a set of attributes or features associated with it. A distance function is a measure of dissimilarity or similarity between two points in the space. It calculates the distance or proximity between points based on their attributes. kNN search is used to identify the k nearest neighbors in **X** to a given query point or a set of query points represented by **Y**. The distance function is used to calculate the distances between the query point(s) and all the points in **X**. The k nearest neighbors are the points in **X** that have the smallest distances to the query point(s).

The kNN search technique is not only used for its practical application but also serves as a benchmark for evaluating the performance of other machine learning algorithms. It provides a baseline for comparing the results obtained from other classification or prediction techniques. The kNN search technique is relatively simple to understand and implement. Its simplicity allows for straightforward comparison and evaluation of the results obtained from other classification techniques or algorithms. By comparing the performance of other algorithms against kNN results, researchers and practitioners can assess the effectiveness of different approaches and make informed decisions. The technique has been used in various areas such as: bioinformatics, image processing and data compression, document retrieval, computer vision, multimedia database, and marketing data analysis.

In brief, kNN classification is one of the most straightforward machine learning methods. The basic idea is to find the known data that is "closest" to the new sample. For more information on kNN, see Chapter 2.

We will use activations from **AlexNet** CNN as the training data for the machine learning model, instead of the entire image. We can get the features at any point in the architecture. In this section, we will get features from the '**fc6**' layer. We can get features using the **activations** function. Using the fully connected layer, as opposed to a convolution layer, decreases the number of features and makes it easier to train the machine learning model.

Feature Extraction for Machine Learning: Flowers

Figure 6.37 shows the code to do the following steps:

1) Defining an image directory followed by an image datastore.
2) Split the data set with **splitEachLabel** function with p = 0.97 for training.
3) Create two augmented image datastores that preprocess **trainImgs** and **testImgs** to be sized 227 × 227. Name the corresponding datastores **trainds** and **testds**, respectively.
4) Load the pretrained CNN AlexNet and store it in a variable named net.
5) Load the training features from **AlexNet**. The training activations are saved in a matrix named **trainingFeatures**, where each row corresponds to a particular image, and each column corresponds to a particular feature.
6) Train a k-nearest neighbors classifier using the **fitcknn** function. We will use the features from **AlexNet** and the training image labels to create the k-nearest neighbors classification model named **knnMdl**.
7) Extract features from the testing images stored in the datastore **testImgs**. Name the features **testFeatures**. The activations need to be stored as rows for the machine learning model, **knnMdl**. Set the '**OutputAs**' option to **"rows"** to accomplish this. The network is stored in variable **net** and we should get the activations from the fully connected layer, '**fc6**'. At the Command prompt, type in: **>>net.Layers** to see the name of each layer.
8) Using **predict** function, **knnMdel** will classify **testFeatures** by predicting the labels. Name the labels **predictedLabels**.
9) Calculate the number of cases where the predicted label is equal to the true label. Divide by the total cases to get the model accuracy as percentage. The correct labels are stored in the categorical array in **testImgs.Labels**. **nnz** function will be used to get the number of nonzero matrix elements where the given condition is met (i.e.; the predicted label is equal to the true label).
10) Use the **confusionchart** function to display the confusion matrix for the true versus predicted labels pertaining to test data.

```
rng(123);
% Create a datastore
ImDir=fullfile('D:\OneDrive_HCT\MatlabBook\MLDL\Flowers');
imds=imageDatastore(ImDir,'IncludeSubfolders',true,'LabelSource','foldernames');
```

Figure 6.37 The code for creating an image datastore, splitting the image data, reading the image features, using **fitcknn** machine learning classification model to be trained and later predict the test data while presenting its accuracy.

```
[trainImgs,testImgs] = splitEachLabel(imds,0.97);
trainds=augmentedImageDatastore([227 227],trainImgs);
testds=augmentedImageDatastore([227 227],testImgs);
net=alexnet;
trainingFeatures = activations(net,trainds,'fc6','OutputAs','rows');
X= trainingFeatures;
Y= trainImgs.Labels;
knnMdl=fitcknn(X,Y);
testFeatures = activations(net,testds,'fc6','OutputAs','rows');
predictedLabels=predict(knnMdl,testFeatures);
numtrue=nnz(testImgs.Labels == predictedLabels);
Accuracy=(numtrue/numel(predictedLabels))*100.0
confusionchart(testImgs.Labels,predictedLabels);
```

Figure 6.37 (Cont'd)

The results are copied from Command Window:

Accuracy = 83.3333

Figure 6.38 shows the confusion matrix for the true versus predicted image labels, using fitcknn classification model.

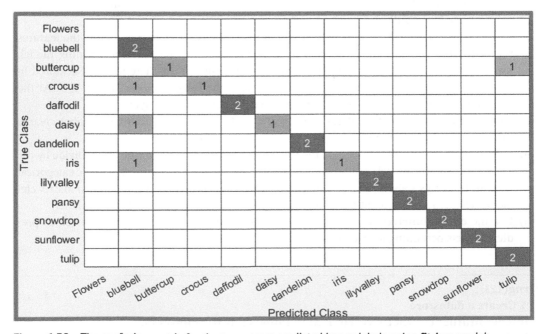

Figure 6.38 The confusion matrix for the true versus predicted image labels, using **fitcknn** model.

If we replace **fitcknn** by **fitcecoc** classification model as in Figure 6.39, an improvement in model predictability can be seen.

```
ecocMdl=fitcecoc(trainingFeatures,trainImgs.Labels);
predictedLabels=predict(ecocMdl,testFeatures);
```

Figure 6.39   Using the same code in Figure 6.37, except for the classification model definition.

The accuracy jumps up to 87.5%, instead of 83.3%.

Moreover, if hyper parameter optimization is automated/enabled for a given classification model, then it will most likely improve the prediction of the given model as in Table 6.8. Please, refer to Chapter 2 for more information on hyper-parameter optimization for each of ML algorithms.

Table 6.8   Automation of hyper parameter optimization for some machine learning classification models.

Classification Model Statement	Prediction Accuracy %
knnMdl = fitcknn(X,Y,'OptimizeHyperparameters',... 'auto','HyperparameterOptimizationOptions',... struct('AcquisitionFunctionName','expected-improvement-plus')) predictedLabels=predict(knnMdl,testFeatures);	91.67
svmtemplate = templateSVM('Standardize',true,'KernelFunction','gaussian'); ecocMdl = fitcecoc(X,Y,'Learners', svmtemplate,'OptimizeHyperparameters', 'auto', ... 'HyperparameterOptimizationOptions', ... struct('AcquisitionFunctionName','expected-improvement-plus')); predictedLabels=predict(ecocMdl,testFeatures);	87.5 CPU time 40 minutes
t = templateTree('Reproducible',true); ensMdl= fitcensemble(X,Y,'OptimizeHyperparameters','auto','Learners',t, ... 'HyperparameterOptimizationOptions',... struct('AcquisitionFunctionName','expected-improvement-plus')) predictedLabels=predict(ensMdl,testFeatures);	70.83 (CPU time 9.35 hr)

NOTE #5

Please, notice that selecting the right layer for features extraction purposes is a matter of trial-and-error procedure. The layer to be chosen is merely governed by maximizing the prediction accuracy for the test images set.

Pattern Recognition Network Generation

```
>>net = patternnet(hiddenSizes,trainFcn,performFcn)
```

returns a pattern recognition neural network with a hidden layer size of **hiddenSizes**, a training function, specified by **trainFcn**, and a performance function, specified by **performFcn**. Table 6.9 shows algorithms that can be used with a pattern recognition network, as quoted from MATLAB offline help.

Table 6.9 Pattern recognition network algorithms.

Training Function	Algorithm	Training Function	Algorithm
'trainlm'	Levenberg-Marquardt	**'traincgf'**	Fletcher-Powell Conjugate Gradient
'trainbr'	Bayesian Regularization	**'traincgp'**	Polak-Ribiére Conjugate Gradient
'trainbfg'	BFGS Quasi-Newton	**'trainoss'**	One Step Secant
'trainrp'	Resilient Backpropagation	**'traingdx'**	Variable Learning Rate Gradient Descent
'trainscg'	Scaled Conjugate Gradient	**'traingdm'**	Gradient Descent with Momentum
'traincgb'	Conjugate Gradient with Powell/Beale Restarts	**'traingd'**	Gradient Descent

The default performance function, specified by **performFcn** is **'crossentropy'**. This argument defines the function used to measure the network's performance. The performance function is used to calculate network performance during training. For a list of functions, in MATLAB Command Window, type in help **nnperformance**.

Pattern recognition networks are feedforward networks that can be trained to classify inputs according to target classes. The target data for pattern recognition networks should consist of vectors of all zero values except for a 1 in element i (i.e., dummy variables), where i is the class it represents.

Figure 6.40 shows the code how to use a pattern recognition network to predict the image label and also to generate the confusion matrix plot for the network being used on some tested images. Notice that the name **net** is reserved for **AlexNet** CNN whereas **net2** is reserved for the pattern recognition network which will be trained on features of the training images, **trainds**, channeled out of the **AlexNet** fully connected layer named **'fc7'** and tested on test mages, **testds**, using **splitEachLabel** function which will split 97% of the image datastore for training and 3% for testing purpose.

```
rng(123);
% Create a datastore
ImDir=fullfile('D:\OneDrive_HCT\MatlabBook\MLDL\Flowers');
imds=imageDatastore(ImDir,'IncludeSubfolders',true,'LabelSource','foldernames');
[trainImgs,testImgs] = splitEachLabel(imds,0.97);
trainds=augmentedImageDatastore([227 227],trainImgs);
testds=augmentedImageDatastore([227 227],testImgs);
net=alexnet;
trainingFeatures = activations(net,trainds,'fc7','OutputAs','rows');
X=trainingFeatures;
```

Figure 6.40 Creation of a pattern recognition network to be trained and tested over the features of an image datastore purged out of **AlexNet** CNN FC layer.

```
Xnorm=zscore(X);
Xnorm=X;
Y= trainImgs.Labels;
YTrain=dummyvar(Y);
testFeatures = activations(net,testds,'fc7','OutputAs','rows');
XTest=zscore(testFeatures);
XTest=testFeatures;
Y2= testImgs.Labels;
YTest=dummyvar(Y2);
net2 = patternnet(24);
net2.trainFcn = 'trainscg';
net2.divideParam.trainRatio = 80/100;
net2.divideParam.valRatio = 10/100;
net2.divideParam.testRatio = 10/100;
[net2,tr] = train(net2, Xnorm', YTrain')
Preds = net2(XTest');
%Confusion matrix plot is generated.
plotconfusion(YTest',Preds)
```

Figure 6.40 (Cont'd)

Figure 6.41 shows the confusion matrix plot with 91.7% testing accuracy and workspace variables. It is worth-mentioning here that without normalizing the extracted features (i.e., **Xnorm=X=trainingFeatures** and **XTest=testFeatures**), the overall accuracy will drop down to 84%.

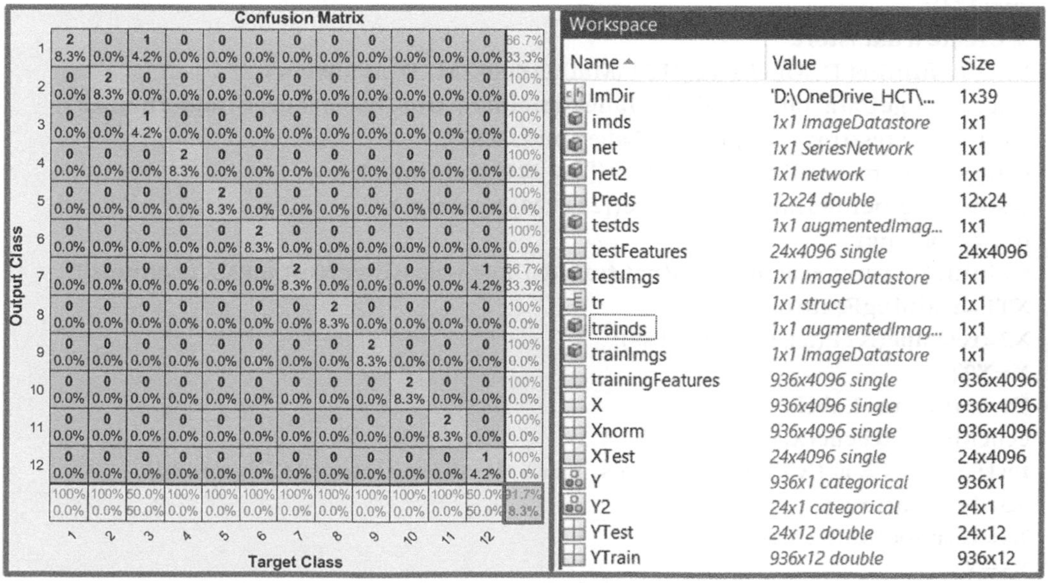

Figure 6.41 The confusion matrix plot for tested images and workspace variables.

The previous procedure, coded in Figure 6.37, can be repeated here for **GoogleNet** CNN, as shown in Figure 6.42. Notice, however, the following differences:

1) The image is resized to [224 224] not [227 227].
2) MATLAB will give the following error if we set the **'OutputAs'** option equal to *'rows'*:

Error using DAGNetwork/activations (line 262)
  To use activations on images larger than the input size of the network, the **"OutputAs"** value must be "**channels.**" For GoogleNet, I also tried using the **'OutputAs'** option equal to *'rows'* and got zero accuracy on platform of R2022a Update 6 version.

---

3) The channel output for the training image set will be 1×1×1000×936 4-D matrix. This 4-D matrix has to be converted into 2-D matrix which accounts for number of trained images and 1000 features for each image. The conversion can be done using **reshape** function. Use **size** function to get dimensions of the 4-D matrix.
   For example, size(XTF) = 1 1 1000 936.
4) On the other hand, the channel output for the test image set will be 1×1×1000×24 4-D matrix. This 4-D matrix has to be converted into 2-D matrix which accounts for number of tested images and 1000 features for each image. The conversion can be done as in step #3.
5) The name of a fully connected layer is **'loss3-classifier'**. You can refer to layer names of a CNN by typing at the Command prompt: >>**net.Layers**. This will enable the user to see the name of each layer and look for a fully connected layer.

```
rng(123);
% Create a datastore
ImDir=fullfile('D:\OneDrive_HCT\MatlabBook\MLDL\Flowers');
imds=imageDatastore(ImDir,'IncludeSubfolders',true,'LabelSource','foldernames');
[trainImgs,testImgs] = splitEachLabel(imds,0.97);
trainds=augmentedImageDatastore([224 224],trainImgs);
testds=augmentedImageDatastore([224 224],testImgs);
net=googlenet;
trainingFeatures = activations(net,trainds, 'loss3-classifier','OutputAs','channels');
XTF=trainingFeatures;
X2=reshape(XTF(:, :, :, :), 1000, 936);
X=X2';
Y= trainImgs.Labels;
ecocMdl=fitcecoc(X,Y);
testFeatures = activations(net,testds, 'loss3-classifier','OutputAs','channels');
X3=testFeatures;
X4=reshape(X3(:, :, :, :), 1000, 24);
X5=X4';
```

Figure 6.42 Creation of machine learning multiclass support vector machines model to be trained and tested over the features of an image datastore purged out of **GoogleNet** CNN FC layer.

```
predictedLabels=predict(ecocMdl,X5);
numtrue=nnz(testImgs.Labels == predictedLabels);
Accuracy=(numtrue/numel(predictedLabels))*100.0
confusionchart(testImgs.Labels,predictedLabels);
```

Figure 6.42 (Cont'd)

The result is copied from Command window:

Accuracy = 83.333%

Figure 6.43 shows the confusion matrix plot of the true versus predicted label for the tested images and generated workspace variables associated with Figure 6.42 code.

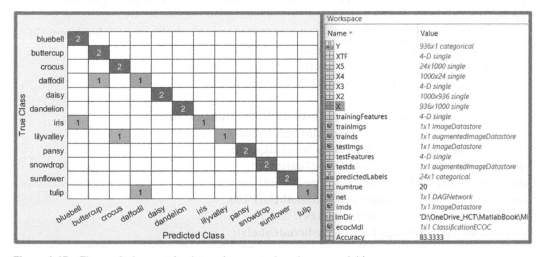

Figure 6.43 The confusion matrix plot and generated workspace variables.

Exercise #1

Repeat **AlexNet**-based tasks, shown earlier in Table 6.8 and in Figure 6.37, but this time for **GoogleNet** CNN to train the given network, test it to predict the labels of a new image test, and estimate the model accuracy and the confusion matrix plot of the true versus predicted label.

Machine Learning Feature Extraction: Spectrograms

The training features for the spectrograms from a cello, flute, and piano have been pre-computed and loaded into the workspace already. The variable is named **trainingFeatures**.

We will use these features and the training image labels to create a Naïve Bayes model. Train a Naïve Bayes classifier using the **fitcnb** function. Name the model **nbMdl**. The training images are stored in a datastore **trainImgs**.

Extract features from the testing images stored in the datastore **testImgs**. Name the features **testFeatures**. The network is stored in variable net and we should get the activations from the '**fc6**' layer.

Then predict the test data using the trained model. Name the output **predictedLabels**. Remember to set the '**OutputAs**' option to '*rows*'.

Calculate the fraction of test images correctly classified. Name the value **accuracy**.

Calculate and display the confusion matrix for the test data.

Figure 6.44 shows the complete code for the afore-mentioned steps.

```
rng(123)
ImDir=fullfile('D:\OneDrive_HCT\MatlabBook\MLDL\MusicProject\Spectro-
grams');
specds=imageDatastore(ImDir,'IncludeSubfolders',true,'LabelSource','foldernames');
[trainImgs,testImgs] = splitEachLabel(specds,0.98);
trainds=augmentedImageDatastore([227 227],trainImgs);
testds=augmentedImageDatastore([227 227],testImgs);
net=alexnet;
trainingFeatures = activations(net,trainds,'fc6','OutputAs','rows');
nbMdl=fitcnb(trainingFeatures,trainImgs.Labels);
testFeatures=activations(net,testds,'fc6',"OutputAs","rows");
predictedLabels=predict(nbMdl,testFeatures);
numtrue=nnz(testImgs.Labels == predictedLabels);
accuracy=(numtrue/numel(predictedLabels))*100;
disp(['Accuracy = ', num2str(accuracy),' %'])
confusionchart(testImgs.Labels,predictedLabels);
```

Figure 6.44 Creation of machine learning Naïve Bayes classification model to be trained and tested over the features of an image datastore purged out of **AlexNet** CNN FC layer.

The result is copied from Command Window:

Accuracy = 69.697%

Figure 6.45 shows the confusion matrix plot of the true versus predicted label for the tested images and generated workspace variables for Figure 6.44 code.

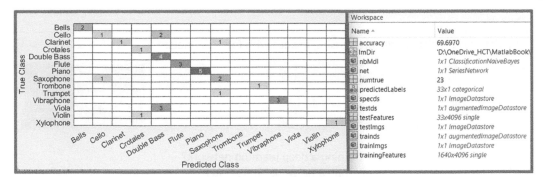

Figure 6.45 The confusion matrix plot and generated workspace variables for Figure 6.44 code.

NOTE #6

Even though **AlexNet** was not earlier trained on non-natural images like spectrograms, the features it learns are still sufficient to train an effective machine learning model.

Quiz #3

In which of the following situation we use classical machine learning to classify features extracted by a CNN?
a. Find a baseline accuracy before training your own deep network
b. The available computing power is insufficient to perform deep learning
c. We do not have time to train a deep network
d. All of the above

Quiz #4

Which command will extract features from the first convolution layer of AlexNet?
a. features = filters (net, catImage, 'conv1')
b. activations = features (net, 'conv1', catImage)
c. features = activations (net, catImage, 'conv1')
d. features = activations (net, 'conv1', catImage)

Quiz #5

The data set consists of images, like the followings:

What is the best approach for training a deep learning network?
a. Transfer learning   b. Train a network from scratch

Quiz #6

The data set consists of remote sensing images with four channels:

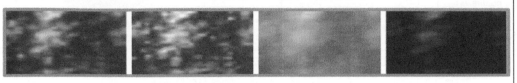

What is the best approach for training a deep learning network?
a. Transfer learning   b. Train a network from scratch

Quiz #7

The data set consists of icons that are 12×12 pixels:
What is the best approach for training a deep learning network?
a. Transfer learning   b. Train a network from scratch

Network Object Prediction Explainers

Toward the end of this chapter, let us explain how pretrained networks make predictions on or recognition for object identity by considering three network object prediction explainers: **occlusionSensitivity**, **imageLIME**, and **gradCAM** function.

Occlusion Sensitivity

An image '**catdog19.jpg**' has been imported. There are two different objects in this image: a cat and dog. Let us resize the image to match **GoogleNet**'s input size of [224 224]. Reassign the output back to **imgres**. We will use **GoogleNet** to classify the image. Name the output **imgpred**. Figure 6.46 shows the code for previous steps.

```
net = googlenet;
[file_name file_path] = uigetfile('*.jpg');
full_file_name = fullfile(file_path, file_name);
cat19=imread(full_file_name);
imshow(cat19)
imgres = imresize(cat19,[224 224]);
imgpred=classify(net,imgres)
```

Figure 6.46   Reading an image file and predicting its class by **GoogleNet** CNN.

The result is copied from Command Window:

```
imgpred = Newfoundland
```

So, **GoogleNet** CNN predicts that the dog class is Newfoundland, while ignoring the cat.

Figure 6.47 (*left*) shows the cat-dog image and workspace variables (*right*).

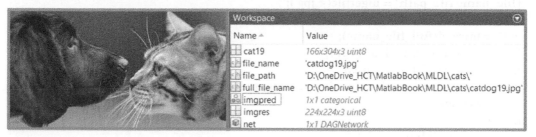

Figure 6.47 The cat-dog image and workspace variables.

The classification is Newfoundland. Why did the network make that prediction? I think it is 50%–50% to say that the classification is either "Newfoundland" or it will be "Newcatland." We can use a technique called *occlusion sensitivity* to visualize which parts of an image are important for a classification.

To calculate the occlusion sensitivity map, we will need three inputs:

>>map = occlusionSensitivity(net,img,label)

map: Sensitivity map that can be visualized with **imshow** and **imagesc** function.
net: A trained CNN that we want to investigate, like **AlexNet** or **GoogleNet**.
img: An image that was imported with **imread** function.
label: One of the class names. The class names can be found using **net.Layers(end).Classes**

Create an occlusion sensitivity map of the cat image. Name the output '**map**'.

To view the sensitivity map, we will overlay map on the cat image. Display the resized cat image.

To overlay the sensitivity map, we can use the **imagesc** function with a low alpha value.

>> imagesc(I,"AlphaData",a)

The alpha value **a** is a number from 0 to 1, where 0 transparent and 1 is opaque.

Enter hold on, then display the sensitivity map with an alpha level of 0.5. After you add the map, enter hold off.

It is easier to see the endpoints of the sensitivity map with the jet colormap, which can be set with the **colormap** function:

>>colormap jet

We can also add a color bar by entering **colorbar** function.

>>colorbar

Figure 6.48 shows the code for the previous steps.

```
net = googlenet;
[file_name file_path] = uigetfile('*.jpg');
full_file_name = fullfile(file_path, file_name);
cat19=imread(full_file_name);
imgres = imresize(cat19,[224 224]);
imgpred=classify(net,imgres)
map = occlusionSensitivity(net,imgres,imgpred);
imshow(imgres)
hold on;
imagesc(map,"AlphaData",0.5)
hold off;
colormap jet
colorbar
```

Figure 6.48   Using the **occlusionSensitivity** and **imagesc** function to visualize the most important features underlying the classification.

Figure 6.49 (left) shows the resized original cat-dog image and the occlusion sensitivity-treated image showing the most important features underlying the judgment made by **GoogleNet** CNN. The reddest areas of the predicted image are around the face of the dog, which explains the "**Newfoundland**" prediction for the dog type.

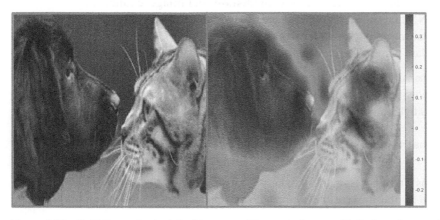

Figure 6.49   (*left*) the resized original image and the occlusion sensitivity-based image (*right*).

Even though **classify** function returns the most likely class, however, one may attempt to select other potential classes. Let us choose or assign the class "**lynx**" or "**tabby**" by changing the third input to the **occlusionSensitivity** function to another prediction to see what is important to another class. Notice that

```
>> scoreMap = occlusionSensitivity(net,X,label);
```

computes a map of the change in classification score for the classes specified by label when parts of the input data **X** are occluded (blocked) with a mask. The change in classification score is relative to the original data without occlusion. The occluding mask is moved across the input data, giving a change in classification score for each mask location. Use an occlusion sensitivity map to identify the parts of your input data that most impact the classification score. Areas in the map with higher positive values correspond to regions of input data that contribute positively to the specified classification label. The network must contain a "**classificationLayer**" as an output layer.

Figure 6.50 shows the code for changing the predicted class to both "**lynx**" and "**tabby**" in the occlusion sensitivity-based based image.

```
net = googlenet;
[file_name file_path] = uigetfile('*.jpg');
full_file_name = fullfile(file_path, file_name);
cat19=imread(full_file_name);
imgres = imresize(cat19,[224 224]);
imgpred=classify(net,imgres)
%map = occlusionSensitivity(net,imgres, "lynx");
map = occlusionSensitivity(net,imgres, "tabby");
imshow(imgres)
hold on;
imagesc(map,"AlphaData",0.5)
hold off;
colormap jet
colorbar
```

Figure 6.50 Changing the predicted class to both "**lynx**" and "**tabby**" in the occlusion sensitivity-based based image.

Figure 6.51 (*left*) shows the occlusion sensitivity-based "**lynx**"-predicted image and (*right*) the occlusion sensitivity-based "**tabby**"-predicted image. **GoogleNet** CNN can barely find the cat, and, overall, the map score is zero or less. The dog face got the lowest (largest negative) score among other zones of the input image.

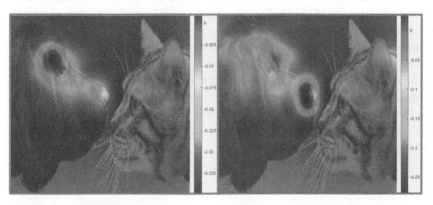

Figure 6.51 (*left*) The occlusion sensitivity-based "**lynx**"-predicted image and (*right*) the occlusion sensitivity-based "**tabby**"-predicted image.

If we want **GoogleNet** CNN to classify the cat instead, we can crop the image to keep only the cat and remove the distracting object, the dog, as shown in Figure 6.52. The image is classified as "**lynx**" cat type.

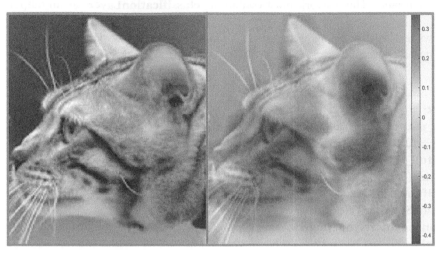

Figure 6.52 The dog is removed (cat only) and hence **GoogleNet** CNN predicts well the image as "**lynx**" cat type.

imageLIME Features Explainer

The command:

```
>>[scoreMap,featureMap,featureImportance] = imageLIME(net,imgres,label);
```

uses the locally interpretable model-agnostic explanation (LIME) technique to compute a map of the importance of the features in the input image **imgres** when the network **net** evaluates the class score (**scoreMap**) for the class given by **label**. We use this function to explain classification decisions and check that the network is focusing on the appropriate features of the image. It also returns a map of the features (**featureMap**) used to compute the LIME results and the calculated importance of each feature (**featureImportance**).

The LIME technique approximates the classification behavior of the net using a simpler, more interpretable model. By generating synthetic data from input **imgres**, classifying the synthetic data using **net**, and then using the results to fit a simple regression model, **imageLIME** function determines the importance of each feature of **imgres** to the network's classification score for class given by **label**. This function requires Statistics and Machine Learning Toolbox™. Figure 6.53 shows the code for reading the image, predicting the object classes using **imageLIME** function, and drawing the road map to important features.

```
net = squeezenet;
[file_name file_path] = uigetfile('*.jpg');
full_file_name = fullfile(file_path, file_name);
cat19=imread(full_file_name);
imgres = imresize(cat19,[227 227]);
%Alternatively, one may resize the image in light of the first input layer.
%inputSize = net.Layers(1).InputSize(1:2);
%imgres = imresize(cat19,inputSize);
label = classify(net,imgres);
[scoreMap,featureMap,featureImportance] = imageLIME(net,imgres,label,'Segment
ation','grid','NumFeatures',64,'NumSamples',3072);
figure
imshow(imgres)
hold on
imagesc(scoreMap,'AlphaData',0.5)
colormap jet
colorbar
```

Figure 6.53 Reading the image, predicting the object classes, and drawing the road map to important features.

Figure 6.54 shows the image with the color map jet showing the most important features or segments in the scanned image. It shows the explained results over the original image with transparency to see areas of the image, largely contributing to the classification score.

Figure 6.54 The score map explaining the network prediction with the dog face getting the highest feature importance.

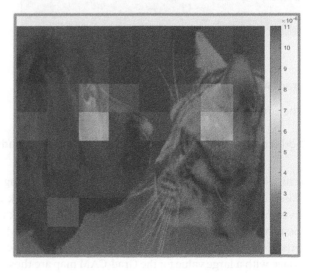

Let us use the feature importance to find the indices of the most important five features.

```
numTopFeatures = 5;
[~,idx] = maxk(featureImportance,numTopFeatures);
% Use the map of the features to mask out the image so only the most important five
%features are visible. Display the masked image.
mask = ismember(featureMap,idx);
maskedImg = uint8(mask).*imgres;
figure
imshow(maskedImg);
```

Figure 6.55 shows the most five important features that contribute to the decision making of the pretrained network regarding the object classification or labeling.

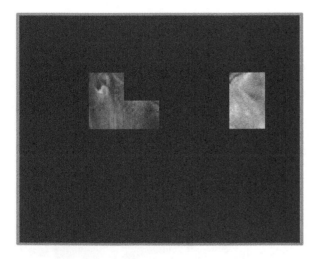

Figure 6.55   The displayed masked image showing the most important five features.

gradCAM Features Explainer

The command:

>>[scoreMap,featureLayer,reductionLayer] = gradCAM(net,imgres,label)

returns the gradient-weighted class activation mapping (Grad-CAM) map of the change in the classification score of image **imgres**, when the network **net** evaluates the class score for the class given by **label**. We use this function to explain network predictions and check that the network is focusing on the right parts of an image. The Grad-CAM interpretability technique uses the gradients of the classification score with respect to the final convolutional feature map. The parts of an image with a large value for the Grad-CAM map are those that significantly affect the network score for that class. It also returns the names of the feature layer **featureLayer** and reduction layer used to compute the Grad-CAM map. Figure 5.56 shows the code for reading an image, resizing it, predicting the label, calculating the score map using **gradCAM**, and imposing it unto the original map.

```
net = squeezenet;
[file_name file_path] = uigetfile('*.jpg');
full_file_name = fullfile(file_path, file_name);
cat19=imread(full_file_name);
imgres = imresize(cat19,[227 227]);
%Alternatively, one may resize the image in light of the first input layer.
%inputSize = net.Layers(1).InputSize(1:2);
%imgres = imresize(cat19,inputSize);
label = classify(net,imgres);
[scoreMap] = gradCAM(net,imgres,label);
figure
imshow(imgres)
hold on
imagesc(scoreMap,'AlphaData',0.5)
colormap jet
colorbar
```

Figure 6.56   Reading an image, resizing it, predicting the label, calculating the score map using **gradCAM**, and imposing it unto the original map.

Figure 6.57 shows which parts of an image are important to the classification decision of a network. The network focuses predominantly on the face of the cat to make the classification decision. If we click on "**label**" variable in workspace, we will see that the label has one entry, that is, 'standard schnauzer' which means that the network recognized that dog type called 'standard schnauzer', a German breed of dog in the Pinscher and Schnauzer group. This is a kind of paradox; the face or even the ear of the cat largely influences the net to believe that the object class is dog-style named 'standard schnauzer'. In conclusion, even with artificial intelligence way of thinking, we may have an embedded artificial stupidity. Since the previous two features explanation methods give a dog-type face recognition, it looks to me that **gradCAM** features explainer has a bug or divergence issue as it is based on the estimation of the gradient-weighted class activation mapping.

Figure 6.57   **gradCAM** features explainer face recognition in telling object class.

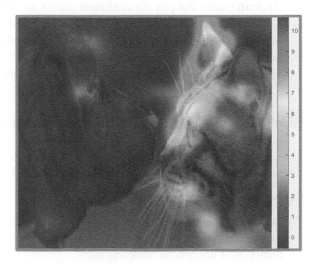

References

1 ImageNet. http://www.image-net.org.

2 Russakovsky, O., Deng, J., Su, H. et al. (2015). ImageNet large scale visual recognition challenge. *International Journal of Computer Vision (IJCV)* 115 (3): 211–252.

HCW 6.1   CNN Retraining for Round Worms Alive or Dead Prediction

In this problem, we are given the images of worms. Use transfer learning to train a deep network that can classify images of roundworms as either alive or dead. (Alive worms are round; dead ones are straight). Conclusion if you are straight, you will be dead.

The images are given in a folder called **WormImages** and the labeling is given in an Excel file where it has the image file name and the worm status. The data set is also available at Wiley book companion.

Carry out the following steps:

1) Get the training images and classes via saving them to an image data store and image data store labels.
2) View the first few images. Use **imshow** function with the second argument to scale the display based on the range of pixel values in the image.
3) Divide data into training (60%) and testing (40%) sets.
4) Create augmented image datastores to resize the images to fit **AlexNet** network. See NOTE #2 for a grayscale image data store.
5) Build a network by starting with a pretrained **AlexNet** network.
6) Consider all AlexNet layers except the last three layers. In addition, add three new layers at the end: One fullyConnectedLayer, one softmaxLayer, and one classificationLayer.
7) Set some training options: The algorithm to 'adam' and 'InitialLearnRate' to 0.0001.
8) Train the network.
9) Evaluate network on test data.
10) Extract features from **'fc6'** layer, and **'OutputAs','rows'** as **X** data. **Y** data will be the labels of the train images set. Use ML **fitcknn** model to fit **X** vs. **Y** data. Calculate the accuracy of **fitcknn** model and plot the confusion matrix for the predicted versus true labels.
11) Convert the extracted features (i.e., **X** data) into 2-d matrix. Convert **Y** data into dummy variables. Create a pattern recognition network and train it over the training data. Calculate the accuracy of the pattern recognition network via plotting the confusion matrix of the predicted versus true labels for the test data. Do not forget to normalize **X** data.

HCW 6.2   CNN Retraining for Food Images Prediction

The Example Food Images data set contains 978 photographs of food in **nine** classes (caesar_salad, caprese_salad, french_fries, greek_salad, hamburger, hot_dog, pizza, sashimi, and sushi).

Carry out the following steps:

1) Download the Example Food Images data set using the downloadSupportFile function and extract the images using the unzip function. This data set is about 77 MB. The data set is also available at Wiley book companion.

```
fprintf("Downloading Example Food Image data set (77 MB)... ")
filename = matlab.internal.examples.downloadSupportFile('nnet', ...
 'data/ExampleFoodImageDataset.zip');
fprintf("Done.\n")
filepath = fileparts(filename);
dataFolder = fullfile(filepath,'ExampleFoodImageDataset');
unzip(filename,dataFolder);
```

The Zip-format file is usually downloaded to a folder location similar to the following location: C:\Users\Kamal\Documents\MATLAB\Examples\R2022a\supportfiles\nnet\data\ ExampleFoodImageDataset.zip

2) Create an images data store.

3) Divide data into training (95%) and testing (5%) sets and prepare the image size that suits GoogleNet network.

4) Build a network by starting with a pretrained **GoogleNet** network.

5) Create a fully connected layer with nine classes and classification layer. Take GoogleNet CNN layers but replace the last fully connected layer by the newly created FC. The name of the last fully connected layer is **"loss3-classifier"** and the name of the output classification layer is **"output."** Replace those two layers by your own, but of the same type, layers.

6) Set some training options: The algorithm to 'adam' and 'InitialLearnRate' to 0.0005.

7) Train the network.

8) Evaluate network on test data.

9) Extract features from **'loss3-classifier'** layer, and **'OutputAs','channels'** as **X** data. **Y** data will be the labels of the train images set. Use ML **fitcecoc** model to fit **X** vs. **Y** data. Calculate the accuracy of **fitcecoc** model and plot the confusion matrix for the predicted versus true labels.

10) Convert the extracted features (i.e., **X** data) into 2-d matrix. Convert **Y** data into dummy variables. Create a pattern recognition network and train it over the training data. Calculate the accuracy of the pattern recognition network via plotting the confusion matrix of the predicted versus true labels for the test data. Do not forget to normalize **X** data.

HCW 6.3   CNN Retraining for Merchandise Data Prediction

The MathWorks Merch data set is a small data set containing 75 images of MathWorks merchandise, belonging to *five different classes (cap, cube, playing cards, screwdriver, and torch)*. You can use this data set to try out transfer learning and image classification quickly.

It can be found at one of the following folder locations:

C:\Program Files\MATLAB\R2022a\examples\deeplearning_shared\data\MerchData.zip
C:\Program Files\MATLAB\R2022a\examples\nnet\data\MerchData.zip
C:\Program Files\MATLAB\R2022a\examples\vision\data\MerchData.zip

The images are of size 227×227×3.

Carry out the following steps:

```
filename = 'MerchData.zip';
dataFolder = fullfile(tempdir,'MerchData');
if ~exist(dataFolder,'dir')
  unzip(filename,tempdir);
end
```

1) Extract the MathWorks Merch data set.
2) Load the data as an image datastore using the **imageDatastore** function and specify the folder containing the image data.
3) Divide data into training (95%) and testing (5%) sets and prepare the image size that suits RestNet-50 network. Notice that RestNet-50 network require an image with 224×224×3 size.
4) Build a network by starting with a pretrained **ResNet-50** network (net=resnet50;).
5) Create a fully connected layer with five classes and classification layer. Take ResNet-50 CNN layers but replace the last fully connected layer by the newly created FC. The name of the last fully connected layer is **"fc1000"** and the name of the output classification layer is **"ClassificationLayer_fc1000."** Replace those two layers by your own, but of the same type, layers.
6) Set some training options.
7) Train the network.
8) Evaluate network on test data.
9) Extract features from **'res4b_branch2b'** layer, and **'OutputAs','channels'** as **X** data. **Y** data will be the labels of the train images set. Use one of ML models to fit **X** vs. **Y** data. Calculate the accuracy of the ML model and plot the confusion matrix for the predicted versus true labels.
10) Convert the extracted features (i.e., **X** data) into 2-d matrix. Convert Y data into dummy variables. Create a pattern recognition network and train it over the train data. Calculate the accuracy of the pattern recognition network via plotting the confusion matrix of the predicted versus true labels for the test data. Do not forget normalize X data.
11) Repeat steps #4 through #10 but this time for **GoogleNet** network. The names of layers in step #5 and #9 are the same as those shown in HCW #2: Food Images.

HCW 6.4 CNN Retraining for Musical Instrument Spectrograms Prediction

The data used for this homework/classwork can be obtained from the University of Iowa Electronic Music Studios. (http://theremin.music.uiowa.edu/index.html). It is also available with Wiley book companion.

The Spectrograms folder contains about 2000 spectrograms. There is one folder for each instrument.

Perform feature extraction using a pretrained network for both **GoogleNet** and **AlexNet** and use a machine learning classification model. Use the **'fc6'** layer of **AlexNet** series network for feature extraction and use the fully connected layer **'loss3-classifier'** of **GoogleNet** DAG network for feature extraction.

Try ML models such as **fitcknn** or **fitcecoc** with and without hyper parameter optimization case to create various machine learning classifiers.

HCW 6.5   CNN Retraining for Fruit/Vegetable Varieties Prediction

The "Fruits" folder contains three subfolders: Training, Validation, and Test; where each contains 24 sub-folders holding the name of a fruit/vegetable. It is also available with Wiley book companion.

Perform feature extraction using a pretrained network for both **GoogleNet** and **AlexNet** and use a machine learning classification model. Use the **'fc6'** layer of **AlexNet** series network for feature extraction and use the fully connected layer **'loss3-classifier'** of **GoogleNet** DAG network for feature extraction.

Try ML models such as **fitcknn** or **fitcecoc** with and without hyper parameter optimization case to create various machine learning classifiers.

7

A Convolutional Neural Network (CNN) Architecture and Training

Convolutional neural networks (CNNs) are a popular type of deep learning tool commonly used for various applications, particularly with images as inputs. However, they are also applicable to other domains like text, signals, and continuous responses. CNNs are inspired by the organization of simple and complex cells in the biological cortex, which respond to specific subregions (receptive regions) of a visual field. Similarly, CNNs are designed to activate based on subregions of an image rather than the entire image. In contrast, the fully connected (FC) layer is responsible for processing the entire image region. This means that neurons in a CNN focus only on their designated subregions and do not handle spatial areas beyond their assigned jurisdiction in the image.

Notice that such subregions might overlap, hence the neurons of a CNN produce spatially correlated outcomes, unlike other types of neural networks, where neurons do not share any connections and produce independent outcomes. Hence, in an FC layer, the number of parameters (weights) can tremendously increase as the size of the input increases. However, in a typical CNN the number of parameters will be reduced as a result of reduced number of connections, shared weights, and down-sampling.

The architecture of a CNN can vary depending on the types and number of layers to include. The types and number of layers to include depend on the intended application or data. if we have categorical responses, then we must have a classification function and a classification layer. On the other hand, if the response is continuous, we must have a regression layer at the end of the network. A small-size network with only one or two convolutional layers might be sufficient to learn a small number of gray scale image data. On the other hand, for more complex data with hundred thousands of colored images, most likely we need a more complicated network with multiple convolutional and fully connected layers.

A CNN passes an image through the network layers and outputs a final class. The network can be composed of tens or hundreds of layers, as each layer learns to detect different image features. Filters are applied to each training image at different levels of resolution, and the output of each convolved image is used as an input to the next layer. The filters can handle very simple features, such as brightness and edges, and very complex features that uniquely define the object as the layers progress. Some commonly used CNN layers are: A typical CNN admits the input image through a set of convolutional filters, each of which activates on (or is triggered by) certain features found in the image. The rectified linear unit (ReLU) layer allows a faster and more efficient training by mapping negative values to zero while retaining positive values as are.

A max-pooling or average-pooling layer simplifies the output by performing nonlinear down-sampling, reducing the number of parameters that the network needs to learn about. A fully connected layer "flattens" the network's 2D spatial features into a 1D vector that represents

Machine and Deep Learning Using MATLAB: Algorithms and Tools for Scientists and Engineers, First Edition.
Kamal I. M. Al-Malah.
© 2024 Kamal I. M. Al-Malah. Published 2024 by John Wiley & Sons, Inc.
Companion Website: www.wiley.com/go/al-malah/machinelearningmatlab

image-level features for classification purposes. Finally, a softmax layer provides the numeric judgment or final decision for each category in the dataset.

A Simple CNN Architecture: The Land Satellite Images

This data set is available at https://www.kaggle.com/datasets/crawford/deepsat-sat6. The original SAT-4 and SAT-6 airborne datasets can also be found at: http://csc.lsu.edu/~saikat/deepsat. Originally, images were extracted from the National Agriculture Imagery Program (NAIP) dataset. The NAIP dataset consists of a total of 330,000 scenes spanning the whole of the Continental United States (CONUS). The authors used the uncompressed digital ortho quarter quad tiles (DOQQs) which are geo TIFF images and the area corresponds to the United States Geological Survey (USGS) topographic quadrangles. The average image tiles are ~6000 pixels in width and ~7000 pixels in height, measuring around 200 megabytes each. The entire NAIP dataset for CONUS is ~65 terabytes. The imagery is acquired at a 1-m ground sample distance (GSD) with a horizontal accuracy that lies within six meters of photo-identifiable ground control points.

The images consist of four bands: red, green, blue, and near infra-red (NIR). In order to maintain the high variance inherent in the entire NAIP dataset, image patches were sampled from a multitude of scenes (a total of 1500 image tiles) covering different landscapes like rural areas, urban areas, densely forested, mountainous terrain, small to large water bodies, agricultural areas, etc. covering the whole state of California.

Once labeled, 28 × 28 non-overlapping sliding window blocks were extracted from the uniform image patch and saved to the dataset with the corresponding label. A 28 × 28 was chosen as the window size to maintain a significantly bigger context, and at the same time not to make it as big as to drop the relative statistical properties of the target class conditional distributions within the contextual window.

The land images made by a satellite are categorized as in the following images. There are six categories in this data set, as shown in Table 7.1.

Table 7.1 Six image classes for land morphology.

Item #	Image Sample	Land Morphology	Item #	Image Sample	Land Morphology
1		Barren Land	4		Road
2		Building	5		Trees
3		Grass Land	6		Water

Displaying Satellite Images

Each individual image has a size of 28 × 28 × 4. We can't visualize all four channels as an image, but we can view the first three together and the fourth separately. Figure 7.1 shows a row which represents one of the six classes listed above.

Figure 7.1 Satellite images view for the first three channels: red, green, and blue.

On the other hand, Figure 7.2 shows the fourth channel which represents the near infra-red (NIR) light spectrum which primarily captures vegetations and can be viewed as a grayscale image.

Figure 7.2 The fourth channel views the captured images using near infrared (NIR) region of the light spectrum.

To classify these images with a pretrained network like **AlexNet**, we would resize the images to 227×227 and remove the fourth channel. This would remove valuable information and significantly increase the training time. Instead, we can solve this problem with a simple custom architecture.

Image datastores are commonly used when images are stored in folders. Another popular way of storing image data is using a 4-D array. This format works well with this data set since the images have four channels. we can use 4-D arrays as input to the **trainNetwork** function.

CNNs use layers specific to image applications. CNNs vary in architecture and depth. For example, **AlexNet** contains 25 layers and **GoogleNet** contains 144. However, all CNNs contain some common layers like an image input layer and convolution layers. Each layer in the network performs some operation on its inputs and outputs a new value.

Here are the sequential steps for creating a simple architecture:

1) The first layer of any convolutional neural network is an image input layer.

>>**imageInputLayer(inputSize)**

This function requires the image size as input. The input size is a three-element vector corresponding to the height, width, and number of channels of that image. Grayscale images have one channel, and color images have three. So, let us Create an image input layer that expects RGB+NIR images sized 28 × 28. Name the layer **inLayer**.

```
inLayer=imageInputLayer([28 28 4]);
```

The first layer defines the input size of the network and normalizes the input images. By default, an image input layer subtracts the mean image of the training data set. This centers the images around zero.

2) Convolution layers learn features in the input image by applying different filters to the image. To create a convolution layer, we need to specify the filter size and the number of filters.

>> **convolution2dLayer([h w],n);**

2-D convolution layers apply sliding filters to the input image. Convolution layers are a key part of the CNN architecture. They rely on the spatial structure of the input image.
   Create a convolution layer with 20 filters each sized 3×3. Name the layer **convLayer**.

```
convLayer=convolution2dLayer([3 3],20);
```

3) Convolution layers are generally followed by rectified linear unit (**ReLU**) and max pooling layer. A **ReLU** layer sets all negative values to zero.

>>**reluLayer();**

The function does not require any inputs.

4) A maximum pooling layer performs down-sampling by dividing the input into rectangular "**pooling**" regions and computing the maximum of each region. Pooling reduces the network complexity and creates a more general network. Use the pool size as input.

>>**maxPooling2dLayer([h w]);**

5) Create a column vector of layers. The first element should be a **ReLU** layer. The second element should be a max pooling layer with a pool size of 3 × 3. Name the array **midLayers**.
We separate each layer with a semicolon (;) to create a column vector.

```
midLayers=[reluLayer();maxPooling2dLayer([3 3])]
```

6) The last three layers of a convolutional neural network are:

>>**fullyConnectedLayer; softmaxLayer; classificationLayer**

Features passing through the network are stored in a collection of matrices until they reach the **fully connected** (**FC**) layer. At the fully connected layer, the input is "flattened" so that it can be mapped to the output classes. This layer is a classical neural network. The output size for this layer is the number of classes for the classification problem. For example, if we were classifying cats and dogs, the output size would be two. In the satellite images case, we have six possible land surfaces.

The **softmax** layer converts the values for each output class into normalized scores using a normalized exponential function. We can interpret each value as the probability that the input image belongs to each class. This layer does not require any input.

The **classification** output layer returns the name of the most likely class. Again, this layer does not require any inputs.

Consequently, create a column vector of output layers named **outLayers**. The first element should be a fully connected layer for six classes. Then add a softmax layer and a classification layer. Separate each layer by a semicolon (;).

```
outLayers=[fullyConnectedLayer(6);softmaxLayer;classificationLayer];
```

To train a network, we need an array of our entire network architecture. The last step is to stack all the layers we have created into a single array. Thus, we create a column vector called layers that contains all the layers created in the previous steps. The order should be as follows:

Figure 7.3 shows the code for the previous steps:

The results are copied from Command Window:

```
inLayer=imageInputLayer([28 28 4]);
convLayer=convolution2dLayer([3 3],20);
midLayers=[reluLayer();maxPooling2dLayer([3 3])];
outLayers=[fullyConnectedLayer(6);softmaxLayer;classificationLayer];
layers=[inLayer;convLayer;midLayers;outLayers]
```

Figure 7.3 Creation of a simple Convolutional Neural Network (CNN).

```
layers = 7×1 Layer array with layers:
    1  "  Image Input          28 ×28 × 4 images with 'zerocenter' normalization
    2  "  Convolution           20 3 × 3 convolutions with stride [1 1] and padding [0 0 0 0]
    3  "  ReLU                  ReLU
    4  "  Max Pooling            3 × 3 max pooling with stride [1 1] and padding [0 0 0 0]
    5  "  Fully Connected       6 fully connected layer
    6  "  Softmax               softmax
    7  "  Classification Output  crossentropyex
```

So far, we have created an architecture that can be used to classify 28×28×4 color images into six classes. Most networks are not this shallow, but this network is sufficient to solve simple image classification problems.

Training Options

Before training, we select training options. There are many options available.

Table 7.2 shows typical training options that can be used in training options, as quoted from MATLAB offline help.

Table 7.2 Typical training options that can be used.

Training Options	Definition	MATLAB Code/Comment
Plots	The plot shows the minibatch loss and accuracy. It includes a stop button that lets the user halt network training at any point.	(**'Plots'**, **'training-progress'**) Plot the progress of the network as it trains. **'none'**: Do not display plots during training.
MaxEpochs	An epoch is the full pass of the training algorithm over the entire training set.	(**'MaxEpochs'**,20) Maximum number of epochs to use for training, specified as a positive integer. The higher value, the longer the network will train, but the higher the accuracy of prediction.
MiniBatchSize	Minibatches are subsets of the training dataset, which are simultaneously processed by GPU. The default is 128.	(**'MiniBatchSize'**,128) The larger the minibatch, the faster the training, but the maximum size will be determined by the GPU memory. If a memory error develops upon training, reduce the minibatch size.
InitialLearnRate	This is a major parameter that controls the speed of training.	A scalar value. The lower the learning rate the more accurate the results, but the longer the network CPU training time.
L2Regularization	Factor for L2 regularizer (weight decay), specified as a nonnegative scalar.	You can specify a multiplier for the L2 regularizer for network layers with learnable parameters.
solverName	Solver for training network. 1) **'sgdm'**: Use the stochastic gradient descent with momentum with the **'Momentum'** name-value pair argument. 2) **'rmsprop'**: Use the RMSProp optimizer with **'SquaredGradientDecayFactor'** name-value pair argument. 3) **'adam'**: Use the Adam optimizer with the **'GradientDecayFactor'** and **'SquaredGradientDecayFactor'** name-value pair arguments, respectively.	

A basic understanding of how training works can help us explore the many training options available.

Mini Batches

To get good results with deep networks, we generally need a lot of training data. Calculating the loss and the gradient by running all the training examples through the network would be a big burden. For this reason, the training set is split up into mini-batches, as shown in Figure 7.4.

Figure 7.4 The concept of mini-batches facilitates the process of spanning the original set of data.

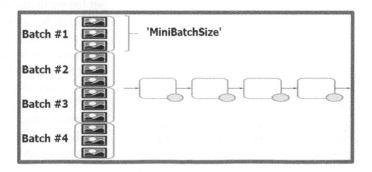

Each iteration of the training is performed with a different mini-batch. For each mini-batch, a CNN will attempt to find the best values of optimizable parameters to maximize the accuracy (or minimize the loss). An iteration is one step taken in the gradient descent algorithm toward minimizing the loss function using a mini-batch. An epoch is the full pass of the training algorithm over the entire training set. A mini-batch is a subset of the training set that is used to evaluate the gradient of the loss function and update the weights. The loss and gradient calculated for each mini-batch is an approximation to the loss and gradient for the full training set, as shown in Figure 7.5. This is known as stochastic gradient descent.

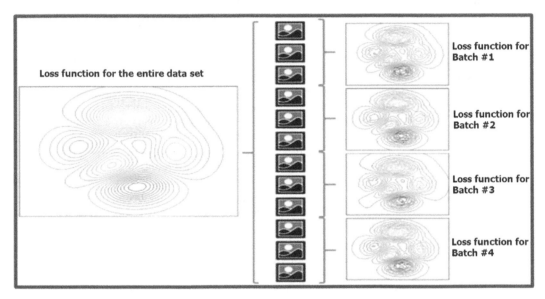

Figure 7.5 The stochastic gradient method based upon the approximate batch-wise gradient methods.

Learning Rates

The gradient descent takes a step in the direction of the negative gradient. The optimizing algorithm moves in $-\nabla E$ direction with the sought of finding the optimum values of model parameters. The size of that step is known as the learning rate. The network training function updates weight and bias values according to gradient descent. The weights and biases are updated in the direction of the negative gradient of the performance function. The learning rate is multiplied times the negative of the gradient to determine the changes to the weights and biases. The larger the learning rate, the bigger the step. If the learning rate is made too large, the algorithm becomes unstable. If the learning rate is set too small, the algorithm takes a longer time to converge. So, we can think of the learning rate as how aggressively we want to modify our optimizable parameters at each step. Figure 7.6 shows how large the jump is from the old to the new values of model parameters. If the jump is relatively large (i.e., red color arrow), then it will be called an aggressive jump. On the other hand, if the jump is relatively short (i.e., green color arrow), then it will be called a cautious or conservative jump.

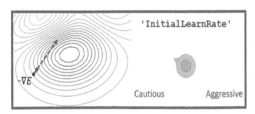

Figure 7.6 The movement in the direction of the negative gradient of the performance function.

We can have a fixed learning rate for the whole training. Alternatively, we can start with a higher learning rate and periodically decrease it as the training progresses, as shown in Figure 7.7.

Figure 7.7 The learning rate for CNN optimizer can be invariant or be held adjustable with training progress.

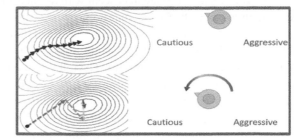

If we plan to change the learning rate with training progress, then a schedule will be made to the learning rate, as explained in Figure 7.8, where the learning rate schedule can be controlled by the indicated options.

Figure 7.8 The learning rate schedule can be controlled by its drop factor and drop period.

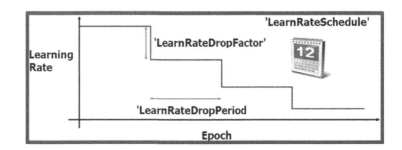

Gradient Clipping

A common problem with using gradients is that the loss function may change rapidly with some parameters but slowly with others, as shown in Figure 7.9. This means some values of the gradient are large and some are small. Very large gradient values can make parameter values change wildly. A simple way to control this is to put a maximum size on the gradient values, using **'GradientThreshold'** option.

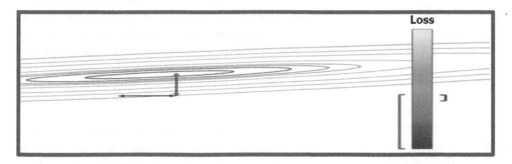

Figure 7.9 The loss function can be sensitive to some parameters over others.

Algorithms

The first input to **trainingOptions** is the name of the training algorithm. We are already familiar with **sgdm**, stochastic gradient descent with momentum, however, there are other algorithms that can be used, as well. By incorporating momentum, **sgdm** enables the optimization process to exhibit smoother and more consistent updates, making it less likely to oscillate or converge slowly. This can lead to faster convergence to a good solution, **especially in high-dimensional and complex optimization problems**. Overall, the main importance of **sgdm** is its ability to accelerate the convergence of the optimization process and improve the chances of finding better solutions by effectively navigating through the parameter space.

Even with a cap limit imposed on the gradient values, using '**GradientThreshold**' option, it turns out that if some values are at the maximum and others are near zero, such a condition can make it hard to choose just the right learning rate. We want smaller steps in the direction of rapid change, but at the same time larger steps in the direction of slow change. A simple fix is to do just that: use a different learning rate for each parameter, as explained in Figure 7.10. Moving parallel to the contours can be aggressive while moving across the contours will be cautious or conservative.

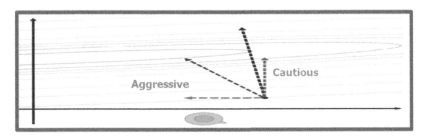

Figure 7.10 Smaller steps are prospected in the direction of rapid change, but at the same time larger steps in the direction of slow change.

The **rmsprop** algorithm keeps a history of the size of the gradient, and uses this to scale the learning rate for each parameter, as explained in Figure 7.11, where the new gradient direction is a combination of the previous two gradient directions. The **rmsprop** algorithm is an optimization algorithm designed to address the limitations of traditional gradient descent methods. It calculates an exponentially weighted average of squared gradients to adapt the learning rate for each parameter. Thus, **rmsprop** uses a moving average of the squared gradients to adjust the learning rate for each parameter individually. It dampens the learning rate for parameters with large gradients and increases it for parameters with small gradients, allowing for faster convergence.

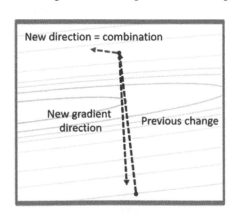

Figure 7.11 The '**rmsprop**' algorithm keeps a history of the size of the gradient, and uses this to scale the learning rate for each parameter.

Thus, **rmsprop** algorithm is particularly well-suited for optimizing neural networks when dealing with **non-stationary (changing over time) or noisy data (containing errors)**.

The adaptive moment estimation **adam** algorithm scales the learning rate for each parameter, like **rmsprop**, but also uses momentum to smooth out the updates, as explained in Figure 7.12.

The **adam** algorithm builds upon the concept of **rmsprop** and incorporates momentum-based updates. In addition to adapting the learning rate like **rmsprop**, **adam** also keeps track of the past gradients' momentum. Thus, **adam** combines the adaptive learning rate approach of **rmsprop** with the momentum update strategy. The momentum term allows the algorithm to maintain a memory of past gradients, which can help accelerate convergence, especially in the presence of **sparse gradients or noisy data**, as is the case with **rmsprop** algorithm.

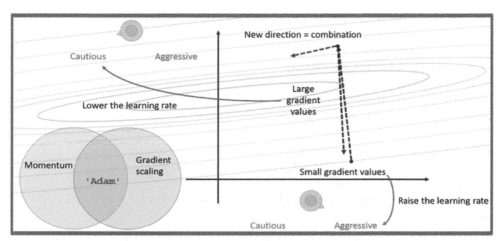

Figure 7.12 The adaptive moment estimation, "**adam**" algorithm scales the learning rate for each parameter, while using momentum to smooth out the updates.

In summary, the performance of these algorithms can vary depending on the specific problem and dataset, so it is often recommended to experiment with different optimization methods to find the most suitable one for a given task.

Training a CNN for Landcover Dataset

Notice that the input image size is 28×28×4. The convolution 2D layer should have 20 filters of size 3×3. The pooling layer should have a pool size of 3×3. There are six classes. Thus, the input of the fully connected layer is set to 6. We can specify any number of settings for training using the **trainingOptions** function. '**MaxEpochs**' and '**InitialLearnRate**' set the maximum number of epochs and the initial learning rate. Create training options named options. Use the stochastic gradient descent with momentum ('**sgdm**'), which is a popular algorithm for CNN training. Set the maximum number of epochs to 5, and set the initial learning rate to 0.0001. Figure 7.13 shows the code for reversing the categories from [0-1]-based values to original class names, defining the layers, setting the training options, and training the customized CNN. Notice that

>> **Ytrncat = onehotdecode(ytrn',mycls3,1);**

decodes each probability vector in **ytrn'** to the most probable class label from the labels specified by **mycls3**. Number *1* specifies the dimension along which the probability vectors are defined. The function decodes the probability vectors into class labels by matching the position of the highest value in the vector with the class label in the corresponding position in **mycls3**. Each probability vector in **ytrn'** is replaced with the value of **mycls3** that corresponds to the highest value in the probability vector.

```
% Loads the landcover dataset.
load sat-6-full.mat;
ytrn=train_y';
mycls= ["building","barren_land","trees","grassland","road","water"];
mycls2 = categorical(mycls);
mycls3=categories(mycls2);
Ytrncat = onehotdecode(ytrn',mycls3,1);
layers=[imageInputLayer([28 28 4]); ...
convolution2dLayer([3 3],20); ...
reluLayer(); maxPooling2dLayer([3 3]); ...
fullyConnectedLayer(6); ...
softmaxLayer(); classificationLayer()];
options=trainingOptions("sgdm","MaxEpochs",5, ...
   "InitialLearnRate",0.0001, 'Plots','training-progress');
landnet = trainNetwork(train_x,Ytrncat,layers,options);
ytest=test_y';
%mycls= ["building","barren_land","trees","grassland","road","water"];
%mycls2 = categorical(mycls);
%mycls3=categories(mycls2);
Ytstcat = onehotdecode(ytest',mycls3,1);
```

Figure 7.13 Fetching response categories, defining CNN layers, and then training the CNN.

Figure 7.14 shows the plot during CNN training phase where both the model accuracy and its loss or plotted versus the number of iterations. The top plot shows the accuracy which is the percentage of training images that the network classified correctly during an iteration. The bottom plot shows the loss. Accuracy does not measure how confident the network is about each prediction. Loss is a measure of how far from a perfect prediction the network is, summed over the set of training images.

The total number of iterations is calculated as number of iterations per epoch times the number of epochs

$$Total\ number\ of\ iterations = 2531 \frac{iterations}{epoch} \times 5\ epoch = 12,655$$

The whole dataset of images was trained for five epochs. This means that the network saw the entire data set five times. Notice that common training options are also shown, like the maximum number of epochs and the learning rate. In Figure 7.13, the default mini-batch size of 128 was used. So, one mini-batch is trained in each iteration. Every point displays the accuracy of classifying 128 different images.

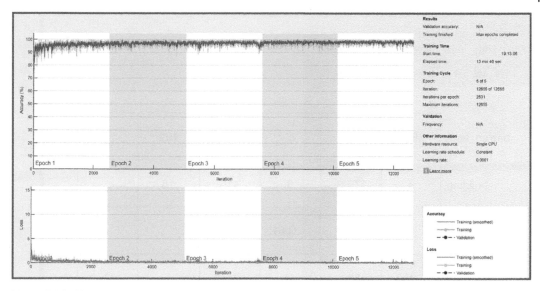

Figure 7.14 The CNN model accuracy and loss as a function of number of iterations.

On the other hand, the test data is stored in a variable **test_x**, and the corresponding response labels are in the one hot decoded variable **Ytstcat**.

We can use the classify function to make predictions with our network. The first input to **classify** function is the trained network name and the second input is the X-data to classify. Use **classify** function to classify the test data **test_x**. Save the predictions in an array named **Ytestpreds**. Display a confusion matrix using the **confusionchart** function. Use the known classes (**Ytstcat'**) as the first input and the predicted classes (**Ytestpreds**) as the second input. Figure 7.15 shows the code for predicting the labels, plotting the confusion matrix, and evaluating the CNN test accuracy.

```
Ytestpreds= classify(landnet,test_x);
confusionchart(Ytstcat',Ytestpreds);
numtrue=nnz(Ytstcat' == Ytestpreds);
accuracy=(numtrue/numel(Ytestpreds))*100;
disp(['Accuracy = ', num2str(accuracy),' %'])
```

Figure 7.15 Predicting the labels, plotting the confusion matrix, and evaluating the CNN test accuracy.

The result is copied from Command Window.

Accuracy = 97.716%.

Figure 7.16 shows the confusion matrix plot for true versus predicted land surface labels.

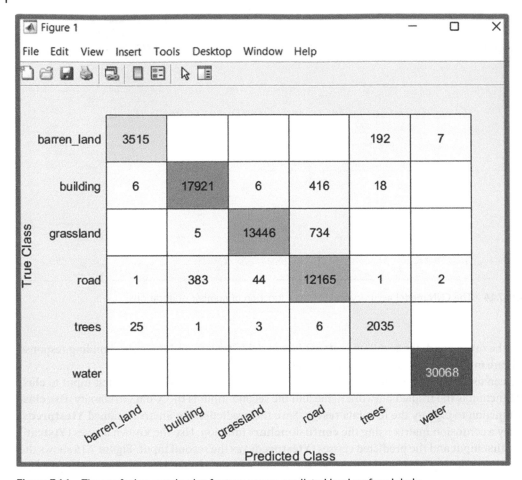

Figure 7.16 The confusion matrix plot for true versus predicted land surface labels.

Exercise

The CNN is about 97.7% accurate on the test set. Which classes does the network most commonly confuse? Does this make sense?

Layers and Filters

When we create a convolution layer, we specify a filter size and the number of filters. These filters extract features found in an input image. Previously, we viewed these extracted features using the **activations** function. Each of the features is created using a different filter. We have seen the result of these filters, but not what the filters look like or how they are learned. We will take a step back to see how convolution layers learn.

When a CNN learns via filters, it involves the process of convolution. Convolutional layers in a CNN apply a set of learnable filters to the input data. These filters are small-sized matrices that are convolved (slid) across the input data, performing element-wise multiplication and summation operations at each position. The resulting values create a feature map, which represents the responses of the filters at different spatial locations in the input. During the training process, the CNN learns the filter weights by adjusting them to minimize a loss function, typically through backpropagation and gradient descent. The filters are updated iteratively, allowing the network to learn the most discriminative and informative patterns from the data. The filters can capture various types of features, such as edges, textures, or more complex structures, depending on the depth and complexity of the CNN.

The activation function is then applied to the outputs of the filters to introduce non-linearities. Common activation functions used in CNNs include Rectified Linear Units (ReLU), sigmoid, or hyperbolic tangent (tanh). These functions allow the network to model non-linear relationships between the input and the learned features, enabling more expressive representations and enhancing the network's ability to generalize to unseen data.

Let us see how an image looks after we apply different filters, like edge detection.

The filters we will use are matrices sized 3 × 3. This is like using a filter sized [3 3] when creating a convolution2dLayer. An RGB image has three channels. To perform convolutions, we need a two-dimensional matrix. We can extract all rows and columns of one of the channels using the colon (:) operator:

>> image(:,:,n)

Extract the first channel of the image **FImage241** to a variable named **RedIm**. Use **imshow** function to display **RedIm**. Figure 7.17 shows the code to load and displays an image.

```
ImDir=fullfile('D:\OneDrive_HCT\MatlabBook\MLDL\Flowers');
%Create an imageDatastore, specifying the read function as a handle to imread
function.
JPGDS = imageDatastore(ImDir, 'IncludeSubfolders',true,'LabelSource',
'foldernames',...
'FileExtensions','.jpg','ReadFcn',@(x) imread(x));
% Access by a reference number of an augmented image datastore.
FImage241=imread(JPGDS.Files{241});
figure(1);
imshow(FImage241)
% Get 1st Channel
RedIm=FImage241(:,:,1);
figure(2);
imshow(RedIm)
```

Figure 7.17 Reading and displaying the original and the red version image.

Figure 7.18 shows the original and red-based image as well as the generated workspace variables.

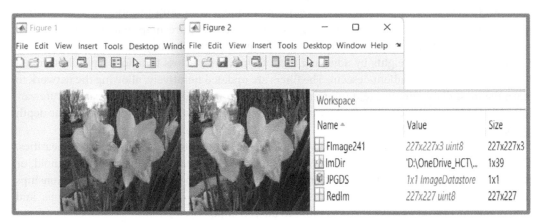

Figure 7.18 The original and red-based image and workspace variables.

We will be using a 3 × 3 matrix called a kernel to filter the image. One of the simplest kernels is the identity kernel.

$$IK = \begin{bmatrix} 1 & 0 & 0 \\ 0 & 1 & 0 \\ 0 & 0 & 1 \end{bmatrix}$$

At every pixel, a convolution using the identity kernel will return the same pixel. The resulting image is therefore the same as the original. We save the identity kernel in a matrix named **IK**.

>>IK = [1 0 0;0 1 0;0 0 1];

Or,

>> IK = eye(3);

We can use **IK** kernel with the **conv2** function to apply the filter **IK** to the image **RedIm**.

>>IKConv=conv2(IK,RedIm);

It is often useful to add a second input argument to **imshow** function when displaying grayscale images. Using empty brackets as the second input will scale the display based on the minimum and maximum values present in the image itself. Display the image **IKConv** using the empty brackets as the second input.

>>imshow(IKConv,[])

Figure 7.19 shows the **IKConv**-treated image and workspace variables.

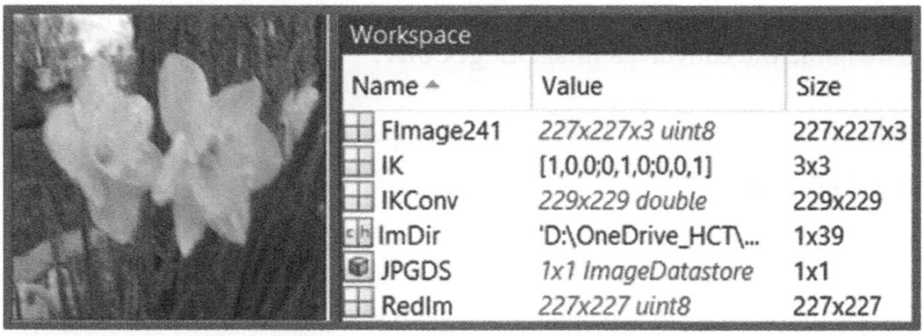

Figure 7.19 The **IKConv**-treated image and workspace variables.

As expected, the image **IKConv** is the same as **RedIm**. Using different kernels will have different effects on our image. We can use the following kernel to blur an image.

$$ONEover9 = \frac{1}{9} \times \begin{bmatrix} 1 & 1 & 1 \\ 1 & 1 & 1 \\ 1 & 1 & 1 \end{bmatrix}$$

At every pixel, this kernel will return an average of the pixel and its immediate neighbors.

Save the blurring kernel in a matrix named **ONEover9**.

Name the convolved image **blurConv**.

ONEover9=(1/9)*ones(3);

blurConv=conv2(ONEover9,RedIm);

imshow(blurConv,[])

Figure 7.20 shows the blurred image by applying **ONEover9** kernel and workspace variables.

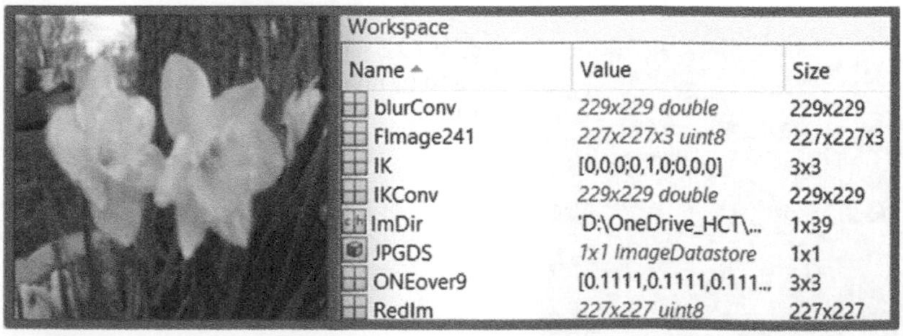

Figure 7.20 The blurred image upon applying **ONEover9** kernel and workspace variables.

Edges are often found as features in early layers of a CNN. We can perform simple edge detection with the kernel:

$$EdgeK = \begin{bmatrix} 0 & 1 & 0 \\ 1 & -4 & 1 \\ 0 & 1 & 0 \end{bmatrix}$$

We save the edge-detecting kernel in a matrix named **EdgeK**. Then we use this kernel to find and display edges in **RedIm**. We name the convolved image **EdgeConv**.

EdgeK=[0 1 0;1 -4 1;0 1 0];

EdgeConv=conv2(EdgeK,RedIm);

imshow(EdgeConv,[])

Figure 7.21 shows the edges in an image as we apply the edge-detecting kernel unto the image and workspace variables.

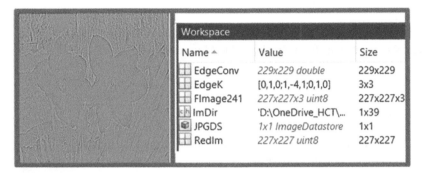

Figure 7.21 Edge detecting kernel being applied to the input image and workspace variables.

We can also use the **imfilter** function to apply the same kernels to the original RGB image. For example:

>>imshow(imfilter(FImage241,EdgeK))

Figure 7.22 shows the filtered image applying **imfilter** function to the original image with **EdgeK** kernel.

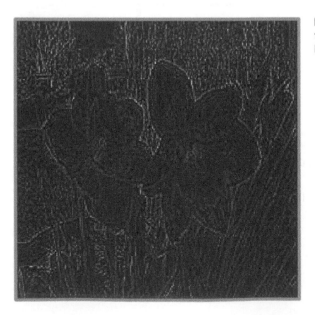

Figure 7.22 The original imaged being filtered by **imfilter** function using one of the kernels.

Filters in Convolution Layers

In the previous section, we applied filters to an input image using convolution. Convolution layers have weights that are used to filter images. Execute the following code:

```
net=alexnet;
layers=net.Layers;
conv=layers(2)
```

Results are copied from Command Window:

```
conv = Convolution2DLayer with properties:
Name: 'conv1'
Hyperparameters
    FilterSize: [11 11]
    NumChannels: 3
    NumFilters: 96
        Stride: [4 4]
    DilationFactor: [1 1]
    PaddingMode: 'manual'
    PaddingSize: [0 0 0 0]
    PaddingValue: 0
Learnable Parameters
    Weights: [11 × 11 × 3 × 96 single]
    Bias: [1 × 1 × 96 single]
```

Notice that the hyperparameters are set when the layer is created. Recall that the first two inputs to **convolution2dLayer** is the filter size and the number of filters. Here, for example, the filter size is set to [11 11] and the number of filters is 96. When we create a convolution layer, the default settings are used for the rest of the hyperparameters, like **Stride**, **PaddingMode**, and **PaddingSize**.

'**NumChannels**' is the number of channels of the input to this convolution layer. In '**conv1**', the input is the image sized 227 × 227 × 3, so the number of channels is 3.

Notice that the learnable parameters are updated during training, but the size of these arrays is calculated from the filter size, number of input channels, and the number of filters, as shown in Figure 7.23. The 11 × 11 size refers to the smallest unit rectangle used by the **convolution2dLayer** to learn image features from the entire 227 × 227 input image.

Figure 7.23 The smallest learning unit rectangle used by **convolution2dLayer** to acquire image features out of the entire 227×227 input image.

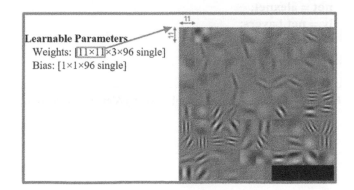

The weights in the first convolution layer can be viewed as an RGB image because there are three channels. We will view filters in **AlexNet** CNN in the next section.

Quiz #1

Given a convolution layer with weights sized 3×3×192×384, what is the number of channels entering the layer?

a) 3   b) 192   c) 384   d) 9

Quiz #2

Given a convolution layer with weights sized 3×3×192×384, what is the filter size?

a) [3 3]   b) [192 192]   c) [192 384]   d) [384 384]

Quiz #3

Given a convolution layer with weights sized 3×3×192×384, what is the number of filters?

a) 3   b) 192   c) 384   d) 9

Viewing Filters: AlexNet Filters

The Weights property of a layer contains that layer's weights. Weights are learned during training.

Since **AlexNet** CNN is a pretrained network, it has already learned weights that will find useful features. In this section, we will inspect these weights in a more detail.

AlexNet's layers are stored in a variable **ly**. Save the weights of the second layer to a variable named **W** and the size of the weight matrix as **WSize**, as shown in Figure 7.24.

```
ImDir=fullfile('D:\OneDrive_HCT\MatlabBook\MLDL\Flowers');
%Create an imageDatastore, specifying the read function as a handle to imread
function.
JPGDS = imageDatastore(ImDir, 'IncludeSubfolders',true,'LabelSource',
'foldernames',...
'FileExtensions',',jpg','ReadFcn',@(x) imread(x));
% Access by a reference number in an augmented image datastore.
FImage241=imread(JPGDS.Files{241});
net = alexnet;
ly = net.Layers;
%Extract weights
W=ly(2).Weights;
WSize=size(W)
```

Figure 7.24   Reading the image file, loading **AlexNet**, and showing the weights of the second layer.

WSize = 11   11   3   96

The weights of the first convolution layer are an array sized $11 \times 11 \times 3 \times 96$. This means that there are 96 different filters with 3 channels. Each of those filters is $11 \times 11 \times 3$, so you can view them as RGB images.

These filters need to be normalized so they can be displayed correctly. The rescale function will scale the weights to be between 0 and 1. View the array of weights using the montage function. Before displaying the weights, we normalize the array with rescale.

>>montage(rescale(W))

Figure 7.25 shows the normalized weights filters (i.e., 96 filters of 11×11 size each).

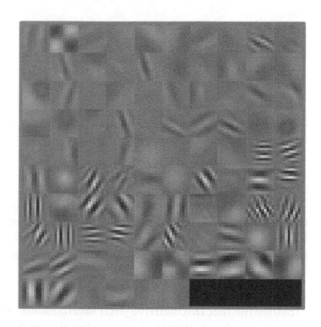

Figure 7.25   The normalized weights filters (96 filters of 11 × 11 size each).

We can view a specific filter by indexing into the weights. We need all rows, columns, and RGB values of a specific filter number.

>>filter = W(:,:,:,n)

Let us extract the tenth filter from the array of weights, then display it. We should rescale this image with rescale before displaying it with **imshow** function.

>>filt10=W(:,:,:,10);

>>imshow(rescale(filt10))

Figure 7.26 shows the normalized tenth filter characterized by red-like spots.

Figure 7.26 The normalized tenth filter, branded by red-like spots.

Let us save the activations from the network net for the image **FImage241** at the layer 'convl' and name the output **actvn**.

>>**actvn = activations(net,FImage241,'convl');**

The activations are sized m×n×96. There is one activation, or feature, for every filter in the first convolution layer. To investigate the tenth filter, we need to get the tenth activation channel. So, let us extract the activations from the tenth filter, rescale this matrix, and then display it with **imshow** function. The complete code is shown in Figure 7.27.

```
ImDir=fullfile('D:\OneDrive_HCT\MatlabBook\MLDL\Flowers');
%Create an imageDatastore, specifying the read function as a handle to imread
function.
JPGDS = imageDatastore(ImDir, 'IncludeSubfolders',true,'LabelSource',
'foldernames',...
'FileExtensions',',jpg','ReadFcn',@(x) imread(x));
% Access by a reference number in an augmented image datastore.
FImage241=imread(JPGDS.Files{241});
net = alexnet;
actvn = activations(net,FImage241,'convl');
filt10b=actvn(:,:,10);
figure; imshow(FImage241, []);
figure; imshow(rescale(filt10b))
```

Figure 7.27 The code for presenting the original image and the 10th filter activated image.

Figure 7.28 shows both the original (*left*) and the orange-like color-activated tenth filter image (*right*).

Figure 7.28 The original (*left*) and the orange-like-color-activated tenth-filter image (*right*). Spots are labeled by either a rectangle or an ellipse.

The orange-like colored spots activate strongly. We can use the previous steps to investigate any of the filters displayed in the montage. For example, the 48th filter finds transitions between blue and orange.

Quiz #4

1) The filter size in a convolution layer is

a) Created by hand, using matrix concatenation and image processing techniques.
b) Learned by the network during training.
c) Set by the user upon creating the 2d convolution layer.

Quiz #5

(Select all that apply) Which of these layers can be assembled together into a deep network?

a) Convolution layer   b) Rectified Linear Unit layer

c) Pooling layer   d) Fully Connected layer

Validation Data

So far, we have evaluated a network using a test data set after training. we can also use a *validation* data set. A validation data set is a separate split of the data used to evaluate the network during training.

- **Training data**: used during training to update weights.
- **Validation data**: used during training to evaluate performance.
- **Testing data**: used after training to evaluate performance.

Validation data is useful to detect if the network is *overfitting*. Although the training loss may be decreasing with training, however, if the validation loss is increasing, we should stop training because the network is learning details about the training data that aren't relevant to new images.

```
>> [trainds,valds,testds] = splitEachLabel(imds,0.8,0.1)
```

where

trainds: **Training images**

valds: **Validation images**

testds: **Testing images**

imds: **Name of the file provided as a string.**

0.8: **80% of the data used to train.**

0.1: **10% of the data is used for validation, and the remaining 10% is used for test.**

There are three training options related to validation.

- 'ValidationData': Validation data and labels.
- 'ValidationFrequency': Number of iterations between each evaluation of the validation data.
- 'ValidationPatience': The number of validations to check before stopping training.

NOTE #1

In addition to the above **splitEachLabel** function, the user can divide the entire image data store into three subsets for training, validation, and testing steps, using **subset** function. See Appendix A.23 for explaining how to split a given image data store into three image data stores.

Fluctuations in the loss from one iteration to another are normal, so we generally don't want to stop training as soon as the validation loss increases. Instead, we perform several validations. If the loss of validation has not reached a new minimum in that time, then we stop the training.

Validation data can be a datastore, table, or cell array. Because the images and labels for this data set are stored in two different variables, it is simplest to create a cell array. A cell array of validation data should have two elements, where the first contains the images and the second contains the labels.

Figure 7.29 shows the code for the following steps:

1) Loading training, validation, and testing data.
2) Creating an image datastore for training, validation, and testing data.
3) Creating CNN layers.
4) Creating validation data.
5) Setting preliminary training options.
6) Training the CNN with the specified layers and training options.
7) Predicting the classification of the test data.
8) Expressing the accuracy of the predicted test data.

```
rng(123);
% Create a datastore
ImDir=fullfile('D:\OneDrive_HCT\MatlabBook\MLDL\Flowers');
imds=imageDatastore(ImDir,'IncludeSubfolders',true,'LabelSource','foldernames');
%Split datastore
[trainIm,valIm,testIm] = splitEachLabel(imds,0.8,0.1);
%Preprocess datastores
trainds=augmentedImageDatastore([28 28],trainIm);
valds=augmentedImageDatastore([28 28],valIm);
testds=augmentedImageDatastore([28 28],testIm);
inLayer=imageInputLayer([28 28 3]);
convLayer=convolution2dLayer([6 6],120);
reluLayer();
midLayers=[reluLayer();maxPooling2dLayer([6 6])];
outLayers=[fullyConnectedLayer(12);softmaxLayer;classificationLayer];
layers=[inLayer;convLayer;midLayers;outLayers];
valData={valds valIm.Labels};
options = trainingOptions('adam','MaxEpochs',40, ...
   'InitialLearnRate',0.0001,'Plots','training-progress', ...
   'ValidationData', valData, ...
  'ValidationFrequency', 40);
mynet2 = trainNetwork(trainds,layers,options);
Ytestpreds= classify(mynet2,testds);
Ytst=testIm.Labels;
confusionchart(Ytst,Ytestpreds);
numtrue=nnz(Ytst == Ytestpreds);
accuracy=(numtrue/numel(Ytestpreds))*100;
disp(['Accuracy = ', num2str(accuracy),' %'])
```

Figure 7.29 Creating and splitting an image datastore and configuring and training a CNN.

Figure 7.30 shows the completion of CNN training and the accuracy of the model using **splitEach-Label** function for image datastore splitting. Workspace variables are also shown.

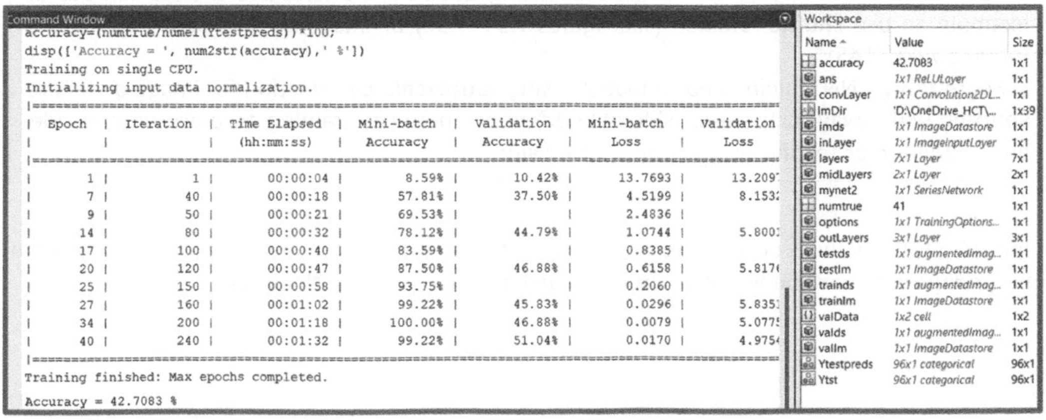

Figure 7.30 The completion of CNN training showing both the training and validation accuracy.

Figure 7.31 shows the CNN training and validation results, using **splitEachLabel** function for image datastore splitting.

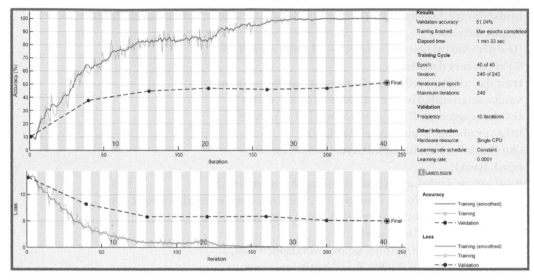

Figure 7.31   CNN training and validation using **splitEachLabel** function of image datastore splitting.

NOTE #2

It should be pointed out here that no matter what we do in terms of maximizing the validation accuracy, the plateau value is still around 50%. For example, Figure 7.32 shows the code for increasing the total number of layers and also inserting, more than once, **dropoutLayer** and **batchNormalizationLayer** within the CNN architecture. MATLAB expects the class label order in training and validation set to be in the same sequence. In the training plot, when there is a mismatch between training and validation set sequence, the **predict()** function computes the probability scores based on the training set sequence and accuracy is computed based on the validation set sequence. This mismatch results in lower accuracy, but actually the CNN model should have trained well, without any over fitting issues. The alternative choices can be, for example, to use **shuffle** function (see figures 7.34–7.36) or image augmentation data store (figures 7.42–7.45).

Regrettably, CNN training and validation using **splitEachLabel** function for image datastore splitting will always cause a considerable difference between training accuracy, on one side, and validation or testing accuracy on another side.

Figure 7.32 shows the code for increasing the total number of layers and also inserting, more than once, **dropoutLayer** and **batchNormalizationLayer** within the CNN layers.

```
rng(123);
% Create a datastore
ImDir=fullfile('D:\OneDrive_HCT\MatlabBook\MLDL\Flowers');
imds=imageDatastore(ImDir,'IncludeSubfolders',true,'LabelSource','foldernames');
%Split datastore
[trainIm,valIm,testIm] = splitEachLabel(imds,0.8,0.1);
%Preprocess datastores
trainds=augmentedImageDatastore([28 28],trainIm);
valds=augmentedImageDatastore([28 28],valIm);
testds=augmentedImageDatastore([28 28],testIm);
layers = [
  imageInputLayer([28 28 3])
  convolution2dLayer(3,8,'Padding',1)
batchNormalizationLayer
  reluLayer
dropoutLayer(0.3)
  maxPooling2dLayer(2,'Strid',1)
  convolution2dLayer(3,16,'Padding',1)
batchNormalizationLayer
  reluLayer
dropoutLayer(0.3)
  maxPooling2dLayer(2,'Stride',1)
  convolution2dLayer(3,32,'Padding',1)
batchNormalizationLayer
  reluLayer
dropoutLayer(0.3)
  fullyConnectedLayer(1024)
batchNormalizationLayer
  reluLayer
dropoutLayer(0.3)
  fullyConnectedLayer(1024)
batchNormalizationLayer
  reluLayer
dropoutLayer(0.3)
  fullyConnectedLayer(1024)
batchNormalizationLayer
  reluLayer
dropoutLayer(0.3)
  fullyConnectedLayer(12)
  softmaxLayer
  classificationLayer];
valData={valds valIm.Labels};
miniBatchSize = 128;
maxEpochs = 40;
```

Figure 7.32 Making CNN deeper and adding two types of filtering layers to cross the gap between training and validation test data, in the form of **dropoutLayer** and **batchNormalizationLayer**.

```
learnRate = 0.0001;
options = trainingOptions('adam',...
   'MiniBatchSize',miniBatchSize,...
   'MaxEpochs',maxEpochs,...
   'InitialLearnRate',learnRate,...
   'ValidationData',valData,...
   'ValidationFrequency',5,...
   'Plots','training-progress',...
   'Shuffle','every-epoch',...
   'Verbose',true);
[net, info] = trainNetwork(trainds, layers, options);
Ytestpreds= classify(net,testds);
Ytst=testIm.Labels;
confusionchart(Ytst,Ytestpreds);
numtrue=nnz(Ytst == Ytestpreds);
accuracy=(numtrue/numel(Ytestpreds))*100;
disp(['Accuracy = ', num2str(accuracy),' %'])
```

Figure 7.32 (Cont'd)

Figure 7.33 shows the training and validation plot for the deepened CNN while using **splitEach-Label** function for image datastore splitting.

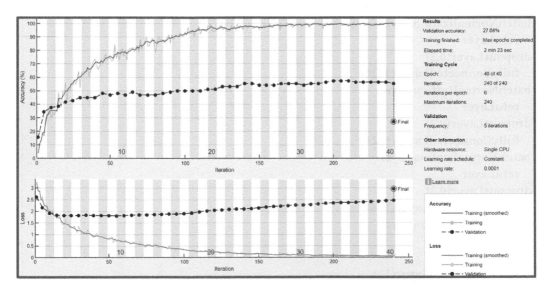

Figure 7.33 The gap between training and validation is still nagging using **splitEachLabel** function for image datastore splitting.

Using shuffle Function

On the other hand, using **shuffle** function to shuffle all entries of an image datastore will give better validation and testing accuracy. The code to use **shuffle** function is shown in Figure 7.34.

```
rng(123);
% Create a datastore
ImDir=fullfile('D:\OneDrive_HCT\MatlabBook\MLDL\Flowers');
imds1=imageDatastore(ImDir,'IncludeSubfolders',true,'LabelSource','foldernames');
trainds=augmentedImageDatastore([18 18],imds1);
rng(123);
imds2=shuffle(imds1);
valds=augmentedImageDatastore([18 18],imds2);
rng(123);
imds3=shuffle(imds1);
testds=augmentedImageDatastore([18 18],imds3);
layers = [
        imageInputLayer([18 18 3])
convolution2dLayer(3,16,'Padding',1)
        reluLayer
        maxPooling2dLayer(2,'Strid',1)
        convolution2dLayer(3,16,'Padding',1)
        reluLayer
        maxPooling2dLayer(2,'Strid',1)
        convolution2dLayer(3,16,'Padding',1)
        reluLayer
maxPooling2dLayer(2,'Stride',1)
convolution2dLayer(3,16,'Padding',1)
        reluLayer
maxPooling2dLayer(2,'Stride',1)
convolution2dLayer(3,16,'Padding',1)
        reluLayer
        maxPooling2dLayer(2,'Stride',1)
        convolution2dLayer(3,32,'Padding',1)
        reluLayer
        fullyConnectedLayer(1024)
        reluLayer
        fullyConnectedLayer(1024)
        reluLayer
        fullyConnectedLayer(1024)
        reluLayer
        fullyConnectedLayer(12)
        softmaxLayer
        classificationLayer];
valData={valds imds2.Labels};
miniBatchSize = 120;
maxEpochs = 20;
learnRate = 0.0001;
```

Figure 7.34 Shuffling the image data store prior to validation and testing step.

```
options = trainingOptions('adam',...
   'MiniBatchSize',miniBatchSize,...
   'MaxEpochs',maxEpochs,...
   'InitialLearnRate',learnRate,...
   'ValidationData',valData,...
   'ValidationFrequency',20,...
   'Plots','training-progress',...
   'Shuffle','every-epoch',...
   'Verbose',true);
[net, info] = trainNetwork(trainds, layers, options);
Ytestpreds= classify(net,testds);
Ytst=imds3.Labels;
confusionchart(Ytst,Ytestpreds);
numtrue=nnz(Ytst == Ytestpreds);
accuracy=(numtrue/numel(Ytestpreds))*100;
disp(['Accuracy = ', num2str(accuracy),' %'])
```

Figure 7.34 (Cont'd)

The accuracy is 100% for all phases: training, validation, and testing. Figure 7.35 shows the training plot accuracy % and loss % for CNN as a function of carried out iterations.

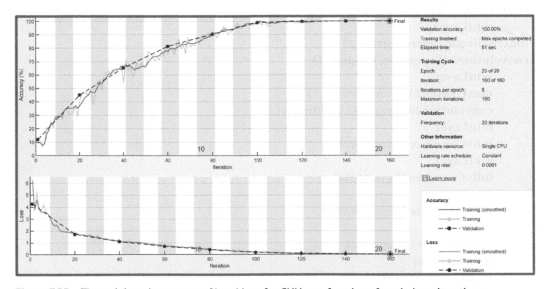

Figure 7.35 The training plot accuracy % and loss for CNN as a function of carried out iterations.

Figure 7.36 shows the confusion matrix plot for true versus predicted image type and workspace variables.

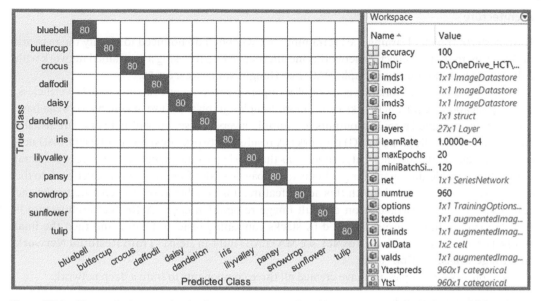

Figure 7.36 The confusion matrix plot for true versus predicted image type and workspace variables.

NOTE #3
Notice that the validation accuracy is similar to the training accuracy. This indicates that the network is not overfitting, which is a good sign.   Moreover, it is possible to stop network training before it reaches the maximum number of epochs using validation patience. When the validation loss does not achieve a new minimum for **n** validations, where **n** is the setting for **'ValidationPatience'**, training will stop.

Improving Network Performance

The first time we train a network, the accuracy is often not adequate. How can we improve the network?

Training Algorithm Options

Modifying the training options is generally the first place to begin improving a network. So far, we have used training options that were provided. See Table 7.2 for other options.

Training Data

If we do not have enough training data, our network may not generalize to new data. If we cannot get more training data, augmentation is a good alternative. In MATLAB, we can use different datastores to augment images.

Architecture

If we perform transfer learning (i.e., retraining), we will often do not need to modify the network architecture to train an effective network. One alternative is to try using a pretrained network with a directed acyclic architecture graph, DAG, like **GoogleNet** or **ResNet-50**.

Notice that using a network that consists of a simple sequence of layers is sufficient. Nevertheless, some applications require networks with a more complex graph structure in which layers can have inputs from multiple layers and outputs to multiple layers. These types of networks are usually called directed acyclic graph (DAG) networks. A residual network (**ResNet**) is a type of DAG network that has residual (or shortcut) connections that can bypass the main network layers. Residual connections enable the parameter gradients to propagate easier from the output layer back to the earlier layers of the network. With this added feature, it is possible to train deeper networks and can result in higher accuracies on difficult image recognition cases. A **ResNet** architecture is mainly made of initial layers, followed by stacks containing residual blocks, and then the final layers. For more information on **ResNet**, see MATLAB | Built-in Help | "**Train Residual Network for Image Classification**" topic.

In the previous examples, we have created all three components to train a deep network:

1) Data with known labels.
2) An array of layers representing the network architecture.
3) Options that control the behavior of the training algorithm.

After the network has been trained, we calculate the accuracy on a test data set. After all, if we are happy with the performance of the network, we can begin using this network in production, as indicated in Figure 7.37.

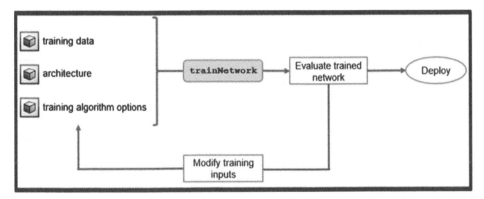

Figure 7.37 CNN options to modify and evaluate the outcome prior to deployment.

Figure 7.38 shows more general guidelines when training a CNN. In addition to these recommendations, we should use a validation data set when possible and use the training progress plot to observe network training patterns. Both accuracy and training time are part of a network's performance. There is always a tradeoff between maintaining a high accuracy and decreasing training time. There is no single "magic" way to set the training options for a deep neural network, but there are some recommended steps to improve our training.

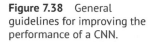

Figure 7.38 General guidelines for improving the performance of a CNN.

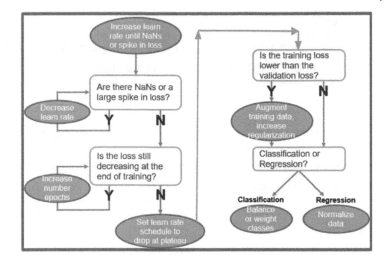

In light of Figure 7.38, it is worth addressing the following points:

- The ideal learning rate is the largest value such that the network trains effectively. To find this learning rate, we should increase the learning rate until we see the loss makes a kick or jump. Then, we decrease the loss until training proceeds appropriately.
- If the learning rate is too small, it will take a very long time to train the network. On the contrary, if the learning rate is too big, the model loss may evaluate to **NaN** or may encounter a large rapid increase in the loss at the beginning of training. If there is a large spike in loss, or loss values are no longer being plotted (**NaN**), our initial learning rate is probably too high. In this case, it is preferred to decrease the learning rate by a power of ten until our loss decreases.
- Increasing the maximum epochs ensures that the network finishes learning. If the accuracy and loss have plateaued for many epochs, we can then decrease the number of epochs to decrease training time.
- Validation criterion exists to prevent overfitting in our network. We can see a difference between training accuracy and validation accuracy. This indicates that the underlying network model begins overfitting. If the validation accuracy begins to plateau over the last epochs, we can limit the training time by reducing the maximum number of epochs.
- Set the learn rate schedule to drop once reaching a plateau value. Once the training has converged, we can set the '**LearnRateSchedule**' and '**learnRateDropPeriod**' to decrease the learning rate at this point. Make sure the training has converged before the learning rate is lowered.
- Validation is key to determining if our network is overfitting. If so, one may attempt to add data augmentation or limit the number epochs. We can also increase the factor **L2** regularization. This option decays the weights.
- If we have very unbalanced classes, try splitting the data so that the training images have an equal number of samples from each class. If regression is sought, data normalization can help stabilize and speed up the network training.
- If we train from scratch, we can try modifying the network architecture to prevent overfitting. Try adding a dropout layer before the last fully connected or convolutional layer. It is also possible to underfit if both the training and validation accuracy are equal. In this case, try increasing the depth of the network (i.e., the number of layers).

Quiz #6

If the validation accuracy is less than the training accuracy, your network may be

a) Overfitting   b) Underfitting   c) Just right

Quiz #7

If the loss is still decreasing at the end of the training, which action is recommended to improve performance?

a) Decrease the learning rate   b) Increase the mini-batch size

c) Increase the number of epochs   d) Decrease the validation frequency

---

Quiz #8

If the loss stops displaying on the training progress plot, which action is recommended to improve performance?

a) Decrease the learning rate   b) Increase the mini-batch size

c) Increase the number of epochs   d) Decrease the validation frequency

---

Image Augmentation: The Flowers Dataset

We can use an **augmentedImageDatastore** function to transform our images with a combination of rotation, reflection, shear, and translation, as shown in Figure 7.39. Augmentation prevents overfitting because the network is unable to memorize the exact details of the training data. When we train the network, the datastore will randomly augment the images using the settings we provide. The transformed images are not stored in memory, which render them as new sets of images from a training point of view.

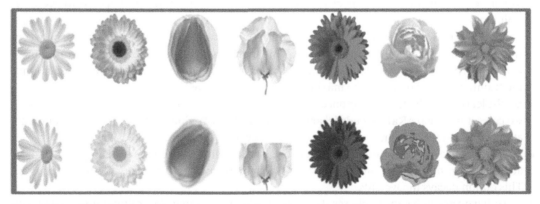

Figure 7.39 Augmentation with a slight modification of the original images (*top*) via an image treatment function will render them as new images (*bottom*).

Let us re-run the flowers image data store. The **imageDataAugmenter** function can be used to choose our augmentation. Possible augmentations include transformations like reflection, translation, and scaling. Generally, we should choose an augmentation that is relevant to your data set.

We will rotate the images a random amount. Create an image data augmenter that rotates an image between 0° (min) and 360° (max) and name the augmenter **aug1**.

>>**aug1=imageDataAugmenter('RandRotation',[min max])**

When we create an augmented image datastore, we need to specify the output image size, the source of the files, and the augmenter.

>>**augmentedImageDatastore(size,ds,'DataAugmentation',aug1);**

Let us create an augmented image datastore of the flowers data set named **augImds1** using our augmenter **aug1**. We set the output size to [200 200].

>>**augImds1=augmentedImageDatastore([200 200],imds,'DataAugmentation',aug1);**

We can read data from an augmented datastore with the read function. Instead of returning one image, read will return a batch of data. Each returned image has a different random augmentation.

Let us save the first batch of augmented data in a table named **data1**.

>>**data1=read(augImds1);**

To display a transformed image, we need to extract one of the images from the table. The variable containing the images is named input.

>>**im = data1.input{n}**

We can display the first image in table data.

>>**imshow(im,[])**

We can stretch an image by setting **'RandXScale'** to [1 2].

Figure 7.40 shows the code for reading an image from the image data store, carrying out image augmentation, and presenting both the original and adulterated image.

```
rng(123);
% Create a datastore
ImDir=fullfile('D:\OneDrive_HCT\MatlabBook\MLDL\Flowers');
imds=imageDatastore(ImDir,'IncludeSubfolders',true,'LabelSource','foldernames');
figure (1);
flwrIm1=imread(imds.Files{1});
imshow(flwrIm1,[])
aug1=imageDataAugmenter('RandRotation',[0 360]);
augImds1=augmentedImageDatastore([200 200],imds,'DataAugmentation',aug1);
```

Figure 7.40 The code for reading an image from the image data store, carrying out image augmentation, and presenting both the original and adulterated image.

```
data1=read(augImds1);
imTbl1=data1.input{1};
figure (2);
imshow(imTbl1,[])
figure (3);
flwrIm2=imread(imds.Files{2});
imshow(flwrIm2,[])
aug2=imageDataAugmenter('RandScale',[1 2]);
augImds2=augmentedImageDatastore([200 200],imds,'DataAugmentation',aug2);
data2=read(augImds2);
imTbl2=data2.input{2};
figure (4);
imshow(imTbl2,[])
```

Figure 7.40 (Cont'd)

Figure 7.41 shows the original (*left*) and the modified image (*middle*) as well as the workspace variables (*right*). The first original image is rotated 90° clockwise and the second original is zoomed in, or scaled by a factor between 1 and 2.

Figure 7.41   The original (*left*) and modified (*middle*) image using one of image augmentation functions. Workspace variables (*right*).

Table 7.3 shows some typical properties of **imageDataAugmenter** function, quoted from MATLAB offline help.

Table 7.3 Typical Properties of **imageDataAugmenter** function.

Property	Value/Description
RandXReflection	Random reflection. false (default) I true. When **RandXReflection** is true (1), each image is reflected horizontally with 50% probability. When **RandXReflection** is false (0), no images are reflected.
RandYReflection	Random reflection. false (default) I true. When **RandYReflection** is true (1), each image is reflected vertically with 50% probability. When **RandYReflection** is false (0), no images are reflected.
RandRotation	Range of rotation, in degrees. [0 0] (default) I 2-element numeric vector I function handle.
	2-element numeric vector. The second element must be larger than or equal to the first element. The rotation angle is picked randomly from a continuous uniform distribution within the specified interval.
RandScale	Range of uniform scaling. [1 1] (default) I 2-element numeric vector I function handle.
	The second element must be larger than or equal to the first element. The scale factor is picked randomly from a continuous uniform distribution within the specified interval.
RandXScale	Range of horizontal scaling. [1 1] (default) I 2-element vector of positive numbers I function handle. The second element must be larger than or equal to the first element. The horizontal scale factor is picked randomly from a continuous uniform distribution within the specified interval. If **RandScale** is specified, then **imageDataAugmenter** will ignore the value of **RandXScale** when scaling images.
RandYScale	Range of vertical scaling. [1 1] (default) I 2-element vector of positive numbers I function handle. 2-element numeric vector. The second element must be larger than or equal to the first element. The vertical scale factor is picked randomly from a continuous uniform distribution within the specified interval. If **RandScale** is specified, then **imageDataAugmenter** will ignore the value of **RandYScale** when scaling images.
RandXShear	Range of horizontal shear; measured as an angle in degrees and is in the range (–90, 90).
	[0 0] (default) I 2-element numeric vector I function handle.
	2-element numeric vector. The second element must be larger than or equal to the first element. The horizontal shear angle is picked randomly from a continuous uniform distribution within the specified interval.
RandYShear	Range of vertical shear; measured as an angle in degrees and is in the range (–90, 90).
	[0 0] (default) I 2-element numeric vector I function handle.
	2-element numeric vector. The second element must be larger than or equal to the first element. The vertical shear angle is picked randomly from a continuous uniform distribution within the specified interval.
RandXTranslation	Range of horizontal translation. Translation distance is measured in pixels.
	[0 0] (default) I 2-element numeric vector I function handle.
	2-element numeric vector. The second element must be larger than or equal to the first element. The horizontal translation distance is picked randomly from a continuous uniform distribution within the specified interval.
RandYTranslation	Range of vertical translation. Translation distance is measured in pixels.
	[0 0] (default) I 2-element numeric vector I function handle.
	2-element numeric vector. The second element must be larger than or equal to the first element. The vertical translation distance is picked randomly from a continuous uniform distribution within the specified interval.

NOTE #4

For further information, see offline built-in MATLAB help: **imageDataAugmenter** function: Configure image data augmentation.

Figure 7.42 shows the creation of two modified augmented image data stores out of the original image data store: **'imds'**, using **'RandScale'** and **'RandXShear'** image data augmenter. These two additionally created augmented image data stores will be utilized as a validation and testing data set.

Moreover, the created CNN along with its layers, number-wise and type-wise, is basically **AlexNet** serial network, except two layers of **batchNormalizationLayer** type are inserted at two positions: One as the sixth layer and another as the twenty-second layer. The goal is to offset the difference between validated and trained accuracy over training iterations. In other words, we try to avoid the overfitting condition where the trained is higher than validated accuracy.

```
rng(123);
% Create a datastore
ImDir=fullfile('D:\OneDrive_HCT\MatlabBook\MLDL\Flowers');
imds=imageDatastore(ImDir,'IncludeSubfolders',true,'LabelSource','foldernames');
trainds=augmentedImageDatastore([227 227],imds);
aug1=imageDataAugmenter('RandScale',[0.8 1.2]);
valds=augmentedImageDatastore([227 227],imds,'DataAugmentation',aug1);
aug2=imageDataAugmenter('RandXShear',[-30 30]);
testds=augmentedImageDatastore([227 227],imds,'DataAugmentation',aug2);
layers = [
imageInputLayer([227 227 3])
convolution2dLayer(11,96,'NumChannels',3,'Stride',4)
reluLayer
crossChannelNormalizationLayer(5,'K',1, 'Alpha', 1.0000e-04,'Beta',0.75);
maxPooling2dLayer(3,'Stride',2)
batchNormalizationLayer
groupedConvolution2dLayer(5,128,2,'Padding',[2 2 2 2])
reluLayer
crossChannelNormalizationLayer(5,'K',1, 'Alpha', 1.0000e-04,'Beta',0.75);
maxPooling2dLayer(3,'Stride',2)
convolution2dLayer(3,384,'NumChannels',256,'Stride',1, 'Padding',[1 1 1 1])
reluLayer
groupedConvolution2dLayer(3,192,2,'Padding',[1 1 1 1])
reluLayer
groupedConvolution2dLayer(3,192,2,'Padding',[1 1 1 1])
reluLayer
maxPooling2dLayer(3,'Stride',2)
fullyConnectedLayer(4096)
reluLayer
dropoutLayer(0.5)
```

Figure 7.42 **AlexNet**-like serial CNN to predict flowers types while using augmented image data stores for validation and testing using **imageDataAugmenter** function.

```
fullyConnectedLayer(4096)
batchNormalizationLayer
reluLayer
dropoutLayer(0.5)
fullyConnectedLayer(12)
softmaxLayer
classificationLayer];
valData={valds imds.Labels};
miniBatchSize = 256;
maxEpochs = 50;
learnRate = 0.00003;
options = trainingOptions('adam',...
  'MiniBatchSize',miniBatchSize,...
  'MaxEpochs',maxEpochs,...
  'InitialLearnRate',learnRate,...
  'ValidationData',valData,...
  'ValidationFrequency',30,...
  'Plots','training-progress',...
  'Shuffle','every-epoch',...
  'Verbose',true);
[net, info] = trainNetwork(trainds, layers, options);
Ytestpreds= classify(net,testds);
Ytst=imds.Labels;
confusionchart(Ytst,Ytestpreds);
numtrue=nnz(Ytst == Ytestpreds);
accuracy=(numtrue/numel(Ytestpreds))*100;
disp(['Accuracy = ', num2str(accuracy),' %'])
```

Figure 7.42 (Cont'd)

Figure 7.43 shows the mini-batch accuracy of the 50th epoch reached 99.61%, the validation accuracy 97.6%, and testing accuracy 97.4%.

```
Initializing input data normalization.
|==================================================================================================|
| Epoch | Iteration | Time Elapsed | Mini-batch | Validation | Mini-batch | Validation | Base Learning |
|       |           | (hh:mm:ss)   | Accuracy   | Accuracy   | Loss       | Loss       | Rate          |
|==================================================================================================|
|    1  |      1    |  00:00:22    |   7.81%    |  20.42%    |  3.3811    |  2.3765    | 3.0000e-05    |
|   10  |     30    |  00:05:27    |  58.20%    |  70.31%    |  1.1767    |  0.9584    | 3.0000e-05    |
|   17  |     50    |  00:08:52    |  71.48%    |            |  0.8182    |            | 3.0000e-05    |
|   20  |     60    |  00:10:52    |  80.47%    |  84.69%    |  0.5196    |  0.4648    | 3.0000e-05    |
|   30  |     90    |  00:16:36    |  91.02%    |  94.48%    |  0.2495    |  0.1630    | 3.0000e-05    |
|   34  |    100    |  00:18:34    |  97.27%    |            |  0.1225    |            | 3.0000e-05    |
|   40  |    120    |  00:22:33    |  98.05%    |  96.15%    |  0.0882    |  0.1113    | 3.0000e-05    |
|   50  |    150    |  00:28:16    |  99.61%    |  97.60%    |  0.0269    |  0.0689    | 3.0000e-05    |
|==================================================================================================|
Training finished: Max epochs completed.
Accuracy = 97.3958 %
```

Figure 7.43 Training, validation and testing accuracy of the trained CNN.

Figure 7.44 shows both the accuracy % and loss as a function of number of iterations for trained and validation data set.

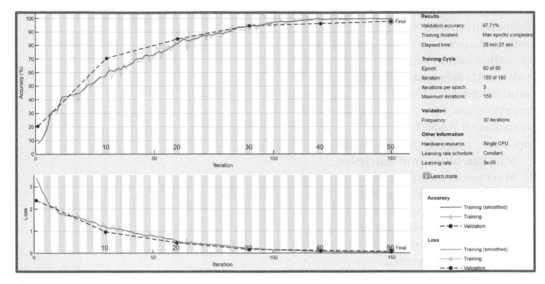

Figure 7.44 The accuracy % and loss for the trained CNN as a function of number of iterations.

Figure 7.45 shows the confusion matrix plot (*left*) of the true versus predicted flower type for the testing augmented image data set and the workspace variables (*right*).

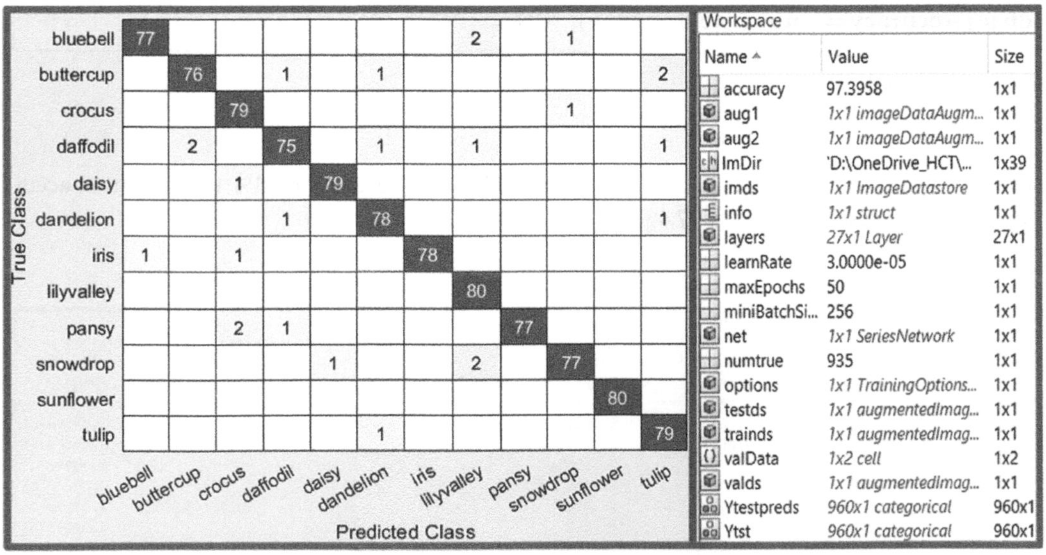

Figure 7.45 The confusion matrix plot for the testing image data set (*left*) and workspace variables (*right*).

Directed Acyclic Graphs Networks

Some networks, such as **AlexNet**, are represented in MATLAB as a column vector of layers. This is called a series architecture. An alternate way to organize layers in a network is called a directed acyclic graph (DAG), such as **GoogleNet**. DAGs have a more complex architecture where layers can have inputs from, or outputs to, multiple layers, as shown in Figure 7.46.

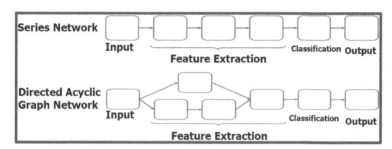

Figure 7.46   The difference between a series and DAG network.

A DAG architecture is defined with layers and connections between these layers. In MATLAB, these are represented in separate network properties. Examples of pretrained DAG networks are: **GoogleNet**, **ResNet-50**, and **SqueezeNet**.

Let us perform transfer learning from **ResNet-50**. We can use the **resnet50** function to create a copy of the pretrained DAG network **ResNet-50**. **ResNet-50** is a pretrained model that has been trained on a subset of the **ImageNet** database and that won the ImageNet Large-Scale Visual Recognition Challenge (ILSVRC) competition in 2015. The model is trained on more than a million images, has 177 layers in total, corresponding to a 50-layer residual network, and can classify images into 1000 object categories (e.g., keyboard, mouse, pencil, and many animals).

NOTE #5

You will get the following error if the specified network is not installed:
    Error using resnet50 (line 61)
    resnet50 requires the Deep Learning Toolbox Model for ResNet-50 Network support package for the pretrained weights. To install this support package, use the Add-On Explorer. To obtain the untrained layers, use resnet50('Weights','none'), which does not require the support package.

To modify the architecture of DAG network, we first need to get a graph of its layers. A layer graph contains both the layers and connections of a DAG.

>>**dagnet = resnet50;**

>>**lgraph = layerGraph(dagnet);**

To modify the image input layer so that we can control the size of the input image, we will create a new image input layer and replace the existing layer.

>>inlayer=imageInputLayer([30 30 3],'Name','input_1');

>>newgraph = replaceLayer(lgraph,'input_1',inlayer);

>>lgraph=newgraph;

As with a series network like **AlexNet**, we need to replace the last fully connected layer and the classification layer in **ResNet-50** to perform transfer learning. Connections between layers in a DAG are defined by each layer's name. When creating a new layer for a DAG network, we should name it by setting the '**Name**' option.

Let us create a fully connected layer with 12 classes for the 12 types of flowers. In the layer creation function, we name the layer '**fc**'. Assign the layer to a variable named **newfc**.

>>newfc= fullyConnectedLayer(12,"Name",'fc');

We can replace a layer using the **replaceLayer** function. The three inputs are the layer graph, the name of the layer to replace, and the variable containing the new layer.

The fully connected layer in **ResNet-50** is named '**fc1000**'. Hence, we use the **replaceLayer** function to replace the last fully connected layer in **lgraph** with **newfc**. Name the updated layer graph **lgraph**. Notice that the classification output layer in **ResNet-50** is named '**ClassificationLayer_fc1000**'. The user may run the following command, as shown below:

>>lgraph

---

lgraph = LayerGraph with properties:
Layers: [177×1 nnet.cnn.layer.Layer]
Connections: [192×2 table]
InputNames: {'input_1'}
OutputNames: {'ClassificationLayer_fc1000'}

>>newgraph = replaceLayer(lgraph,'fc1000',newfc);

>>lgraph=newgraph;

We create and name a new classification layer and set the '**Name**' option to '**flwr**'. Replace the classification layer in **lgraph** and name the final architecture **lgraph**.

>>ncl=classificationLayer("Name",'flwr');

>>newcl=replaceLayer(lgraph,'ClassificationLayer_fc1000',ncl);

>>lgraph=newcl;

Figure 7.47 shows the code for carrying out the cosmetic surgical operations needed to modify originally **ResNet-50** DAG layers so that we can later use in transfer learning.

```
rng(123);
% Create a datastore
ImDir=fullfile('D:\OneDrive_HCT\MatlabBook\MLDL\Flowers');
imds=imageDatastore(ImDir,'IncludeSubfolders',true,'LabelSource','foldernames');
trainds=augmentedImageDatastore([30 30],imds);
aug1=imageDataAugmenter('RandScale',[0.8 1.2]);
valds=augmentedImageDatastore([30 30],imds,'DataAugmentation',aug1);
aug2=imageDataAugmenter('RandXShear',[-30 30]);
testds=augmentedImageDatastore([30 30],imds,'DataAugmentation',aug2);
valData={valds imds.Labels};
miniBatchSize = 40;
maxEpochs = 20;
learnRate = 0.0008;
options = trainingOptions('adam',...
   'MiniBatchSize',miniBatchSize,...
   'MaxEpochs',maxEpochs,...
   'InitialLearnRate',learnRate,...
   'ValidationData',valData,...
   'ValidationFrequency',20,...
   'Plots','training-progress',...
   'Shuffle','every-epoch',...
   'Verbose',true);
dagnet = resnet50;
lgraph = layerGraph(dagnet);
inlayer=imageInputLayer([30 30 3],'Name','input_1');
newgraph = replaceLayer(lgraph,'input_1',inlayer);
lgraph=newgraph;
newfc= fullyConnectedLayer(12,"Name",'fc');
newgraph = replaceLayer(lgraph,'fc1000',newfc);
lgraph=newgraph;
ncl=classificationLayer("Name",'flwr');
newcl=replaceLayer(lgraph,'ClassificationLayer_fc1000',ncl);
lgraph=newcl;
[dagnet, info] = trainNetwork(trainds, lgraph, options);
Ytestpreds= classify(dagnet,testds);
Ytst=imds.Labels;
confusionchart(Ytst,Ytestpreds);
numtrue=nnz(Ytst == Ytestpreds);
accuracy=(numtrue/numel(Ytestpreds))*100;
disp(['Accuracy = ', num2str(accuracy),' %'])
```

Figure 7.47 Re-using **resnet50** DAG network with modification of the input image layer, the fully connected, and classification layer to suit training the network using the 12-class flowers image datastore.

Figure 7.48 shows the mini-batch accuracy of the 20th epoch reached 100.0%, the validation accuracy 90.1%, and testing accuracy 90.1%.

```
Training on single CPU.
Initializing input data normalization.
|=========================================================================================================
| Epoch  | Iteration | Time Elapsed | Mini-batch | Validation | Mini-batch | Validation | Base Learning
|        |           | (hh:mm:ss)   | Accuracy   | Accuracy   | Loss       | Loss       | Rate
|=========================================================================================================
|   20 |      460 |   00:11:54 |   100.00% |    87.08% |   0.1716 |   0.5496 |    0.000
|   20 |      480 |   00:12:26 |   100.00% |    90.10% |   0.0472 |   0.3571 |    0.000
|=========================================================================================================
Training finished: Max epochs completed.
Accuracy = 90.1042 %
```

Figure 7.48   Training, validation and testing accuracy of the re-trained ResNet-50 DAG network.

Figure 7.49 shows both the accuracy % and loss as a function of number of iterations for trained and validation data set.

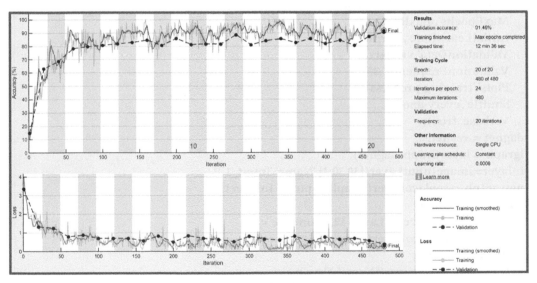

Figure 7.49   The accuracy % and loss for the trained **ResNet-50** DAG network as a function of number of iterations.

Figure 7.50 shows the confusion matrix plot (*left*) of the true versus predicted flower type for the testing augmented image data set and the workspace variables (*right*).

Confusion matrix plot (True Class vs Predicted Class):

True \ Pred	bluebell	buttercup	crocus	daffodil	daisy	dandelion	iris	lilyvalley	pansy	snowdrop	sunflower	tulip
bluebell	67		2				4	7				
buttercup		74		2			1	1				2
crocus	3		73		1		1		1	1		
daffodil		2		69		2	2			1		4
daisy					79					1		
dandelion		3		1		75	1					
iris	3		3				71			3		
lilyvalley				1				77		2		
pansy	2		1	1	3		2	1	70			
snowdrop			1		2		1	1		75		
sunflower		5		1				1			73	
tulip	2	1		6		3	1	4			1	62

Workspace

Name ▲	Value	Size
accuracy	90.1042	1x1
aug1	1x1 imageDataAugm...	1x1
aug2	1x1 imageDataAugm...	1x1
dagnet	1x1 DAGNetwork	1x1
ImDir	'D:\OneDrive_HCT\...	1x39
imds	1x1 ImageDatastore	1x1
info	1x1 struct	1x1
inlayer	1x1 ImageInputLayer	1x1
learnRate	8.0000e-04	1x1
lgraph	1x1 LayerGraph	1x1
maxEpochs	20	1x1
miniBatchSi...	40	1x1
ncl	1x1 ClassificationOut...	1x1
newcl	1x1 LayerGraph	1x1
newfc	1x1 FullyConnectedL...	1x1
newgraph	1x1 LayerGraph	1x1
numtrue	865	1x1
options	1x1 TrainingOptions...	1x1
testds	1x1 augmentedImag...	1x1
trainds	1x1 augmentedImag...	1x1
valData	1x2 cell	1x2
valds	1x1 augmentedImag...	1x1
Ytestpreds	960x1 categorical	960x1
Ytst	960x1 categorical	960x1

Figure 7.50 The confusion matrix plot for the testing image data set (*left*) and workspace variables (*right*).

NOTE #6

We can view the architecture of a network using the layer graph as input to the plot function.
>>lgraph=layerGraph(dagnet);>>plot(lgraph)
    Check out the Deep Learning Tips and Tricks page in the documentation for more general advice on improving the accuracy of the deep network.

Deep Network Designer (DND)

Before we move to DND platform, let us run the following code to populate the workspace with variables needed for training, validation, and testing, as shown in Figure 7.51.

```
rng(123);
% Create a datastore
ImDir=fullfile('D:\OneDrive_HCT\MatlabBook\MLDL\Flowers');
imds=imageDatastore(ImDir,'IncludeSubfolders',true,'LabelSource','foldernames');
trainds=augmentedImageDatastore([30 30],imds);
aug1=imageDataAugmenter('RandScale',[0.8 1.2]);
valds=augmentedImageDatastore([30 30],imds,'DataAugmentation',aug1);
aug2=imageDataAugmenter('RandXShear',[-30 30]);
testds=augmentedImageDatastore([30 30],imds,'DataAugmentation',aug2);
valData={valds imds.Labels};
```

Figure 7.51 Creating workspace variables to be later used by DND app.

Let us show how to use MATLAB Deep Network Designer. At the prompt, simply type in

>>deepNetworkDesigner

Or,

From **APPS | APPS | MACHINE LEARNING AND DEEP LEARNING** menu, click on "**Deep Network Designer**" icon to open DND app. Figure 7.52 shows the main window of DND app. On the left, we have the layer library where we can drag different types of layers and release the mouse in the design working area or canvas. We can then reconnect the layers in serial, parallel mode, or both. In the middle, we can flip from "**Designer**" to "**Data**" or "**Training**" tab window. We will show shortly how to deal with such tab windows. On the right, we have "**Properties**" pane where the user can carry out modification for the highlighted or selected layer from the middle "**Designer**" tab window.

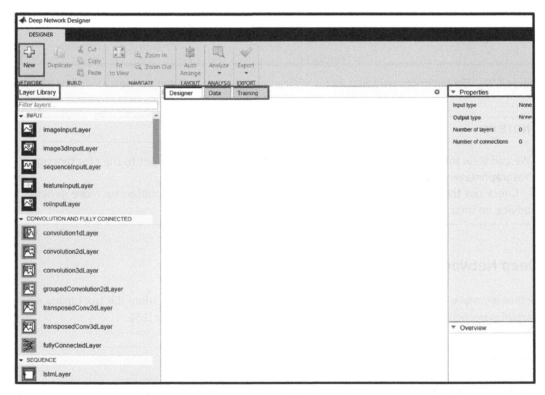

Figure 7.52 DND main window to handle network layers creation and modification.

Figure 7.53 shows upon clicking "**New**" button different forms or templates are available for the user to select from. The user may select to create a set of layers from scratch, retrained, a template, or a workspace layer variable.

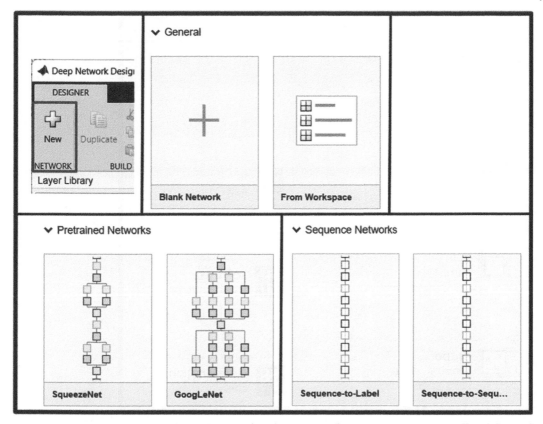

Figure 7.53 Different forms for creating a set of layers from scratch, retrained layers, a template, or workspace layer variable.

Let us select "**Sequence-to-Label**" network and Figure 7.54 shows the created set of network layers. This template will serve as a starting set of layers where will modify some of them and add additional types of layers. A "**Sequence-to-Label**" network will be dealt with specifically in Chapter 9.

NOTE #7

We can highlight one icon at a time or select a group of icons by bracketing (or squaring) them in a drag (left-mouse) mode. Once highlighted, the user may use keyboard arrows and/or the mouse itself to move the selected icon(s).

We can highlight or select a layer and carry out modification using the "**Properties**" right pane or delete it and replace it by a new type selected from "**Layer Library**" left pane. Let us replace the first layer by first deleting it and second adding a new layer of "**imageInputLayer**" type. Figure 7.55 shows the selection of the image input layer type from left pane and modification of the input image size utilizing the "**Properties**" right pane.

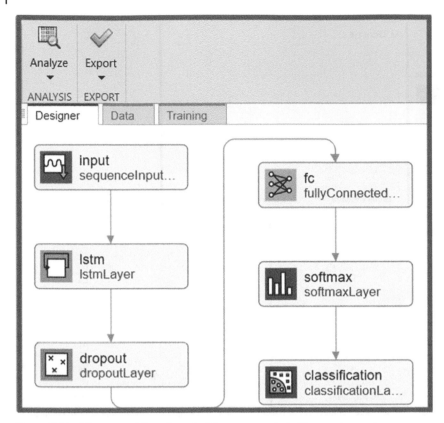

Figure 7.54 The created flowchart for "**Sequence-to-Label**" network template.

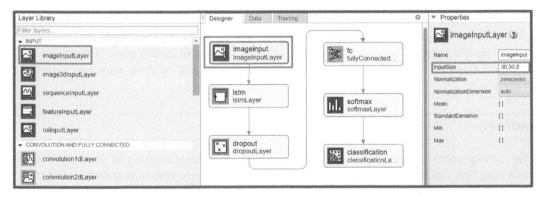

Figure 7.55 Insertion of an image input layer type at the start and modifying it size properties.

The second change deals with replacement of the long short-term memory (**lstmLayer**) by a "**convolution2dLayer**"-type. The default filter size (3×3), number of filters (32), and stride (1,1) value will be retained for the "**convolution2dLayer.**" The **lstmLayer** will be covered in a more detail in Chapter 9.

The third change is insertion of three successive layers right after the **convolution2dLayer**, as shown in Figure 7.56. These three layers are: **Batch Normalization** layer, **ReLU** layer, and

Average Pooling layer. The default values of settings, selected by DND app, for those three layers will be retained. In addition, two properties for two layers were modified as also shown on the right side of Figure 7.56.

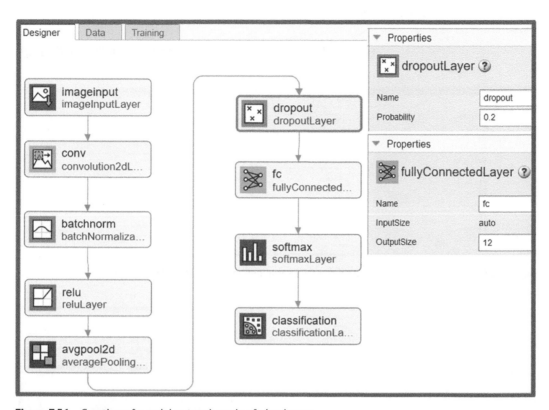

Figure 7.56 Creation of a serial network made of nine layers.

Click on "**Analyze**" button to check for potential errors and warnings. Figure 7.57 shows that upon analysis, zero errors and warnings were found.

At this stage after assuring a bug-free set of layers, we may proceed and export the layers to workspace environment, as shown in Figure 7.58. The created "**layers_1**" variable can be seen under workspace environment, where it says that it is 9×1 layers. This can be modified by prompt commands and then can be imported back into DND platform. So does the case where the user may change the set of layers, in terms of number and type, using DND platform, and re-export to workspace environment.

Let us move now to "**DATA**" tab window where we will import training and validations data sets into DND platform. Figure 7.59 shows that we imported both "**trainds**" and "**valds**" augmented image datastore.

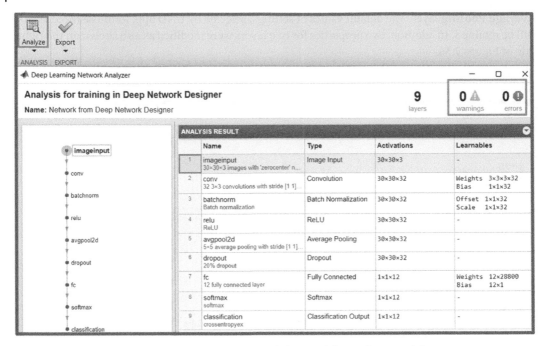

Figure 7.57 Results of analysis of the constructed serial network layers for potential errors.

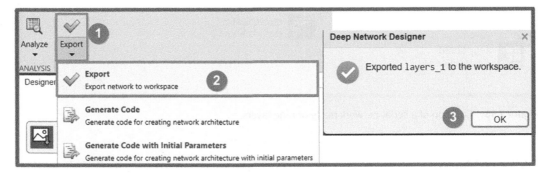

Figure 7.58 Saving the set of layers to workspace environment.

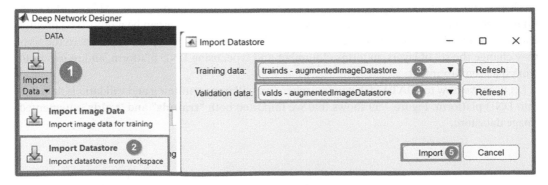

Figure 7.59 Training and validation data sets are imported into DND platform.

Figure 7.60 show a preview for the first five images of both training and validation image data stores.

Figure 7.60 A preview for the first five imported images for training and validation image data stores.

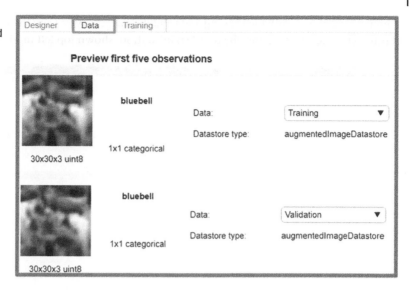

Figure 7.61 shows the training option upon clicking "**Training Options**" under "**Training**" tab window.

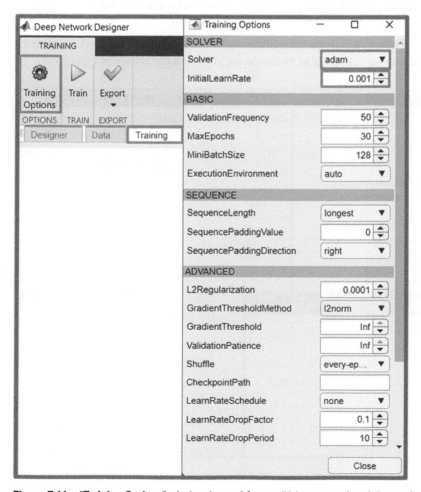

Figure 7.61 "**Training Options**" window is used for modifying network training options.

Figure 7.62 shows both the training and validation accuracy (%) and the model loss as a function of number of iterations, upon clicking "Train" button, shown top left in Figure 7.61.

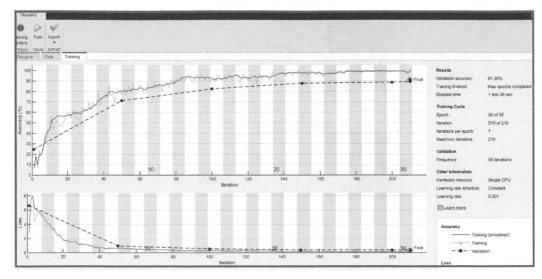

Figure 7.62 The model accuracy (%) and loss as a function of carried out iterations for both training and validation step.

Figure 7.63 (*top*) shows how to export the trained network and its results to the workspace environment. Details on the trained network and its results are shown at the *bottom* of Figure 7.63.

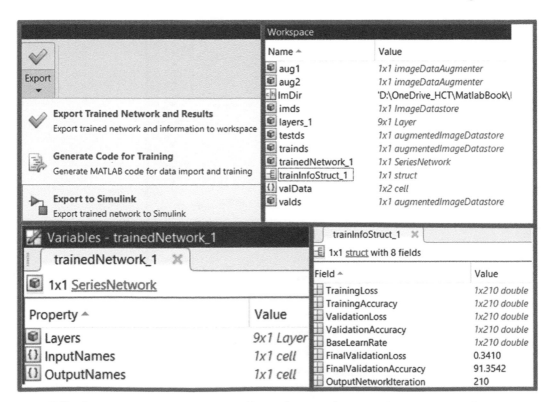

Figure 7.63 Exporting the trained network and its results to workspace environment (top) and more details are shown (bottom).

Figure 7.64 shows the code on how to utilize the DND-generated trained network called **trainedNetwork_1**.

```
Ytestpreds= classify(trainedNetwork_1,testds);
Ytst=imds.Labels;
confusionchart(Ytst,Ytestpreds);
numtrue=nnz(Ytst == Ytestpreds);
accuracy=(numtrue/numel(Ytestpreds))*100;
disp(['Accuracy = ', num2str(accuracy),' %'])
```

Figure 7.64   DND-generated trained network called **trainedNetwork_1**.

Copied from Command Window: Accuracy = 87.6042%

Figure 7.65 shows the confusion matrix plot for the test image data store, named **testds**.

True Class \ Predicted Class	bluebell	buttercup	crocus	daffodil	daisy	dandelion	iris	lilyvalley	pansy	snowdrop	sunflower	tulip
bluebell	78						1	1				
buttercup		68				5	1	1				5
crocus	6		66		2	1	1	2	1			1
daffodil		3		61		7		2		1		6
daisy			2		75					2		1
dandelion		2		1		75	1	1				
iris	3	1	2				68	3	1	1		1
lilyvalley			1		1			72	1	3		2
pansy	3			1				3	69	1		3
snowdrop		1	1					8		70		
sunflower		1		3		2					73	1
tulip		3		2		4		4		1		66

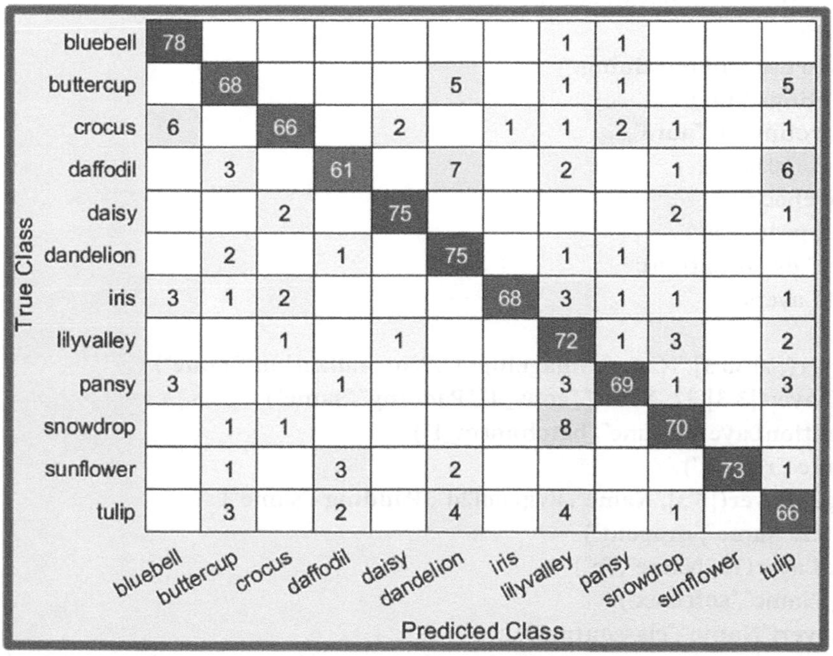

Figure 7.65   The confusion matrix plot for the test image data store, named **testds**.

Under "**Training**" tab window, clicking on "**Export**" button followed by "**Generate Code for Training**" option will create MATLAB code for data import and training in the form of ***.mlx** file.

NOTE #8

MATLAB® stores live scripts and functions using the Live Code file format in a file with a ***.mlx** extension. The Live Code file format uses Open Packaging Conventions technology, which is an extension of the zip file format. Code and formatted content are stored in an XML document

(Continued)

NOTE (Continued)

separate from the output using the Office Open XML (ECMA-376) format. Benefits of live code file format are: Interoperable (or interchangeable) across locales, extensible (or supporting the range of formatting options offered by Microsoft Word, and both forward and backward compatible at the same time, with future and previous versions of MATLAB.

Figure 7.66 shows the main portion of the DND-created *.mlx file, where the user has the option to make changes for number of layers and their types as well as tuning-up training options.

```
trainingSetup = load("D:\OneDrive_HCT\MatlabBook\MLDL\
params_2021_12_23__20_22_11.mat");
%Import training and validation data.
dsTrain = trainingSetup.dsTrain;
dsValidation = trainingSetup.dsValidation;
%Set Training Options
%Specify options to use when training.
opts = trainingOptions("adam",...
   "ExecutionEnvironment","auto",...
   "InitialLearnRate",0.001,...
   "Shuffle","every-epoch",...
   "Plots","training-progress",...
   "ValidationData",dsValidation);
%Create Array of Layers
layers = [
   imageInputLayer([30 30 3],"Name","imageinput","Normalization","none")
   convolution2dLayer([3 3],32,"Name","conv_1","Padding","same")
   batchNormalizationLayer("Name","batchnorm_1")
   reluLayer("Name","relu_1")
   averagePooling2dLayer([5 5],"Name","avgpool2d","Padding","same")
   dropoutLayer(0.2,"Name","dropout")
   fullyConnectedLayer(12,"Name","fc")
   softmaxLayer("Name","softmax")
   classificationLayer("Name","classoutput")];
%Train the network using the specified options and training data.
[net, traininfo] = trainNetwork(dsTrain,layers,opts);
```

Figure 7.66 DND-generated code for network layers and training options.

At the end, one can see that DND app can do a lot of work on the behalf of the user or coder.

Semantic Segmentation

Semantic segmentation associates each pixel of an image with a class label, such as flower, person, road, sky, or car. Use the Image Labeler and the Video Labeler app to interactively label pixels and export the label data for training a neural network.

Analyze Training Data for Semantic Segmentation

To train a semantic segmentation network, we need a collection of images and its corresponding collection of pixel-labeled images. A pixel labeled image is an image where every pixel value represents the categorical label of that pixel.

1. Image Dataset Location

Let us use a small set of images and their corresponding pixel labeled images:

```
dataDir = fullfile(toolboxdir('vision'),'visiondata');
imDir = fullfile(dataDir,'building');
pxDir = fullfile(dataDir,'buildingPixelLabels');
```

2. Create an Image Data Store

Load the image data using an **imageDatastore** function. An image datastore can efficiently represent a large collection of images because images are only read into memory when needed.

```
imds = imageDatastore(imDir);
```

3. Display an Image

Read and display the fifth image.

```
I = readimage(imds,5);
figure
imshow(I)
```

Figure 7.67 shows the fifth image out of the five-image data store.

Figure 7.67   The fifth image out of the five-image data store.

4. Define Class Names and Create the Pixel Label Data Store

Let us create the pixel label images using **pixelLabelDatastore** function to define the mapping between label IDs and categorical names. **pixelLabelDatastore** function is used to create a datastore specifically designed for handling pixel-wise labeled data. This function is commonly used in computer vision tasks, such as semantic segmentation or image annotation, where each pixel in an image is assigned a specific label or category. **pixelLabelDatastore** function is a convenient way to handle large datasets of labeled images efficiently, as it allows for on-the-fly loading and processing of data during training, without loading all the data into memory at once.

In the given dataset, the labels are "**sky,**" "**grass,**" "**building,**" and "**sidewalk.**" The label IDs for these classes are 1, 2, 3, 4, respectively.

```
classNames = ["sky" "grass" "building" "sidewalk"];
pixelLabelID = [1 2 3 4];
pxds = pixelLabelDatastore(pxDir,classNames,pixelLabelID);
%Read the fifth pixel-labeled image.
C = readimage(pxds,5);
```

The output C is a categorical matrix where C(i,j) is the categorical label of pixel I(i,j).

5. Overlay Pixel Labels on an Image

Overlay the pixel labels on the image to see how different parts of the image are labeled.

```
B = labeloverlay(I,C);
figure
imshow(B)
```

Figure 7.68 shows the overlay of pixel labels unto the original image. As one can see that there are four colors, characterizing four different zones or sub-areas out of the entire area of the image. Thus, each color represents the jurisdictional area of each of the four types of classes.

Figure 7.68 The overlay of pixel label unto the original image, showing the jurisdictional area of each class type.

The categorical output format simplifies tasks that require doing things by class names. For example, we can create a binary mask of just the building, as shown in Figure 7.69.

```
buildingMask = C == 'building';
figure
imshowpair(I, buildingMask,'montage')
```

Figure 7.69   The pixel labels are reduced to "**building**" where the right image becomes binary (i.e., building or none).

Figure 7.70 shows workspace variables for the previously executed commands.

Workspace		
Name ▲	**Value**	**Size**
B	*480x640x3 uint8*	480x640x3
buildingMask	*480x640 logical*	480x640
C	*480x640 categorical*	480x640
classNames	*1x4 string*	1x4
dataDir	'C:\Program Files\MATLAB\R2022a\toolbox\vision\visiondata'	1x56
I	*480x640x3 uint8*	480x640x3
imDir	'C:\Program Files\MATLAB\R2022a\toolbox\vision\visiondata\building'	1x65
imds	*1x1 ImageDatastore*	1x1
pixelLabelID	[1,2,3,4]	1x4
pxDir	'C:\Program Files\MATLAB\R2022a\toolbox\vision\visiondata\buildingPixelLabels'	1x76
pxds	*1x1 PixelLabelDatastore*	1x1

Figure 7.70   The workspace variables for previously executed commands.

Create a Semantic Segmentation Network

A common pattern in semantic segmentation networks requires the down-sampling of an image between convolutional and **ReLU** layers, and then up-sampling the output to match the input size. This operation is analogous to the standard scale-space analysis using image pyramids. During this process however, a network performs the operations using non-linear filters optimized for a specific set of classes that we want to segment.

1. Create an Image Input Layer

A semantic segmentation network starts with an **imageInputLayer**, which defines the smallest image size the network can process. Most semantic segmentation networks are fully convolutional, which means they can process images that are larger than the specified input size. Here, an image size of [32 32 1] is used for the network to process 32×32 gray images.

```
inputSize = [32 32 1];
imgLayer = imageInputLayer(inputSize);
```

2. Create a 2-D Convolution and ReLU Layer

Start with the convolution and **ReLU** layer. The convolution layer padding is selected such that the output size of the convolution layer is the same as the input size. This makes it easier to construct a network because the input and output sizes between most layers remain the same as the processed image moves through the network.

```
filterSize = 1;
numFilters = 32;
conv = convolution2dLayer(filterSize,numFilters,'Padding',1);
relu = reluLayer();
```

3. Create a Downsampling Layer

The downsampling is performed using a max pooling layer. A 2-D max pooling layer performs downsampling by dividing the input into rectangular pooling regions, **'poolSize'**, then computing the maximum of each **'poolSize'** region. On the other hand, the **'Stride'** is the step size for traversing the input vertically and horizontally, specified as a vector of two positive integers [*a b*], where *a* is the vertical step size and *b* is the horizontal step size. The default **'Stride'** values are [1,1]. When creating the layer, one can also specify **'Stride'** as a scalar to use the same value for both dimensions.

```
poolSize = 2;
maxPoolDownsample2x = maxPooling2dLayer(poolSize,'Stride',2);
```

NOTE #9

If the dimensions of **'Stride'** are less than those of **'PoolSize'**, then the pooling regions overlap. The padding dimensions **'PaddingSize'** must be less than the pooling region dimensions, **'PoolSize'**. The **'PaddingSize'** with a non-negative integer **p** means that padding of size **p** is added to all the edges of the input. The default **'PaddingSize'** is zero.

---

4. Downsampling Layer Stacking

Stack the convolution, **ReLU**, and max pooling layers to create a network that downsamples its input.

```
downsamplingLayers = [
  conv
  relu
  maxPoolDownsample2x
  conv
  relu
  maxPoolDownsample2x
  ]
```

The results are copied from Command Window:
  downsamplingLayers = 6×1 Layer array with layers:

```
1 "    Convolution 32 1×1 convolutions with stride [1 1] and padding [1 1 1 1]
2 "    ReLU ReLU
3 "    Max Pooling 2×2 max pooling with stride [2 2] and padding [0 0 0 0]
4 "    Convolution 32 1×1 convolutions with stride [1 1] and padding [1 1 1 1]
5 "    ReLU ReLU
6 "    Max Pooling 2×2 max pooling with stride [2 2] and padding [0 0 0 0]
```

5. Create Upsampling Layer

The upsampling is done using the transposed convolution layer (also commonly referred to as "**deconv**" or "**deconvolution**" layer). When a transposed convolution is used for upsampling, it performs the upsampling and the filtering at the same time.

Create a transposed convolution layer to upsample.

```
filterSize = 4;
transposedConvUpsample2x = transposedConv2dLayer(4,numFilters,'Stride',2,'Crop
ping',1);
```

The '**Cropping**' parameter is set to 1 to make the output size equal twice the input size. Cropping is a technique used to extract a smaller region or patch from a larger image. In the context of CNNs, cropping refers to the operation of removing a portion of an input image to focus on a specific region of interest. This technique offers several advantages: Reduction of irrelevant information, enhanced spatial resolution, improved computational efficiency, and data augmentation.

6. Upsampling Layer Stacking

Stack the transposed convolution and **ReLU** layers. An input to this set of layers is upsampled.

```
upsamplingLayers = [
  transposedConvUpsample2x
  relu
  transposedConvUpsample2x
  relu
  ]
```

Results are copied from Command Window.

upsamplingLayers = 4×1 Layer array with layers:

1	" Transposed Convolution	32 4×4 transposed convolutions with stride [2 2] and cropping [1 1 1 1]
2	" ReLU	ReLU
3	" Transposed Convolution	32 4×4 transposed convolutions with stride [2 2] and cropping [1 1 1 1]
4	" ReLU	ReLU

7. Create a Pixel Classification Layer

The final set of layers are responsible for making pixel classifications. These final layers process an input that has the same spatial dimensions (height and width) as the input image. However, the number of channels (third dimension) is larger and is equal to number of filters in the last transposed convolution layer. This third dimension needs to be squeezed down to the number of classes we wish to segment. This can be done using a 1×1 convolution layer whose number of filters equal the number of classes, e.g., 2 **["triangle" "background"]**.

Create a convolution layer to combine the third dimension of the input feature maps down to the number of classes.

```
numClasses = 2;
conv1x1 = convolution2dLayer(1,numClasses);
```

In addition to the 1×1 convolution layer, the softmax and pixel classification layer are added. These two layers combine to predict the categorical label for each image pixel.

```
finalLayers = [
  conv1x1
  softmaxLayer()
  pixelClassificationLayer()
  ]
```

Results are copied from Command Window:

finalLayers = 3×1 Layer array with layers:

1	" Convolution	2 1×1 convolutions with stride [1 1] and padding [0 0 0 0]
2	" Softmax	softmax
3	" Pixel Classification Layer	Cross-entropy loss

8. Stack All Layers

Stack all the layers to complete the semantic segmentation network.

```
SemSegNet = [
  imgLayer
```

```
    downsamplingLayers
    upsamplingLayers
    finalLayers
]
```

Figure 7.71 shows the architecture of semantic segmentation network, **SemSegNet**.

```
SemSegNet = 14×1 Layer array with layers:
1    ''    Image Input              32×32×1 images with 'zerocenter' normalization
2    ''    Convolution              32 1×1 convolutions with stride [1  1] and padding [1  1  1  1]
3    ''    ReLU                     ReLU
4    ''    Max Pooling              2×2 max pooling with stride [2  2] and padding [0  0  0  0]
5    ''    Convolution              32 1×1 convolutions with stride [1  1] and padding [1  1  1  1]
6    ''    ReLU                     ReLU
7    ''    Max Pooling              2×2 max pooling with stride [2  2] and padding [0  0  0  0]
8    ''    Transposed Convolution   32 4×4 transposed convolutions with stride [2  2] and cropping [1  1  1  1]
9    ''    ReLU                     ReLU
10   ''    Transposed Convolution   32 4×4 transposed convolutions with stride [2  2] and cropping [1  1  1  1]
11   ''    ReLU                     ReLU
12   ''    Convolution              2 1×1 convolutions with stride [1  1] and padding [0  0  0  0]
13   ''    Softmax                  softmax
14   ''    Pixel Classification Layer   Cross-entropy loss
```

Figure 7.71   The command-created semantic segmentation network, **SemSegNet**.

This network is ready to be trained using **trainNetwork** function.
Figure 7.72 shows workspace variables related to executing previous code.

Workspace		
Name ▲	Value	Size
conv	*1x1 Convolution2DLayer*	1x1
conv1x1	*1x1 Convolution2DLayer*	1x1
downsamplingLayers	*6x1 Layer*	6x1
filterSize	4	1x1
finalLayers	*3x1 Layer*	3x1
imgLayer	*1x1 ImageInputLayer*	1x1
inputSize	[32,32,1]	1x3
maxPoolDownsample2x	*1x1 MaxPooling2DLayer*	1x1
numClasses	2	1x1
numFilters	32	1x1
poolSize	2	1x1
relu	*1x1 ReLULayer*	1x1
SemSegNet	*14x1 Layer*	14x1
transposedConvUpsample2x	*1x1 TransposedConvolution2DLayer*	1x1
upsamplingLayers	*4x1 Layer*	4x1

Figure 7.72   The workspace variables related to the previously executed code.

Figure 7.73(left) shows DND-constructed semantic segmentation network, which is exactly similar to the previously command-created network, **SemSegNet**. The properties of layers requiring input parameters are also shown (**right**). Under **"ANALYSIS"** tab, do not forget to click on **"Analyze"** | **"Analyze"** sub menu to assure that the DND-created network is bug-free. You may also export to Workspace environment and save the created variable as a *.mat file. The **SemSegNet** set of layers will be used in the next section.

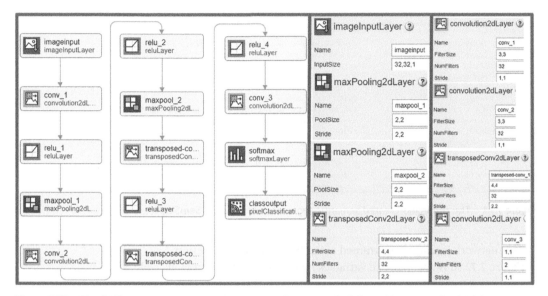

Figure 7.73 The DND-constructed semantic segmentation network (*left*) and its properties (*right*).

Train and Test the Semantic Segmentation Network

1. Import a Data Set

The **triangleImages** main folder (C:\Program Files\MATLAB\R2022a\toolbox\vision\vision-data\triangleImages) contains 200 training images and 100 test images with ground truth labels. Define the main location folder of the data set for training and testing.

```
dataSetDir = fullfile(toolboxdir('vision'),'visiondata','triangleImages');
```

2. Define the Location Folder of the Train Images

```
trainImagesDir = fullfile(dataSetDir,'trainingImages');
```

3. Create an imageDatastore Object Holding the Train Images

```
imdstrn = imageDatastore(trainImagesDir);
```

4. Define the location of the ground truth labels associated with the train images.

```
trainLabelsDir = fullfile(dataSetDir,'trainingLabels');
```

5. Define the Class Names and their Associated Label IDs

The label IDs are the pixel values used in the image files to represent each class.

```
classNames = ["triangle" "background"];
labelIDs = [255 0];
```

6. Pixel Label Data Store Creation for Training Images

Create a **pixelLabelDatastore** object holding the ground truth pixel labels for the training images.

```
pxdsTruth = pixelLabelDatastore(trainLabelsDir,classNames,labelIDs);
```

7. Splitting Training Folder into Training and Validation

```
imdstrn.Files=fullfile(imdstrn.Files);
r= size(imdstrn.Files,1);
P = 0.90;
idx = randperm(r);
trainData = imdstrn.Files(idx(1:round(P*r)), :);
trainData = imageDatastore(trainData);
trainLbl = pxdsTruth.Files(idx(1:round(P*r)), :);
trainLbl = pixelLabelDatastore(trainLbl,classNames,labelIDs);
ValData = imdstrn.Files(idx(round(P*r)+1:end), :);
ValData=imageDatastore(ValData);
ValLbl = pxdsTruth.Files(idx(round(P*r)+1:end), :);
ValLbl= pixelLabelDatastore(ValLbl,classNames,labelIDs);
dsval = combine(ValData, ValLbl);
dstrn = combine(trainData,trainLbl);
```

8. Training Options

```
miniBatchSize = 40;
maxEpochs = 30;
learnRate = 0.0005;
options = trainingOptions('adam',...
   'MiniBatchSize',miniBatchSize,...
   'MaxEpochs',maxEpochs,...
   'InitialLearnRate',learnRate,...
   'Plots','training-progress',...
   'Shuffle','every-epoch',...
   'ValidationData',dsval,...
   'ValidationFrequency',10,...
```

```
  'Verbose',true);
[SerNet, info] = trainNetwork(dstrn, SemSegNet, options);
```

SerNet is now a trained series CNN network. Figure 7.74 shows both training and validation accuracy % and loss for the examined series network.

We need to test predictability of the previously trained network.

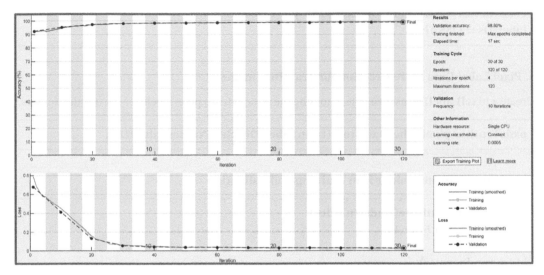

Figure 7.74 The training and validation accuracy % and loss for the examined series network.

9. Define the Location Folder of the Test Images

```
testImagesDir = fullfile(dataSetDir,'testImages');
```

10. Create an imageDatastore Object Holding the Test Images

```
imdstst = imageDatastore(testImagesDir);
```

11. Define the Location of the Ground Truth Labels Associated with Test Images

```
testLabelsDir = fullfile(dataSetDir,'testLabels');
```

12. Pixel Label Data Store Creation for Test Images
Create a **pixelLabelDatastore** object holding the ground truth pixel labels for the test images.

```
pxdsTruth2 = pixelLabelDatastore(testLabelsDir,classNames,labelIDs);
```

13. Run a Semantic Segmentation Classifier
Run the network on the test images. Predicted labels are written to disk in a temporary directory and returned as a **pixelLabelDatastore** object.

```
pxdsResults2 = semanticseg(imdstst,SerNet,"WriteLocation",tempdir);
```

14. Evaluate the Quality of the Prediction

The predicted labels are compared to the ground truth labels. While the semantic segmentation metrics are being computed, progress is printed to Command Window.

```
metrics = evaluateSemanticSegmentation(pxdsResults2,pxdsTruth2);
```

Semantic segmentation accuracy is a measure of how accurately an algorithm or model can assign semantic labels to each pixel in an image. Several parameters are commonly used to evaluate the accuracy of semantic segmentation:

1) Global Classification Accuracy: This parameter measures the overall accuracy of the segmentation model by comparing the predicted labels to the ground truth labels. It calculates the percentage of correctly classified pixels in the entire image.
2) Mean Accuracy for All Classes: This metric calculates the average accuracy across all classes. It considers the accuracy of each individual class separately and then computes the mean accuracy. It provides a balanced measure of segmentation performance across different classes.
3) Mean Intersection over Union (IoU) for All Classes: Intersection over Union is a commonly used evaluation metric for semantic segmentation. It measures the overlap between the predicted segmentation and the ground truth segmentation for each class. The IoU is calculated as the intersection of the predicted and ground truth regions divided by their union. Mean IoU is then computed by averaging the IoU scores across all classes.
4) Mean Boundary F-1 Score for All Classes: The boundary F-1 score evaluates the quality of the predicted segmentation boundaries. It measures the harmonic mean of precision and recall for the predicted boundaries compared to the ground truth boundaries. The F-1 score provides a balance between precision and recall and is commonly used to evaluate the performance of boundary-aware segmentation algorithms. Mean boundary F-1 score is calculated by averaging the F-1 scores across all classes.

Figure 7.75 shows semantic segmentation results in terms of indicated accuracy parameters, like the global classification accuracy, the mean accuracy for all classes, the mean intersection over union for all classes, and the mean boundary F-1 score for all classes. The boundary F1 (BF) contour matching score indicates how well the predicted boundary of each class aligns with the true boundary. It is recommended to use the BF score if someone wants a metric that tends to correlate better with human qualitative assessment than the IoU metric.

```
Evaluating semantic segmentation results
----------------------------------------
* Selected metrics: global accuracy, class accuracy, IoU, weighted IoU, BF score
* Processed 100 images.
* Finalizing... Done.
* Data set metrics:

    GlobalAccuracy    MeanAccuracy    MeanIoU    WeightedIoU    MeanBFScore
    _____    _____    _____    _____    _____

       0.98874           0.93807       0.88589      0.9788         0.77858
```

Figure 7.75 Semantic segmentation results in terms of indicated accuracy parameters.

NOTE #10

For each class, IoU is the ratio of correctly classified pixels to the total number of ground truth and predicted pixels in that class. In other words,
  IoU score = TP / (TP + FP + FN)
  As shown in Figure 7.76, the sketch describes the true positives (TP), false positives (FP), and false negatives (FN).

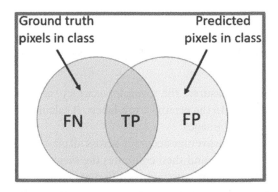

Figure 7.76 The IoU is defined as the ratio of the common area, TP, over the total area.

On the other hand, the following command:

---

metrics.ClassMetrics

---

will give the metrics of each class in terms of the classification accuracy, the intersection over union (IoU), and the boundary F-1 score for each class in the data set.
  The following results are copied from Command Window:

Class	Accuracy	IoU	MeanBFScore
triangle	0.88224	0.78351	0.63457
background	0.9939	0.98826	0.92258

15. Confusion Matrix

---

metrics.ConfusionMatrix

The confusion matrix is copied from Command Window:

class	triangle	background
triangle	4173	557
background	596	97074

16. Normalized Confusion Matrix

The normalized confusion chart using the following command:

```
cm = confusionchart(metrics.ConfusionMatrix.Variables, ...
  classNames, Normalization='row-normalized');
cm.Title = 'Normalized Confusion Matrix (%)';
```

Figure 7.77 shows a plot for the normalized confusion matrix with percentile distribution of data between matched and mismatched predicted semantic segmentation against the ground truth segmentation.

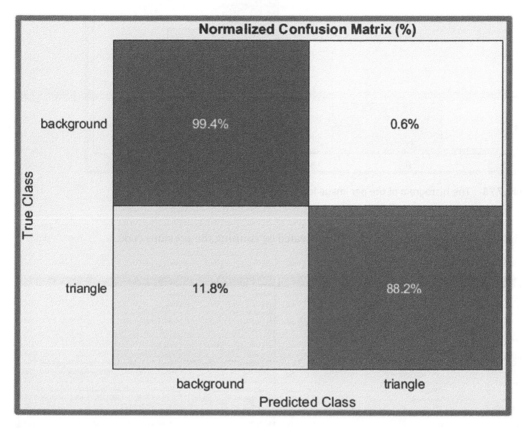

Figure 7.77 The confusion matrix for the normalized confusion matrix with percentile distribution of data between matched and mismatched predicted cases.

17. Inspect an Image Metric

Visualize the histogram of the per-image intersection over union (IoU).

```
imageIoU = metrics.ImageMetrics.MeanIoU;
figure
histogram(imageIoU)
title('Image Mean IoU')
```

Figure 7.78 shows the histogram of the per image IoU.

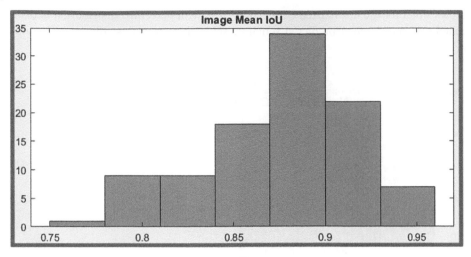

Figure 7.78 The histogram of the per image IoU.

Figure 7.79 shows workspace variables created by running the previous code.

Name ^	Value	Size
classNames	1x2 string	1x2
cm	1x1 ConfusionMatrixChart	1x1
dataSetDir	'C:\Program Files\MATLAB\R2022a\toolbox\vision\vis...	1x71
dstrn	1x1 CombinedDatastore	1x1
dsval	1x1 CombinedDatastore	1x1
idx	1x200 double	1x200
imageIoU	100x1 double	100x1
imdstrn	1x1 ImageDatastore	1x1
imdstst	1x1 ImageDatastore	1x1
info	1x1 struct	1x1
labelIDs	[255,0]	1x2
learnRate	5.0000e-04	1x1
maxEpochs	30	1x1
metrics	1x1 semanticSegmentationMetrics	1x1
miniBatchSize	40	1x1
options	1x1 TrainingOptionsADAM	1x1

Name ^	Value	Size
P	0.9000	1x1
pxdsResults2	1x1 PixelLabelDatastore	1x1
pxdsTruth	1x1 PixelLabelDatastore	1x1
pxdsTruth2	1x1 PixelLabelDatastore	1x1
r	200	1x1
SemSegNet	14x1 Layer	14x1
SerNet	1x1 SeriesNetwork	1x1
testImagesDir	'C:\Program Files\MATLAB\R2022a\toolbox\vision\vis...	1x82
testLabelsDir	'C:\Program Files\MATLAB\R2022a\toolbox\vision\vis...	1x82
trainData	1x1 ImageDatastore	1x1
trainImagesDir	'C:\Program Files\MATLAB\R2022a\toolbox\vision\vis...	1x86
trainLabelsDir	'C:\Program Files\MATLAB\R2022a\toolbox\vision\vis...	1x86
trainLbl	1x1 PixelLabelDatastore	1x1
ValData	1x1 ImageDatastore	1x1
ValLbl	1x1 PixelLabelDatastore	1x1

Figure 7.79 Command-generated workspace variables.

HCW 7.1 CNN Creation for Round Worms Alive or Dead Prediction

In this problem, we are given the images of worms. Use transfer learning to train a deep network that can classify images of roundworms as either alive or dead. (Alive worms are round; dead ones are straight). Conclusion if you are straight, you will be dead.

The images are given in a folder called **WormImages** and the labeling is given in an Excel file where it has the image file name and the worm status. Data is also available with Wiley web companion.

Carry out the following steps:

1) Get the training images and classes via saving them to an image data store and image data store labels.
2) Create an augmented image datastore for training.

3) Use **shuffle** function to create two more data sets for validation and testing.
4) Create a CNN from scratch by creating appropriate number and type of layers.
5) Set some training options.
6) Train the network.
7) Evaluate network on test data, like calculating the accuracy % and plotting confusion chart for true versus predicted value of the test data.
8) Use **augmentedImageDatastore** function to transform data store images with a combination of rotation, reflection, shear, and/or translation. Create three augmented image data stores for training, validation, and testing.
9) Go to step #4 and continue until step #7. See which way is a more efficient in terms of overall accuracy and CPU time for training, validation, and testing steps.

HCW 7.2   CNN Creation for Food Images Prediction

The Example Food Images data set contains 978 photographs of food in **nine** classes (Caesar-salad, caprese-salad, french-fries, Greek-salad, hamburger, hot-dog, pizza, sashimi, and sushi).

Carry out the following steps:

1) Download the Example Food Images data set using the downloadSupportFile function and extract the images using the unzip function. This data set is about 77 MB. The data is also available with Wiley web companion.

```
fprintf("Downloading Example Food Image data set (77 MB)... ")
filename = matlab.internal.examples.downloadSupportFile('nnet', ...
  'data/ExampleFoodImageDataset.zip');
fprintf("Done.\n")
filepath = fileparts(filename);
dataFolder = fullfile(filepath,'ExampleFoodImageDataset');
unzip(filename,dataFolder);
```

The Zip-format file is usually downloaded to a folder location similar to the following location: C:\Users\Kamal\Documents\MATLAB\Examples\R2022a\supportfiles\nnet\data\ExampleFood ImageDataset.zip

2) Get the training images and classes via saving them to an image data store and image data store labels.
3) Create an augmented image datastore for training.
4) Use **shuffle** function to create two more data sets for validation and testing.
5) Create a CNN from scratch by creating appropriate number and type of layers.
6) Set some training options.
7) Train the network.
8) Evaluate network on test data, like calculating the accuracy % and plotting confusion chart for true versus predicted value of the test data.
9) Use **augmentedImageDatastore** function to transform data store images with a combination of rotation, reflection, shear, and/or translation. Create three augmented image data stores for training, validation, and testing.
10) Go to step #5 and continue until step #8. See which way is a more efficient in terms of overall accuracy and CPU time for training, validation, and testing steps.

HCW 7.3   CNN Creation for Merchandise Data Prediction

The MathWorks Merch data set is a small data set containing 75 images of MathWorks merchandise, belonging to *five different classes (cap, cube, playing cards, screwdriver, and torch)*. You can use this data set to try out transfer learning and image classification quickly.

It can be found at one of the following folder locations:

C:\Program Files\MATLAB\R2022a\examples\deeplearning_shared\data\MerchData.zip
C:\Program Files\MATLAB\R2022a\examples\nnet\data\MerchData.zip
C:\Program Files\MATLAB\R2022a\examples\vision\data\MerchData.zip

The images are of size 227 × 227 × 3.

Carry out the following steps:

1) Extract the MathWorks Merch data set.

```
filename = 'MerchData.zip';
dataFolder = fullfile(tempdir,'MerchData');
if ~exist(dataFolder,'dir')
  unzip(filename,tempdir);
end
```

2) Load the data as an image datastore using the **imageDatastore** function and specify the folder containing the image data and their classes.
3) Create an augmented image datastore for training.
4) Use **shuffle** function to create two more data sets for validation and testing.
5) Create a CNN from scratch by creating appropriate number and type of layers.
6) 6. Set some training options.
7) Train the network.
8) Evaluate network on test data, like calculating the accuracy % and plotting confusion chart for true versus predicted value of the test data.
9) Use **augmentedImageDatastore** function to transform data store images with a combination of rotation, reflection, shear, and/or translation. Create three augmented image data stores for training, validation, and testing.
10) Go to step #5 and continue until step #8. See which way is a more efficient in terms of overall accuracy and CPU time for training, validation, and testing steps.

HCW 7.4   CNN Creation for Musical Instrument Spectrograms Prediction

The data used for this homework/classwork can be obtained from the University of Iowa Electronic Music Studios. (http://theremin.music.uiowa.edu/index.html)

The Spectrograms folder contains about 2000 spectrograms. There is one folder for each instrument. The data is also available with Wiley web companion.

Carry out the following steps:

1) Load the data as an image datastore using the **imageDatastore** function and specify the folder containing the image data and their classes.
2) Split the data so that 80% of the images are used for training, 10% are used for validation, and the rest are used for testing, using **splitEachLabel** function. If changing the image size, make augmented image datastores.
3) Create a network from scratch with different number and types of layers.
4) Set the training options.
5) Train the network.
6) Evaluate network on test data, like calculating the accuracy % and plotting confusion chart for true versus predicted value of the test data.
7) Use **shuffle** function to create two more data sets for validation and testing and continue starting from step #3 up to #6.
8) Use **augmentedImageDatastore** function to transform data store images with a combination of rotation, reflection, shear, and/or translation. Create three augmented image data stores for training, validation, and testing.
9) Go to step #3 and continue until step #6.

HCW 7.5   CNN Creation for Chest X-ray Prediction

The data used for this homework/classwork can be obtained from one of the following web sites (size ~ 1.12 GB):
   https://drive.google.com/drive/folders/1tthDMG019wk4Oq4ar6EHAy20-xl1cVra
   https://bimcv.cipf.es/bimcv-projects/padchest/
   https://www.kaggle.com/datasets/paultimothymooney/chest-xray-pneumonia (size ~ 2GB)
   https://datasets.activeloop.ai/docs/ml/datasets/chest-x-ray-image-dataset/

Carry out the following steps:

1) Load the data as an image datastore using the **imageDatastore** function and specify the folder containing the image data and their classes.
2) Create three augmented image data stores for training, validation, and testing using **'ColorPreprocessing'**, set to **'gray2rgb'**. See Ch.5 | NOTE #2.
3) Create a network from scratch with different number and types of layers.
4) Set the training options.
5) Train the network.
6) Evaluate network on test data, like calculating the accuracy % and plotting confusion chart for true versus predicted value of the test data.

HCW 7.6   Semantic Segmentation Network for CamVid Dataset

CamVid is a popular dataset for semantic segmentation tasks, consisting of images captured from a vehicle-mounted camera. Here's how you can access and use the CamVid dataset.

Download the dataset: Visit the official CamVid dataset website (http://mi.eng.cam.ac.uk/research/projects/VideoRec/CamVid/) and navigate to the "Downloads" section. Download the CamVid dataset, which contains labeled images and corresponding annotations. A softcopy is available for book users at the book web companion. There are 32 different pixel classes with their class names and their digital IDs. It covers 32 different object classes, including various road and traffic-related objects. The dataset provides a good starting point for training and testing semantic segmentation networks on smaller-scale tasks.

Carry out the following steps:

1) Set the path to the extracted dataset folder.
2) Specify the image and annotation (pixel label) paths.
3) Create **imageDatastor**e objects for images and annotations
4) Read both LabelID and class names from the attached Excel/Text file.
5) Create the pixel label data store using **pixelLabelDatastore** function.
6) Display a sample image and annotations.
7) Create a binary mask of just the car. Use **"Car"** class and show the masked image.
8) Create a semantic segmentation network from scratch, made of adequate type and number of layers.
9) Split the entire data set into three subsets: Training, validation, and testing.
10) Set the training options.
11) Train the network using **trainNetwork** function to create a trained series CNN network.
12) Run the network on the test images using **semanticseg** function. Predicted labels are written to disk in a temporary directory and returned as a **pixelLabelDatastore** object.
13) Evaluate the quality of the prediction using **evaluateSemanticSegmentation** function to report different metrics.

8

Regression Classification

Object Detection

A convolutional neural network (CNN) is often used for image classification. Generally speaking, classification refers to the task of using data and known responses to build a predictive model that predicts a discrete response for new data. For image classification, the input data is the image and the known response is the label of the image subject. On the other hand, regression is another task that can be accomplished with deep learning. Regression refers to assigning continuous response values to data, instead of discrete classes. One example of image regression is correcting rotated images. The input data is a rotated image, and the known response is the angle of rotation.

Preparing Data for Regression

To perform transfer learning for regression, we can use the **trainNetwork** function with the data, architecture, and training options as input.

NOTE #1
We cannot use an **imageDatastore** to store training data when performing regression because the labels are not categorical. One convenient alternative is to use the table datatype. The first variable of the table should contain the file directories and names of each image. The second variable should contain the responses. In this case, hand-written synthetic digit images will have one class or output, that is, the rotated angle.

Modification of CNN Architecture from Classification to Regression

Let us load **AlexNet** and save the network layers in a variable named **layers**.

>>**net = alexnet;**

>>**layers = net.Layers**

Figure 8.1 shows the first and last portion of AlexNet serial network where the last three layers are to be replaced later.

Machine and Deep Learning Using MATLAB: Algorithms and Tools for Scientists and Engineers, First Edition.
Kamal I. M. Al-Malah.
© 2024 Kamal I. M. Al-Malah. Published 2024 by John Wiley & Sons, Inc.
Companion Website: www.wiley.com/go/al-malah/machinelearningmatlab

```
layers =

25×1 Layer array with layers:

    1   'data'      Image Input    227×227×3 images with 'zerocenter' normalization
    2   'conv1'     Convolution    96 11×11×3 convolutions with stride [4  4] and padding [0  0  0  0]
    3   'relu1'     ReLU           ReLU

   23   'fc8'       Fully Connected              1000 fully connected layer
   24   'prob'      Softmax                      softmax
   25   'output'    Classification Output        crossentropyex with 'tench' and 999 other classes
```

Figure 8.1 A portion of **AlexNet** CNN where it shows the last three layers used for image classification.

The last three layers of **AlexNet** are specific to image classification. Because we need the network to perform regression, we can delete these three layers before replacing them with the correct layers.

The last **n** elements of an array are easily removed with the end keyword.

>>**layers(end-n+1:end) = []**

For example, to remove the last three layers of **AlexNet** serial network, we key in:

>>**layers(end-2:end) = []**

Let us repeat the previous steps for **GoogleNet** DAG network.

Load **GoogleNet** to a variable named **net2** and save the network layers in a variable named **layers**.

>>**net2 = googlenet;**

>**layers2 = net2.Layers**

Figure 8.2 shows the last portion of the 144-layer **GoogleNet** DAG network where the last three layers are to be replaced later.

```
layers2 =

144×1 Layer array with layers:

    1   'data'                Image Input          224×224×3 images with 'zerocenter' normalization
  142   'loss3-classifier'    Fully Connected      1000 fully connected layer
  143   'prob'                Softmax              softmax
  144   'output'              Classification Output  crossentropyex with 'tench' and 999 other classes
```

Figure 8.2 The last three layers of **GoogleNet** network are to be replaced later.

Let us remove the last three layers.

>>**layers(end-2:end)=[];**

For regression problems, the last two layers must be a fully connected layer and a regression layer. The corresponding functions are

>>fullyConnectedLayer(outputSize)

and

>>regressionLayer()

The regression response is one numeric value; one for each rotated angle. This means that the fully connected layer should have one output.

We only need to add two more layers to our architecture because regression networks do not need a softmax layer. We can get the layer graph using the **layerGraph** function.

Let us extract the layer graph into a variable named **lgraph**.

>>lgraph=layerGraph(net);

We need to modify the final three layers of **GoogleNet** to suit our data set. When we create a fully connected layer, we should specify the number of output classes and a **"Name"**.

>>fc = fullyConnectedLayer(numClasses,"Name","lyname")

Then we can use the **replaceLayer** function to modify **GoogleNet**'s architecture. To replace a layer, we need to know layer's name, which can be found by looking at the output of **net. Layers**.

>>lg = replaceLayer(lg,"NameToReplace",newlayer)

Let us create a new fully connected layer that has one class and is named **"new_fc."** Name the variable **fc**. Then replace the fully connected layer in **lgraph**.

In **GoogleNet**, the fully connected layer is named **"loss3-classifier."**

>>fc = fullyConnectedLayer(1, "Name","new_fc");

>>lgraph = replaceLayer(lgraph,"loss3-classifier",fc)

Regression networks do not need a softmax layer, so we need to remove the softmax layer from the layer graph.

To remove a layer from a network, we can use the **removeLayers** function.

>>lg = removeLayers(lg,"NameToRemove")

In GoogleNet, the softmax layer is named **"prob."** Let us remove the **softmax** layer from **lgraph**.

>>lgraph = removeLayers(lgraph,"prob");

For regression problems, the output layer should be a regression layer instead of a classification layer.

>>regressionLayer("Name","lyname")

Let us create a regression output layer that is named "**new_out.**" Name the variable **out**. Then replace the final layer in **lgraph**.

In **GoogleNet**, the classification layer is named "**output.**"

>>**out = regressionLayer("Name","new_out")**

>>**lgraph = replaceLayer(lgraph,"output",out)**

The modified network has all the layers it needs, but since we removed the softmax layer, the last two layers need to be connected.

Connect the layers "**new_fc**" and "**new_out.**"

>>**lgraph = connectLayers(lgraph,"new_fc","new_out");**

The cosmetic surgery is complete! These layers can be used with the **trainNetwork** function to train a regression network. Just like with classification, we also would need training options and training data.

Root-Mean-Square Error

In classification, a prediction is either correct or incorrect. The effectiveness of a prediction with a regression network must be calculated differently.

When a regression network is trained, the root-mean-square error (RMSE) is calculated instead of accuracy.

To find the error for one image, we can just find the difference between the known value and the predicted value.

>>**predictedColor = predict(cnnet,testImage);**

>>**err = testColor - predictedColor;**

To find the error for multiple images, calculate RMSE.

>>**rmse = sqrt(mean(err.^2));**

AlexNet-Like CNN for Regression: Hand-Written Synthetic Digit Images

The data set contains synthetic images of handwritten digits together with the corresponding angles (in degrees) by which each image is rotated. Let us load the training and validation images as 4-D arrays using **digitTrain4DArrayData** and **digitTest4DArrayData**. The outputs **YTrain** and **YValidation** are the rotation angles in degrees. The training and validation data set both contain 5000 images.

When training neural networks, it is often helpful to make sure that the data is normalized at all stages of the network. Normalization helps stabilize and speed up network training using gradient descent. If the data is poorly scaled, then the loss can become NaN and the network parameters can diverge during training. One way of treating data include rescaling the data so that its range is [0,1]. Another way is normalizing plus standardizing the data; i.e., with zero mean and 1 standard deviation, σ, (i.e., 99.7 % of data will fall within the range $\pm 3\sigma$).

- Input data. Scale the predictors before we input them to the network. In this example, the input images are already scaled to the range [0,1].
- Layer outputs. We can normalize the outputs of each convolutional and fully connected layer by using a batch normalization layer.
- Responses. If the batch normalization layers are used to normalize the layer outputs toward the end of the network, then the predictions of the network will be normalized when training starts. If the response has a very different scale from these predictions, then network training can fail to converge. If the response is poorly scaled, then response normalization/scaling will be better tested to see if the network training improves. If the response variable is normalized/scaled before training, then we must transform back the predictions of the trained network to obtain the predictions of the original response.

If the distribution of the input or response is very uneven or skewed, we can also perform non-linear transformations (for example, taking logarithms) to the data before training the network.

Figure 8.3 shows the type and number of layers for the modified AlexNet (*left*) and the original AlexNet CNN (*right*).

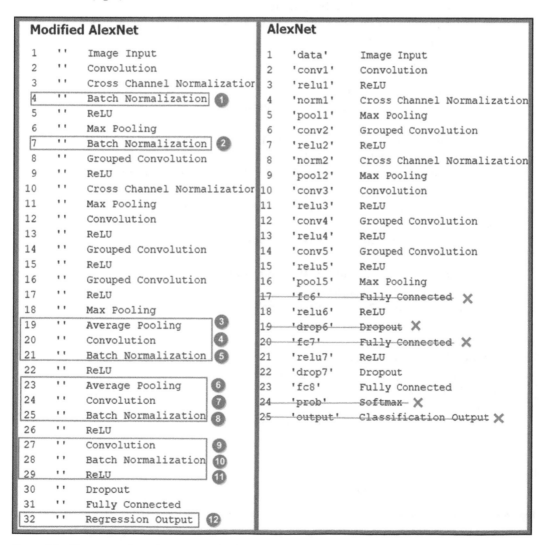

Figure 8.3 The type and number of layers for the modified AlexNet (*left*) and the original AlexNet CNN (*right*).

Figure 8.4 shows the code for using **AlexNet** serial network but with modifications indicated in Figure 8.3, including the number of layers and their types. The training options are also shown at the end.

We let it train for 30 epochs. We set the initial learn rate to 0.001 and lower the learning rate after 25 epochs. We monitor the network accuracy during training by specifying validation data and validation frequency. The software trains the network on the training data and calculates the accuracy on the validation data at regular intervals during training. The validation data is not used to update the network weights. We turn on the training progress plot, and turn off the command window output.

```
[XTrain,~,YTrain] = digitTrain4DArrayData;
[XValidation,~,YValidation] = digitTest4DArrayData;
layers = [
    imageInputLayer([28 28 1])
    convolution2dLayer(3,8,'Padding','same')
    crossChannelNormalizationLayer(5,'K',1, 'Alpha', 1.0000e-04,'Beta',0.75);
    batchNormalizationLayer
    reluLayer
    maxPooling2dLayer(3,'Stride',2)
    batchNormalizationLayer
    groupedConvolution2dLayer(5,128,2,'Padding', 'same')
    reluLayer
    crossChannelNormalizationLayer(3,'K',1, 'Alpha', 1.0000e-04,'Beta',0.75)
    maxPooling2dLayer(3,'Stride',2, 'Padding', 'same')
    convolution2dLayer(3,384,'NumChannels',256,'Stride',1, 'Padding', 'same')
    reluLayer
    groupedConvolution2dLayer(3,192,2,'Padding', 'same')
    reluLayer
    groupedConvolution2dLayer(3,192,2,'Padding', 'same')
    reluLayer
    maxPooling2dLayer(3,'Stride',2, 'Padding', 'same')
    averagePooling2dLayer(2,'Stride',2)
    convolution2dLayer(3,16,'Padding','same')
    batchNormalizationLayer
    reluLayer
    averagePooling2dLayer(2,'Stride',2)
    convolution2dLayer(3,32,'Padding','same')
    batchNormalizationLayer
    reluLayer
    convolution2dLayer(3,32,'Padding','same')
    batchNormalizationLayer
    reluLayer
    dropoutLayer(0.2)
    fullyConnectedLayer(1)
```

Figure 8.4 Training **AlexNet**-like serial network and the training options.

```
      regressionLayer];
miniBatchSize  = 128;
validationFrequency = floor(numel(YTrain)/miniBatchSize);
options = trainingOptions('adam', ...
   'MiniBatchSize',miniBatchSize, ...
   'MaxEpochs',30, ...
   'InitialLearnRate',1e-3, ...
   'LearnRateSchedule','piecewise', ...
   'LearnRateDropFactor',0.1, ...
   'LearnRateDropPeriod',25, ...
   'Shuffle','every-epoch', ...
   'ValidationData',{XValidation,YValidation}, ...
   'ValidationFrequency',validationFrequency, ...
   'Plots','training-progress', ...
   'Verbose',false);
net2 = trainNetwork(XTrain,YTrain,layers,options);
```

Figure 8.4 (Cont'd)

Figure 8.5 shows both RMSE and loss (MSE) values for the trained and validated CNN.

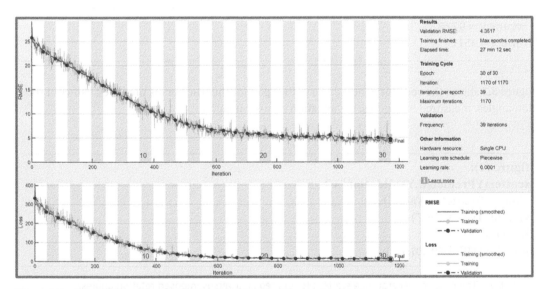

Figure 8.5   The RMSE and loss (MSE) for both CNN training and validation step.

Figure 8.6 shows the code continuation of Figure 8.4 code, where the accuracy and RMSE are both calculated in light of deviation from the true values for the validation data set. Both the predicted and true values represent the rotation angles, measured in degrees, imposed on the handwritten digital images. We use **predict** function to predict the angles of rotation of the validation images.

The performance of the model is examined by calculating the percentage of predictions within an acceptable error margin and the root-mean-square error (RMSE) of the predicted and actual angles of rotation. The number of correct predictions is evaluated so long as the error fits within an acceptable error margin from the true angles. We set the threshold to be 10 degrees.

```
YPredicted = predict(net2,XValidation);
predictionError = YValidation - YPredicted;
%Set a threshold of 10 degrees. Calculate the percentage of predictions within this
threshold.
thrshld = 10;
numCorrect = sum(abs(predictionError) < thrshld);
numValidationImages = numel(YValidation);
accuracy = (numCorrect/numValidationImages)*100.0
squares = predictionError.^2;
rmse = sqrt(mean(squares))
```

Figure 8.6 The code for calculating the model accuracy and RMSE using the validation data set.

The results are copied from Command Window.

```
accuracy = 97.68
rmse = 4.3517
```

Figure 8.7 shows the code for generating a scatter plot for the true versus predicted values of rotation angles, measured in degrees, of validation data-set images.

```
figure (2);
scatter(YPredicted,YValidation,'+')
xlabel("Predicted Value")
ylabel("True Value")
hold on
plot([-60 60], [-60 60],'r--')
```

Figure 8.7 The code for generating a scatter plot of predicted versus imposed rotation angles.

Figure 8.8 is the generated plot of true versus predicted rotation angles of validation data set images. A perfect model occurs when all datapoints fall unto the diagonal.

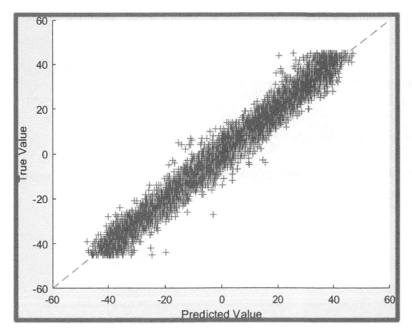

Figure 8.8 The true versus predicted rotation angle of the validation data-set of images.

Figure 8.9 shows the code for displaying the original digits with their corrected as model-predicted rotations. We can use montage function to display the digits together in a single image. Fifty-six (56) sample digits are rotated according to their predicted angles of rotation using **imrotate** function. We display the original (rotated) digits with their corrected rotations so that they become unrotated. We can use **montage** function to display the digits together in a single image.

```
idx = randperm(numValidationImages,56);
for i = 1:numel(idx)
   image = XValidation(:,:,:,idx(i));
   predictedAngle = YPredicted(idx(i));
   imagesRotated(:,:,:,i) = imrotate(image,predictedAngle,'bicubic','crop');
end
figure (3);
subplot(1,2,1)
montage(XValidation(:,:,:,idx))
title('Original')
subplot(1,2,2)
montage(imagesRotated)
title('Corrected')
```

Figure 8.9 The code for displaying 56 images as both original and correctly predicted images.

Figure 8.10 (*left*) shows the originally rotated digit images and their correctly predicted counterpart images (*right*).

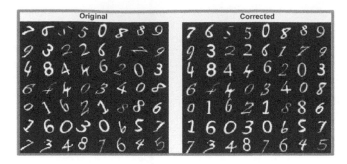

Figure 8.10 The original digit images (*left*) and their correctly predicted counterpart images (*right*).

Figure 8.11 shows MATLAB workspace variables created by running the previous code.

Workspace					Workspace				
Name ▲	Value	Size	Min	Max	Name ▲	Value	Size	Min	Max
accuracy	0.9768	1x1	0.9768	0.9768	predictedA...	-37.8682	1x1	-37.86...	-37.8682
i	56	1x1	56	56	predictionE...	5000x1 single	5000x1	-24.32...	23.3444
idx	1x56 double	1x56	48	4653	rmse	4.3517	1x1	4.3517	4.3517
image	28x28 double	28x28	0	0.8941	squares	5000x1 single	5000x1	5.6571...	591.7658
imagesRot...	4-D double	4-D	-0.0768	1.083/	thrshld	10	lxl	10	10
layers	32x1 Layer	32x1			validationFr...	39	1x1	39	39
miniBatchSi...	128	1x1	128	128	XTrain	4-D double	4-D	<Too ...	<Too mar
net2	1x1 SeriesNetwork	1x1			XValidation	4-D double	4-D	<Too ...	<Too mar
numCorrect	4884	1x1	4884	4884	YPredicted	5000x1 single	5000x1	-47.48...	46.9699
numValidat...	5000	1x1	5000	5000	YTrain	5000x1 double	5000x1	-45	45
options	1x1 TrainingOptionsADAM	1x1			YValidation	5000x1 double	5000x1	-45	45

Figure 8.11 MTALB workspace variables.

A New CNN for Regression: Hand-Written Synthetic Digit Images

Create New Network Layers

To solve the regression problem, we create the layers of the network and include a regression layer at the end of the network. The first layer defines the size and type of the input data. The input images are $28 \times 28 \times 1$. We create an image input layer of the same size as the training images. The middle layers of the network define the core architecture of the network, where most of the computation and learning take place. The final layers define the size and type of output data. For regression problems, a fully connected layer must precede the regression layer at the end of the network. We create a fully connected output layer of size 1 and a regression layer. Combine all the layers together in an array named **layers**.

Figure 8.12 shows the code for loading the predictor and response data set, creating a CNN from scratch, specifying the training options, and training the network.

```
[XTrain,~,YTrain] = digitTrain4DArrayData;
[XValidation,~,YValidation] = digitTest4DArrayData;
layers = [
  imageInputLayer([28 28 1])
  convolution2dLayer(3,8,'Padding','same')
  batchNormalizationLayer
  reluLayer
  averagePooling2dLayer(2,'Stride',2)
  convolution2dLayer(3,16,'Padding','same')
  batchNormalizationLayer
  reluLayer
  averagePooling2dLayer(2,'Stride',2)
  convolution2dLayer(3,32,'Padding','same')
  batchNormalizationLayer
  reluLayer
  convolution2dLayer(3,32,'Padding','same')
  batchNormalizationLayer
  reluLayer
  dropoutLayer(0.2)
  fullyConnectedLayer(1)
  regressionLayer];
miniBatchSize  = 128;
validationFrequency = floor(numel(YTrain)/miniBatchSize);
options = trainingOptions('sgdm', ...
  'MiniBatchSize',miniBatchSize, ...
  'MaxEpochs',30, ...
  'InitialLearnRate',1e-3, ...
  'LearnRateSchedule','piecewise', ...
  'LearnRateDropFactor',0.1, ...
  'LearnRateDropPeriod',25, ...
  'Shuffle','every-epoch', ...
  'ValidationData',{XValidation,YValidation}, ...
  'ValidationFrequency',validationFrequency, ...
  'Plots','training-progress', ...
  'Verbose',false);
net = trainNetwork(XTrain,YTrain,layers,options);
```

Figure 8.12   The code for creating a CNN from scratch, specifying training options, and training the network.

Figure 8.13 shows both RMSE and loss (MSE) of the trained CNN network. Again, the software trains the network on the training data and calculates the accuracy on the validation data at regular intervals during training.

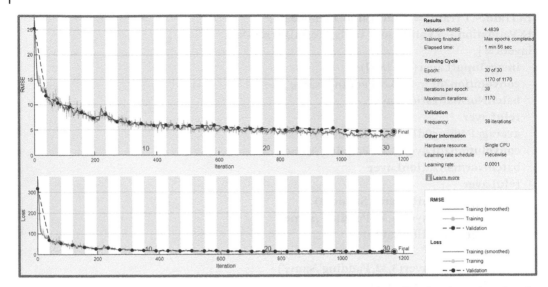

Figure 8.13 RMSE and loss (MSE) values of the trained CNN model for the predicted and actual angles of rotation.

To examine the details of the network architecture contained in the **Layers** property of net, at the command run:

>>net.Layers

Figure 8.14 shows the network layer architecture.

```
18×1 Layer array with layers:
1   'imageinput'        Image Input           28×28×1 images with 'zerocenter' normalization
2   'conv_1'            Convolution           8 3×3×1 convolutions with stride [1  1] and padding 'same'
3   'batchnorm_1'       Batch Normalization   Batch normalization with 8 channels
4   'relu_1'            ReLU                  ReLU
5   'avgpool2d_1'       Average Pooling       2×2 average pooling with stride [2  2] and padding [0  0  0  0]
6   'conv_2'            Convolution           16 3×3×8 convolutions with stride [1  1] and padding 'same'
7   'batchnorm_2'       Batch Normalization   Batch normalization with 16 channels
8   'relu_2'            ReLU                  ReLU
9   'avgpool2d_2'       Average Pooling       2×2 average pooling with stride [2  2] and padding [0  0  0  0]
10  'conv_3'            Convolution           32 3×3×16 convolutions with stride [1  1] and padding 'same'
11  'batchnorm_3'       Batch Normalization   Batch normalization with 32 channels
12  'relu_3'            ReLU                  ReLU
13  'conv_4'            Convolution           32 3×3×32 convolutions with stride [1  1] and padding 'same'
14  'batchnorm_4'       Batch Normalization   Batch normalization with 32 channels
15  'relu_4'            ReLU                  ReLU
16  'dropout'           Dropout               20% dropout
17  'fc'                Fully Connected       1 fully connected layer
18  'regressionoutput'  Regression Output     mean-squared-error with response 'Response'
```

Figure 8.14 The CNN network layer architecture, created from scratch.

Figure 8.15 shows the code for testing the performance of the network by evaluating the accuracy on the validation data.

```
YPredicted = predict(net,XValidation);
thrshld = 10;
numCorrect = sum(abs(predictionError) < thrshld);
numValidationImages = numel(YValidation);
accuracy = (numCorrect/numValidationImages)*100.0
squares = predictionError.^2;
rmse = sqrt(mean(squares))
```

Figure 8.15 Writing down the definition of CNN model accuracy and RMSE.

Results are copied from Command Window.

```
accuracy = 96.90
rmse = 4.4839
```

Figure 8.16 shows the code for displaying the original (rotated) 56 sample images and their correctly predicted counterpart images such that they become unrotated.

```
idx = randperm(numValidationImages,56);
for i = 1:numel(idx)
   image = XValidation(:,:,:,idx(i));
   predictedAngle = YPredicted(idx(i));
   imagesRotated(:,:,:,i) = imrotate(image,predictedAngle,'bicubic','crop');
end
figure (2);
subplot(1,2,1)
montage(XValidation(:,:,:,idx))
title('Original')
subplot(1,2,2)
montage(imagesRotated)
title('Corrected')
```

Figure 8.16 The code for displaying sample images with their original (*rotated*) and their counterparts as correctly predicted unrotated form.

Figure 8.17 (*left*) shows the originally rotated digit images and their correctly predicted counterpart images (*right*).

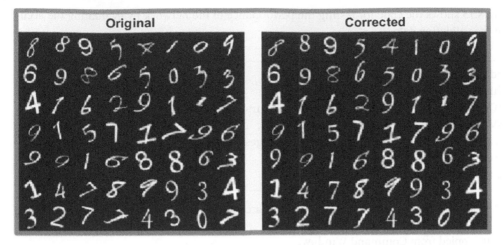

Figure 8.17 The original digit image (*left*) and its correctly predicted counterpart image (*right*).

Quiz #1

(Select all that apply) Which of the following performance indicators are displayed when training a regression network?

a)Loss   b) Accuracy   c) RMSE

Quiz #2

In image regression, the response to an image is a continuous value.

a) True   b) False

Quiz #3

(Select all that apply) Which of the following layers in **AlexNet** network are specific to classification?

a) The last fully connected layer   b) The last dropout layer

c) The softmax layer   d) The classification output layer

Deep Network Designer (DND) for Regression

In Chapter 7 we introduced DND platform. This section shows how to use DND app to construct and train an image-to-image regression network for super resolution. Spatial resolution is the number of pixels used to construct a digital image. An image with a high spatial resolution is composed of a greater number of pixels and as a result the image contains more details. *Super resolution is the process of taking a low-resolution input image and upscaling it into a higher resolution output image.* When we work with image data, we might reduce the spatial resolution to decrease the size of the data, at the cost of image quality. To recover the lost information, we can train a deep learning network to predict the missing details of an image. In this section, we recover 28×28-pixel images from images that were initially compressed down to 7×7 pixels.

Loading Image Data

This example uses the digits data set, which consists of 10,000 synthetic grayscale images of hand-written digits. Each image is $28 \times 28 \times 1$ pixels.

Load the data and create an image datastore made of $28 \times 28 \times 1$-pixel images.

```
dataFolder = fullfile(toolboxdir('nnet'),'nndemos','nndatasets','DigitDataset');
imds = imageDatastore(dataFolder, ...
   'IncludeSubfolders',true, ....
   'LabelSource','foldernames');
```

We use the shuffle function to shuffle the data prior to training.

```
imds = shuffle(imds);
```

We use the **splitEachLabel** function to divide the image datastore into three image datastores containing images for training, validation, and testing.

```
[imdsTrain,imdsVal,imdsTest] = splitEachLabel(imds,0.7,0.15,0.15,'randomized');
```

Rescale the data in each image to the range [0,1]. Notice that

```
>>B = rescale(A);
```

will scale the entries of **A** to the interval [0,1]. The output array **B** is the same size as A.

Scaling helps stabilize and speed up network training using gradient descent. If the data is initially poorly scaled, then the loss can become NaN and the network parameters can diverge during training.

```
imdsTrain = transform(imdsTrain,@(x) rescale(x));
imdsVal = transform(imdsVal,@(x) rescale(x));
imdsTest = transform(imdsTest,@(x) rescale(x));
```

Generating Training Data

We create a training data set by generating pairs of images consisting of low-resolution images and the corresponding high-resolution images. To train a network to perform image-to-image regression, the images need to be pairs consisting of an input and a response where both images are of the same size. Generate the training data by downgrading each image to 7×7 pixels and then upgrading to 28×28 pixels. Using the pairs of transformed and original images, the network can learn how to map between the two different levels of resolutions.

```
targetSize = [7,7];
imdsInputTrain = transform(imdsTrain,@(x) imresize(x,targetSize,
'method','bilinear'));
imdsInputVal = transform(imdsVal,@(x) imresize(x,targetSize, 'method','bilinear'));
imdsInputTest = transform(imdsTest,@(x) imresize(x,targetSize,
'method','bilinear'));
targetSize = [28,28];
imdsInputTrain = transform(imdsInputTrain,@(x) {imresize(x,targetSize,
'method','bilinear')});
imdsInputVal = transform(imdsInputVal,@(x) {imresize(x,targetSize,
'method','bilinear')});
imdsInputTest = transform(imdsInputTest,@(x) {imresize(x,targetSize,
'method','bilinear')});
```

We then use the **combine** function to combine the low- and high-resolution images into a single datastore. The output of the **combine** function is a **CombinedDatastore** object.

```
dsTrain = combine(imdsInputTrain,imdsTrain);
dsVal = combine(imdsInputVal,imdsVal);
dsTest = combine(imdsInputTest,imdsTest);
```

Creating a Network Architecture

We create the network architecture using the **unetLayers** function. This function provides a network suitable for semantic segmentation that can be easily adapted for image-to-image regression. Create a network with input size 28 × 28 × 1 pixels.

```
layers = unetLayers([28,28,1],3,'EncoderDepth',2);
```

returns a U-Net network. **unetLayers** includes a pixel classification layer in the network to predict the categorical label for every pixel in an input image. The second input is the number of classes or number of responses, which is one in our case. But the number of classes has a minimum value of 2. However, it can be modified as in Figure 8.19. See **Note #2**, below.

We use **unetLayers** to create the U-Net network architecture. We must train the network using **trainNetwork** function. The encoder depth is specified as a positive integer. U-Net is composed of both an encoder subnetwork and a corresponding decoder subnetwork. The depth of these networks determines the number of times the input image is down-sampled or up-sampled during processing. The encoder network down-samples the input image by a factor of 2^D, where D is the value of '**EncoderDepth**'. On the contrary, the decoder network up-samples the encoder network output by a factor of 2^D.

NOTE #2

Upon double clicking on **layers** icon, found in workspace variables, one will be able to see that "**layers**" variable is 1×1 **LayerGraph** type and is made of 34 layers. Moreover, the number of classes in the semantic segmentation is specified as integer greater than 1. Yet, it can be modified as in Figure 8.19 via modifying the number of filters for the last convolution 2-D layer.

Let us edit the network for image-to-image regression using DND app.

>>deepNetworkDesigner(layers);

In the Designer working area, replace the softmax and pixel classification layer by a regression layer from the Layer Library, as shown in Figure 8.18.

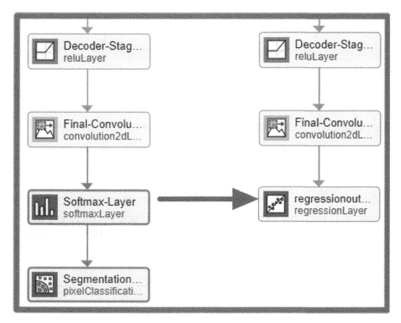

Figure 8.18 Replacement of the last two layers by a regression layer.

Highlight the final convolutional 2-D layer and set the "**NumFilters**" property to *1*, as shown in Figure 8.19. We have one single output that represents the regressed (best) value for estimating the rotated angle in this example.

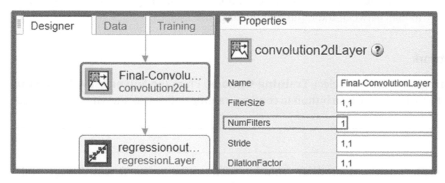

Figure 8.19 The number of filters is set to 1 (equal to number of possible output values) for the final convolutional 2-D layer.

Under "**Designer**" tab window, click on "**Analyze**" button from the top toolbar followed by "**Analyze**" icon to let DND app carry out an initial assessment for the architecture of the proposed layers. Be sure to end up with zero bugs (errors and warnings) as shown in Figure 8.20; otherwise fix the bug if any exists.

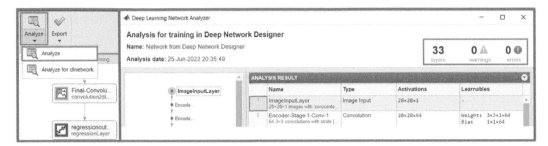

Figure 8.20 An initial assessment for the architecture of layers is highly recommended to avoid any potential bugs and headaches.

NOTE #3

Under "**Design**" tab window, one may click on "**Export**" button to save a copy of the latest version of the created layers to workspace environment. Then the user may save the workspace variable in a file for a future reference. The network layers can be re-imported into DND environment and training can proceed from this point onward.

Importing Data

Import the training and validation data into DND app. Under the **Data** tab window, click **Import Data > Import Datastore** and select **dsTrain** as the training data and **dsVal** as the validation data. Import both datastores by clicking on **Import** button, as shown in Figure 8.21. Notice that DND app displays the pairs of images in the combined datastore. The up-scaled low-resolution input images are on the left, and the original high-resolution response images are on the right. The network learns how to map between the input and the response images.

Training the Network

Under the **Training** tab windows, select **Training Options**. From the **Solver** list, select **adam**. Set **MaxEpochs** to 10. Click on **Close** button to confirm training options, as shown in Figure 8.22.

NOTE #4

If the default ('**sgdm**') solver is used, the solution (network parameters) will diverge and training will stop as a result of having loss values of NaN. The following warning is issued upon using the default solver:

Warning: Training stopped at iteration 4 because training loss is NaN. Predictions using the output network might contain NaN values.

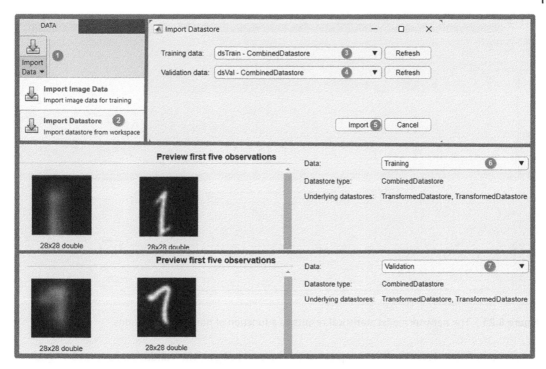

Figure 8.21 Importing training and validation data sets into DND environment. Preview for some low- and high-resolution images is shown for both data sets.

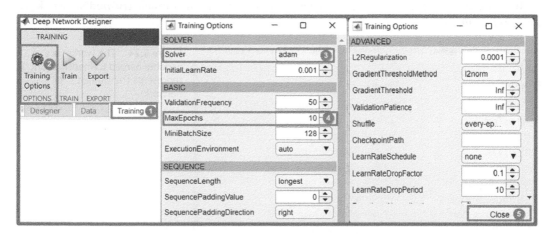

Figure 8.22 The chance to modify training options prior to carrying out network training.

Train the network on the combined datastore by clicking on **Train** button at the top toolbar. Wait until DND finishes the training session and shows the results in the form of RMSE and loss (MSE) as a function of number of iterations for both training and validation data sets, as shown in Figure 8.23.

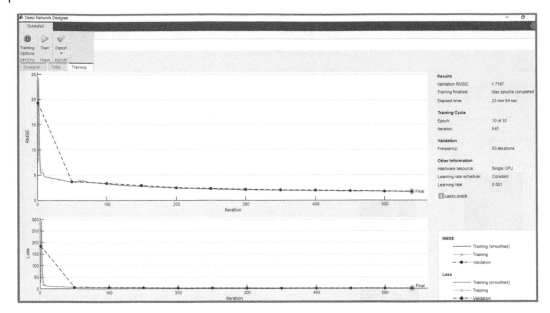

Figure 8.23 The network model statistical results as a function of number of iterations.

As the network learns how to map between the two images, the validation root mean squared error (RMSE) decreases.

Under "**Training**" tab window, if we click on "**Export**" button followed by the first choice, that is, "**Export Trained Network and Results**" option, we will be able to export the trained network to the workspace environment. The trained network is stored in the variable **trainedNetwork_1**, as shown in Figure 8.24.

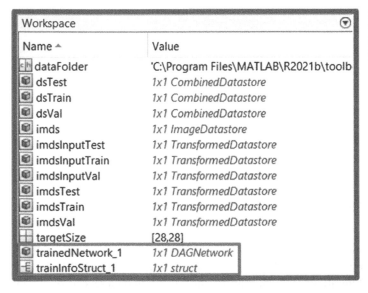

Figure 8.24 The workspace variable named **trainedNetwork_1** represents the trained CNN network.

On the other hand, if the user clicks on the second choice, that is, "**Generate Code for Training**" option, then a *.mlx file will be generated. MATLAB® stores live scripts and functions using the Live Code file format in a file with a *.mlx extension. More on *.mlx file has been explained in

Chapter 7. Figure 8.25 shows the DND-generated *.mlx code for training and validating the CNN network, the network architecture, and the training options. The user may attempt to re-run the code as it stands which basically reproduces the same previous results or make some changes or modifications to the code and re-run the modified code for the sake of improving the training and/ or validation performance. In Figure 8.25, I removed the annotated comment statements for the sake of making the story short.

```
trainingSetup = load("D:\OneDrive_HCT\MatlabBook\MLDL\params_2021_12_25.
mat");
dsTrain = trainingSetup.dsTrain;
dsValidation = trainingSetup.dsValidation;
opts = trainingOptions("adam",...
   "ExecutionEnvironment","auto",...
   "InitialLearnRate",0.001,...
   "MaxEpochs",10,...
   "Shuffle","every-epoch",...
   "Plots","training-progress",...
   "ValidationData",dsValidation);

lgraph = layerGraph();

%Add layer branches

tempLayers = [
  imageInputLayer([28 28 1],"Name","ImageInputLayer")
  convolution2dLayer([3 3],64,"Name","Encoder-Stage-1-Conv-1","Padding","same","We
ightsInitializer","he")
  reluLayer("Name","Encoder-Stage-1-ReLU-1")
  convolution2dLayer([3 3],64,"Name","Encoder-Stage-1-Conv-2","Padding","same","We
ightsInitializer","he")
  reluLayer("Name","Encoder-Stage-1-ReLU-2")];
lgraph = addLayers(lgraph,tempLayers);
tempLayers = [
  maxPooling2dLayer([2 2],"Name","Encoder-Stage-1-MaxPool","Stride",[2 2])
  convolution2dLayer([3 3],128,"Name","Encoder-Stage-2-Conv-1","Padding","same","W
eightsInitializer","he")
  reluLayer("Name","Encoder-Stage-2-ReLU-1")
  convolution2dLayer([3 3],128,"Name","Encoder-Stage-2-Conv-2","Padding","same","W
eightsInitializer","he")
  reluLayer("Name","Encoder-Stage-2-ReLU-2")];
lgraph = addLayers(lgraph,tempLayers);
tempLayers = [
  dropoutLayer(0.5,"Name","Encoder-Stage-2-DropOut")
  maxPooling2dLayer([2 2],"Name","Encoder-Stage-2-MaxPool","Stride",[2 2])
  convolution2dLayer([3 3],256,"Name","Bridge-Conv-1","Padding","same","WeightsIni
tializer","he")
  reluLayer("Name","Bridge-ReLU-1")
```

Figure 8.25 DND-generated code for training and validating CNN network with editable layer architecture and training options.

```
  convolution2dLayer([3 3],256,"Name","Bridge-Conv-2","Padding","same","WeightsIni
tializer","he")
  reluLayer("Name","Bridge-ReLU-2")
  dropoutLayer(0.5,"Name","Bridge-DropOut")
  transposedConv2dLayer([2 2],128,"Name","Decoder-Stage-1-UpConv","BiasLearnRat
eFactor",2,"Stride",[2 2],"WeightsInitializer","he")
  reluLayer("Name","Decoder-Stage-1-UpReLU")];
lgraph = addLayers(lgraph,tempLayers);
tempLayers = [
  depthConcatenationLayer(2,"Name","Decoder-Stage-1-DepthConcatenation")
  convolution2dLayer([3 3],128,"Name","Decoder-Stage-1-Conv-1","Padding","same","W
eightsInitializer","he")
  reluLayer("Name","Decoder-Stage-1-ReLU-1")
  convolution2dLayer([3 3],128,"Name","Decoder-Stage-1-Conv-2","Padding","same","W
eightsInitializer","he")
  reluLayer("Name","Decoder-Stage-1-ReLU-2")
  transposedConv2dLayer([2 2],64,"Name","Decoder-Stage-2-UpConv","BiasLearnRate
Factor",2,"Stride",[2 2],"WeightsInitializer","he")
  reluLayer("Name","Decoder-Stage-2-UpReLU")];
lgraph = addLayers(lgraph,tempLayers);
tempLayers = [
  depthConcatenationLayer(2,"Name","Decoder-Stage-2-DepthConcatenation")
  convolution2dLayer([3 3],64,"Name","Decoder-Stage-2-Conv-1","Padding","same","We
ightsInitializer","he")
  reluLayer("Name","Decoder-Stage-2-ReLU-1")
  convolution2dLayer([3 3],64,"Name","Decoder-Stage-2-Conv-2","Padding","same","We
ightsInitializer","he")
  reluLayer("Name","Decoder-Stage-2-ReLU-2")
  convolution2dLayer([1 1],1,"Name","Final-ConvolutionLayer","Padding","same","We
ightsInitializer","he")
  regressionLayer("Name","regressionoutput")];
lgraph = addLayers(lgraph,tempLayers);
% Clean up helper variable
clear tempLayers;

% Connect all the branches of the network to create the network graph.

lgraph = connectLayers(lgraph,"Encoder-Stage-1-ReLU-2","Encoder-Stage-1-MaxPool");
lgraph = connectLayers(lgraph,"Encoder-Stage-1-ReLU-2","Decoder-Stage-2-Dep-
thConcatenation/in2");
lgraph = connectLayers(lgraph,"Encoder-Stage-2-ReLU-2","Encoder-Stage-2-DropOut");
lgraph = connectLayers(lgraph,"Encoder-Stage-2-ReLU-2","Decoder-Stage-1-Dep-
thConcatenation/in2");
lgraph = connectLayers(lgraph,"Decoder-Stage-1-UpReLU","Decoder-Stage-1-Dep-
thConcatenation/in1");
lgraph = connectLayers(lgraph,"Decoder-Stage-2-UpReLU","Decoder-Stage-2-Dep-
thConcatenation/in1");

%Train the network using the specified options and training data.

[net, traininfo] = trainNetwork(dsTrain,lgraph,opts);
```

Figure 8.25 (Cont'd)

Test Network

Evaluate the performance of the network using the test data. Using **predict** function, we can test if the network can produce a high resolution-image from a low-resolution input image that was not included in the training set. Figure 8.26 shows the code for comparing a triplicate for each of six images borrowed from three batches: The low-resolution, the predicted, and the response (true).

```
ypred = predict(trainedNetwork_1,dsTest);
for i = 1:6
  I(1:2,i) = read(dsTest);
  I(3,i) = {ypred(:,:,:,i)};
end
%Compare the input, predicted, and response images.
subplot(1,3,1)
imshow(imtile(I(1,:),'GridSize',[6,1]))
title('Input')
subplot(1,3,2)
imshow(imtile(I(3,:),'GridSize',[6,1]))
title('Predict')
subplot(1,3,3)
imshow(imtile(I(2,:),'GridSize',[6,1]))
title('Response')
```

Figure 8.26 Code for comparing a triplicate of an image borrowed from a low-resolution, predicted, and the response pool.

Figure 8.27 shows the first six tested images where the triplicate is made of the low-resolution, the predicted, and the response image.

Figure 8.27 A triplicate for each of the six images borrowed from the low-resolution, predicted, and true pool.

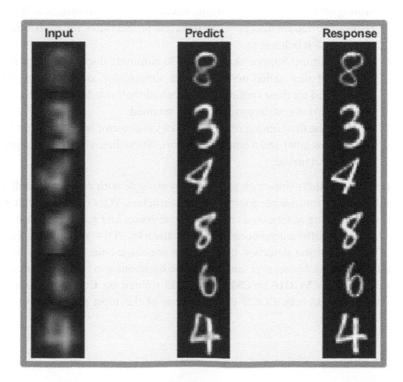

It can be seen that the network successfully produces high-resolution from low-resolution inputs. The network in this example is very simple and highly linked to the digits data set. For an example showing how to create a more complex image-to-image regression network for everyday images, see MATLAB built-in help | Single Image Super-Resolution Using Deep Learning.

YOLO Object Detectors

YOLO (**Y**ou **O**nly **L**ook **O**nce) is a popular object detection algorithm known for its real-time performance and high accuracy. It's a deep learning-based approach that can detect and localize objects in images and videos by dividing the input image into a grid and predicting bounding boxes and class probabilities for objects within each grid cell.

Here's a step-by-step overview of how a YOLO object detector works:

1) Input Image: The YOLO algorithm takes an input image or a frame from a video as its input.
2) Preprocessing: The input image is typically resized to a fixed size to ensure consistency throughout the detection process. It is then divided into a grid of cells, usually with a size of, for example, 13×13, 26×26, or 52×52.
3) Anchor Boxes: Before training the YOLO model, a set of anchor boxes is predefined. These anchor boxes represent different aspect ratios and sizes of objects that the model will try to detect. Each grid cell is responsible for predicting bounding boxes associated with these anchor boxes.
4) Convolutional Neural Network (CNN): YOLO utilizes a CNN as its backbone architecture. This CNN processes the input image and extracts meaningful features at different scales. The output of the CNN is a feature map.
5) Bounding Box and Class Predictions: Each grid cell in the feature map is responsible for predicting bounding boxes and class probabilities. For each grid cell, the model predicts multiple bounding boxes, each associated with a confidence score representing the likelihood of containing an object. These bounding boxes are defined as offsets to the dimensions of the anchor boxes. The model also predicts the class probabilities for each bounding box, indicating the object class it belongs to.
6) Non-Maximum Suppression (NMS): To eliminate duplicate or overlapping detections, a post-processing step called non-maximum suppression is performed. NMS filters out bounding boxes based on their confidence scores and their overlap with other boxes. Only the most confident and non-overlapping boxes are retained.
7) Output: The final output of the YOLO object detector is a set of bounding boxes, each associated with a class label and a confidence score. These bounding boxes represent the detected objects in the input image.

However, YOLO's drawback is that it may struggle with detecting small objects compared to algorithms using multi-scale approaches. Nonetheless, YOLO remains widely used in various applications, including autonomous driving, surveillance, and real-time video analysis.

There are different versions of YOLO networks. YOLO version 4 is featured in this section.

YOLO v4 object detection network is a one-stage object detection network and is composed of three parts: *backbone, neck, and head.* The backbone can be a pretrained convolutional neural network such as **VGG16** or **CSPDarkNet53** trained on **C**ommon **O**bjects in **Co**ntext (COCO) or ImageNet data sets. COCO dataset is one of the most popular open-source object recognition databases used to train deep learning programs. COCO has hundreds of thousands of images with millions of learnable objects. On the other hand, ImageNet (www.image-net.org) dataset is organized according to the WordNet hierarchy, where each node of the hierarchy is depicted by hundreds to thousands of images.

The YOLO v4 network consists of three main components: the backbone, the neck, and the head. The backbone serves as the feature extraction network, generating feature maps from the input images. Connecting the backbone and the head, the neck comprises a Spatial Pyramid Pooling (SPP) module and a Path Aggregation Network (PAN). Its role is to combine the feature maps from various layers of the backbone and pass them as inputs to the head. Lastly, the head takes in the aggregated features and performs the task of predicting bounding boxes, objectness scores, and classification scores. The YOLO v4 network uses a one-stage object detector, such as YOLO v3, as detection heads. Figure 8.28 shows a schematic for YOLO v4 skeletal architecture.

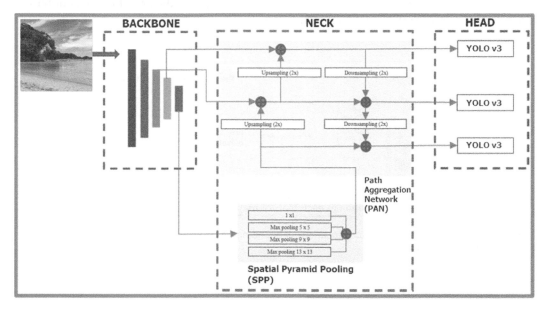

Figure 8.28   Yolo v4 architecture showing the main three skeletal parts.

The YOLO v4 network uses **CSPDarkNet-53** as the backbone for extracting features from the input images. The backbone in YOLO v4 has five residual block modules that produce feature maps. These feature maps are then combined with the neck of the network. The neck contains an SPP module, which merges the max-pooling results of the low-resolution feature map to capture important features. The SPP module uses different kernel sizes like 1×1, 5×5, 9×9, and 13×13 for max-pooling, with a stride of 1. Combining the feature maps helps improve the network's accuracy in detecting small objects by expanding the receptive field of the backbone features. The fused feature maps from the SPP module are then joined with the high-resolution feature maps using a PAN. The PAN employs up-sampling and down-sampling operations to create pathways for combining low-level and high-level features.

The PAN module produces a collection of combined feature maps that are used for making predictions. In YOLO v4, there are three detection heads, each functioning as a YOLO v3 network to generate the final predictions. The YOLO v4 network produces feature maps of different sizes (19 × 19,

38×38, and 76×76) to predict the bounding boxes, classification scores, and objectness scores. In contrast, the tiny YOLO v4 network is a lighter version of YOLO v4, designed with fewer layers. It uses a feature pyramid network as the neck and includes two YOLO v3 detection heads. The network generates feature maps of size 13×13 and 26×26 for making predictions.

YOLO v4 uses anchor boxes to detect classes of objects in an image. Similar to YOLO v3, YOLO v4 predicts these three attributes for each anchor box:

- Intersection over union (**IoU**): Predicts the objectness score of each anchor box.
- Anchor box offsets: Refines the anchor box position.
- Class probability: Predicts the class label assigned to each anchor box.

In Figure 8.29, we can see dotted lines representing predefined anchor boxes placed at various positions on a feature map. After applying offsets, the anchor boxes' locations are adjusted and refined. The colored anchor boxes indicate those that have been matched with a specific class. It's important to note that during network training, we need to specify the predefined anchor boxes, also referred to as a priori boxes, as well as the classes.

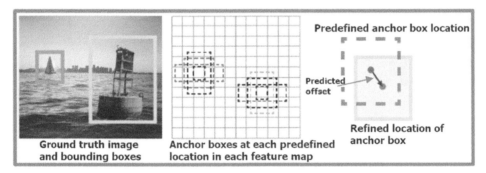

Figure 8.29 An image with predefined locations of bounding boxes to be fed to and processed by the YOLO training network.

NOTE #5

The **IoU** is the ratio between the area of intersection (common area) between the proposed anchor box and the ground-truth box to the area of union of the two boxes. Thus, the first box is the anchor generated by the region proposal and the second is the ground-truth box. The **IoU** value ranges from 0.0 to 1.0. When there is no intersection, the IoU is 0.0. As the two boxes get closer to one another, the **IoU** value will increase until it hits 1.0 (100% common overlap area).

For each rectangular sliding (small size probing) window of size n×n, where n varies for each CNN, R region proposals are created. Each proposal is parametrized according to a reference box which is called an anchor box. The two parameters of the anchor boxes are: Scale and Aspect Ratio.

Object Detection Using YOLO v4

Object detection using deep learning provides a fast and accurate means to predict the location of an object within an image. Deep learning is an instrumental machine learning technique by which the object detector automatically learns image features required for detection tasks. Several techniques for object detection using deep learning are available such as Faster R-CNN, you only look once (YOLO) v2, YOLO v3, YOLO v4, and single shot detection (SSD). Applications for object detection include, but are not limited to: Image classification, scene understanding, self-driving vehicles, and surveillance. Let us use **trainYOLOv4ObjectDetector** object to create a you only look once version 4 (YOLO v4) one-stage object detector for detecting objects in an image. Using this object, we can:

1) Predict objects in a new image using a pretrained YOLO v4 object detector, previously trained on COCO dataset: This means that we can use a pre-existing YOLO v4 model that has been trained on the COCO (Common Objects in Context) dataset, which is a widely used benchmark for object detection. The pretrained model has already learned to recognize a wide range of objects from various categories. We can then apply this pretrained model to new images and use it to predict the objects present in those images. The pretrained model comes with its own set of learned weights and parameters that enable it to detect objects based on the features it has learned during training.

2) Retrain a YOLO v4 object detector using any pretrained or untrained YOLO v4 deep learning network: This point refers to the ability to customize or fine-tune a YOLO v4 object detector using transfer learning. Transfer learning involves taking a pretrained model and further training it on a different dataset, either using the same network architecture or modifying it according to specific requirements. In this case, we have the option to retrain a YOLO v4 model using either a pretrained YOLO v4 model (which might have been trained on a different dataset) or an untrained YOLO v4 model. This process allows us to adapt the model's knowledge to a more specific domain or dataset, potentially improving its performance on our specific task or set of objects.

COCO-Based Creation of a Pretrained YOLO v4 Object Detector

Figure 8.30 shows the code for creating a pretrained YOLO v4 object detector using YOLO v4 deep learning networks trained on COCO dataset.

```
% Specify the name of a pretrained YOLO v4 deep learning network.
name = "tiny-yolov4-coco";
% Create YOLO v4 object detector using the pretrained YOLO v4 network.
detector = yolov4ObjectDetector(name);
% Display and inspect the properties of the YOLO v4 object detector.
disp(detector)
% Use analyzeNetwork to display the YOLO v4 network architecture and get information
% about the network layers.
analyzeNetwork(detector.Network)
%Detect objects in an unknown image by using the pretrained YOLO v4 object detector.
img = imread("highway.png");
[bboxes,scores,labels] = detect(detector,img);
%Display the detection results.
detectedImg = insertObjectAnnotation(img,"Rectangle",bboxes,labels);
figure
imshow(detectedImg)
```

Figure 8.30   Creating a YOLO object detector, analyzing it, and using it to detect some image objects.

The following results are copied from Command Window, in response to **disp(detector)**:

yolov4ObjectDetector with properties:
Network: [1×1 dlnetwork]
AnchorBoxes: {2×1 cell}
ClassNames: {80×1 cell}
InputSize: [416 416 3]
ModelName: 'tiny-yolov4-coco'

In response to **analyzeNetwork(detector.Network)**, Figure 8.31 shows a portion of the analyzed tiny YOLO v4 network architecture and its 74 layers with six million learnable objects.

Deep Learning Network Analyzer						
Analysis for dlnetwork usage			6M total learnables	74 layers	0 ⚠ warnings	0 ⓘ errors
Name: Unnamed Network						

ANALYSIS RESULT

input_1
conv_2
bn_2
leaky_2
conv_3
bn_3
leaky_3
conv_4
bn_4
leaky_4
slice_5
conv_6

	Name	Type	Activations
1	input_1 416×416×3 images	Image Input	416(S) × 416(S) × 3(C) × 1(B)
2	conv_2 32 3×3×3 convolutions with stride [2 2] a	Convolution	208(S) × 208(S) × 32(C) × 1(B)
3	bn_2 Batch normalization with 32 channels	Batch Normalization	208(S) × 208(S) × 32(C) × 1(B)
4	leaky_2 Leaky ReLU with scale 0.1	Leaky ReLU	208(S) × 208(S) × 32(C) × 1(B)
5	conv_3 64 3×3×32 convolutions with stride [2 2] ...	Convolution	104(S) × 104(S) × 64(C) × 1(B)
6	bn_3 Batch normalization with 64 channels	Batch Normalization	104(S) × 104(S) × 64(C) × 1(B)
7	leaky_3 Leaky ReLU with scale 0.1	Leaky ReLU	104(S) × 104(S) × 64(C) × 1(B)
8	conv_4 64 3×3×64 convolutions with stride [1 1] ...	Convolution	104(S) × 104(S) × 64(C) × 1(B)
9	bn_4 Batch normalization with 64 channels	Batch Normalization	104(S) × 104(S) × 64(C) × 1(B)

Figure 8.31 A portion of the tiny YOLO v4 trained network which has 74 layers.

Figure 8.32 shows the predicted (or recognized) objects within the image such that a label and a bounding box are shown around them.

Figure 8.32 The original highway image with the detected, labeled, and box-bound objects.

> **NOTE #6**
>
> In a typical **[bboxes,scores,labels]** output, the **bboxes** column contains M-by-4 matrices, of M bounding boxes for the objects found in the image. Each row contains a bounding box as a 4-element vector in the format [x,y,width,height] with pixel units. The **scores** column returns the class-specific confidence scores for each bounding box, and the **labels** column is a categorical array of labels assigned to the bounding boxes. The labels for object classes are pre-defined earlier during training.
>
> The objectness score uses the **IoU** value (see **NOTE #5**) to determine whether a positive or a negative objectness score is assigned to an anchor, using the following guidelines:
>
> a) An anchor that has an **IoU** value higher than 0.7 is assigned a positive objectness label. Others with **IoU** values < 0.7 are arbitrarily assigned negative score values.
> b) If there is no anchor with an **IoU** value higher than 0.7, then a positive label will be assigned to the anchor(s) with the highest **IoU** value, given that its **IoU** value is still > 0.5. Others will be assigned negative score values.
> c) A negative objectness score is arbitrarily assigned to a non-positive anchor when the **IoU** values for all ground-truth boxes is less than 0.3. A negative objectness score means the anchor is classified as background.
> d) Anchors that are neither positive nor negative are not used in the training classifier. This represents the case $0.3 \leq IoU \leq 0.5$. This is the twilight or uncertainty zone between night and light. Or, between the background (or black ground) and a detected object.

Fine-Tuning of a Pretrained YOLO v4 Object Detector

Let us outline how to fine-tune a pretrained YOLO v4 object detector for detecting vehicles in an image. We will use a tiny YOLO v4 network trained on COCO dataset.

Load a pretrained YOLO v4 object detector and inspect its properties.

```
detector = yolov4ObjectDetector("tiny-yolov4-coco")
```

The number of anchor boxes must be same as that of the number of output layers in the YOLO v4 network. The tiny YOLO v4 network contains two output layers. Use **analyzeNetwork** to display the YOLO v4 network architecture and get information about the network layers, as given by the following command:

```
analyzeNetwork(detector.Network)
```

Figure 8.33 shows the output layers as open-ended lines.

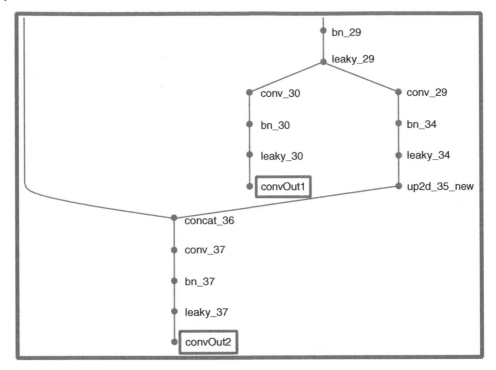

Figure 8.33 Implementing **analyzeNetWork(detector.Network)** shows the two terminal outputs.

Alternatively, we can directly use:

>>detector.Network

Figure 8.34 shows the two output layers upon executing **detector.Network** command.

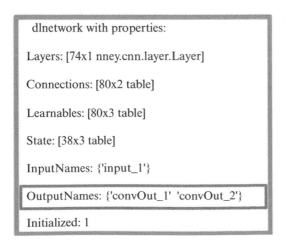

dlnetwork with properties:

Layers: [74x1 nney.cnn.layer.Layer]

Connections: [80x2 table]

Learnables: [80x3 table]

State: [38x3 table]

InputNames: {'input_1'}

OutputNames: {'convOut_1' 'convOut_2'}

Initialized: 1

Figure 8.34 Implementing **detector.Network** shows the two terminal outputs.

1. Prepare Training Data

Load a *.mat file containing information about the vehicle dataset to use for training. The information stored in the *.mat file is a table. The first column contains the training images and the remaining column contains the labeled bounding boxes.

```
data = load("vehicleTrainingData.mat");
trainingData = data.vehicleTrainingData;
```

2. Specify the directory in which training samples are stored. Add a full path to the file names in training data.

```
dataDir = fullfile(toolboxdir('vision'),'visiondata');
trainingData.imageFilename = fullfile(dataDir,trainingData.imageFilename);
```

3. Create an imageDatastore using the files from the table.

```
imds = imageDatastore(trainingData.imageFilename);
```

4. Create a boxLabelDatastore using the label columns from the table.

```
blds = boxLabelDatastore(trainingData(:,2));
```

5. Combine the datastores.

```
ds = combine(imds,blds);
```

6. Specify the input size to use for resizing the training images. The size of the training images must be a multiple of 32 when we use the tiny-yolov4-coco and csp-darknet53-coco pretrained YOLO v4 deep learning networks. We must also resize the bounding boxes based on the specified input size.

```
inputSize = [224 224 3];
```

7. Resize and rescale the training images and the bounding boxes using the **preprocessData** helper function (Appendix A.18A or Appendix A.18.B). Also, convert the preprocessed data to a datastore object using the transform function.

```
trainingDataForEstimation = transform(ds,@(data)preprocessData(data,inputSize));
```

8. Estimate the anchor boxes from the training data. We must assign the same number of anchor boxes to each output layer in the YOLO v4 network. YOLO detectors rely on anchor boxes to find bounding boxes. If we know the dimensions of the bounding boxes, we can pass those directly to the network architecture, but that isn't the case. Thus, the **estimateAnchorBoxes** function will find suitable anchor boxes for a data set, then such anchor boxes will be sorted out by their areas

```
numAnchors = 6;
[anchors, meanIoU] = estimateAnchorBoxes(trainingDataForEstimation,numAnch
ors);
area = anchors(:,1).*anchors(:,2);
[~,idx] = sort(area,"descend");
anchors = anchors(idx,:);
anchorBoxes = {anchors(1:3,:);anchors(4:6,:)};
```

NOTE #7

I tried numAnchors=4 and the trained network managed to label the truck only. I also tried numAnchors=8 and it managed to label both the truck and car. There has to be a *minimum even number* to be assigned so that we can divide it by 2 for having the same number of anchor boxes assigned to each output layer of this lightweight version (i.e., tiny YOLO v4 network). Keep in mind that the larger the number of objects originally labeled within a given image, the larger the number of anchor boxes will be.

9. Specify the class names and configure the pretrained YOLOv4 deep learning network to retrain for the new dataset using **yolov4ObjectDetector** function.

```
classes = {'vehicle'};
detector = yolov4ObjectDetector("tiny-yolov4-coco",classes,anchorBoxes,
InputSize=inputSize);
```

10. Specify the training options and retrain the pretrained YOLO v4 network on the new dataset using the **trainYOLOv4ObjectDetector** function.

In the below code, when **BatchNormalizationStatistics** is set to "**batch**," the mean and variance statistics are computed based on the current batch of training out of the entire data set. This approach is simple and computationally efficient but can introduce some batch-to-batch variability, especially when the batch size is small. The statistics are not stored and are recalculated for each batch during both training and inference. On the other hand, when **BatchNormalizationStatistics** is set to "**moving**" the mean and variance statistics are accumulated and updated over multiple batches. This approach maintains a moving average of the statistics, allowing for a more stable estimation of the mean and variance. During training, the statistics are updated after each batch, and during inference, the accumulated moving averages are used. This method can help reduce the batch-to-batch variability and improve the overall performance of the model. Thus, setting **BatchNormalizationStatistics** to "**moving**" has a few advantages:

1) Improved generalization: The moving average statistics capture a broader representation of the data distribution over time, which can lead to better generalization. It helps the model to normalize the activations consistently, even when the current batch might have atypical data.

2) Smoother training dynamics: The moving average statistics provide a smoother update signal during training. This can help stabilize the training process, avoid drastic changes in the model's behavior, and lead to faster convergence.

3) Better performance on small batches: When working with small batch sizes, setting **BatchNormalizationStatistics** to "**moving**" can be particularly beneficial. It mitigates the impact of batch-to-batch variability, allowing for more reliable and accurate estimation of the mean and variance.

```
options = trainingOptions("sgdm", ...
   InitialLearnRate=0.001, ...
   MiniBatchSize=16,...
   MaxEpochs=40, ...
   BatchNormalizationStatistics="moving",...
   ResetInputNormalization=false,...
   VerboseFrequency=30);
trainedDetector = trainYOLOv4ObjectDetector(ds,detector,options);
```

11. Read a test image to later detect objects

```
I = imread('highway.png');
```

12. Use the fine-tuned YOLO v4 object detector to detect vehicles in a test image and display the detection results.

```
[bboxes, scores, labels] = detect(trainedDetector,I,Threshold=0.05);
detectedImg = insertObjectAnnotation(I,"Rectangle",bboxes,labels);
figure;
imshow(detectedImg)
```

Figure 8.35 shows the entire code for the previous 12 steps.

```
data = load("vehicleTrainingData.mat");
trainingData = data.vehicleTrainingData;
dataDir = fullfile(toolboxdir('vision'),'visiondata');
trainingData.imageFilename = fullfile(dataDir,trainingData.imageFilename);
imds = imageDatastore(trainingData.imageFilename);
blds = boxLabelDatastore(trainingData(:,2));
ds = combine(imds,blds);
inputSize = [224 224 3];
trainingDataForEstimation = transform(ds,@(data)preprocessData(data,inputSize));
numAnchors = 6;
[anchors, meanIoU] = estimateAnchorBoxes(trainingDataForEstimation,numAnchors);
```

Figure 8.35 The code for re-training a pretrained tiny YOLO v4 object detector and presenting results.

```
area = anchors(:,1).*anchors(:,2);
[~,idx] = sort(area,"descend");
anchors = anchors(idx,:);
anchorBoxes = {anchors(1:3,:);anchors(4:6,:)};
classes = {'vehicle'};
detector = yolov4ObjectDetector("tiny-yolov4-coco",classes,anchorBoxes,InputSize=i
nputSize);
options = trainingOptions("sgdm", ...
   InitialLearnRate=0.001, ...
   MiniBatchSize=16,...
   MaxEpochs=40, ...
   BatchNormalizationStatistics="moving",...
   ResetInputNormalization=false,...
   VerboseFrequency=30);
trainedDetector = trainYOLOv4ObjectDetector(ds,detector,options);
I = imread('highway.png');
[bboxes, scores, labels] = detect(trainedDetector,I,Threshold=0.05);
detectedImg = insertObjectAnnotation(I,"Rectangle",bboxes,labels);
figure
imshow(detectedImg)
```

Figure 8.35 (Cont'd)

Figure 8.36 (*left*) shows the model training loss versus epochs or iterations and the image with objects being labeled and box bounded (*right*).

Epoch	Iteration	TimeElapsed	LearnRate	TrainingLoss
2	30	00:00:54	0.001	10.879
4	60	00:01:43	0.001	8.0307
5	90	00:02:26	0.001	3.2769
7	120	00:03:06	0.001	0.80468
8	150	00:03:47	0.001	1.1949
37	690	00:16:46	0.001	0.12071
38	720	00:17:32	0.001	0.11075
40	750	00:18:18	0.001	0.14077

Figure 8.36 Retraining a pretrained network showing the training loss (*left*) and the recognized objects (*right*).

Evaluating an Object Detector

We can visually evaluate the effectiveness of an object detector, but if we want to test multiple images it can be useful to calculate a numeric metric. There are many ways we could calculate the effectiveness of the network, but we should account for both the precision of the bounding box and the correctness of the label. Figure 8.37 shows what potential errors are encountered in terms of either mis-location of the bounding box (*left*), or even mislabeling the object itself (*right*).

Figure 8.37   Incorrect bounding box (*left*) or incorrect label (*right*) as potential errors.

Figure 8.38 shows the code for creating a tiny YOLO v4 object detector, reading an image, borrowing the pretrained detector's classes, to predict the objects (i.e., persons) as well as showing the enclosed object, its predicted label, and its objectness score. Notice that the tiny version of YOLO v4 object detector has 80 different built-in classes; one of them is **"person"**.

```
yv4objdetect = yolov4ObjectDetector("tiny-yolov4-coco");
image = imread('visionteam.jpg');
classNames = yv4objdetect.ClassNames;
executionEnvironment = 'auto';
[bboxes, scores, labels] = detect(yv4objdetect, image);
annotations = string(labels) + ": " + string(scores);
image = insertObjectAnnotation(image, 'rectangle', bboxes, annotations);
figure, imshow(image)
```

Figure 8.38   Code for predicting objects in an image using a pretrained object detector.

Figure 8.39 shows the predicted objects associated with their labels and objectness scores.

Figure 8.39   Box bounded objects, their labels, and scores as predicted by the trained tiny YOLO v4 detector.

Figure 8.40 shows workspace variables pertinent to running the code of figures 8.35 and 8.38.

Name ▲	Value	Size
anchorBoxes	*2x1 cell*	2x1
anchors	*6x2 double*	6x2
area	*[4828;18228;1386;21600;11536;736]*	6x1
bboxes	*[139.8195,86.6172,94.4769,80.5360;100.8866,104....*	2x4
blds	*1x1 boxLabelDatastore*	1x1
classes	*1x1 cell*	1x1
data	*1x1 struct*	1x1
dataDir	*'C:\Program Files\MATLAB\R2022a\toolbox\visi...*	1x56
detectedImg	*240x320x3 uint8*	240x320x3
detector	*1x1 yolov4ObjectDetector*	1x1
ds	*1x1 CombinedDatastore*	1x1
I	*240x320x3 uint8*	240x320x3
idx	*[4;2;5;1;3;6]*	6x1
imds	*1x1 ImageDatastore*	1x1
inputSize	*[224,224,3]*	1x3
labels	*2x1 categorical*	2x1
meanIoU	0.8500	1x1
numAnchors	6	1x1
options	*1x1 TrainingOptionsSGDM*	1x1
scores	*[0.5543;0.1502]*	2x1
trainedDetector	*1x1 yolov4ObjectDetector*	1x1
trainingData	*295x2 table*	295x2
trainingDataForEstimat...	*1x1 TransformedDatastore*	1x1

Figure 8.40 Workspace variables created by running the code of figures 8.35 and 8.38.

Object Detection Using R-CNN Algorithms

R-CNN is a deep learning approach that combines two stages to perform object detection. In the first stage, the algorithm proposes a set of rectangular regions within an image that are likely to contain objects. These regions are potential candidates for further analysis. In the second stage, each proposed region is individually processed and classified to determine the specific object present in that region.

The applications of R-CNN object detectors are diverse and include autonomous driving, smart surveillance systems, and facial recognition. These systems leverage R-CNN to detect and classify objects in real-time scenarios.

Furthermore, the Computer Vision Toolbox™ provides object detectors for various algorithms, including R-CNN, Fast R-CNN, and Faster R-CNN. These detectors facilitate the implementation of object detection tasks using these specific approaches. Thus, models for object detection using regions with CNNs are based on the following three processes:

- Find regions in the image that might contain an object. These regions are called *region proposals*.
- Extract CNN features from the region proposals.
- Classify the objects using the extracted features.

There are three variants of an R-CNN. Each variant attempts to optimize, speed up, or enhance the results of one or more of these processes.

R-CNN

The R-CNN detector first generates region proposals using an algorithm such as Edge Boxes. The proposal regions are cropped out of the image and resized. These region proposals are candidates that might have objects within them. The number of such regions is typically several thousands.

Then, the CNN classifies the cropped and resized regions. Finally, the region proposal bounding boxes are refined by a support vector machine (SVM) that is trained using CNN features, where it uses a pre-trained SVM algorithm to classify the region proposal to either the background or one of the object classes.

Use the **trainRCNNObjectDetector** function to train an R-CNN object detector. The function returns an **rcnnObjectDetector** object that detects objects in an image. Figure 8.41 shows a schematic for the three major components of an R-CNN. Notice that an R-CNN typically has two output layers: Classification for objects detection and regression for defining the bounding box surrounding the detected object.

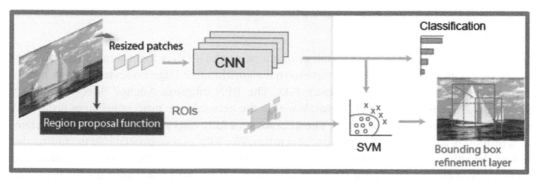

Figure 8.41   R-CNN three major components: Creation of a region proposal, feature extraction, and classification

Fast R-CNN

The Fast R-CNN detector, similar to the R-CNN detector, employs an algorithm such as Edge Boxes to generate region proposals. However, unlike the R-CNN detector, which crops and resizes these proposals, the Fast R-CNN detector processes the entire image. While an R-CNN detector must classify every region, Fast R-CNN performs pooling of CNN features corresponding to each region proposal. The efficiency of Fast R-CNN surpasses that of R-CNN due to the shared computations for overlapping regions. To achieve this, a novel layer called ROI (region of interest) pool is introduced to extract feature vectors of equal length from all proposals (ROIs) within the same image. Fast R-CNN shares computations, particularly convolutional layer calculations, across all proposals (ROIs) instead of conducting separate calculations for each proposal. This is accomplished by utilizing the ROI Pooling layer, thereby enhancing the speed of Fast R-CNN in comparison to R-CNN. The feature map from the last convolutional layer is fed to an ROI Pooling layer, simply to extract a fixed-length feature vector out of each region proposal.

Use the **trainFastRCNNObjectDetector** function to train a Fast R-CNN object detector. The function returns a **fastRCNNObjectDetector** that detects objects from an image. Figure 8.42 shows a schematic for the major components of a Fast R-CNN. Again, notice that a Fast R-CNN typically has two output layers: Classification for objects detection and regression for defining the bounding box surrounding the detected object.

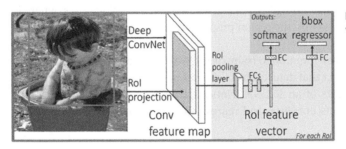

Figure 8.42 Fast R-CNN components with ROI pool inclusion.

Faster R-CNN

In the Faster R-CNN detector, a region proposal network (RPN) is incorporated into the network itself, eliminating the need for an external algorithm like Edge Boxes to generate region proposals. This is illustrated in Figure 8.43. The RPN employs Anchor Boxes for Object Detection. By generating region proposals within the network, the process becomes faster and can be finely tuned to the input data. The RPN acts as a fully convolutional network and generates proposals with various scales and aspect ratios. The RPN plays a crucial role in guiding the object detection (Fast R-CNN) by providing information about where to focus attention. Essentially, it informs the Fast R-CNN component which regions of the image are likely to contain objects of interest.

Instead of employing various images with different sizes or multiple filters with different dimensions, the Faster R-CNN detector utilizes anchor boxes. An anchor box is a predefined reference box with a specific scale and aspect ratio. By using multiple reference anchor boxes, a range of scales and aspect ratios can be represented within a single region. This collection of reference anchor boxes forms a pool.

Each region in the image is then associated with each reference anchor box, enabling the detection of objects at different scales and aspect ratios. This approach allows the sharing of convolutional computations between the region proposal network (RPN) and the Fast R-CNN component. Consequently, this sharing of computations leads to a reduction in computational time, making the detection process faster and more efficient.

Use the **trainFasterRCNNObjectDetector** function to train a Faster R-CNN object detector. The function returns a **fasterRCNNObjectDetector** that detects objects potentially found in a test image.

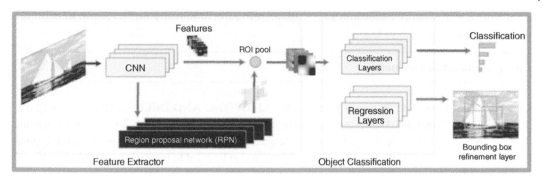

Figure 8.43 Faster R-CNN components with RPN inclusion.

Transfer Learning (Re-Training)

We can use a pretrained convolution neural network (CNN) as the basis for an R-CNN detector, also referred to as transfer learning. See Pretrained Deep Neural Networks (Deep Learning Toolbox). Any of the following networks can be used with **trainRCNNObjectDetector**, **trainFastRCNNObjectDetector**, or **trainFasterRCNNObjectDetector** function:

- 'alexnet'
- 'vgg16'
- 'vgg19'
- 'resnet50'
- 'resnet101'
- 'inceptionv3'
- 'googlenet'
- 'inceptionresnetv2'
- 'squeezenet'

Notice that all of the above CNNs require installation of MATLAB Deep Learning Toolbox.

On the other hand, we can also design a custom model based on a pretrained image classification CNN. See the upcoming three sections about R-CNN, Fast R-CNN, and Faster R-CNN creation and training.

R-CNN Creation and Training

Let us create the series of R-CNNs, starting with R-CNN, passing through Fast R-CNN, and finally reaching at Faster R-CNN. We will take **resnet50** out of the above listed CNNs as a demonstration on how to re-use the base CNN, modify its layer architecture, retrain it, and finally test it.

1. % Load pretrained ResNet-50.

```
net = resnet50();
```

2. % Convert the network layers into a layer graph (or graph of network layers) object to manipulate the layers.

```
lgraph = layerGraph(net);
```

3. From APPS menu | APPS group | MACHINE LEARNING AND DEEP LEARNING category, click on "**Deep Network Designer**" icon. The main window will pop-up. From top left menu, click on "**New**" button. Select import from workspace. Choose "**lgraph**" layer, as shown in Figure 8.44. Notice that "**Zoom In**" and "**Zoom Out**" button can be used to magnify or shrink the overall size, respectively.

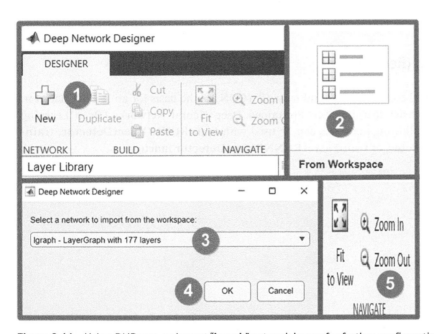

Figure 8.44   Using DND app, to import "**lgraph**" network layers for further configuration.

4. As shown in Figure 8.45, replace the last three layers, on the left side, by the new layers of the same type, on the right side. The properties of the newly added layer are also shown on the right-side of Figure 8.45. The replacement is a must to allow editing the properties of those layers.

NOTE #8

The number of Fully Connected (**FC**) output classes, modified from 3 to 2 in this case, is subject to change later, depending upon training and/or testing a new set of images or a single image. MATLAB will warn the user if there is a mismatch between the entered number of classes and the number of classes available in an examined image data store. Number 2 here accounts for the number of object classes, including the background as a class. For example, Figure 8.47 shows **stopSigns** table with two columns: The first is for the image file name and the second is for the original ground-truth bounding box specifications. Thus, we have two classes: One class, that is, vehicle and another class reserved for the background, other than the object itself.

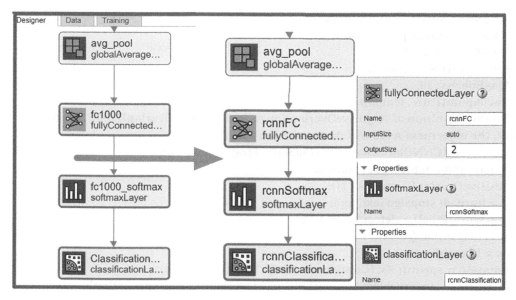

Figure 8.45 Replacing the last three layers by the same type to facilitate layers editing. Their properties are shown on the right-side, as well.

At this stage, the modified **resnet50** CNN network layers are ready to be used in training an R-CNN object detector.

5. From the top toolbar, click on "**Analyze**" button and select the first choice and the "**Deep Learning Network Analyzer**" window pops-up, where it tells the user if there are bugs. Be sure to have no errors.

6. Export the created R-CNN layer graph to workspace environment. Click on "**Export**" button from the top toolbar and select the first choice, that is, "**Export network to workspace.**" A pop-up message saying that **lgraph_1** (1×1 LayerGraph) variable was exported to the workspace, which can be utilized shortly. It is better to save this workspace variable into a *.mat file so that it can be reused by loading it to into the workspace and then importing it into DND app. For example, I saved it as **RCNNLayer.mat** which includes **lgraph_1** layer graph.

The network layers are ready to be trained using **trainRCNNObjectDetector**. Figure 8.46 shows the code for loading a data set of images, shuffling the individual images, creating training options, creating an R-CNN object detector, and finally testing the re-trained R-CNN object detector versus an image containing a traffic sign.

```
data = load('rcnnStopSigns.mat', 'stopSigns', 'layers');
stopSigns = data.stopSigns;
stopSigns.imageFilename = fullfile(toolboxdir('vision'),'visiondata', ...
   stopSigns.imageFilename);
rng(0);
shuffledIdx = randperm(height(stopSigns));
```

Figure 8.46 Using the customized **resnet50** CNN to suit R-CNN object detector for both retraining and testing. **lgraph_1** layer graph is used for R-CNN training object detector.

```
stopSigns = stopSigns(shuffledIdx,:);
options = trainingOptions('sgdm', ...
  'MiniBatchSize', 10, ...
  'InitialLearnRate', 1e-3, ...
  'MaxEpochs', 10, ...
  'CheckpointPath', tempdir);
% For the definition of 'NegativeOverlapRange' and 'PositiveOverlapRange', see NOTE
#6 on the objectness score which uses the IoU range [0 1].
rcnn = trainRCNNObjectDetector(stopSigns, lgraph_1, options, ...
  'NegativeOverlapRange', [0 0.1], ...
  'PositiveOverlapRange', [0.7 1]);
img = imread('stopSignTest.jpg');
[bbox, score, label] = detect(rcnn, img);
[score, idx] = max(score);
bbox = bbox(idx, :);
annotation = sprintf('%s: (Confidence = %f)', label(idx), score);
detectedImg = insertObjectAnnotation(img, 'rectangle', bbox, annotation);
figure
imshow(detectedImg)
```

Figure 8.46 (Cont'd)

Figure 8.47 shows the training results in terms of model accuracy and loss (*left*) and the workspace variables (*right*).

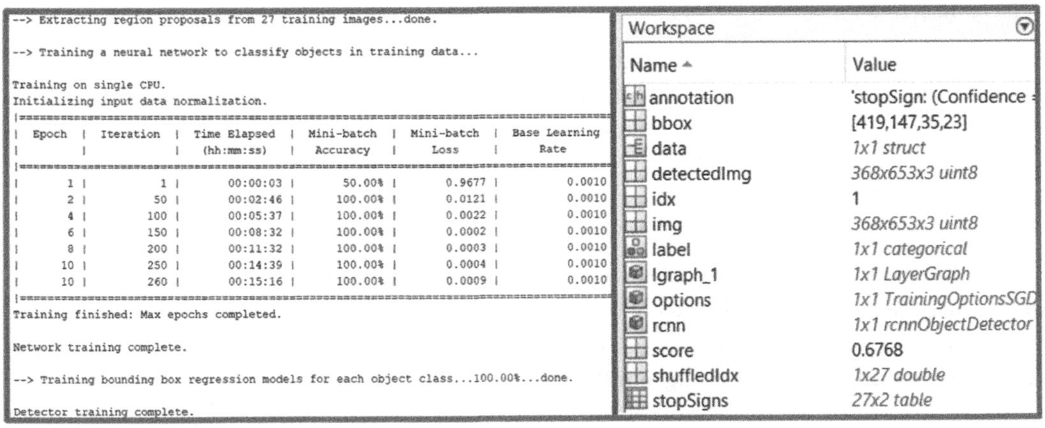

```
--> Extracting region proposals from 27 training images...done.

--> Training a neural network to classify objects in training data...

Training on single CPU.
Initializing input data normalization.
|==========================================================================================|
|  Epoch  |  Iteration  |  Time Elapsed  |  Mini-batch  |  Mini-batch  |  Base Learning  |
|         |             |   (hh:mm:ss)   |   Accuracy   |     Loss     |      Rate       |
|==========================================================================================|
|       1 |           1 |       00:00:03 |       50.00% |       0.9677 |          0.0010 |
|       2 |          50 |       00:02:46 |      100.00% |       0.0121 |          0.0010 |
|       4 |         100 |       00:05:37 |      100.00% |       0.0022 |          0.0010 |
|       6 |         150 |       00:08:32 |      100.00% |       0.0002 |          0.0010 |
|       8 |         200 |       00:11:32 |      100.00% |       0.0003 |          0.0010 |
|      10 |         250 |       00:14:39 |      100.00% |       0.0004 |          0.0010 |
|      10 |         260 |       00:15:16 |      100.00% |       0.0009 |          0.0010 |
|==========================================================================================|
Training finished: Max epochs completed.

Network training complete.

--> Training bounding box regression models for each object class...100.00%...done.

Detector training complete.
```

Workspace

Name	Value
annotation	'stopSign: (Confidence
bbox	[419,147,35,23]
data	1x1 struct
detectedImg	368x653x3 uint8
idx	1
img	368x653x3 uint8
label	1x1 categorical
lgraph_1	1x1 LayerGraph
options	1x1 TrainingOptionsSGD
rcnn	1x1 rcnnObjectDetector
score	0.6768
shuffledIdx	1x27 double
stopSigns	27x2 table

Figure 8.47 The training results of R-CNN object detector (*left*) and workspace variables (*right*).

Figure 8.48 shows that **restnet50**-customized trained R-CNN object detector managed to find the traffic sign and embrace it with a box.

Figure 8.48 The box-embraced traffic sign is detected by **restnet50**-customized R-CNN object detector.

Fast R-CNN Creation and Training

Let us continue with **restnet50**-customized layer graph. We can continue with the previous layer graph of network layers via loading the saved *.mat file and importing into workspace platform, or start from scratch.

1) Add a box regression layer to learn a set of box offsets to apply to the region proposal boxes. The learned offsets transform the region proposal boxes so that they are closer to the original ground truth bounding box. This transformation helps improve the localization performance of **Fast R-CNN**. The box regression layers are composed of a fully connected layer followed by an R-CNN box regression layer. The fully connected layer is configured to output a set of 4 box offsets for each labeled class. Thus, for **rcnnBoxFC** layer, we have the **OutputSize** = 4×1 class = 4. The background class is excluded because the background bounding boxes are not refined. We have one ground-truth labeled class, that is, the traffic sign. Hence, for **rcnnFC** fully connected layer, the **OutputSize** = number of classes + 1 for background = 2. Notice that such two numbers are adjustable parameters, which means we can change them upon the ground truth training data being handled. For instance, in the below running examples, we have only one class that is either a stop-sign or a vehicle, but not both. Figure 8.49 shows the addition of a regression parallel to the classification route.

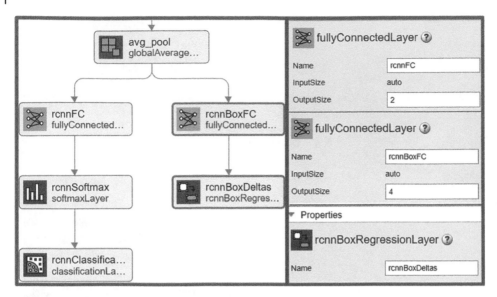

Figure 8.49 Adding, in parallel to the classification route, the regression route which accounts for bounding boxes.

2) The next step is to choose which layer in the network to use as the feature extraction layer. This layer will be connected to the ROI max pooling layer which will pool features for classifying the pooled regions. Selecting a feature extraction layer requires empirical evaluation. For **ResNet-50**, a typical feature extraction layer is the output of the 4-th block of convolutions, which corresponds to the layer named **activation40_relu**. In order to insert the ROI max pooling layer, first we need to disconnect the layers attached to the feature extraction layer: **res5a_branch2a** and **res5a_branch1**, as shown in Figure 8.50.

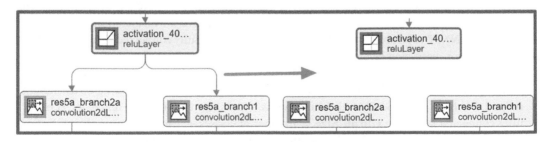

Figure 8.50 Dislodging the below connected two layers from **activation40_relu** layer.

3) We add an ROI max pooling layer, **roiMaxPooling2dLayer**([14,14],'Name','**roiPool**') and ROI input layer, **roiInputLayer**('Name','**roiInput**'). Connect the two newly added ROI-type layers as shown in Figure 8.51. The properties for the newly added layers are shown on the right-side of Figure 8.51. Of course, the user can search for a layer type by navigating through the Layer Library, the left pane of DND main window.

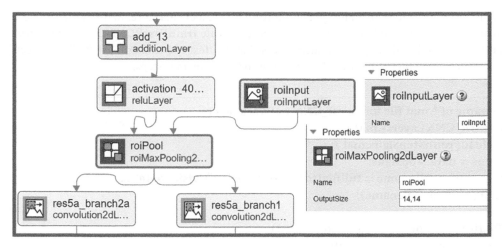

Figure 8.51 Adding and connecting the two ROI-type layers to the rest of layers top and bottom.

4) From the top toolbar, click on "**Analyze**" button and select the first choice. Figure 8.52 shows the "**Deep Learning Network Analyzer**" window where it tells that there are no bugs. The warning says we cannot use it for a typical **trainNetwork** function; however, it can be used for R-CNN-based train functions as indicated in the body of the message itself.

Figure 8.52 The deep learning network analysis showing no bugs with the created layer graph.

5) Export the created R-CNN network layers to workspace environment. Click on "**Export**" button from the top toolbar and select the first choice, that is, *Export network to workspace*. A pop-up message saying that **lgraph_1** (1×1 LayerGraph) variable was exported to the workspace, which can be utilized shortly. It is better to save this workspace variable into a *.mat file so that it can be reused by loading it into the workspace and then importing it into DND app. For example, I saved it to a file named **FastR_CNNLayers1.mat**. The network is ready to be trained using **trainFastRCNNObjectDetector**.

6) Load the data in the form of an image data store and box label data store. Reuse of the created network layers. Define options for training purposes. Create **trainFastRCNNObjectDetector** to be trained while utilizing the created layers. Finally, test the trained Fast R-CNN object detector on a given image. Figure 8.53 shows the code for previous steps.

```
% The name of *.mat file for FastR_CNN network layers.
load('FastR_CNNLayers1.mat');
data = load('rcnnStopSigns.mat');
stopSigns = data.stopSigns;
stopSigns.imageFilename = fullfile(toolboxdir('vision'),'visiondata', ...
  stopSigns.imageFilename);
rng(0);
shuffledIdx = randperm(height(stopSigns));
stopSigns = stopSigns(shuffledIdx,:);
imds = imageDatastore(stopSigns.imageFilename);
blds = boxLabelDatastore(stopSigns(:,2));
ds = combine(imds, blds);
ds = transform(ds,@(data)preprocessData(data,[112 112 3]));
options = trainingOptions('adam', ...
  'MiniBatchSize', 10, ...
  'InitialLearnRate', 1.5e-3, ...
  'MaxEpochs', 10, ...
  'LearnRateSchedule','piecewise', ...
  'LearnRateDropFactor',0.1, ...
  'LearnRateDropPeriod',25, ...
  'Shuffle','every-epoch', ...
  'CheckpointPath', tempdir);
% For the definition of 'NegativeOverlapRange' and 'PositiveOverlapRange', see NOTE
#6 on the objectness score which uses the IoU range [0 1].
frcnn = trainFastRCNNObjectDetector(ds, fastrcnn,  options, ...
  'NegativeOverlapRange', [0 0.25], ...
  'PositiveOverlapRange', [0.6 1]);
img = imread('stopSignTest.jpg');
[bbox, score, label] = detect(frcnn, img);
detectedImg = insertObjectAnnotation(img,'rectangle',bbox,score);
figure
imshow(detectedImg)
```

Figure 8.53 Creation of an image and box label data store. Reuse of the created network layers with training options to **trainFastRCNNObjectDetector** and test it later.

Figure 8.54 (*left*) shows the mini-batch loss, accuracy, and RMSE with number of epochs or iterations and workspace variables (*right*).

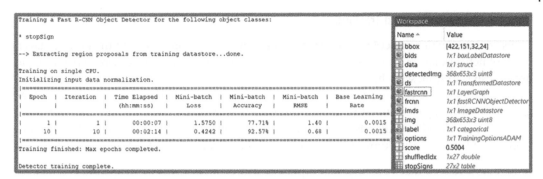

Figure 8.54 The mini-batch loss, accuracy, and RMSE with number of epochs or iterations (*left*) and workspace variables (*right*).

Figure 8.55 shows how the stop sign within the image was detected, framed in, and emblemed by an objectness score value of 0.5.

Figure 8.55 The Fast R-CNN object detector spotted the stop sign with a box and objectness score.

NOTE #9

The efficiency of spotting the traffic sign within an image is a function of the training accuracy, the solver (i.e., '**sgdm**', '**rmsprop**', or '**adam**'), the image size (degree of resolution) of both trained and tested images, the '**NegativeOverlapRange**', and the '**PositiveOverlapRange**'. Consequently, it requires more than one attempt to find the best settings for the afore-mentioned factors.

Faster R-CNN Creation and Training

Let us continue with **restnet50**-customized layer graph. We can continue with the previous layer graph of network layers via loading a saved *.mat file and importing into workspace platform, or else we can start from scratch.

Faster R-CNN utilizes a region proposal network (RPN) to create region proposals. The RPN accomplishes this by predicting the class (either "object" or "background") and box offsets for a collection of predefined bounding box templates called "anchor boxes." The size of anchor boxes is determined based on prior knowledge of object scale and aspect ratio in the training dataset.

1) Figure 8.56 shows that we changed the input image layer to accept the new **InputSize**: 56×56×3. This will facilitate the process of testing images later as a large-size of an input image requires a large size of memory to handle testing on a single CPU or else requires GPU hardware. Usage of GPU requires Parallel Computing Toolbox™.

Figure 8.56 Resizing the input image layer to accept 56×56×3 image size.

2) Figure 8.57 shows the deletion of the old Fast R-CNN **roiInputLayer** and insertion of eight layers right beneath "**activation_40_relu**" reluLayer. The inter-layer connection is also shown, starting from the added layer #1 down to the added layer #8. Layer #6 called the region proposal layer is connected to **roiMaxPooling2dLayer**. Of course, **roiMaxPooling2dLayer** is originally found in the old scheme of Fast R-CNN (see Figure 8.51). So, "**activation_40_relu**" reluLayer has now two outputs: One passes through this relatively long-chain added branch and another directly goes into **roiMaxPooling2dLayer**. Layers prior to "**activation_40_relu**" reluLayer and layers just after **roiMaxPooling2dLayer** are the same as those of the old scheme of Fast R-CNN.

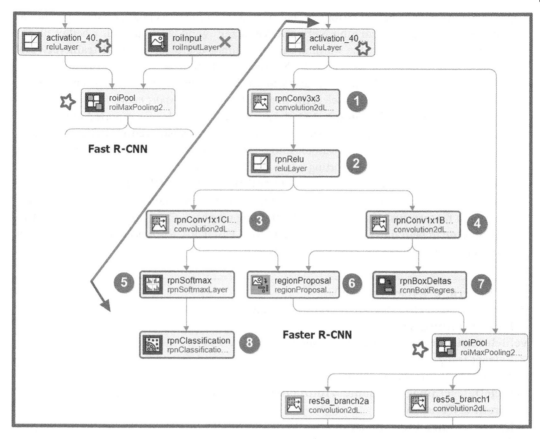

Figure 8.57 Modification of Fast R-CNN layers to Faster R-CNN layers.

Properties of each of the eight added layers, as well as, of **roiMaxPooling2dLayer** are shown in Figure 8.58.

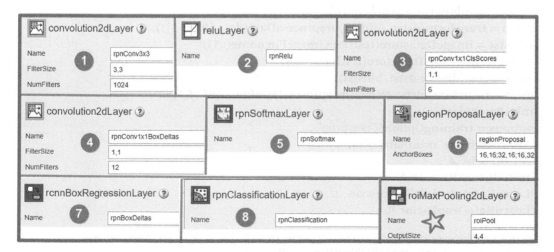

Figure 8.58 Properties of the eight added layers as well as of **roiMaxPooling2dLayer.**

3) From the top toolbar, click on "**Analyze**" button and select the first choice. As shown earlier in Figure 8.52, the "**Deep Learning Network Analyzer**" window should tell whether or not there are no errors. Again, the warning says we cannot use it for a typical **trainNetwork** function; however, it can be used for R-CNN-based train functions.

4) Export the created R-CNN network layers to workspace environment. Click on "**Export**" button from the top toolbar and select the first choice, that is, *Export network to workspace*. A pop-up message saying that **lgraph_1** (1×1 LayerGraph) variable was exported to the workspace platform, which can be utilized shortly. It was renamed to **fasterrcnn2** (1×1 LayerGraph). It is better to save this workspace variable into a *.mat file so that it can be reused by loading it into the workspace and then importing it into DND app. I saved it as '**FasterR_CNNLayers2. mat**'. The network is ready to be trained using **trainFasterRCNNObjectDetector**.

5) Load the data in the form of an image data store and box label data store. Reuse the created layer graph **fasterrcnn2**. Define options for training purposes. Create **trainFasterRCNNObjectDetector** to be trained. Finally, test the trained Faster R-CNN object detector on a given image. Figure 8.59 shows the code for previous steps.

```
load('FasterR_CNNLayers2.mat')
data = load('fasterRCNNVehicleTrainingData.mat');
% Extract training set
vehicleDataset = data.vehicleTrainingData;
% Extract image path
dataDir = fullfile(toolboxdir('vision'),'visiondata');
vehicleDataset.imageFilename = fullfile(dataDir, vehicleDataset.imageFilename);
%% train CNN network
% Divide the data into two parts
% Top 97 % for training, the last 3 % for testing
ind = round(size(vehicleDataset,1) * 0.97);
trainData = vehicleDataset(1 : ind, :);
testData = vehicleDataset(ind+1 : end, :);
imdstrn = imageDatastore(trainData.imageFilename(:,1));
bldstrn = boxLabelDatastore(trainData(:,2));
dstrn = combine(imdstrn, bldstrn);
dstrn = transform(dstrn,@(data)preprocessData(data,[95 95 3]));
imdstst = imageDatastore(testData.imageFilename(:,1));
bldstst = boxLabelDatastore(testData(:,2));
dstst = combine(imdstst, bldstst);
dstst = transform(dstst,@(data)preprocessData(data,[95 95 3]));
miniBatchSize  = 1;
options = trainingOptions('rmsprop', ...
'MiniBatchSize',miniBatchSize, ...
'MaxEpochs',4, ...
'InitialLearnRate',1e-4, ...
'LearnRateSchedule','piecewise', ...
'LearnRateDropFactor',0.1, ...
```

Figure 8.59 Creation of an image and box label data store. Re-use of the created network layers with training options to **trainFasterRCNNObjectDetector** and test it later.

```
'LearnRateDropPeriod',20, ...
'Shuffle','every-epoch', ...
'Plots','training-progress', ...
'Verbose',false);
% For the definition of 'NegativeOverlapRange' and 'PositiveOverlapRange', see NOTE
#6 on the objectness score which uses the IoU range [0 1].
detector = trainFasterRCNNObjectDetector(dstrn, fasterrcnn2, options, ...
'NegativeOverlapRange', [0 0.2], ...
'PositiveOverlapRange', [0.5 1]);
m=size(testData,1);
for i = 1:m
I = imread(testData.imageFilename{i});
[bboxes, scores, labels] = detect(detector, I);
annotation = sprintf('%s: (Confidence = %f)', labels, scores);
detectedImg = insertObjectAnnotation(I, 'rectangle', bboxes, annotation);
%subplot(m,3,i)
figure (i)
imshow(detectedImg)
end
```

Figure 8.59 (Cont'd)

Figure 8.60 shows the training performance in terms of Faster R-CNN model accuracy, RMSE, and loss as a function of number of epochs or iterations.

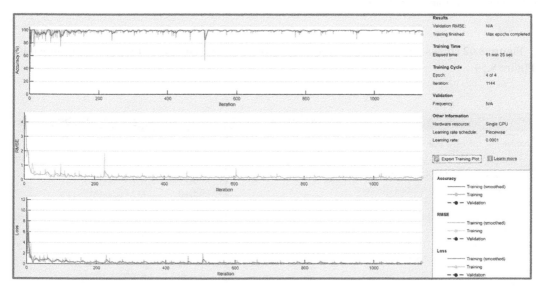

Figure 8.60 The training performance in terms of **restnet50**-customized **Faster R-CNN** model accuracy, RMSE, and loss as a function of number of epochs or iterations.

Figure 8.61 shows the testing results for nine images where the trained **restnet50**-customized **trainFasterRCNNObjectDetector** managed to spot the location of vehicle objects within the tested images. Obviously, the accuracy is not 100% in recognizing all the vehicle objects found in a test image.

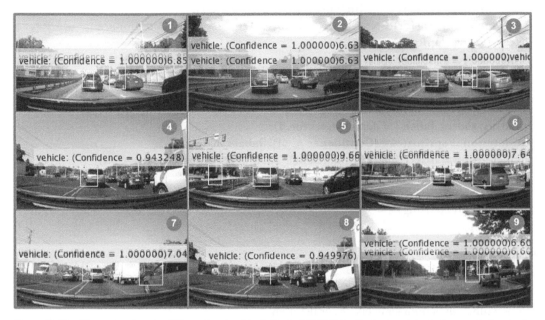

Figure 8.61 The trained **restnet50**-customized **trainFasterRCNNObjectDetector** detected some, but not all, vehicle objects within a tested image.

Figure 8.62 shows the workspace variables generated as a result of executing the code shown in Figure 8.59.

Workspace		Workspace	
Name ▲	Value	Name ▲	Value
annotation	'vehicle: (Confidence = 1.000000)6.608174e-01: (Co	I	128x228x3 uint8
bboxes	[141,67,19,30;156,56,17,31]	imdstrn	1x1 ImageDatastore
bldstrn	1x1 boxLabelDatastore	imdstst	1x1 ImageDatastore
bldstst	1x1 boxLabelDatastore	ind	286
data	1x1 struct	labels	2x1 categorical
dataDir	'C:\Program Files\MATLAB\R2022a\toolbox\vision\	m	9
detectedImg	128x228x3 uint8	miniBatchSi...	1
detector	1x1 fasterRCNNObjectDetector	options	1x1 TrainingOptionsRMSProp
dstrn	1x1 TransformedDatastore	scores	[0.6608;0.6281]
dstst	1x1 TransformedDatastore	testData	9x2 table
fasterrcnn2	1x1 LayerGraph	trainData	286x2 table
i	9	vehicleData...	295x2 table

Figure 8.62 The workspace variables generated by executing the code of Figure 8.59.

NOTE #10

The user may attempt to train one of three R-CNN-based object detectors using either or both of the two ways:

1) a set of layers created either by running his/her own code or using DND platform
2) an image and box label data store other than those of the running examples

Then MATLAB may give the following error message:
  Error using trainFasterRCNNObjectDetector. Invalid network. Caused by:
  Layer '**rcnnBoxDeltas**': The input size must be 1×1×4. This R-CNN box regression layer expects the third input dimension to be 4 times the number of object classes the network should detect (1 classes). See the documentation for more details about creating Fast or Faster R-CNN networks.
  Layer '**rcnnClassification**': The input size must be 1×1×2. The classification layer expects the third input dimension to be the number of object classes the network should detect (1 classes) plus 1. The additional class is required for the "background" class. See the documentation for more details about creating Fast or Faster R-CNN networks.
  The above message means that the user must change the output size of the two fully connected (**FC**) layers; namely, **rcnnFC** and **rcnnBoxFC**, as shown in Figure 8.49 and in light of NOTE #8.

---

evaluateDetectionPrecision Function for Precision Metric

In previous classification models, we calculated accuracy based only on misclassification. Now we should consider both the bounding box and the label. The function **evaluateDetectionPrecision** calculates precision metrics using an overlap threshold between the predicted and true bounding boxes. Precision is a ratio of true positive instances to all positive instances of objects in the detector. For a multiclass detector, the function returns the average precision (ap) as a vector of scores for each object class in the order specified by the ground truth data. Figure 8.63 shows the code on how to use **evaluateDetectionPrecison** function.

>>[ap,recall,precision] = evaluateDetectionPrecision(detectionResults, groundTruthData)

   returns the average precision, ap, of the **detectionResults** compared to the **groundTruthData**. We can use the average precision to measure the performance of an object detector. A plot of data points for plotting the precision–recall curve can be drawn. Notice that **detector** in this example is a faster R-CNN object detector. It can also be a YOLO object detector, as well.

```
load('FasterR_CNNLayers3_4.mat', 'fasterrcnn3')
data = load('fasterRCNNVehicleTrainingData.mat');
% Extract training set
vehicleDataset = data.vehicleTrainingData;
% Extract image path
```

Figure 8.63 **evaluateDetectionPrecision** function to calculate precision metric of a trained object detector.

```
dataDir = fullfile(toolboxdir('vision'),'visiondata');
vehicleDataset.imageFilename = fullfile(dataDir, vehicleDataset.imageFilename);
% train CNN network
% Divide the data into three parts
% Top 60% for training, 15 % for validation, and the last 25 % for testing.
r=size(vehicleDataset,1);
P = 0.60 ;
idx = randperm(r)  ;
trainData = vehicleDataset(idx(1:round(P*r)), :);
ValData = vehicleDataset(idx(round(P*r)+1:round((P+0.15)*r)), :);
testData=vehicleDataset(idx(round((P+0.15)*r)+1:end), :);
imdstrn = imageDatastore(trainData.imageFilename(:,1));
bldstrn = boxLabelDatastore(trainData(:,2));
dstrn = combine(imdstrn, bldstrn);
imdsval = imageDatastore(ValData.imageFilename(:,1));
bldsval = boxLabelDatastore(ValData(:,2));
dsval = combine(imdsval, bldsval);
imdstst = imageDatastore(testData.imageFilename(:,1));
bldstst = boxLabelDatastore(testData(:,2));
dstst = combine(imdstst, bldstst);
miniBatchSize  = 1;
valData=dsval;
options = trainingOptions('adam', ...
'MiniBatchSize',miniBatchSize, ...
'MaxEpochs',3, ...
'InitialLearnRate',1e-4, ...
'LearnRateSchedule','piecewise', ...
'LearnRateDropFactor',0.1, ...
'LearnRateDropPeriod',10, ...
'Shuffle','every-epoch', ...
'ValidationData',valData,...
'ValidationFrequency',20,...
'Plots','training-progress', ...
'Verbose',false);
detector = trainFasterRCNNObjectDetector(dstrn, fasterrcnn3, options, ...
'NegativeOverlapRange', [0 0.3], ...
'PositiveOverlapRange', [0.5 1]);
results = struct;
    for i = 1: size(trainData,1);
        I = imread(trainData.imageFilename{i});
        [bboxes, scores, labels] = detect(detector, I);
        results(i).Boxes = bboxes;
        results(i).Scores = scores;
        results(i).Labels = labels;
    end
results = struct2table(results);
GTResults = trainData(:,2);
```

Figure 8.63 (Cont'd)

```
[ap, recall, precision] = evaluateDetectionPrecision(results, GTResults);
figure;
plot(recall, precision);
xlabel('Recall');
ylabel('Precision')
grid on;
title(sprintf('Average Precision for trainData = %.2f', ap))
results2 = struct;
  for i2 = 1: size(testData,1);
    I2 = imread(testData.imageFilename{i2});
    [bboxes2, scores2, labels2] = detect(detector, I2);
    results2(i2).Boxes = bboxes2;
    results2(i2).Scores = scores2;
    results2(i2).Labels = labels2;
  end
results2 = struct2table(results2);
GTResults2 = testData(:,2);
%The average precision is used to evaluate the detection effect
[ap2, recall2, precision2] = evaluateDetectionPrecision(results2, GTResults2);
figure;
plot(recall2, precision2)
xlabel('Recall');
ylabel('Precision')
grid on;
title(sprintf('Average Precision for testData = %.2f', ap2));
```

Figure 8.63 (Cont'd)

Figure 8.64 shows the training and validation results in terms of accuracy %, RMSE, and loss for the selected faster R-CNN.

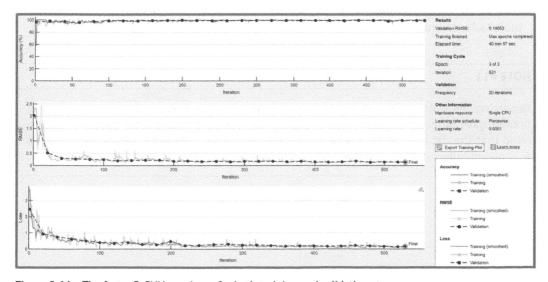

Figure 8.64   The faster R-CNN goodness for both training and validation step.

Figure 8.65 shows the plot of precision (y) versus recall (x) for both training (*top*) and test (*bottom*) data set, based on evaluation of the results against the ground truth data, **blds**, and the generated workspace variables (*right*).

Here are the advantages of using a precision-recall plot for both training and test data:

1) Performance Assessment: The plot allows us to assess the overall performance of our detection algorithm by considering both precision and recall simultaneously. Precision measures the accuracy of positive predictions, while recall quantifies the ability of the algorithm to correctly identify positive instances. By examining the plot, we can visually analyze the algorithm's behavior across different thresholds and understand how it performs in terms of precision and recall.

2) Threshold Selection: The plot helps us select an appropriate threshold for our detection algorithm. As we vary the classification threshold, the precision and recall values change accordingly. The plot allows us to identify the threshold that strikes the desired balance between precision and recall, based on our specific application requirements. For example, if we prioritize minimizing false positives, we can choose a threshold that maximizes precision while still maintaining an acceptable level of recall.

3) Comparing Models: The plot enables us to compare the performance of different detection models or algorithms. By plotting multiple curves on the same precision-recall plot, we can easily compare their performance. This comparison can help us identify which model performs better in terms of precision and recall, and select the most suitable algorithm for our application.

4) Analysis of Imbalanced Data: The precision-recall plot is particularly useful when dealing with imbalanced datasets, where the number of negative instances far exceeds the positive instances. In such cases, accuracy alone might not provide a complete picture of the algorithm's performance. The precision-recall plot allows us to evaluate the algorithm's ability to correctly identify the positive instances (recall) while considering the impact of false positives (precision).

5) Performance Monitoring: By generating separate precision-recall plots for training and test data, we can monitor the algorithm's performance during training and evaluate its generalization ability. If the training data precision-recall curve significantly outperforms the test data curve, it indicates potential overfitting, suggesting the need for regularization techniques or model adjustments.

NOTE #11

Repeating the training procedure with the same data sets and also with the same training options gave different average precision results for both **trainData** and **testData** set. Keep in mind that the data distribution is randomly selected for each partition. Definitely, changing the trained object detector itself will affect the average precision, as well.

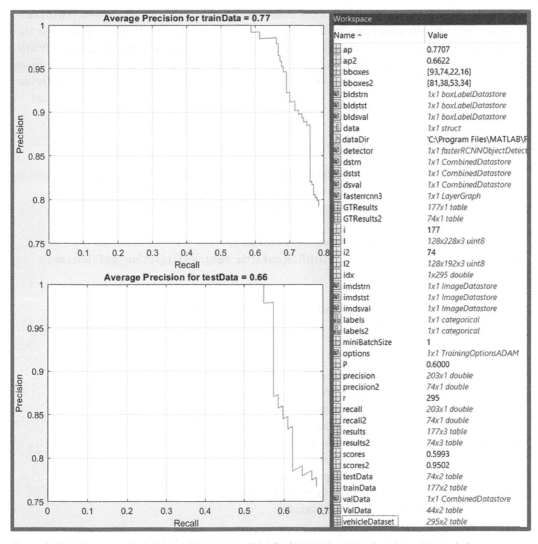

Figure 8.65 The plot of precision (*y*) versus recall (*x*) for both training (*top*) and test (*bottom*) data set, based on evaluation of the results against the ground truth data, **blds**. Created workspace variables are shown (*right*).

evaluateDetectionMissRate for Miss Rate Metric

We should also consider the case when the detector fails to find an object. This is called the miss rate. We can calculate miss rate metrics using **evaluateDetectionMissRate** function.

>>[logAverageMissRate,fppi,missRate]= evaluateDetectionMissRate(detectionResults, groundTruthData)

It returns the log-average miss rate of the **detectionResults** compared to **groundTruthData**, which is used to measure the performance of the object detector. For a multiclass detector, the log-average miss rate is a vector of scores for each object class in the order specified by **groundTruth-Data**. It also returns data points for plotting the log miss rate versus false positives per image (FPPI) curve, using input arguments. Figure 8.66 shows the code for exploiting **evaluateDetec-tionMissRate** function. Again, any faster R-CNN or YOLO object detector can be used in replacement of **detector** variable.

```
load('FasterR_CNNLayers2.mat')
data = load('fasterRCNNVehicleTrainingData.mat');
% Extract training set
vehicleDataset = data.vehicleTrainingData;
% Extract image path
dataDir = fullfile(toolboxdir('vision'),'visiondata');
vehicleDataset.imageFilename = fullfile(dataDir, vehicleDataset.imageFilename);
% train CNN network
% Divide the data into three parts
% Top 60% for training, 15 % for validation, and the last 25 % for testing
r=size(vehicleDataset,1);
P = 0.60 ;
idx = randperm(r) ;
trainData = vehicleDataset(idx(1:round(P*r)), :);
ValData = vehicleDataset(idx(round(P*r)+1:round((P+0.15)*r)), :);
testData=vehicleDataset(idx(round((P+0.15)*r)+1:end), :);
imdstrn = imageDatastore(trainData.imageFilename(:,1));
bldstrn = boxLabelDatastore(trainData(:,2));
dstrn = combine(imdstrn, bldstrn);
dstrn = transform(dstrn,@(data)preprocessData(data,[300 300 3]));
imdsval = imageDatastore(ValData.imageFilename(:,1));
bldsval = boxLabelDatastore(ValData(:,2));
dsval = combine(imdsval, bldsval);
dsval = transform(dsval,@(data)preprocessData(data,[300 300 3]));
imdstst = imageDatastore(testData.imageFilename(:,1));
bldstst = boxLabelDatastore(testData(:,2));
dstst = combine(imdstst, bldstst);
dstst = transform(dstst,@(data)preprocessData(data,[300 300 3]));
miniBatchSize  = 2;
valData=dsval;
options = trainingOptions('adam', ...
'MiniBatchSize',miniBatchSize, ...
'MaxEpochs',4, ...
'InitialLearnRate',1e-4, ...
'LearnRateSchedule','piecewise', ...
'LearnRateDropFactor',0.1, ...
```

Figure 8.66 Exploiting **evaluateDetectionMissRate** function for evaluating the miss rate of a trained object detector.

```
'LearnRateDropPeriod',10, ...
'Shuffle','every-epoch', ...
'ValidationData',valData,...
'ValidationFrequency',20,...
'Plots','training-progress', ...
'Verbose',false);
detector = trainFasterRCNNObjectDetector(dstrn, fasterrcnn2, options, ...
'NegativeOverlapRange', [0 0.2], ...
'PositiveOverlapRange', [0.5 1]);
results = struct;
  for i = 1: size(trainData,1);
    I = imread(trainData.imageFilename{i});
    [bboxes, scores, labels] = detect(detector, I);
    results(i).Boxes = bboxes;
    results(i).Scores = scores;
    results(i).Labels = labels;
  end
results = struct2table(results);
GTResults = trainData(:,2);
[am, fppi, missRate] = evaluateDetectionMissRate(results,GTResults);
%Plot log-miss-rate (y) vs. FPPI (x) curve, using a base-10 logarithmic scale for both
axes.
figure
loglog(fppi, missRate);
grid on
title(sprintf('log Average Miss Rate for trainData = %.2f', am))
results2 = struct;
  for i2 = 1: size(testData,1);
    I = imread(testData.imageFilename{i2});
    [bboxes, scores, labels] = detect(detector, I);
    results2(i2).Boxes = bboxes;
    results2(i2).Scores = scores;
    results2(i2).Labels = labels;
  end
results2 = struct2table(results2);
GTResults2 = testData(:,2);
[am2, fppi2, missRate2] = evaluateDetectionMissRate(results2,GTResults2);
%Plot log-miss-rate (y) vs. FPPI (x) curve, using a base-10 logarithmic scale for both
axes.
figure
loglog(fppi2, missRate2);
grid on
title(sprintf('log Average Miss Rate for testData= %.2f', am2))
```

Figure 8.66 (Cont'd)

Figure 8.67 shows the training and validation results in terms of accuracy %, RMSE, and loss for the selected faster R-CNN.

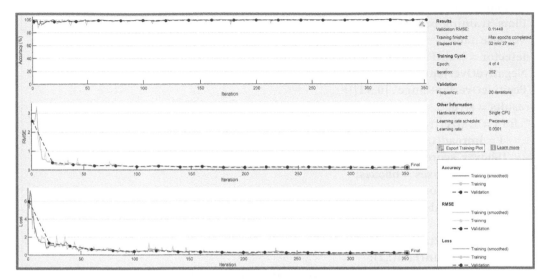

Figure 8.67   The faster R-CNN goodness for both training and validation step.

Figure 8.68 shows the log average miss rate for both training (*top*) and test (*bottom*) data store, and the created workspace variables (*right*). Notice that zero log average miss rate value means a perfect match between predicted and ground truth data.

Here are the advantages of using the log miss rate (y) versus FPPI (x) curve:

1) Comprehensive Performance Evaluation: The log miss rate-FPPI curve provides a comprehensive evaluation of a detection algorithm's performance across different thresholds or confidence levels. The log miss rate measures the logarithm of the ratio between false negatives and the total number of ground truth positives. The FPPI quantifies the average number of false positives per image. By analyzing the curve, we can assess the algorithm's performance at various operating points.

2) Visualizing Detection Trade-offs: The curve helps us understand the trade-off between the false positive rate (FPPI) and the false negative rate (miss rate) of the algorithm. It provides a visual representation of the algorithm's behavior as we adjust the detection threshold or confidence level. This allows us to identify the threshold that achieves the desired balance between false positives and false negatives based on our specific application requirements.

3) Threshold Selection: The log miss rate-FPPI curve assists in selecting an appropriate threshold for the detection algorithm. By analyzing the curve, we can identify the threshold that corresponds to a desired miss rate or FPPI level. For example, if our application demands a low false positive rate, we can choose a threshold that achieves a low FPPI while still maintaining an acceptable miss rate.

4) Performance Comparison: The curve facilitates the comparison of different detection models or algorithms. By plotting multiple curves on the same graph, we can compare their performance at different thresholds or operating points. This comparison helps us determine which algorithm

performs better in terms of miss rate and false positive rate, enabling us to select the most suitable model for our application.

5) Performance Monitoring: Generating separate log miss rate-FPPI curves for training and test datasets allows us to monitor the algorithm's performance during training and assess its generalization ability. If the training data curve significantly outperforms the test data curve, it suggests potential overfitting, indicating the need for regularization techniques or model adjustments.

6) Analysis of Imbalanced Data: The log miss rate-FPPI curve is particularly valuable when dealing with imbalanced datasets, where the number of positive instances is much smaller than the negatives. In such cases, traditional performance metrics like accuracy might not adequately represent the algorithm's performance. The curve provides insights into the trade-off between false positives and false negatives, allowing us to evaluate the algorithm's ability to detect positive instances while controlling false positives.

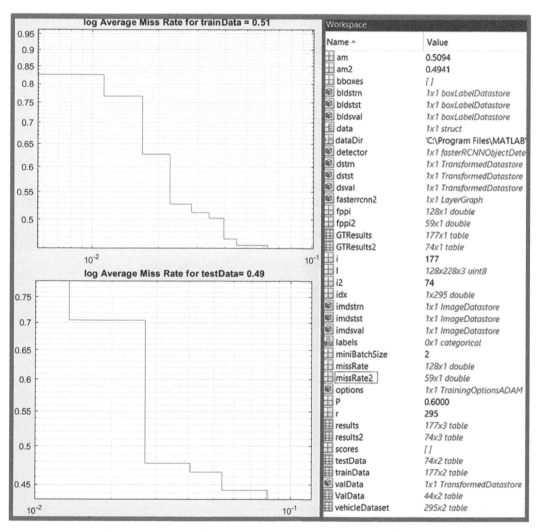

Figure 8.68 The plot of the log miss rate (y) versus FPPI (x) curve for both training (*top*) and test (*bottom*) data set, based on evaluation of the results against the ground truth data, **blds**. Created workspace variables shown *right*.

Quiz #4

An object detector has been trained to identify ducks and penguins. After performing object detection on an image of four ducks, the red-arrow marked image is:

Which of the following describes the issue with the detection of the left duck?

a) Inaccurate bounding box location   b) Inaccurate label

c) Detector failed to find an object   d) None of the above.

Quiz #5

An object detector has been trained to identify ducks and penguins. After performing object detection on an image of four ducks, the red-arrow marked image is:

Which of the following describes the issue with the detection of the left duck?

a) Inaccurate bounding box location   b) Inaccurate label

c) Detector failed to find an object   d) None of the above.

Quiz #6

An object detector has been trained to identify ducks and penguins. After performing object detection on an image of four ducks, the red-arrow marked image is:

Which of the following describe the issue with the detection of the fourth duck? (Select all that apply)

a) Inaccurate bounding box location   b) Inaccurate label

c) Detector failed to find an object   d) None of the above.

Quiz #7

An object detector has been trained to identify ducks and penguins. After performing object detection on an image of four ducks, the red-arrow marked image is:

Which of the following describes the issue with the detection of the third duck?

a) Inaccurate bounding box location   b) Inaccurate label

c) Detector failed to find an object   d) None of the above.

Quiz #8

We would like to detect the location of pedestrians in images from a camera mounted in a car. Which of the following types of deep networks would be most appropriate?

a) CNN   b) YOLO   c) LSTM

HCW 8.1   Testing yolov4ObjectDetector and fasterRCNN Object Detector

Refer to Figure 8.30.

a. Create a yolov4ObjectDetector based on

1. 'tiny-yolov4-coco'
2. 'csp-darknet53-coco'

Use the attached images for testing purpose: cat1.jpeg, cat2.jpeg, cat3.jpeg, and cat4.jpeg. Resize the given images to [416 416 3]. Upon testing, you may need to use a threshold value between 0.0 and 0.05 as in the following statement:

```
[bboxes,scores,labels] = detect(detector41,img,Threshold=0.03);
```

b. Create a fasterRCCN object detector using the following statements:

```
data = load('fasterRCNNVehicleTrainingData.mat', 'detector');
detector = data.detector;
```

Use the attached images for testing purpose: audi-798530__340.jpg, car-1093927__340.jpg, car-5548242__340.jpg, and vw-bus-1845719__340.jpg. Resize the given images to [512 512 3]. Upon testing, you may need to use a threshold value between 0.0 and 0.05, as in part a.

NOTE: Cat and car images are available with Wiley web companion.

HCW 8.2   Creation of Two CNN-based yolov4ObjectDetectors

Refer to Figure 8.35. Create two customized versions of yolov4ObjectDetector. The first one will be **restnet50**-based and another will be **GoogleNet**-based.

a. For the sake of helping the user, use the following code and insert within the code of Figure 8.35 by replacing the old code:

```
detector = yolov4ObjectDetector("tiny-yolov4-coco",classes,anchorBoxes,InputSize=
inputSize);
```

by the new code:

```
net = resnet50;
%Convert the network to a layer graph and remove the layers used for classification
using removeLayers.
lgraph = layerGraph(net);
lgraph = removeLayers(lgraph,'ClassificationLayer_fc1000');
```

```
imageSize = lgraph.Layers(1).InputSize;
layerName = lgraph.Layers(1).Name;
newinputLayer = imageInputLayer(imageSize,'Normalization','none','Name',layerName);
lgraph = replaceLayer(lgraph,layerName,newinputLayer);
%Convert the network to a dlnetwork object.
dlnet = dlnetwork(lgraph);
%Select two feature extraction layers in the base network to serve as the source for
detection subnetwork. The number of feature layers = number of anchorBoxes.
featureLayer = ["activation_22_relu","activation_40_relu"];
detector = yolov4ObjectDetector(dlnet,classes,anchorBoxes,'ModelName','RestNet50
YOLO v4','DetectionNetworkSource',featureLayer);
```

Carry out the training for this customized version of yolov4ObjectDetector using the following data set:

```
data = load("vehicleTrainingData.mat");
trainingData = data.vehicleTrainingData;
dataDir = fullfile(toolboxdir('vision'),'visiondata');
trainingData.imageFilename = fullfile(dataDir,trainingData.imageFilename);
imds = imageDatastore(trainingData.imageFilename);
blds = boxLabelDatastore(trainingData(:,2));
ds = combine(imds,blds);
```

Test the trained model using the image called ('highway.png'). Upon testing, you may need to use a threshold value between 0.0 and 0.05, as in part a of HCW 8.1.

b. Repeat step **a.** but this time for **GoogleNet**, instead of **restnet50** CNN. To get to know the name of potential layers of **GoogleNet**, where you can extract features from, use the following code:

```
analyzeNetwork(net)
```

Or:

```
analyzeNetwork(dlnet)
```

Upon testing, you may need to use a threshold value between 0.0 and 0.05, as in part a of HCW 8.1.

HCW 8.3 Creation of GoogleNet-Based Fast R-CNN Object Detector

Refer to "**R-CNN Creation and Training**" and "**Fast R-CNN Creation and Training**" section. Use DND app. Create a fast R-CNN. Use the following data to train your created fast R-CNN object detector:

```
data = load("vehicleTrainingData.mat");
trainingData = data.vehicleTrainingData;
dataDir = fullfile(toolboxdir('vision'),'visiondata');
```

```
trainingData.imageFilename = fullfile(dataDir,trainingData.imageFilename);
imds = imageDatastore(trainingData.imageFilename);
blds = boxLabelDatastore(trainingData(:,2));
ds = combine(imds,blds);
```

Test the created detector using the image called ('**highway.png**'). Upon testing, you may need to use a threshold value between 0.0 and 0.05, as in part a of HCW 8.1.

HCW 8.4   Creation of a GoogleNet-Based Faster R-CNN Object Detector

Refer to "**R-CNN Creation and Training**," "**Fast R-CNN Creation and Training**," and "**Faster R-CNN Creation and Training**," section. Create a faster R-CNN. Notice that you may use DND app to continue creating a GoogleNet-based faster R-CNN object detector, or use the following code to create a faster R-CNN layer graph which will be used in training trainFasterRCNNObjectDetector.

```
inputImageSize = [224 224 3];
%Specify the number of objects to detect.
numClasses = 1; % For our problem, it will be 'vehicle' class.
%Use a pretrained GoogleNet network as the base network for the Faster R-CNN net-
work. You must download GoogleNet support package.
net = 'googlenet';
%Specify the network layer to use for feature extraction. You can use the analyzeNet-
work function to see all the layer names in a network.
featureLayer = 'pool4-3x3_s2';
%Specify the anchor boxes. You can also use the estimateAnchorBoxes function to
estimate anchor boxes from your training data. See Figure 8.35 in "Fine-Tuning of a
Pretrained YOLO v4 Object Detector" section.
% The arbitrarily assigned anchorBoxes values will definitely affect the performance
of the training step and may not converge if off values are selected outside the image
dimensions.
anchorBoxes = [64,64; 28,28; 45,45];
%Create a Faster R-CNN object detection network layer graph.
lgraph = fasterRCNNLayers(inputImageSize,numClasses,anchorBoxes, ...
            net,featureLayer)
```

Use the same data set shown in HWC 8.3 for training purpose. Use **lgraph** in the training statement, as shown below:

```
detector = trainFasterRCNNObjectDetector(dstrn, lgraph, options, ...
'NegativeOverlapRange', [0 0.35], ...
'PositiveOverlapRange', [0.55 1]);
```

Show test images annotated by both the predicted label and its bounding box. Upon testing, you may need to use a threshold value between 0.0 and 0.05, as in part a of HCW 8.1.

HCW 8.5   Calculation of Average Precision and Miss Rate Using GoogleNet-Based Faster R-CNN Object Detector

Use the following data set for creation of Google-based faster R-CNN layer graph which will be used in training the faster R-CNN object detector. Test the trained object detector and evaluate the average precision and average miss rate of **'vehicle'** class for both the training and test data set. Upon testing, you may need to use a threshold value between 0.0 and 0.05, as in part a of HCW 8.1.

```
%unzip vehicleDatasetImages.zip
data = load("vehicleDatasetGroundTruth.mat");
vehicleDataset = data.vehicleDataset;
rng("default");
r=size(vehicleDataset,1);
P = 0.80 ;
idx = randperm(r) ;
trainData = vehicleDataset(idx(1:round(P*r)), :);
ValData = vehicleDataset(idx(round(P*r)+1:round((P+0.1)*r)), :);
testData=vehicleDataset(idx(round((P+0.1)*r)+1:end), :);
imdstrn = imageDatastore(trainData.imageFilename(:,1));
bldstrn = boxLabelDatastore(trainData(:,2));
dstrn = combine(imdstrn, bldstrn);
imdsval = imageDatastore(ValData.imageFilename(:,1));
bldsval = boxLabelDatastore(ValData(:,2));
dsval = combine(imdsval, bldsval);
imdstst = imageDatastore(testData.imageFilename(:,1));
bldstst = boxLabelDatastore(testData(:,2));
dstst = combine(imdstst, bldstst);
```

HCW 8.6   Calculation of Average Precision and Miss Rate Using GoogleNet-Based yolov4 Object Detector

Use the data set shown in HCW 8.5 problem for creation of Google-based yolov4 object detector layer graph which will be used in training the yolov4 object detector. Test the trained object detector and evaluate the average precision and average miss rate of **'vehicle'** class for both the training and test data set. Upon testing, you may need to use a threshold value between 0.0 and 0.05, as in part a of HCW 8.1.

HCW 8.7   Faster RCNN-based Car Objects Prediction and Calculation of Average Precision for Training and Test Data

Acknowledgement: Images in this database were taken from the TU-Darmstadt, Caltech, TU-Graz and UIUC databases. Additional images were provided by INRIA. Funding was provided by PASCAL.

Publications: M. Everingham, A. Zisserman, C. K. I. Williams, L. Van Gool, et al. The 2005 PASCAL Visual Object Classes Challenge. In Machine Learning Challenges. Evaluating Predictive Uncertainty, Visual Object Classification, and Recognising Textual Entailment., eds. J. Quinonero-Candela, I. Dagan, B. Magnini, and F. d'Alche-Buc, LNAI 3944, pages 117-176, Springer-Verlag, 2006.

The following data set can be downloaded from: http://host.robots.ox.ac.uk/pascal/VOC/data bases.html#VOC2005_1.

http://www.pascal-network.org/challenges/VOC/databases.html

In addition, it is available with Wiley's web companion. You can use any data set combined with its bounding box data. For this problem, use the data in the folder called: voc2005_1\PNGImages\ TUGraz_cars. On the other hand, the bounding box data are given within a text file for each image residing in the same folder name but with annotation: voc2005_1\Annotations\TUGraz_cars.

Our first **challenging** task is to extract the numerical bounding box four-data points from the text file. Here is the code to use for bounding box data extraction out of the given text file.

```
%Set the folder path and file extension
folderPath = 'D:\Downloads\voc2005_1\Annotations\TUGraz_cars';
fileExtension = '*.txt';
% Get a list of text files in the folder
fileList = dir(fullfile(folderPath, fileExtension));
% Initialize empty arrays to store the box labels and corresponding file paths
boxLabels = {};
filePaths = {};
% Iterate over each text file and extract the box labels
for i = 1:numel(fileList)
    % Read the text file
    filePath = fullfile(folderPath, fileList(i).name);
    fileData = fileread(filePath);
% Extract the bounding box information using regular expressions
    matches = regexp(fileData, 'Bounding box for object \d+ "([^"]+)" \(Xmin, Ymin\) -
\(Xmax, Ymax\) : \((\d+), (\d+)\) - \((\d+), (\d+)\)', 'tokens');
% Initialize an empty cell array to store the box labels for this file
    fileLabels = {};
% Process each match and extract the bounding box coordinates
    for j = 1:numel(matches)
        match = matches{j};
% Extract the label name and bounding box coordinates
        labelName = match{1};
        xmin = str2double(match{2});
        ymin = str2double(match{3});
        xmax = str2double(match{4});
        ymax = str2double(match{5});
% Ensure the bounding box coordinates are valid
        if ~isnan(xmin) && ~isnan(ymin) && ~isnan(xmax) && ~isnan(ymax)
            % Store the bounding box coordinates as separate variables
            boxLabel = {xmin, ymin, xmax, ymax};
            % Store the bounding box label in the fileLabels cell array
            fileLabels{end+1} = boxLabel;
        end
```

```
      end
% Store the file path and box labels for this file
   filePaths{end+1} = filePath;
   boxLabels{end+1} = fileLabels;
% Print the extracted bounding box coordinates and label names
   fprintf('File: %s\n', fileList(i).name);
   for k = 1:numel(fileLabels)
      label = fileLabels{k};
      fprintf('Bounding Box: [%d, %d, %d, %d]\n', label{:});
   end
end
% Check the dimensions of the bounding box data
sizes = cellfun(@(x) size(x, 2), boxLabels);
fprintf('Bounding Box Sizes: %s\n', mat2str(sizes));
% Create a table with box labels and file paths
data = table(filePaths', boxLabels', 'VariableNames', {'FilePath', 'BoxLabels'});
bL=boxLabels';
% Create an empty cell array to store the converted values
convertedLabels = cell(size(bL));
% Iterate over each cell in bL and extract the 1x4 double array
for i = 1:numel(bL)
   % Access the nested cell array within the current cell
   nestedCell = bL{i};
   % Check if the nested cell is empty
   if ~isempty(nestedCell)
      % Access the 1x4 cell of numerical values within the nested cell
      innerCell = nestedCell{1};
      % Convert the 1x4 cell of numerical values into a 1x4 double array
      convertedLabels{i} = cell2mat(innerCell);
   end
end
blT=cell2table(convertedLabels);
blT.Properties.VariableNames{1} = 'cars';
iT=cell2table(imds.Files);
xtemp=[iT blT];
% Example cell array with bounding box dimensions
boundingBoxes = xtemp{:,2};
% Find empty rows using cellfun and isempty
emptyRows = cellfun(@isempty, boundingBoxes);
% Remove empty rows from the cell array
nxtemp = xtemp(~emptyRows,1:2);
blT2=nxtemp(:,2);
```

Once you manage to extract the bounding box data, carry out partition of entire TUGraz_cars data into training, validation, and test set. Use a faster RCNN object detector to train it on the training together with the validation data. Try the trained and validated model on the test data set and calculate the average precision (AP) for both training and test data set.

9

Recurrent Neural Network (RNN)

A Recurrent Neural Network (RNN) is a type of artificial neural network designed to process sequential data by maintaining an internal memory or state. Unlike traditional feedforward neural networks, which process data in a single pass, RNNs can take into account the previous steps in the sequence, making them well-suited for tasks that involve sequential or time-dependent data.

The key feature of an RNN is its ability to use its internal memory to process sequences of inputs of varying lengths. It achieves this by applying the same set of weights to each element of the sequence while maintaining a hidden state that represents the context from previous inputs. This hidden state is updated at each step, allowing the network to capture the dependencies and patterns in the sequence.

The basic building block of an RNN is a recurrent neuron, which takes an input along with the previous hidden state and produces an output and a new hidden state. The output can be used for prediction or further processing, and the new hidden state becomes the input for the next step in the sequence. This recursive structure enables the RNN to pass information from one step to another, allowing it to model long-term dependencies.

RNNs have been successfully applied to various tasks involving sequential data, such as natural language processing, speech recognition, machine translation, sentiment analysis, and time series forecasting. However, traditional RNNs can struggle to capture long-term dependencies due to the "vanishing gradient" problem, where the gradients that flow backward in time can diminish exponentially. To address this, variants of RNNs, such as Long Short-Term Memory (LSTM) and Gated Recurrent Unit (GRU), have been developed, which incorporate additional mechanisms to alleviate the vanishing gradient problem and enhance the memory capabilities of the network.

Long Short-Term Memory (LSTM) and BiLSTM Network

In LSTM (Long Short-Term Memory) training using MATLAB, different types of sequences can be used, depending on the specific problem we are addressing. Here are some common types of sequences used in LSTM training:

1) Time Series Data: LSTM models are particularly effective in handling time series data, such as stock prices, weather data, or sensor readings over time. In MATLAB, we can represent time series data as sequences, where each element represents a specific point in time.

Machine and Deep Learning Using MATLAB: Algorithms and Tools for Scientists and Engineers, First Edition.
Kamal I. M. Al-Malah.
© 2024 Kamal I. M. Al-Malah. Published 2024 by John Wiley & Sons, Inc.
Companion Website: www.wiley.com/go/al-malah/machinelearningmatlab

2) Natural Language Processing (NLP) Sequences: LSTM models are widely used in NLP tasks, such as text classification, sentiment analysis, and language generation. In MATLAB, we can represent text data as sequences of words or characters, where each element in the sequence corresponds to a word or character in the text.

3) Image Sequences: LSTM models can also be applied to sequential image data, such as video or image sequences. In MATLAB, we can represent image sequences as a series of image frames, where each element in the sequence represents a single frame.

4) Sensor Data Sequences: LSTM models can be used to analyze sequential sensor data, such as data collected from IoT devices or wearable sensors. In MATLAB, we can represent sensor data as sequences, where each element represents a specific sensor reading or measurement.

An LSTM layer learns long-term dependencies between time steps in time series and sequence data. The state of the layer consists of the hidden state (also known as the output state) and the cell state. The hidden state at time step t contains the output of the LSTM layer for this time step. The cell state contains information learned from the previous time steps. At each time step, the layer adds information to or removes information from the cell state. The layer controls these updates using gates.

An LSTM can be depicted as an RNN but with a memory pool that will do two tasks:

1) Short-term state: Keeps the output at the current time step.

2) Long-term state: Stores, reads, and rejects items dedicated for the long-term while passing through the network.

The decision of reading, storing, and writing is based on activation functions the output of which varies between [0, 1].

On the other hand, a BiLSTM operates in both forward and backward direction, where it learns the sequence of the provided input (forward) and learns the reverse of the same sequence (backward). Toward the end of the learning process, both have to be merged via a mechanism to combine both. This merge step can be done via Sum, Multiplication, Averaging, and Concatenation (default).

Table 9.1 shows components controlling the cell state and hidden state of the layer.

Table 9.1 Components that control the cell and hidden state of the layer.

Component	Purpose
Input gate (i)	It takes as input the current input to the LSTM cell and the previous hidden state. It applies a sigmoid activation function to these inputs to produce a gate activation vector. The gate activation vector is essentially a set of values between 0 and 1 that determine the extent to which new information should be allowed into the memory cell.
Forget gate (f)	It takes as input the current input to the LSTM cell and the previous hidden state. Similar to the input gate, it applies a sigmoid activation function to these inputs, producing a gate activation vector. This gate activation vector determines the extent to which the LSTM should forget or retain information from the previous time step.
	So, it acts as a filter that decides which information from the previous memory cell state should be discarded. It does this by element-wise multiplying the gate activation vector with the previous memory cell state. The resulting product determines the amount of information to be forgotten.

(Continued)

Table 9.1 (Continued)

Component	Purpose
Cell candidate (g)	The cell candidate is obtained by applying a hyperbolic tangent (tanh) activation function to a linear combination of the current input and the previous hidden state. It captures the new information that could potentially be stored in the memory cell. The cell candidate represents the new information that the LSTM cell is considering to incorporate into the memory cell state. It is generated regardless of whether this information will be stored or discarded.
Output gate (o)	It takes as input the current input to the LSTM cell and the previous hidden state. It applies a sigmoid activation function to these inputs, producing a gate activation vector. This gate activation vector determines the extent to which the LSTM should allow the information from the memory cell to pass through to the output.
	The output gate acts as a filter that decides how much information from the memory cell should be included in the output of the LSTM layer. It does this by element-wise multiplying the gate activation vector with the current memory cell state, which produces the output of the LSTM cell.

Figure 9.1 shows the flow of data at time step t. The diagram highlights how the gates forget, update, and output the cell and hidden states.

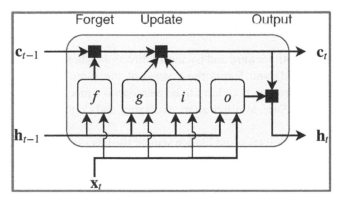

Figure 9.1 A snapshot for flow of data at time step t. The gates control the forget, update, and output steps between the cell (**c**) and hidden (**h**) states.

The first step in the LSTM RNN is to decide which information should be omitted from the cell in that particular time step. The sigmoid function determines this. It looks at the previous state (h_{t-1}) along with the current input x_t and computes the function.

The learnable weights of an LSTM RNN are the input weights **W** (InputWeights), the recurrent weights **R** (RecurrentWeights), and the bias **b** (Bias). The matrices **W**, **R**, and **b** are concatenations of the input weights, the recurrent weights, and the bias of each component, respectively. These matrices are concatenated as follows:

$$W = \begin{bmatrix} W_i \\ W_f \\ W_g \\ W_o \end{bmatrix}, \quad R = \begin{bmatrix} R_i \\ R_f \\ R_g \\ R_o \end{bmatrix}, \quad b = \begin{bmatrix} b_i \\ b_f \\ b_g \\ b_o \end{bmatrix}$$

where $i, f, g,$ and o denote the input gate, forget gate, cell candidate, and output gate, respectively. The cell state at time step t is given by:

$$c_t = f_t \odot c_{t-1} + i_t \odot g_t$$

where $\odot$ denotes the Hadamard product (element-wise multiplication of vectors).

The hidden state at time step t is given by:

$$h_t = o_t \odot \sigma_c(c_t)$$

where σ_c denotes the state activation function. The **lstmLayer** function, by default, uses the hyperbolic tangent function (**tanh**) to compute the cell state activation function. Table 9.2 shows formulas describing the components at time step t, as quoted from MATLAB offline help.

Table 9.2 Formulas for describing the components at time step t.

Component	Formula
Input gate	$i_t = \sigma_g\left(W_i X_t + R_i h_{t-1} + b_i\right)$
Forget gate	$f_t = \sigma_g\left(W_f X_t + R_f h_{t-1} + b_f\right)$
Cell candidate	$g_t = \sigma_c\left(W_g X_t + R_g h_{t-1} + b_g\right)$
Output gate	$o_t = \sigma_g\left(W_o X_t + R_o h_{t-1} + b_o\right)$

In these calculations, σ_g denotes the gate activation function. The **lstmLayer** function, by default, uses the sigmoid function given by $\sigma_g(x) = \dfrac{1}{\left(1 + e^{-x}\right)}$ to compute the gate activation function.

To explain the mechanism, let's use an example with two sentences:

1) "George is good in PlayStation games. Sam is good in boxing."
2) "Sam watches TV boxing match. He likes to watch the world championship heavyweight boxing."

In an LSTM layer, the forget gate identifies a change in context after the first full stop. It compares the current input sentence with the previous one. Since the next sentence focuses on Sam, information about George is considered for removal. The forget gate uses a sigmoid function to decide which values should be kept (1) or discarded (0).

Let us consider the possible scenarios based on the values of the sigmoid gate and the tanh function in an LSTM cell:

1) Sigmoid Gate Output: 0, Tanh Value: −1
   - The sigmoid gate output of 0 suggests forgetting the information.
   - The tanh value of −1 indicates a strong negative influence or reset of the information.
   - Net Output: The information from the cell state is completely ignored or reset.
2) Sigmoid Gate Output: 0, Tanh Value: 0
   - The sigmoid gate output of 0 suggests forgetting the information.
   - The tanh value of 0 indicates no influence or importance assigned to the information.
   - Net Output: The information from the cell state is completely ignored or forgotten.

3) Sigmoid Gate Output: 0, Tanh Value: 1
   - The sigmoid gate output of 0 suggests forgetting the information.
   - The tanh value of 1 indicates a positive influence or importance assigned to the information.
   - Net Output: The information from the cell state is completely ignored or forgotten.
4) Sigmoid Gate Output: 1, Tanh Value: −1
   - The sigmoid gate output of 1 suggests retaining all the information.
   - The tanh value of -1 indicates a strong negative influence or reset of the information.
   - Net Output: The information from the cell state is reset or suppressed, but not completely ignored.
5) Sigmoid Gate Output: 1, Tanh Value: 0
   - The sigmoid gate output of 1 suggests retaining all the information.
   - The tanh value of 0 indicates no influence or importance assigned to the information.
   - Net Output: The information from the cell state is retained as is.
6) Sigmoid Gate Output: 1, Tanh Value: 1
   - The sigmoid gate output of 1 suggests retaining all the information.
   - The tanh value of 1 indicates a positive influence or importance assigned to the information.
   - Net Output: The information from the cell state is retained as is.

All in all, an LSTM RNN does the following tasks: How much history data should be retained versus deleted; prioritize the current information in light of the retained history data; and finally picks up the most important prioritized piece of information and pushes it as an output.

Figure 9.2 illustrates the architecture of a simple LSTM network for classification. The network starts with a sequence input layer followed by an LSTM layer. To predict class labels, the network ends with a fully connected layer, a softmax layer, and a classification output layer.

Figure 9.2 A schematic for the backbone of an LSTM RNN for classification.

Figure 9.3 illustrates the architecture of a simple LSTM network for regression. The network starts with a sequence input layer followed by an LSTM layer. The network ends with a fully connected layer and a regression output layer.

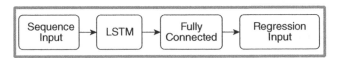

Figure 9.3 A schematic for the backbone of an LSTM RNN for regression.

Figure 9.4 illustrates the architecture of an LSTM network for video *classification*. To input image sequences to the network, use a sequence input layer. To use convolutional layers to extract features, that is, to apply the convolutional operations to each frame of the videos independently, use a sequence folding layer followed by the convolutional layers, and then a sequence unfolding layer. To use the LSTM layers to learn from sequences of vectors, use a flatten layer followed by the LSTM and output layers.

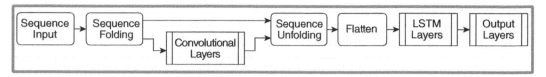

Figure 9.4 A schematic for the backbone of an LSTM RNN for video classification.

Explanation for each of the appearing layers in an LSTM RNN is show in Table 9.3, quoted from MATLAB offline help.

Table 9.3 Explanation for each layer appearing in LSTM RNN.

Layer	Description
sequenceInputLayer	A sequence input layer inputs sequence data to a network.
lstmLayer	An LSTM layer learns long-term dependencies between time steps in time series and sequence data.
fullyConnectedLayer	A fully connected layer multiplies the input by a weight matrix and then adds a bias vector.
softmaxLayer	A softmax layer applies a softmax function to the input. The softmax function is the output unit activation function after the last fully connected layer for multi-class classification problems. The softmax function is also known as the normalized exponential and can be considered the multi-class generalization of the logistic sigmoid function.
classificationLayer	A classification layer computes the cross-entropy loss for classification and weighted classification tasks with mutually exclusive classes. The layer infers the number of classes from the output size of the previous layer. For example, to specify the number of classes K of the network, you can include a fully connected layer with output size K and a softmax layer before the classification layer.
bilstmLayer	A bidirectional LSTM (BiLSTM) layer learns bidirectional long-term dependencies between time steps of time series or sequence data. These dependencies can be useful when you want the network to learn from the complete time series at each time step.
regressionLayer	A regression layer computes the half-mean-squared-error loss for regression tasks. For sequence-to-sequence regression networks, the loss function of the regression layer is the half-mean-squared-error of the predicted responses for each time step, not normalized by R, the number of responses.
sequenceFoldingLayer	A sequence folding layer converts a batch of image sequences to a batch of images. Use a sequence folding layer to perform convolution operations on time steps of image sequences independently.
sequenceUnfoldingLayer	A sequence unfolding layer restores the sequence structure of the input data after sequence folding.

(Continued)

Table 9.3 (Continued)

Layer	Description
convolution1dLayer convolution2dLayer convolution3dLayer	A 1-D convolutional layer applies sliding convolutional filters to 1-D input. The layer convolves the input by moving the filters along the input and computing the dot product of the weights and the input, then adding a bias term. A 2-D convolutional layer applies sliding convolutional filters to 2-D input. The layer convolves the input by moving the filters along the input vertically and horizontally and computing the dot product of the weights and the input, and then adding a bias term. A 3-D convolutional layer applies sliding cuboidal convolution filters to 3-D input. The layer convolves the input by moving the filters along the input vertically, horizontally, and along the depth, computing the dot product of the weights and the input, and then adding a bias term.
flattenLayer	A flatten layer collapses the spatial dimensions of the input into the channel dimension.

NOTE #1

The dimensions that the layer convolves over depends on the layer input:

For time series and vector sequence input (data with three dimensions corresponding to the channels, observations, and time steps), the layer convolves over the time dimension.

For 1-D image input (data with three dimensions corresponding to the spatial pixels, channels, and observations), the layer convolves over the spatial dimension.

For 1-D image sequence input (data with four dimensions corresponding to the spatial pixels, channels, observations, and time steps), the layer convolves over the spatial dimension.

For 2-D image input (data with four dimensions corresponding to pixels in two spatial dimensions, the channels, and the observations), the layer convolves over the spatial dimensions.

For 2-D image sequence input (data with five dimensions corresponding to the pixels in two spatial dimensions, the channels, the observations, and the time steps), the layer convolves over the two spatial dimensions.

For 3-D image input (data with five dimensions corresponding to pixels in three spatial dimensions, the channels, and the observations), the layer convolves over the spatial dimensions.

For 3-D image sequence input (data with six dimensions corresponding to the pixels in three spatial dimensions, the channels, the observations, and the time steps), the layer convolves over the spatial dimensions.

Train LSTM RNN Network for Sequence Classification

NOTE #2
To train an LSTM, the data needs to follow a specific format. The input data should be organized as a cell array with one column. Each column represents a time step within a sample. It's important to note that each sample can have a varying number of time steps, but be careful as sequences with different lengths will be padded. Excessive padding can hinder training progress. The rows in the data correspond to the feature dimension of each sample. This could be data from various sensors or different letters in a vocabulary. It's necessary for all samples to have the same number of rows.

Let us load the Japanese Vowels data set, as described in [1] and [2]. **XTrain** is a cell array containing 270 sequences of varying length with 12 features corresponding to linear predictive coding cepstral (LPCC) coefficients. LPCC coefficients are a feature representation commonly used in speech and audio processing and they are derived from Linear Predictive Coding (LPC), which is a technique used to model the spectral envelope of a speech signal.

In LPC analysis, the speech signal is modeled as a linear combination of past samples, where the coefficients of this linear combination are estimated using methods like the autocorrelation method or the covariance method. These coefficients represent the vocal tract characteristics of the speech signal. The LPCC coefficients are obtained by taking the cepstrum of the LPC coefficients. The cepstrum represents the spectral envelope of the LPC coefficients in the quefrency domain. The quefrency domain is derived from the cepstrum, which is the result of taking the inverse Fourier Transform of the logarithm of the power spectrum of a signal. The cepstrum is primarily used for analyzing the periodicity or repetition within a signal. The term "quefrency" is used to refer to the time-domain representation in the cepstral domain.

LPCC coefficients are often used as features for speech recognition, speaker recognition, and other speech processing tasks. They provide a compact representation of the spectral properties of speech signals, allowing for efficient analysis and classification.

Y is a categorical vector of labels 1, 2, ..., 9. The entries in **XTrain** are matrices with 12 rows (one row for each feature) and a varying number of columns (one column for each time step). From a speech analysis point of view, speech is composed of an excitation sequence linearly convolved with the impulse response of the vocal tract transfer function. The process of cepstral deconvolution is an attempt to deconvolve the excitation signal from the vocal tract transfer function without making any judicial assumptions needed for linear prediction. At the end of the tunnel, a transfer function, with both zeros and poles, can be concluded and be used in system modeling. Here are the steps of reading the Japanese vowels data, training the RNN model, loading the test set, and classifying the sequences into speakers.

1) Load Data and Show a Plot

Load data and show a plot for the first observation in terms of features versus time steps.

```
[XTrain,YTrain] = japaneseVowelsTrainData;
%Visualize the first time series in a plot. Each line corresponds to a feature.
figure;
plot(XTrain{1}')
```

(Continued)

(Continued)

title("Training Observation 1")
numFeatures = size(XTrain{1},1);
legend("Feature " + string(1:numFeatures),'Location','northeastoutside')

Figure 9.5 shows the first observation of Japanese vowels data for the 12 features as a function of time step.

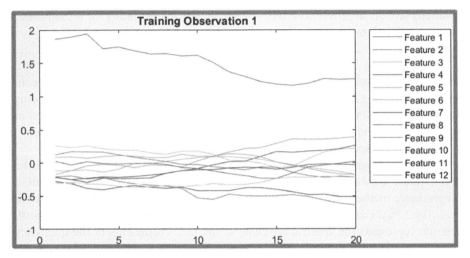

Figure 9.5 Features from 1 up to 12 for the first observation with 20 time-steps for each.

2) Define the LSTM Network Architecture
Specify the input size as 12 (the number of features of the input data). Specify an LSTM layer to have 100 hidden units and to output the last element of the sequence. Furthermore, specify nine classes by including a fully connected layer of size 9, followed by a softmax and a classification layer.

NOTE #3

The number of hidden units is read-only, also known as the hidden size, specified as a positive integer. The number of hidden units corresponds to the amount of information remembered between time steps (the hidden state). The hidden state can contain information from all previous time steps, regardless of the sequence length. If the number of hidden units is too large, then the layer might overfit the training data. This value can vary from dozens to thousands. The hidden state does not limit the number of time steps that are processed in an iteration. To split sequences into smaller sequences upon using **trainNetwork** function, use the **"SequenceLength"** training option.

inputSize = 12;
numHiddenUnits = 100;
numClasses = 9;

```
layers = [ ...
  sequenceInputLayer(inputSize)
  lstmLayer(numHiddenUnits,'OutputMode','last')
  fullyConnectedLayer(numClasses)
  softmaxLayer
  classificationLayer]
```

The output is shown below, copied from Command Window:

```
layers = 5 × 1 Layer array with layers:
  1  "  Sequence Input        Sequence input with 12 dimensions
  2  "  LSTM                   LSTM with 100 hidden units
  3  "  Fully Connected        9 fully connected layer
  4  "  Softmax                softmax
  5  "  Classification Output  crossentropyex
```

3) Specify the Training Options

Specify the solver as '**adam**' and '**GradientThreshold**' as 1. Set the mini-batch size to 27 and set the maximum number of epochs to 70.

Because the mini batches are small with short sequences, the CPU is better suited for training. Set '**ExecutionEnvironment**' to '*cpu*'. To train on a GPU, if available, set '**ExecutionEnvironment**' to '*auto*' (the default value).

```
maxEpochs = 70;
miniBatchSize = 27;
options = trainingOptions('adam', ...
  'ExecutionEnvironment','cpu', ...
  'MaxEpochs',maxEpochs, ...
  'MiniBatchSize',miniBatchSize, ...
  'GradientThreshold',1, ...
  'Verbose',false, ...
  'Plots','training-progress');
```

NOTE #4

In general, RNNs may encounter one of the following gradients-based critical issues:

1) Vanishing gradient problem: The gradient becomes too small, hence, the parameter updates become insignificant. This renders the learning of long data sequences a tedious task.
2) Exploding gradient problem: On the contrary, if the slope tends to grow exponentially instead of decaying, then it will end up with an exploding gradient condition. This condition arises when large error gradients accumulate, resulting in very large updates to RNN model weights during the training process, resulting in poor performance or bad accuracy.
3) The "**GradientThreshold**" is Inf (default) | positive scalar. If the gradient exceeds the value of *GradientThreshold*, then the gradient is clipped according to the GradientThresholdMethod training option. The "gradient explosion" is indicated by a training loss that goes to NaN

(Continued)

NOTE #4 (Continued)

---

or Inf. Gradient clipping helps prevent gradient explosion by stabilizing the training at higher learning rates and in the presence of outliers. Gradient clipping enables networks to be trained faster and does not usually impact the accuracy of the learned task. For more information, see MATLAB | Built in Help | "**Gradient Clipping**".

---

4) Train the LSTM Network

---

net = trainNetwork(XTrain,YTrain,layers,options);

Figure 9.6 shows the LSTM RNN training results, expressed in terms of accuracy and loss.

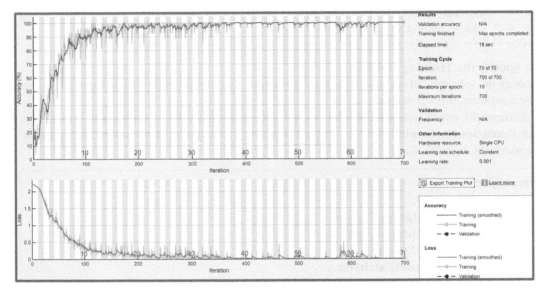

Figure 9.6 The LSTM RNN training results, expressed in terms of accuracy and loss.

5) Load the Test Set

[XTest,YTest] = japaneseVowelsTestData;

---

6) Classify the Test Data Set

Specify the same mini-batch size used for training.

YPred = classify(net,XTest,'MiniBatchSize',miniBatchSize);

---

7) Estimation of Accuracy

Calculate the classification accuracy of the predictions.

---

acc = (sum(YPred == YTest)./numel(YTest))*100;

---

acc = 95.9459

Improving LSTM RNN Performance

We learned about improving performance of CNNs in previous chapters (see Chapter 7). Most of these guidelines are still applicable to training LSTM RNNs. Figure 9.7 shows the relevant steps we can take to improve accuracy of a sequence classification network. Sequences can be normalized using a variety of methods. Sequence length and padding are specific to LSTM RNNs.

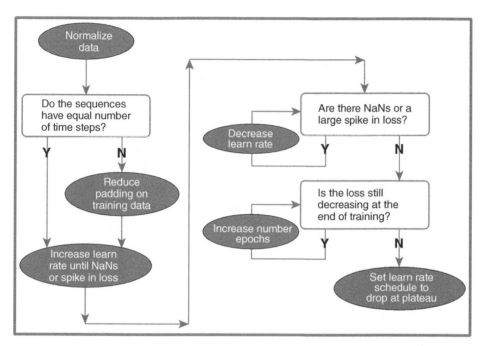

Figure 9.7   How to improve accuracy of an LSTM RNN.

Sequence Length

Sequences may contain any number of time steps. Nevertheless, we should be cautious if sequences have different lengths. During training, the sequences in each mini-batch are padded (expanded) with a number, usually zero, to equalize the lengths. A network cannot distinguish between values created for padding and values that are originally part of the sequence. We should minimize the amount of padding by sorting data by sequence length and carefully choosing the mini-batch size.

Consider eighteen sequences with varying lengths. If we train a network with this data and nine sequences per mini-batch, a substantial amount of padding is added to most of the sequences, as shown in Figure 9.8.

On the other hand, sorting the data by sequence length reduces padding significantly. If we sort training data, we should also set the "**Shuffle**" training option to "*never*". Notice that there is still too much padding on the shorter sequences in each mini-batch, as shown in Figure 9.9.

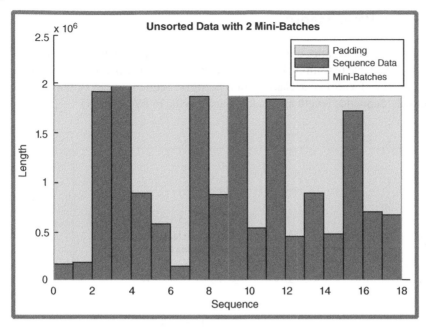

Figure 9.8 The "**MiniBatchSize**" is set to 9, which means 2 mini-batches are needed for 18 sequences.

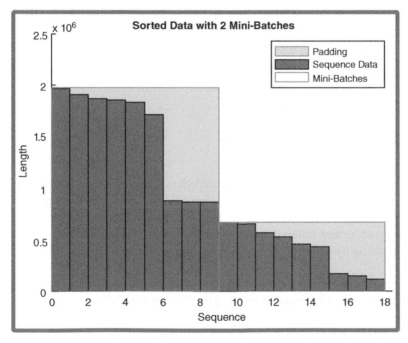

Figure 9.9 The "**MiniBatchSize**" is still 9 but sequences are sorted by length; hence, less padding.

Setting the "**MiniBatchSize**" to *6* will further reduce padding, as shown in Figure 9.10. However, decreasing the mini-batch size can increase the network training time and number of iterations because the computing resource will process less data at a time.

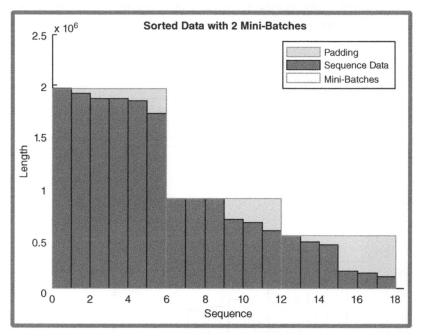

Figure 9.10   Setting the "**MiniBatchSize**" to *6* will reduce padding yet increasing the training CPU time.

As we continue to decrease the "**MiniBatchSize**" value down to *3*, the amount of padding will further decrease; however, more iterations and CPU time are needed to train the network for the same number of epochs, as shown in Figure 9.11.

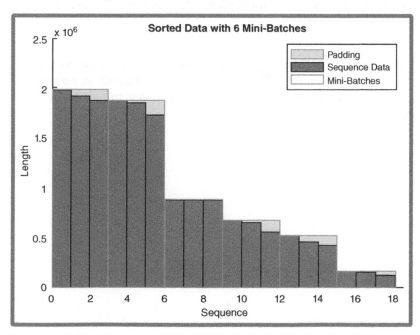

Figure 9.11   Less padding can be made by further reduction of "**MiniBatchSize**" value down to *3* while a longer computational time is needed.

Moreover, we can also use the "**SequenceLength**" option to modify the padding behavior. By default, the sequences are padded to have the same length as the "*longest*" sequence. This behavior is shown in previous figures 9.8–9.11. Figure 9.12 shows the use of the "*shortest*" option to trim longer sequences to the same length as the shortest sequence. This option has no padding; however, it may end up with removing important data from sequences.

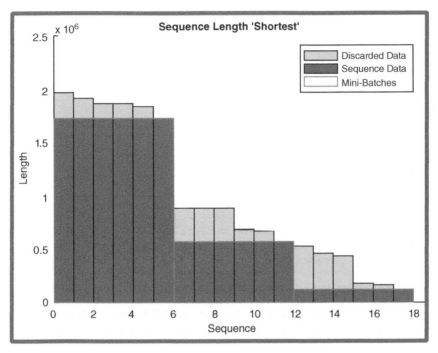

Figure 9.12 Setting the "**MiniBatchSize**" to *6* and "**SequenceLength**" to "*shortest*" will minimize padding but with a potential loss of data from longer sequences.

Finally, if sequences are too long to fit in memory, we can also specify a threshold to split the sequences on the length basis. The sequences will be padded before they are split, so we may still have padding in data, as shown in Figure 9.13. The split value is set to $5 \times 10^5 = 0.5 \times 10^6$. The "**MiniBatchSize**" is still set to *6*.

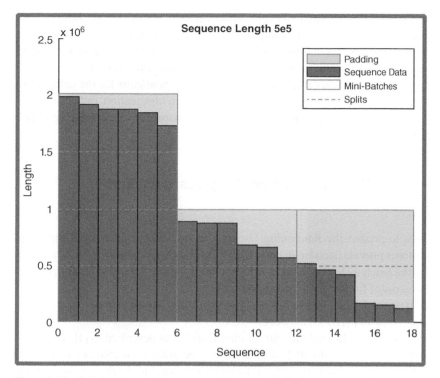

Figure 9.13 Splitting long sequences to avoid memory-limited processor problems, with "**MiniBatchSize**" set to 6.

Classifying Categorical Sequences

Training an LSTM RNN requires the sequences to be numeric. If sequences are categorical, how can we train a deep network?

Categorical sequences could be a sequence of the weather, DNA, or music notes, as shown in Figure 9.14.

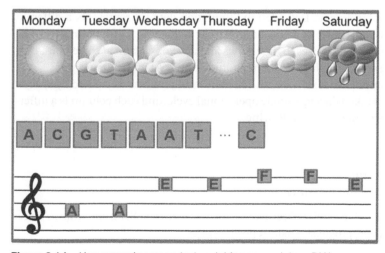

Figure 9.14 Non-numeric categorical variables as weekdays, DNA sequences, or music notes.

One option is to assign a number to each category. However, this results in imposing a false numerical structure on the observations. For example, if you assign the numbers 1 through 4 to four categories in a predictor, it implies that the distance between the categories 1 and 4 is longer than the distance between the categories 3 and 4. Instead, we use dummy predictors.

Instead of assigning a number to each category, we create *dummy predictors* for the categories. Each dummy predictor can have only two values: 0 or 1. For any given observation, only one of the dummy predictors can have the value equal to 1. See Chapter 3 | **Accommodating Categorical Data: Creating Dummy Variables** section.

Sequence-to-Sequence Regression Using Deep Learning: Turbo Fan Data

This section shows how to predict the Remaining Useful Life (RUL) of engines by using deep learning. The RUL estimates provide decision makers with information that allows them to change operational characteristics (such as load) which in turn may prolong the life of the component. It also allows planners to account for upcoming maintenance [3].

To train a deep neural network to predict numeric values from time series or sequence data, we can use a long short-term memory (LSTM) network.

We will use the Turbofan Engine Degradation Simulation Data Set as described in [3]. The tutorial trains an LSTM network to predict the RUL of an engine (predictive maintenance), measured in cycles, given time series data representing various sensors in the engine. Each sequence varies in length and corresponds to a full Run to Failure (RTF) instance. Both the training and the test data are sub-divided into four data sets with varying size of sequences.

The engine is operating normally at the start of each time series and develops a fault at some point during the series. In the *training set*, the fault grows in magnitude until system failure. In the *test set*, the time series ends some time prior to system failure. The objective of the competition is to predict the number of remaining operational cycles before failure (RUL) in the test set, i.e., the number of operational cycles after the last cycle that the engine will continue to operate. Also provided a vector of true RUL values for the sake of assessing the predicted RUL test data.

1) Download Data
Each time series of the Turbofan Engine Degradation Simulation data set represents a different engine. Each engine starts with unknown degrees of initial wear and manufacturing variation. The engine is operating normally at the start of each time series and develops a fault at some point during the series. In the training set, the fault grows in magnitude until system failure occurs.

The data contains a ZIP-compressed text files with 26 columns of numbers, separated by spaces. Each row is a snapshot of data taken during a single operational cycle, and each column is a different variable. The columns correspond to the following:

- Column 1: Unit number
- Column 2: Time in cycles
- Columns 3–5: Operational settings
- Columns 6–26: Twenty-one sensor measurements

Download and extract the Turbofan Engine Degradation Simulation Data Set from: https://ti.arc.nasa.gov/tech/dash/groups/pcoe/prognostic-data-repository[4].

Alternatively, download from:
https://www.kaggle.com/datasets/behrad3d/nasa-cmaps

2) Create a Directory to Store the Turbofan Engine Degradation Simulation Data Set

```
dataFolder = fullfile(tempdir,"turbofan");
if ~exist(dataFolder,'dir')
   mkdir(dataFolder);
end
```

3) Prepare Training Data

Load the data using the **processTurboFanDataTrain** function attached to this example. The **processTurboFanDataTrain** function extracts the data from **filenamePredictors** and returns the cell arrays **XTrain** and **YTrain**, which contain the training predictor and response sequences. See Appendix A19 for **processTurboFanDataTrain** function.

```
filename = "CMAPSSData.zip";
unzip(filename,dataFolder)
% Reading the training failure data, train_FD.txt
filenamePredictors = fullfile(dataFolder,"train_FD.txt");
[XTrain,YTrain] = processTurboFanDataTrain(filenamePredictors);
```

Notice here that **XTrain** is 260×1 cell array which means each entry, of the 260 entries, is made of a fixed number of features equal to 24 and a variable sequence length (i.e., time step or stamp). On the other hand, **YTrain** is also 260×1 cell array with one fixed feature for each entry but with a variable sequence length. Notice that **XTrain** 260 data points have all the same number of features, that is, 24 each.

4) Remove Features with Constant Values

Features that remain constant for all time steps can negatively impact the training. Find the rows of data that have the same minimum and maximum values and remove the rows, accordingly.

```
%min(A,[],2) is a column vector containing the minimum value of each row in A.
m = min([XTrain{:}],[],2);
M = max([XTrain{:}],[],2);
idxConstant = M == m;
for i = 1:numel(XTrain)
   XTrain{i}(idxConstant,:) = [];
end
```

Notice here that **idxConstant** values are all zero, which means no feature removal takes place. This step is not needed for such **XTrain** data, but it is good to keep this practice for other case studies.

numFeatures = size(XTrain{1},1)

5) Normalize Training Predictors

Normalize the training predictors to have zero mean and unit variance. To calculate the mean and standard deviation over all observations, concatenate the sequence data horizontally.

```
mu = mean([XTrain{:}],2); % mean of each row.
% To maintain the default normalization while specifying the dimension of opera-
tion, set
% the second argument to zero.
sig = std([XTrain{:}],0,2); % the standard of row data around zero mean.
for i = 1:numel(XTrain)
    XTrain{i} = (XTrain{i} - mu) ./ sig; % normalization step.
end
```

6) Clip Responses

To learn more from the sequence data when the engines are close to failing, clip the responses at the threshold *150*. This makes the network treat instances with higher RUL values as equal.

```
thrsh = 150;
for i = 1:numel(YTrain)
    YTrain{i}(YTrain{i} > thrsh) = thrsh;
end
```

7) Prepare Data for Padding

To minimize the amount of padding added to the mini-batches, sort the training data by sequence length. Then, choose a mini-batch size which divides the training data evenly and reduces the amount of padding in the mini-batches. Sort the training data by sequence length.

```
for i=1:numel(XTrain)
    sequence = XTrain{i};
    sequenceLengths(i) = size(sequence,2); %equal to number of columns per
sequence.
end
[sequenceLengths,idx] = sort(sequenceLengths,'descend');
XTrain = XTrain(idx);
YTrain = YTrain(idx);
%View the sorted sequence lengths in a bar chart.
figure
bar(sequenceLengths)
xlabel("Sequence")
ylabel("Length")
title("Sorted Data")
```

Figure 9.15 shows the plot of the sequence length for the 260 sequences while being sorted from longest down to shortest.

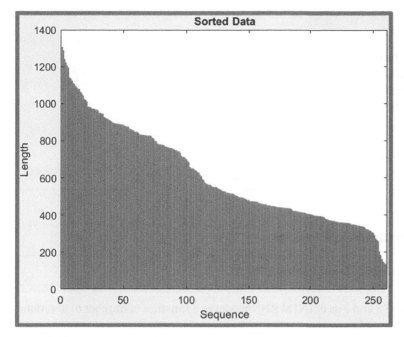

Figure 9.15   Arrangement of sequences from the longest to shortest.

8) Define Network Architecture

Define the network architecture. Create an LSTM network that consists of an LSTM layer with *200* hidden units, followed by a fully connected layer of size *100* and a dropout layer with dropout probability *0.2*.

```
numResponses = size(YTrain{1},1); % For  YTrain, # of responses is 1.
numHiddenUnits = 200;
numFeatures = size(XTrain{1},1); % For Xtrain, # of features is 24.
layers = [ ...
  sequenceInputLayer(numFeatures)
  bilstmLayer(numHiddenUnits,'OutputMode','sequence')
  fullyConnectedLayer(100)
  bilstmLayer(numHiddenUnits,'OutputMode','sequence')
  fullyConnectedLayer(100)
  dropoutLayer(0.2)
  fullyConnectedLayer(numResponses)
  regressionLayer];
```

9) Training Options

Specify the training options. Train for *40* epochs with mini-batches of size *13* using the solver **'adam'**. Specify the learning rate *0.0003*. To prevent the gradients from exploding, set the gradient threshold to *10*. To keep the sequences sorted by length, set **'Shuffle'** to *'never'*.

```
maxEpochs = 40;
miniBatchSize = 13;
options = trainingOptions('adam', ...
   'MaxEpochs',maxEpochs, ...
   'MiniBatchSize',miniBatchSize, ...
   'InitialLearnRate',0.0003, ...
   'GradientThreshold',10, ...
   'Shuffle','never', ...
   'Plots','training-progress',...
   'Verbose',0);
```

10) Train the Network
Train the network using **trainNetwork** function.

```
net = trainNetwork(XTrain,YTrain,layers,options);
```

Figure 9.16 shows both RMSE and loss of LSTM RNN model as a function of number of iterations or epochs.

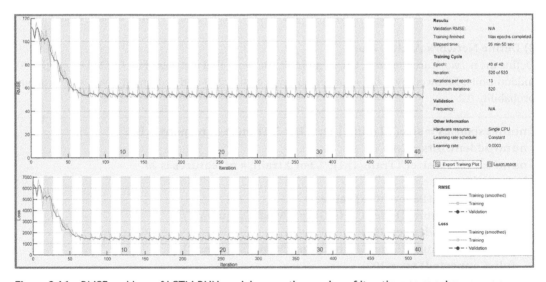

Figure 9.16   RMSE and loss of LSTM RNN model versus the number of iterations or epochs.

11) Test the Network
Prepare the test data using the **processTurboFanDataTest** function attached to this example. The **processTurboFanDataTest** function extracts the data from **filenamePredictors** (test failure data, test_FD001.txt) and **filenameResponses** (true RUL values, RUL_FD001.txt) and returns the cell arrays **XTest** and **YTest**, which contain the test predictor and response sequences, respectively. Notice that there are four groups of test data that can be examined and one of them is tested here. See Appendix A20 for **processTurboFanDataTest** function.

```
filenamePredictors = fullfile(dataFolder,"test_FD001.txt");
filenameResponses = fullfile(dataFolder,"RUL_FD001.txt");
[XTest,YTest] = processTurboFanDataTest(filenamePredictors,filenameResponses);
```

12) Remove Features with Constant Values

Remove features with constant values using **idxConstant** calculated from the training data. Normalize the test predictors using the same parameters as in the training data. Clip the test responses at the same threshold used for the training data.

```
for i = 1:numel(XTest)
  %XTest{i}(idxConstant,:) = [];
  %No constant value features found.
  XTest{i} = (XTest{i} - mu) ./ sig;
  YTest{i}(YTest{i} > thrsh) = thrsh;
end
```

13) Predict RUL

Make predictions on the test data using predict. To prevent the function from adding padding to the data, specify the mini-batch size 1.

```
YPred = predict(net,XTest,'MiniBatchSize',1);
```

The LSTM network makes predictions on the partial sequence one step at a time. At each time step, the network predicts using the value at this time step, and the network state calculated from the previous time steps only. The network updates its state between each prediction. The predict function returns a sequence of these predictions. *The last element of the prediction corresponds to the predicted RUL for the partial sequence.*

Let us use both the last element and the mean value of the predicted RUL and compare it with the true RUL that can be retrieved from **filenameResponses** variable. The **filenameResponses** refers to **RUL_FD001.txt** file, as shown below. We have selected the per cent relative error (PRE) of the predicted RUL value to be $\leq$ 10%. However, it is left to the engineering maintenance department to decide on the threshold value of PRE.

```
YRUL = readmatrix(filenameResponses);
for i = 1:numel(YTest)
YTST(i)=mean(YTest{i});
YPRD(i)=mean(YPred{i});
end
for i = 1:numel(YTest)
YPRE(i)=(abs(YPRD(i) - YTST(i))./YTST(i))*100;
end
YPREMean=mean(YPRE)
YPRELT10=numel(YPRE(YPRE<=10.0))
```

See the last case in Table 9.4, which matches with the above code.

This means that 11% out of the predicted 100 data points fall within PRE range between [0,10%]. The prediction is said to be accurate and accepted within engineering measurements and standards.

Table 9.4 shows the different combinations of the predicted and reference Y data and the corresponding mean PRE and (PRE < 10%) cases. The last case has the lowest PRE in terms of the lowest mean PRE and with adequate cases having PRE < 10%. This is expected as the previous training takes place between XTrain and YTrain data, not between XTrain and YRUL data. Notice that both YTrain and YTest are sequences with variable lengths; whereas YRUL is a single value for each row (i.e., iteration).

Table 9.4 Different combinations of predicted and reference datum for Y.

$Y_{Predict}$ for i^{th} iteration	$Y_{Reference}$ for i^{th} iteration	PRE_{Mean} for all 100 Test Data	(PRE < 10 %) Cases
YPred{i}(end)	YRUL(i)	89.4%	4%
mean(YPred{i})	YRUL(i)	148.3%	17%
YPred{i}(end)	YTest{i}(end)	89.5%	4%
mean(YPred{i})	mean(YTest{i})	29.4%	11%

Based on the best case (last case in Table 9.4), we can calculate the root-mean-square error (RMSE) of the predictions and visualize the prediction error in a histogram.

```
for i = 1:numel(YTest)
YTST(i)=mean(YTest{i});
YPRD(i)=mean(YPred{i});
end
figure
rmse = sqrt(mean((YPRD - YTST).^2))
histogram(YPRD - YTST)
title("RMSE = " + rmse)
ylabel("Frequency")
xlabel("Error")
```

Figure 9.17 shows the frequency or distribution of the difference between the predicted and the test last value of sequences. The overall RMSE is 43.3.

Figure 9.18 shows workspace variables where one can see the PRE values, based on the mean YPred and mean YTest value, with an upper threshold PRE value of 10%.

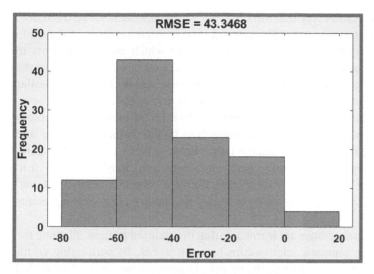

Figure 9.17 The histogram of prediction error based on the mean value of the test versus the predicted sequences.

Workspace		Workspace	
Name ▲	Value	Name ▲	Value
dataFolder	"C:\Users\Kamal\AppData\...	sequence	24x316 double
filename	"CMAPSSData.zip"	sequenceLengths	1x260 double
filenamePredictors	"C:\Users\Kamal\AppData\...	sig	24x1 double
filenameResponses	"C:\Users\Kamal\AppData\...	thrsh	150
i	100	XTest	100x1 cell
idx	1x260 double	XTrain	100x1 cell
idxConstant	24x1 logical	YPRD	1x100 single
layers	8x1 Layer	YPrdct	1x100 double
m	24x1 double	YPRE	1x100 single
M	24x1 double	YPred	100x1 cell
maxEpochs	30	YPredM	86.0003
miniBatchSize	5	YPRELT10	11
mu	24x1 double	YPREM	89.4869
net	1x1 SeriesNetwork	YPREMean	29.3882
numFeatures	24	YRUL	100x1 double
numHiddenUnits	200	YTest	100x1 cell
numResponses	1	YTrain	100x1 cell
options	1x1 TrainingOptionsADAM	YTST	1x100 double
rmse	43.3468		

Figure 9.18 Workspace variables showing PRE and RMSE value, based on the mean YPred and mean YTest value.

Classify Text Data Using Deep Learning: Factory Equipment Failure Text Analysis – 1

This section shows how to classify text data using a deep learning LSTM network. Text data is naturally sequential. A piece of text is a sequence of words, which might have dependencies among them. To learn and use long-term dependencies to classify sequence data, we use an LSTM neural network. An LSTM network is an RNN type that can learn long-term dependencies between time steps of sequence data. To input text to an LSTM network, first we convert the text data into numeric sequences. We can achieve this using a word encoding which maps documents to sequences of numeric indices. For better results, we also include a word embedding layer in the network. Word embeddings map words in a vocabulary to numeric vectors rather than scalar indices. These embeddings capture semantic details of the words, so that words with similar meanings have similar vectors. The following example demonstrates how LSTM models can represent relationships between words using vector arithmetic: "Rome is to Italy as Paris is to France" and describes this relationship using the equation: France = Italy − Rome + Paris. This equation demonstrates a simple way to capture the relationship between "Rome" and "Italy" and apply it to "Paris" and "France" using vector arithmetic. By subtracting the vector representation of "Rome" from "Italy" and then adding the vector representation of "Paris," we obtain a vector that represents "France."

In an LSTM model, the word embeddings are learned during the training process, allowing the model to capture and generalize semantic relationships between words. By performing vector arithmetic operations on these word embeddings, the model can generate new word vectors that capture similar semantic relationships. This capability of LSTM models to represent and manipulate semantic relationships through vector arithmetic is a powerful aspect of their ability to understand and generate natural language.

There are four steps involved in training and using the LSTM network in this section:

- Import and preprocess the data.
- Convert the words to numeric sequences using a word encoding.
- Create and train an LSTM network with a word embedding layer.
- Classify new text data using the trained LSTM network.

1) Import Data
Import the factory reports data. This data contains labeled textual descriptions of factory events. To import the text data as strings, specify the text type to be '**string**'.

```
filename = "factoryReports.csv";
data = readtable(filename,'TextType','string');
% Get top rows of a table, timetable, or tall array
head(data)
```

Here are the heads of the table as shown in Figure 9.19. The description refers to the insinuation of an event, usually an infringement of a safety rule, how it is categorized, level of danger, remedy step, and the cost of potential damage.

Description	Category	Urgency	Resolution	Cost
"Items are occasionally getting stuck in the scanner spools."	"Mechanical Failure"	"Medium"	"Readjust Machine"	45
"Loud rattling and banging sounds are coming from assembler pistons."	"Mechanical Failure"	"Medium"	"Readjust Machine"	35
"There are cuts to the power when starting the plant."	"Electronic Failure"	"High"	"Full Replacement"	16200
"Fried capacitors in the assembler."	"Electronic Failure"	"High"	"Replace Components"	352
"Mixer tripped the fuses."	"Electronic Failure"	"Low"	"Add to Watch List"	55
"Burst pipe in the constructing agent is spraying coolant."	"Leak"	"High"	"Replace Components"	371
"A fuse is blown in the mixer."	"Electronic Failure"	"Low"	"Replace Components"	441
"Things continue to tumble off of the belt."	"Mechanical Failure"	"Low"	"Readjust Machine"	38

Figure 9.19 Heads of the table, which include description, category, urgency, resolution, and cost.

2) Transformation to Categorical Classes

Our goal is to classify events by the label in the "**Category**" column. To divide the data into classes, we need to convert these labels to categorical.

```
data.Category = categorical(data.Category);
```

3) Histogram of Class Distribution

View the distribution of the classes in the data using a histogram.

```
figure
histogram(data.Category);
xlabel("Class")
ylabel("Frequency")
title("Class Distribution")
```

Figure 9.20 shows the distribution of 480 data points among the four types of categorical classes.

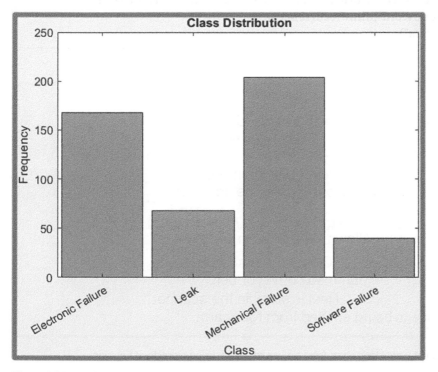

Figure 9.20   Distribution of the data among the failure types.

4) Data Partition

Partition the data into a training partition and a held-out partition for validation. Specify the holdout percentage to be 20%.

```
cvp = cvpartition(data.Category,'Holdout',0.2);
dataTrain = data(training(cvp),:);
dataValidation = data(test(cvp),:);
%Extract the text data and labels from the partitioned tables.
textDataTrain = dataTrain.Description;
textDataValidation = dataValidation.Description;
YTrain = dataTrain.Category;
YValidation = dataValidation.Category;
```

5) Creation of a Word Cloud from a Text Data

To check that the data is correctly imported, visualize the training text data using a word cloud. Figure 9.21 shows the **wordcloud** output. A **wordcloud** function creates word cloud chart from a given text data.

```
figure
% number of occurrences for the unique words.
[numOccurrences,uniqueWords] = histcounts(categorical(textDataTrain));
wordcloud(uniqueWords,numOccurrences);
title("Training Data")
```

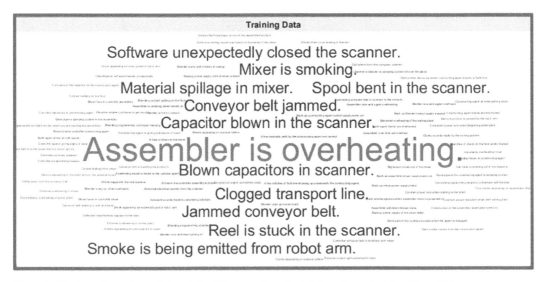

Figure 9.21 Creation of word cloud chart from a given text data, using **wordcloud** function.

6) Preprocess Text Data

We create a function that tokenizes, converts to lower case, and removes the punctuation of the text data. The function is called **preprocessText**. See Appendix A21.

Preprocess the training data and the validation data using the **preprocessText** function.

```
documentsTrain = preprocessText(textDataTrain);
documentsValidation = preprocessText(textDataValidation);
```

NOTE #5

If you attempt to run the above code, you may have the following error message: 'tokenizedDocument' requires Text Analytics Toolbox. As it says, you must install "**Text Analytic Toolbox**".

Notice that the tokenized document is a document represented as a collection of words (or tokens) which is used for text analysis. We can use tokenized documents to:

- Detect complex tokens in a text, such as web addresses, emoticons, emoji, and hashtags.
- Remove words such as stop words using the **removeWords** or **removeStopWords** functions.
- Perform word-level preprocessing tasks such as stemming or lemmatization using the **normalizeWords** function.
- Analyze word and n-gram frequencies using **bagOfWords** and **bagOfNgrams** objects.
- Add sentence and part-of-speech details using the **addSentenceDetails** and **addPartOfSpeechDetails** functions.
- Add entity tags using the **addEntityDetails** function.
- Add grammatical dependency details using the **addDependencyDetails** function.
- View details about the tokens using the **tokenDetails** function.
- For more examples, see MATLAB | Built-in Help | **tokenizedDocument**.

The function supports English, Japanese, German, and Korean text. To learn how to use **tokenizedDocument** for other languages, see Language Considerations.

---

7) View the Tokenized Train Data

View the first few preprocessed training documents. Thus, a sentence or a phrase is split into a number of tokens. If we examine **documentsTrain** variable in Workspace window, we will notice that it is made of 384 × 1 string array.

```
documentsTrain(1:5)
5 × 1 tokenizedDocument:
   9 tokens: items are occasionally getting stuck in the scanner spools
  10 tokens: loud rattling and banging sounds are coming from assembler pistons
  10 tokens: there are cuts to the power when starting the plant
   5 tokens: fried capacitors in the assembler
   4 tokens: mixer tripped the fuses
```

8) Convert Document to Sequences

To input the documents into an LSTM network, we use a word encoding to convert the documents into sequences of numeric indices. To create a word encoding, use the **wordEncoding** function. Again, it requires "**Text Analytic Toolbox**" installation.

```
enc = wordEncoding(documentsTrain);
```

Notice that **enc** is 1×1 **wordEncoding** struct which has two properties: '**NumWords**' with a value of *432* and '**Vocabulary**' with a value of *1×432* string array. If we explore '**Vocabulary**' vector, we will see that it is made of 432 non-repeating, unique words.

9) Sequence Length and Padding

The next conversion step is to pad and truncate documents so they are all the same length. The **trainingOptions** function provides options to pad and truncate input sequences automatically. However, these options are not well suited for sequences of word vectors. Instead, we manually pad and truncate the sequences. If we left-pad and truncate the sequences of word vectors, then the training might improve.

To pad and truncate the documents, first we choose a target length, and then truncate documents that are longer than it and left-pad documents that are shorter than it. For best results, the target length should be short without discarding large amounts of data. To find a suitable target length, view a histogram of the training document lengths.

```
documentLengths = doclength(documentsTrain);
figure
histogram(documentLengths)
title("Document Lengths")
xlabel("Length")
ylabel("Number of Documents")
```

Figure 9.22 shows the word length distribution. Most of the training documents have fewer than 10 tokens. Hence, we will use this as the target length for truncation and padding. Convert the documents to sequences of numeric indices using **doc2sequence** function. To truncate or left-pad the sequences to have length *10* or *11*. Set the '**Length**' option to *10*.

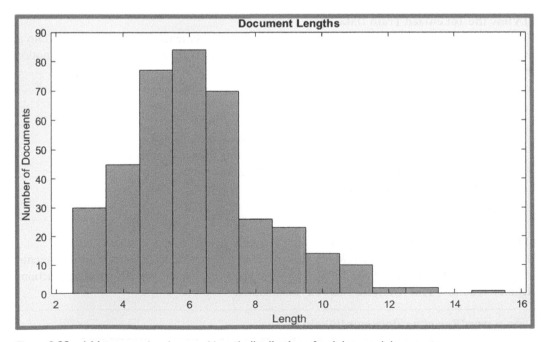

Figure 9.22 A histogram showing word length distribution of training word documents.

```
sequenceLength = 10;
XTrain = doc2sequence(enc,documentsTrain,'Length',sequenceLength);
%Convert the validation documents to sequences using the same options.
XValidation = doc2sequence(enc,documentsValidation,'Length',sequenceLength);
```

10) Create and Train LSTM Network

Define an LSTM network architecture. To input sequence data into the network, include a sequence input layer and set the input size to *1*. Next, include a word embedding layer of dimension *50* and the same number of words as the word encoding. Next, include an LSTM layer and set the number of hidden units to *80*. To use the LSTM layer for a sequence-to-label classification problem, set the output mode to *'last'*. Finally, add a fully connected layer with the same size as the number of classes, a softmax layer, and a classification layer.

```
inputSize = 1;
embeddingDimension = 50;
numHiddenUnits = 80;
numWords = enc.NumWords;
numClasses = numel(categories(YTrain));
layers = [ ...
    sequenceInputLayer(inputSize)
    wordEmbeddingLayer(embeddingDimension,numWords)
    lstmLayer(numHiddenUnits,'OutputMode','last')
    fullyConnectedLayer(2*numWords);
    fullyConnectedLayer(numClasses)
    softmaxLayer
    classificationLayer]
```

NOTE #6

A word embedding layer maps a sequence of word indices to embedding vectors and learns the word embedding during training. This layer requires Deep Learning Toolbox installation. Both **embeddingDimension** and **numWords** input parameters of the word embedding layer are specified as a positive integer. The value of **embeddingDimension** trades off between model under-fitting and over-fitting case. On the other hand, if the number of unique words in the training data is greater than **numWords**, then the layer will map the out-of-vocabulary words to the same vector. Thus, **numWords** value should be at least equal to the number of unique words in the train documents.

---

11) Specify Training Options

By default, **trainNetwork** uses a GPU if one is available. Otherwise, it uses the CPU. To specify the execution environment manually, use the '**ExecutionEnvironment**' name-value pair argument of **trainingOptions**. Training on a CPU can take significantly longer than training on a GPU. Training with a GPU requires Parallel Computing Toolbox™ and a supported GPU device.

```
options = trainingOptions('adam', ...
   'MiniBatchSize',16, ...
   'GradientThreshold',2, ...
   'Shuffle','never', ...
   'ValidationData',{XValidation,YValidation}, ...
   'Plots','training-progress', ...
   'Verbose',false);
```

12) Train the LSTM network

```
net = trainNetwork(XTrain,YTrain,layers,options);
```

Figure 9.23 shows the training model accuracy and loss as a function of the number of iterations or epochs.

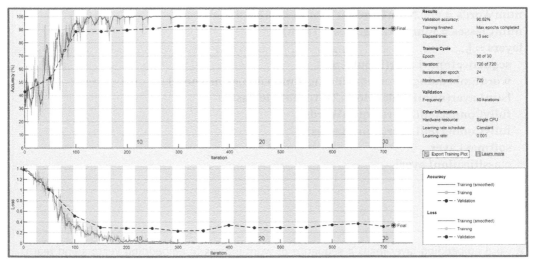

Figure 9.23 The LSTM RNN training model accuracy and loss as a function of the number of iterations or epochs.

13) Predict Using New Data
Classify the event type of three new reports. Create a string array containing the new reports.

```
reportsNew = [ ...
   "Coolant is pooling underneath sorter."
   "Sorter blows fuses at start up."
   "There are some very loud rattling sounds coming from the assembler."];
```

14) Preprocess the Test Text Data

Preprocess the test text data, using the preprocessing steps as in the training documents.

```
documentsNew = preprocessText(reportsNew);
```

15) Convert the Test Text Data to Sequences

Convert the test text data to sequences using **doc2sequence** function with the same options as when creating the training sequences. We use the original **enc,** 1×1 wordEncoding struct, created in step #8.

```
XNew = doc2sequence(enc,documentsNew,'Length',sequenceLength);
```

16) Classify the New Sequences

Classify the new sequences using the trained LSTM network.

```
labelsNew = classify(net,XNew)
```

The results are copied from Command Window.

```
labelsNew = 3 × 1 categorical array
    Leak
    Electronic Failure
    Mechanical Failure
```

Figure 9.24 shows workspace variables.

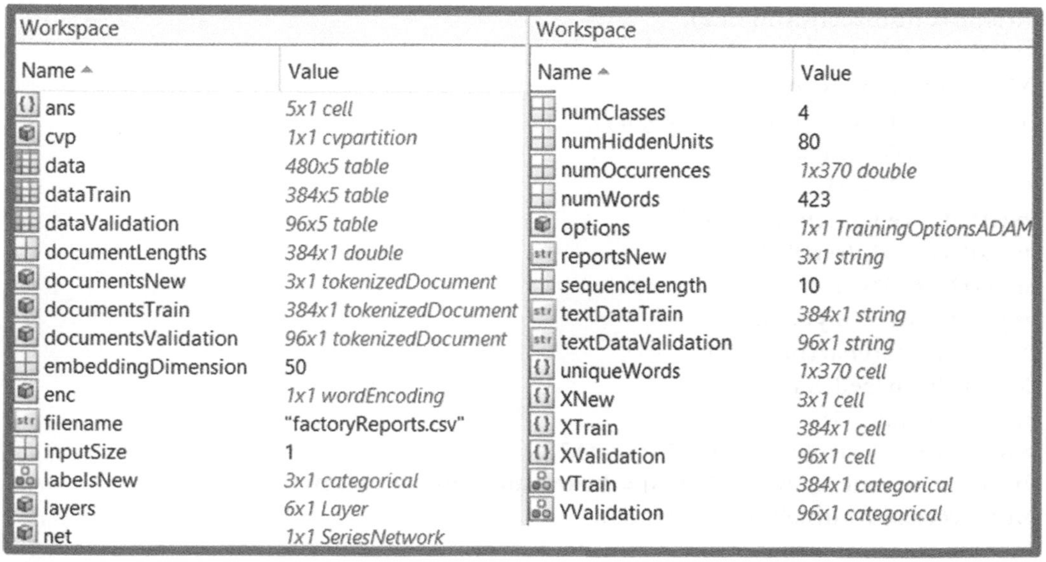

Workspace		Workspace	
Name ▲	Value	Name ▲	Value
ans	5x1 cell	numClasses	4
cvp	1x1 cvpartition	numHiddenUnits	80
data	480x5 table	numOccurrences	1x370 double
dataTrain	384x5 table	numWords	423
dataValidation	96x5 table	options	1x1 TrainingOptionsADAM
documentLengths	384x1 double	reportsNew	3x1 string
documentsNew	3x1 tokenizedDocument	sequenceLength	10
documentsTrain	384x1 tokenizedDocument	textDataTrain	384x1 string
documentsValidation	96x1 tokenizedDocument	textDataValidation	96x1 string
embeddingDimension	50	uniqueWords	1x370 cell
enc	1x1 wordEncoding	XNew	3x1 cell
filename	"factoryReports.csv"	XTrain	384x1 cell
inputSize	1	XValidation	96x1 cell
labelsNew	3x1 categorical	YTrain	384x1 categorical
layers	6x1 Layer	YValidation	96x1 categorical
net	1x1 SeriesNetwork		

Figure 9.24 The workspace variables.

Classify Text Data Using Deep Learning: Factory Equipment Failure Text Analysis – 2

Figure 9.25 shows an alternative code to carry out training for an LSTM RNN and then to test the trained RNN, using factory equipment failure data.

```
filename = "factoryReports.csv";
tbl = readtable(filename,'TextType','string');
numObservations = size(tbl,1);
numObservationsTrain = floor(0.7*numObservations);
numObservationsVal = numObservations - numObservationsTrain;
idx = randperm(numObservations);
idxTrain = idx(1:numObservationsTrain);
idxVal = idx(numObservationsTrain+1:end);
tblTrain = tbl(idxTrain,:);
tblVal = tbl(idxVal,:);
txtDTrain = tblTrain.Description;
YTrain = tblTrain.Category;
Ytrncat=categorical(tblTrain.Category);
TF = (txtDTrain(:,1)=="");
txtDTrain(TF) = [];
p = [".","?","!",",",";",":"];
txtDTrain = replace(txtDTrain,p," ");
txtDTrain = strip(txtDTrain);
mytxt= txtDTrain;
mytxt=lower(mytxt);
Xtnum=uint8(char(mytxt));
Xtemp4=num2cell(Xtnum,2);
txtDVal = tblVal.Description;
YVal = tblVal.Category;
YValcat=categorical(tblVal.Category);
TF2 = (txtDVal(:,1)=="");
txtDVal(TF2) = [];
txtDVal = replace(txtDVal,p," ");
txtDVal = strip(txtDVal);
mytxt2= txtDVal;
mytxt2=lower(mytxt2);
XVal2=uint8(char(mytxt2));
XVal4=num2cell(XVal2,2);
numFeatures = 1;
numClasses = size(categories(Ytrncat),1);
[numOccurrences,uniqueWords] = histcounts(categorical(mytxt));
numWords=2*numel(uniqueWords);
```

Figure 9.25 An alternative code to carry out training for an LSTM RNN, using factory equipment failure data and then test the trained RNN with a new report.

```
numHiddenUnits = numWords;
layers = [
  sequenceInputLayer(numFeatures)
  wordEmbeddingLayer(500, numWords,"Name","word-embedding")
  bilstmLayer(numHiddenUnits,"Name","biLSTM",'OutputMode','last')
     batchNormalizationLayer
  bilstmLayer(numHiddenUnits,"Name","biLSTM2",'OutputMode','last')
     fullyConnectedLayer(numWords,"Name","fc_1")
  batchNormalizationLayer
     reluLayer
     dropoutLayer(0.2)
  fullyConnectedLayer(numClasses,"Name","fc_2")
  softmaxLayer
  classificationLayer];
miniBatchSize = 30;
maxEpochs = 60;
options = trainingOptions('adam', ...
  'MiniBatchSize',miniBatchSize, ...
  'MaxEpochs',maxEpochs,...
  'Shuffle','never', ...
  'InitialLearnRate',0.0008, ...
  'Plots','training-progress', ...
       'ValidationData',{XVal4,YValcat}, ...
       'validationFrequency', floor(numel(YTrain)/miniBatchSize), ...
  'Verbose',false);
net = trainNetwork(Xtemp4,Ytrncat,layers,options);
reportsNew = [ ...
  "Coolant is pooling underneath sorter."
  "Sorter blows fuses at start up."
  "There are some very loud rattling sounds coming from the assembler."];
TF3 = (reportsNew(:,1)=="");
reportsNew(TF3) = [];
p = [".","?","!",",",";",":"];
reportsNew = replace(reportsNew,p," ");
reportsNew= strip(reportsNew);
mytxt4= reportsNew;
mytxt4=lower(mytxt4);
Xtnum6=uint8(char(mytxt4));
Xtemp6=num2cell(Xtnum6,2);
YPred = classify(net,Xtemp6,'MiniBatchSize',miniBatchSize)
```

Figure 9.25 (Cont'd)

Figure 9.26 shows the RNN training model accuracy and loss as a function of number of iterations (or epochs).

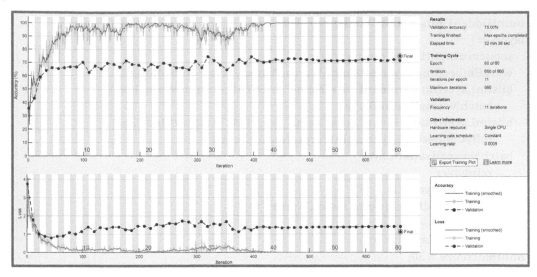

Figure 9.26 The RNN training model accuracy and loss as a function of number of iterations.

The result is copied from Command Window. The third statement is categorized as software failure. Apparently, the training step has to be improved and more words are to be added for the new report text.

YPred = 3×1 categorical array
    Leak
    Electronic Failure
    Software Failure

Figure 9.27 shows the generated workspace variables for the code of Figure 9.26.

Workspace		Workspace	
Name ▲	Value	Name ▲	Value
filename	"factoryReports.csv"	tbl	480x5 table
idx	1x480 double	tblTrain	336x5 table
idxTrain	1x336 double	tblVal	144x5 table
idxVal	1x144 double	TF	336x1 logical
layers	12x1 Layer	TF2	144x1 logical
maxEpochs	60	TF3	3x1 logical
miniBatchSize	30	txtDTrain	336x1 string
mytxt	336x1 string	txtDVal	144x1 string
mytxt2	144x1 string	uniqueWords	1x323 cell
mytxt4	3x1 string	validationFrequency	3
net	1x1 SeriesNetwork	Xtemp4	336x1 cell
numClasses	4	Xtemp6	3x1 cell
numFeatures	1	Xtnum	336x82 uint8
numHiddenUnits	646	Xtnum6	3x66 uint8
numObservations	480	XVal2	144x71 uint8
numObservationsTrain	336	XVal4	144x1 cell
numObservationsVal	144	YPred	3x1 categorical
numOccurrences	1x323 double	YTrain	336x1 string
numWords	646	Ytrncat	336x1 categorical
options	1x1 TrainingOptionsADAM	YVal	144x1 string
p	1x6 string	YValcat	144x1 categorical
reportsNew	3x1 string		

Figure 9.27 Workspace variables showing the accuracy %.

Word-by-Word Text Generation Using Deep Learning – 1

This section shows how to train a deep learning LSTM network to generate text word-by-word. Information is being created everywhere at a large scale on social media (i.e., WhatsApp, Twitter, and Facebook). Such text messages are highly unstructured. There is a need to create significant bits of knowledge from this information. Natural Language Processing (NLP) algorithms are used to manage human languages. Computational techniques represent and analyze language automatically. Learning sentence representation is fundamental to numerous natural language applications. Such models try to learn the fixed-length feature vector which encodes the semantic and syntactic properties of sentences. One of the popular approaches to train a sentence model is the encoder-decoder framework using Recurrent Neural Network (RNN) [3].

Deep learning technology has made a great progress, and text generation based on deep learning has received extensive attention. In the form of complex and diverse information, text information is a mainstream form of data, and its number is growing very fast. NLP algorithms are used to quickly and accurately extract and use effective information from massive text data [4].

ChatGPT (https://chat.openai.com/) stands for "Chat Generative Pre-trained Transformer." The "Chat" part signifies its focus on engaging in conversations and providing interactive dialogue. "GPT" refers to the underlying architecture, the Generative Pre-trained Transformer, which is a deep learning model known for its ability to generate coherent and contextually relevant text.

ChatGPT is an advanced language model developed by OpenAI that uses the GPT-3.5 architecture. It is designed to engage in human-like conversations and assist users in a wide range of tasks and inquiries. Trained on a diverse dataset, ChatGPT can provide information, answer questions, generate text, offer suggestions, and engage in interactive discussions, enabling users to interact with it in a natural and conversational manner, making it a powerful tool for communication and information retrieval. See my articles posted at: https://icarus-ai.enki.tech/blog/. Such articles are useful in helping beginners master the AI-based tool.

To train a deep learning network for word-by-word text generation, train a sequence-to-sequence LSTM network to predict the next word in a sequence of words. To train the network to predict the next word, specify the responses to be the input sequences shifted by one time step.

MATLAB reads a text from a website. It reads and parses the HTML code to extract the relevant text, then uses a custom mini-batch datastore **documentGenerationDatastore** to input the documents to the network as mini-batches of sequence data. The datastore converts documents to sequences of numeric word indices.

The deep learning network is an LSTM network that contains a word embedding layer. A mini-batch datastore is an implementation of a datastore with support for reading data in batches. We can use a mini-batch datastore as a source of training, validation, test, and prediction data sets for deep learning applications. Use mini-batch datastores to read out-of-memory data or to perform specific preprocessing operations when reading batches of data.

1) Load Training Data
Load the training data. Read the HTML code from one of MATLAB's public pages on deep learning. Obviously, you may read from any other source available on the world wide web.

```
url = 'https://americanliterature.com/author/hans-christian-andersen/short-story/
the-snow-queen';
code = webread(url);
```

2) Parse HTML Code

The HTML code contains the relevant text inside <p> (paragraph) elements. Extract the relevant text by parsing the HTML code using **htmlTree** function and then finding all the elements with element name "p". Notice that **htmlTree** is used to create a tree structure from an HTML string. It parses the HTML string and creates a hierarchical tree structure, with each node in the tree representing an HTML element. It can also be used to traverse the tree structure and extract data from the HTML elements. In general, **htmlTree** object represents a parsed HTML element or node. It extracts parts of interest using **findElement** function or **Children** property, and it can also extract text using **extractHTMLText** function.

The **findElement** function is used to search for specific HTML elements in a parsed HTML document. This function takes an HTML Document Object Model (DOM) element and a search criterion as input and returns an array of HTML elements that match the search criterion. In a typical HTML document, DOM is a hierarchical representation of an HTML document that allows programs, like MATLAB, to access and manipulate the content and structure of the document. An HTML DOM element is a part of an HTML document, such as a tag or an attribute, that can be accessed and manipulated using DOM. Each element in DOM is represented as an object, and these objects can be manipulated using JavaScript, or in the case of MATLAB, using HTML parsing functions.

Below, we will have an HTML main tree called **tree**. The **subtrees** variable returns the elements in tree matching the "p" selector. If we double-click on **subtrees** variable, we will notice that it is a hierarchical structure basically made of branches (parents) and leaves (children).

```
tree = htmlTree(code);
selector = "p";
subtrees = findElement(tree,selector);
```

3) Extract the Data

Extract the text data from the HTML **subtrees** using **extractHTMLText**.

```
textData = extractHTMLText(subtrees);
```

4) Remove Empty Paragraphs

Remove the empty paragraphs and view the first 10 remaining paragraphs.

```
textData(textData == "") = [];
textData(1:10)
```

Figure 9.28 shows a portion of the first non-empty extracted paragraphs.

```
10×1 string array

"FIRST STORY. Which Treats of a Mirror and of the Splinters"
"Now then, let us begin. When we are at the end of the story, we shall know more than
"Once upon a time there was a wicked sprite, indeed he was the most mischievous of al
""That's glorious fun!" said the sprite. If a good thought passed through a man's min
"SECOND STORY. A Little Boy and a Little Girl"
"In a large town, where there are so many houses, and so many people, that there is n
"The children's parents had large wooden boxes there, in which vegetables for the kit
""It is the white bees that are swarming," said Kay's old grandmother."
""Do the white bees choose a queen?" asked the little boy; for he knew that the honey
""Yes," said the grandmother, "she flies where the swarm hangs in the thickest cluste
```

Figure 9.28 Non-empty paragraphs, as extracted by **extractHTMLText** function.

5) Visualize the Text Data

Visualize the text data in a word cloud.

```
figure
wordcloud(textData);
title("The Snow Queen")
```

Figure 9.29 shows the word cloud of the extracted text.

Figure 9.29 The word cloud of extracted text.

6) Prepare Data for Training

Create a datastore that contains the data for training using **documentGenerationDatastore** function. Here is an outline:

1) When tokenizing a document, or a piece of text, a special token called the "start of text" token is assigned an index of 0. This token represents the beginning of the document.
2) The 'start of text' token is given by the string "**startOfText**". The actual representation of the "start of text" token in the tokenized sequence is the string "**startOfText**". This string is used as a placeholder to identify the beginning of the document during tokenization.
3) On the other hand, the words in the response (i.e., Y variable) are tokenized into categorical sequences, where each word is represented by a unique token. However, the tokens are shifted by one position compared with the original predictor (i.e., X variable) words. This means that the token for the first word in the response (Y) corresponds to the second word in the original predictor (X), the token for the second word corresponds to the third word, and so on. This shifting ensures that the tokenized sequences align correctly with the original text when generating or processing responses.

Tokenize the text data using **tokenizedDocument** function. We can customize the mini-batch datastore specified by **documentGenerationDatastore.m** to fit our own data by customizing the functions. This m-file can be found in Appendix A22. For an example showing how to create one's own custom mini-batch datastore, see MATLAB | Built-in Help | Develop Custom Mini-Batch Datastore.

```
documents = tokenizedDocument(textData);
```

Create a document generation datastore using the tokenized documents.

```
ds = documentGenerationDatastore(documents);
```

Using MATLAB **documentGenerationDatastore** function, the predictor and response data are embedded within the **documents** that we provide as input to the function. The **documents** input is expected to be a cell array of character vectors or a string array representing the text data. Each element of the **documents** array represents a document or a text sample.

When we create a **documentGenerationDatastore** using **documentGenerationDatastore(documents)**, the function automatically assumes that the last element of each document is the response or target data, and the preceding elements are the predictor or input data. We can access the predictor and response data from the **documentGenerationDatastore** object using the **read** function. For instance, we can retrieve the predictor and response data as follows:

```
ds = documentGenerationDatastore(documents);
while hasdata(ds)
    data = read(ds);
    predictorData = data(:,1); % Access the predictor data
    responseData = data(:,2); % Access the response data
% Perform further processing or analysis on the predictor and response data
end
reset(ds); %if the user would like to re-run read statement.
```

To reduce the amount of padding added to the sequences, sort the documents in the datastore by sequence length.

```
ds = sort(ds);
```

7) Create and Train LSTM Network

Define the LSTM network architecture. To input sequence data into the network, include a sequence input layer and set the input size to *1*. Next, include a word embedding layer of dimension *200* and the same number of words as the word encoding. Next, include an **LSTM** layer and specify the hidden size to be *200*. Finally, add a fully connected (**FC**) layer with the same size as the number of classes, a **softmax** layer, and a **classification** layer. The number of classes is the number of words in the vocabulary plus an extra class for the "**end of text**" class.

```
inputSize = 1;
embeddingDimension = 200;
numWords = numel(ds.Encoding.Vocabulary);
numClasses = numWords + 1;
layers = [
  sequenceInputLayer(inputSize)
  wordEmbeddingLayer(embeddingDimension,numWords)
  lstmLayer(200)
  dropoutLayer(0.2)
  fullyConnectedLayer(numClasses)
  softmaxLayer
  classificationLayer];
```

Specify the training options. Specify the solver to be '**adam**'. Train for *300* epochs with learn rate *0.01*. Set the mini-batch size to *32*. To keep the data sorted by sequence length, set the '**Shuffle**' option to '*never*'. To monitor the training progress, set the '**Plots**' option to '*training-progress*'. To suppress verbose output, set '**Verbose**' to *false*.

```
options = trainingOptions('adam', ...
  'MaxEpochs',300, ...
  'InitialLearnRate',0.01, ...
  'MiniBatchSize',32, ...
  'Shuffle','never', ...
  'Plots','training-progress', ...
  'Verbose',false);
```

Train the network using **trainNetwork** function.

```
net = trainNetwork(ds,layers,options);
```

Figure 9.30 shows RNN model accuracy and loss as a function of number of iterations.

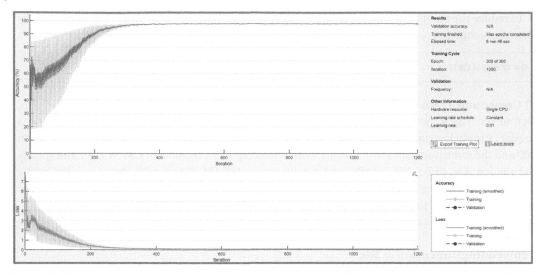

Figure 9.30 RNN model accuracy and loss as a function of number of iterations.

8) Generate New Text

Generate the first word of the text by sampling a word from a probability distribution according to the first words of the text in the training data. Generate the remaining words by using the trained LSTM network to predict the next time step using the current sequence of generated text. Keep generating words one-by-one until the network predicts the "end of text" word. To make the first prediction using the network, input the index that represents the "start of text" token. Find the index by using **word2ind** function with the word encoding used by the document datastore.

```
enc = ds.Encoding;
wordIndex = word2ind(enc,"startOfText")
```

For the remaining predictions, sample the next word according to the prediction scores of the network. The prediction scores represent the probability distribution of the next word. Sample the words from the vocabulary given by the class names of the output layer of the network.

```
vocabulary = string(net.Layers(end).Classes);
```

Make predictions word by word using **predictAndUpdateState** function. For each prediction, input the index of the previous word. Stop predicting when the network predicts the end of text word or when the generated text is 800 characters long. For large collections of data, long sequences, or large networks, predictions on GPU are usually faster to compute than predictions on CPU. Otherwise, predictions on CPU are usually faster to compute. For single time step predictions, use the CPU. To use the CPU for prediction, set the 'ExecutionEnvironment' option of **predictAndUpdateState** function to '*cpu*'.

```
generatedText = "";
maxLength = 800;% It can be adjusted by the user.
```

```
while strlength(generatedText) < maxLength
  % Predict the next word scores.
  [net,wordScores] = predictAndUpdateState(net,wordIndex,'ExecutionEnvironment',
'cpu');
   % Sample the next word.
   newWord = datasample(vocabulary,1,'Weights',wordScores);
   % Stop predicting at the end of text.
   if newWord == "EndOfText"
     break
   end
   % Add the word to the generated text.
   generatedText = generatedText + " " + newWord;
   % Find the word index for the next input.
   wordIndex = word2ind(enc,newWord);
end
```

The generation process introduces whitespace characters between each prediction, which means that some punctuation characters appear with unnecessary spaces before and after. Reconstruct the generated text by removing the spaces before and after the appropriate punctuation characters.

9) Refinement of Text

Remove the spaces that appear *before* the specified punctuation characters.

```
punctuationCharacters = ["." ";" "," """ ")" ":" "?" "!"];
generatedText = replace(generatedText," " + punctuationCharacters,punctuationCha
racters);
```

Remove the spaces that appear *after* the following leading punctuation characters: ["(" """].

```
punctuationCharacters = ["(" """];
generatedText = replace(generatedText,punctuationCharacters + " ",punctuation-
Characters)
```

The generated text is copied from Command Window.

```
generatedText = "It was quite early; she kissed her old grandmother, who was still asleep, put
on her red shoes, and went alone to the river."
```

Below is a portion of "**code**" text, where it shows the relevant text. The "**code**" variable is created by: >>**code=webread(url);**

```
<p>It was quite early; she kissed her old grandmother, who was still asleep, put on her red
shoes, and went alone to the river. </p>
```

10) Regeneration of Text

To generate multiple pieces of text, reset the network state between generations using **resetState** function. Otherwise, it will generate nothing as the status, from a previous run, brings the cursor to the end of the scanned text.

net = resetState(net);

After reset, start then from step #8 above to generate a new text.

Figure 9.31 shows workspace variables, including the URL address and a portion of the generated text.

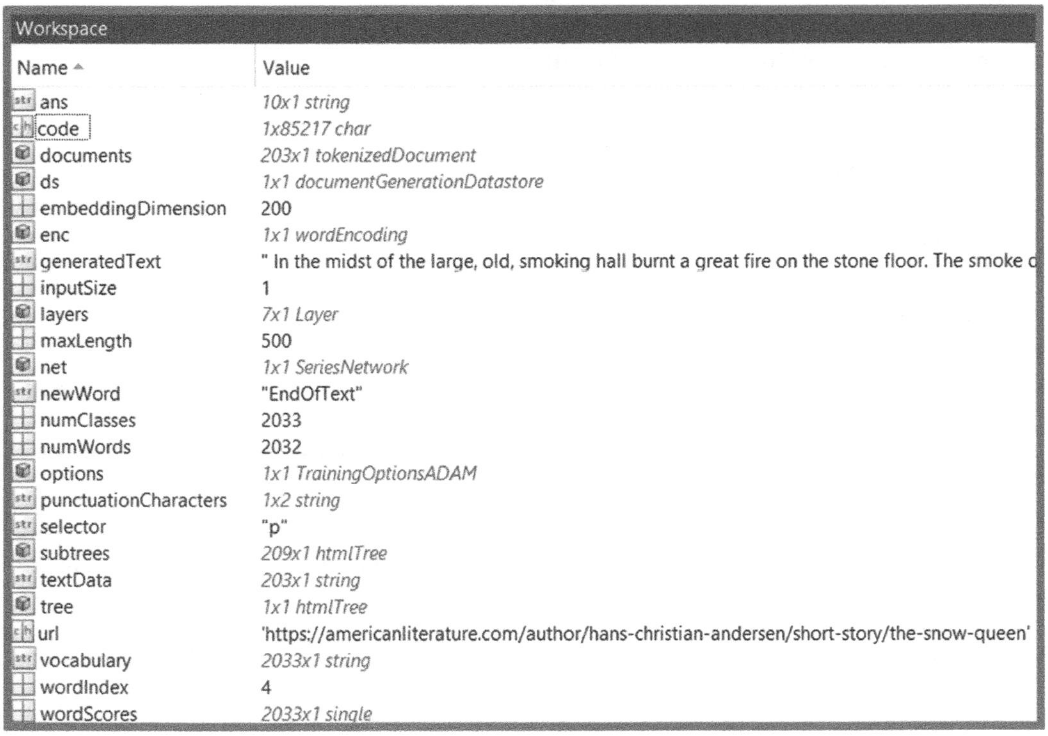

Figure 9.31 Workspace variables, including the URL address and a portion of the generated text.

NOTE #7

Using the above outlined method, some Uniform Resource Locaters (URL) have Hyper-Text Markup Language (HTML) pages where upon extraction gives an error; for example, the URL's HTML page used in the next section will give an error as shown below:
  Error using trainNetwork
  Index exceeds the number of array elements. Index must not exceed 7.
  Caused by:
    Error using documentGenerationDatastore/read
    Index exceeds the number of array elements. Index must not exceed 7.
  Do not panic; use the second approach that is covered in the following section.

Word-by-Word Text Generation Using Deep Learning – 2

Figure 9.32 shows an alternative code for performing the same duty as in the previous section.

Alternatively, one may read text data from a text file, instead of reading from a web source, and continue afterward. The insertion where to read from a text file is commented within the code of Figure 9.32.

```
url='https://americanliterature.com/childrens-stories/jack-and-the-beanstalk';
code = webread(url);
tree = htmlTree(code);
selector = "p";
subtrees = findElement(tree,selector);
textData = extractHTMLText(subtrees);
textData(textData == "") = [];
% Alternatively, one may read a text data from a text file, instead from a web source,
as shown here in the %commented statements, and then continue afterward.
%filename = "sonnets.txt";
%textData = fileread(filename);
%textData = replace(textData," ","");
%textData = split(textData,[newline newline]);
startOfTextCharacter = compose("\x0002");
whitespaceCharacter = compose("\x00B7");
endOfTextCharacter = compose("\x2403");
newlineCharacter = compose("\x00B6");
textData = startOfTextCharacter + textData;
textData = replace(textData,[" " newline],[whitespaceCharacter newlineCharacter]);
uniqueCharacters = unique([textData{:}]);
numUniqueCharacters = numel(uniqueCharacters);
numDocuments = numel(textData);
XTrain = cell(1,numDocuments);
YTrain = cell(1,numDocuments);
for i = 1:numel(textData)
  characters = textData{i};
  sequenceLength = numel(characters);
  % Get indices of characters.
  [~,idx] = ismember(characters,uniqueCharacters);
  % Convert characters to vectors.
  X = zeros(numUniqueCharacters,sequenceLength);
  for j = 1:sequenceLength
    X(idx(j),j) = 1;
  end
  % Create vector of categorical responses with end of text character.
  charactersShifted = [cellstr(characters(2:end)')' endOfTextCharacter];
  Y = categorical(charactersShifted);
  XTrain{i} = X;
  YTrain{i} = Y;
```

Figure 9.32 An alternative code for extracting and generating a text from an HTML page.

```
end
inputSize = size(XTrain{1},1);
numHiddenUnits = 300;
numClasses = numel(categories([YTrain{:}]));
embeddingDimension = 300;
tokendocs = preprocessText(textData);
enc = wordEncoding(tokendocs);
numWords = enc.NumWords;
layers = [
  sequenceInputLayer(inputSize)
  lstmLayer(numHiddenUnits,'OutputMode','sequence')
  fullyConnectedLayer(numClasses)
  softmaxLayer
  classificationLayer];
options = trainingOptions('adam', ...
  'MaxEpochs',600, ...
  'InitialLearnRate',0.01, ...
  'GradientThreshold',2, ...
  'MiniBatchSize',77,...
  'Shuffle','every-epoch', ...
  'Plots','training-progress', ...
  'Verbose',false);
net = trainNetwork(XTrain,YTrain,layers,options);
generatedText = generateText(net,uniqueCharacters,startOfTextCharacter, ...
newlineCharacter,whitespaceCharacter,endOfTextCharacter)
```

Figure 9.32   (Cont'd)

Figure 9.33 shows the RNN model accuracy and loss versus the number of iterations.

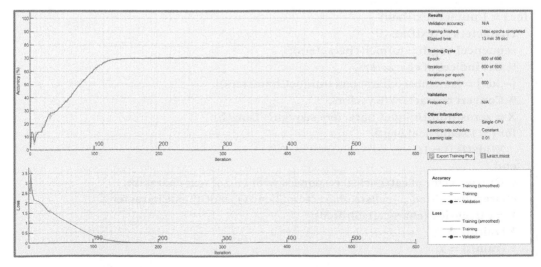

Figure 9.33   The RNN model accuracy and loss versus the number of iterations.

The generated text is copied from Command Window.

---

generatedText =
"While he was eating, the giant came home. The giant was very big and looked very fearsome. Jack was terrified and went and hid inside. The giant cried, 'Fee-fi-fo-fum, I smell the blood of an Englishman. Be he alive, or be he dead, I'll grind his bones to make my bread!' The wife said, 'There is no boy in here!' So, the giant ate his food and then went to his room. He took out his sacks of gold coins, counted them and kept them aside. Then he went to sleep. In the night, Jack crept out of his hi"

Below is a portion of the HTML-extracted text and how it looks, which pertains to the above generated text paragraph.

"While·he·was·eating,·the·giant·came·home.··The·giant·was·very·big·and·looked·very·fearsome. ·Jack·was·terrified·and·went·and·hid·inside.··The·giant·cried,·'Fee-fi-fo-fum,·I·smell·the·blood· of·an·Englishman.··Be·he·alive,·or·be·he·dead,·I'll·grind·his·bones·to·make·my·bread!'··The·wife ·said,·'There·is·no·boy·in·here!'·So,·the·giant·ate·his·food·and·then·went·to·his·room.··He·took· out·his·sacks·of·gold·coins,·counted·them·and·kept·them·aside.··Then·he·went·to·sleep.··In·the· night,·Jack·crept·out·of·his·hiding·place,·took·one·sack·of·gold·coins·and·climbed·down·the·be anstalk.··At·home,·he·gave·the·coins·to·his·mother.··His·mother·was·very·happy·and·they·lived· well·for·sometime."

Figure 9.34 shows the workspace variables, including the URL address and a portion of the generated text.

Name ▲	Value		Name ▲	Value
characters	'DJack·and·his·mother·were·now·very·rich·and·they·lived·happily·ever·after.'		numUniqueCharacters	49
charactersShifted	1x74 string		numWords	90
code	1x35711 char		options	1x1 TrainingOptionsADAM
embeddingDimension	300		selector	"p"
enc	1x1 wordEncoding		sequenceLength	74
endOfTextCharacter	"\"		startOfTextCharacter	"□"
generatedText	"While he was eating, the giant came home. The giant was very big and I		subtrees	9x1 htmlTree
i	7		textData	7x1 string
idx	1x74 double		tokendocs	7x1 tokenizedDocument
inputSize	49		tree	1x1 htmlTree
j	74		uniqueCharacters	'DL-.?ABCDEFHULOSTWYabcdefghiklmnopqrstuvwxy."'"'
layers	5x1 Layer		url	'https://americanliterature.com/childrens-stories/jack-and-the-beanstalk'
net	1x1 SeriesNetwork		whitespaceCharacter	"."
newlineCharacter	"¶"		X	49x74 double
numClasses	49		XTrain	1x7 cell
numDocuments	7		Y	1x74 categorical
numHiddenUnits	300		YTrain	1x7 cell

Figure 9.34 Workspace variables, including the URL address and a portion of the generated text.

Train Network for Time Series Forecasting Using Deep Network Designer (DND)

In this section, we show how to forecast time series data by training an LSTM network in DNN. DNN allows us to interactively create and train deep neural networks for sequence classification and regression tasks. To forecast the values of future time steps of a sequence, we can train a sequence-to-sequence regression LSTM network, where the responses are the training sequences with values shifted by one time step. That is, at each time step of the input sequence, the LSTM network learns to predict the value of the next time step. We will use the data set **chickenpox_ dataset**. We then create and train an LSTM network to forecast the number of chickenpox cases, given the number of cases in previous months.

1) Load Sequence Data

To begin, we need to load the provided sample data known as "**chickenpox_dataset**". This dataset consists of a solitary time series, where each time step represents a month, and the values represent the number of cases. The output is in the form of a cell array, where each element represents a single time step. Our task is to transform the data into a row vector of type double by reshaping it accordingly. The data set is available from within MATLAB itself.

```
data = chickenpox_dataset;
data = [data{:}];
figure
plot(data)
xlabel("Month")
ylabel("Cases")
title("Monthly Cases of Chickenpox")
```

Figure 9.35 shows the monthly cases of chickenpox.

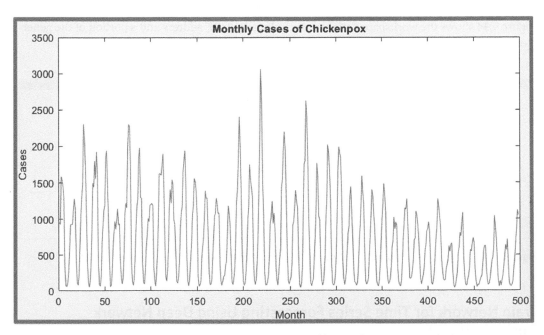

Figure 9.35 Chickenpox cases as a function of time, expressed in months

2) Data Partition

Partition the training and test data. Train on the first 80% of the sequence, 10% for validation, and test on the last 10%.

```
num1 = floor(0.8*numel(data));
num2 = floor(0.9*numel(data));
dataTrain = data(1:num1);
dataVal=data(num1+1:num2);
dataTest = data(num2+1:end);
```

3) Standardize Data

For a better fit and to avoid divergence upon training, we will standardize both the training and validation data to have zero mean and unit variance, based on what we statistically call population properties. For prediction using **dataTest** set, we must standardize the test data using the same parameters as those of the population (i.e., entire) data.

```
mu = mean(data);
sig = std(data);
dataTrainStandardized = (dataTrain - mu) / sig;
dataValStandardized = (dataVal - mu) / sig;
```

4) Prepare Predictors and Responses

To forecast the values of future time steps of a sequence, specify the responses as the training sequences with values shifted by one time step. That is, at each time step of the input sequence, the LSTM network learns to predict the value of the next time step.

```
XTrain = dataTrainStandardized(1:end-1);
YTrain = dataTrainStandardized(2:end);
XVal = dataValStandardized(1:end-1);
YVal = dataValStandardized(2:end);
```

To train the network using DNN, we convert the training data to a datastore object. We use **arrayDatastore** to convert the training data predictors and responses into **ArrayDatastore** objects. We then use **combine** function to combine the two datastores.

```
adsXTrain = arrayDatastore(XTrain);
adsYTrain = arrayDatastore(YTrain);
cdsTrain = combine(adsXTrain,adsYTrain);
adsXVal = arrayDatastore(XVal);
adsYVal = arrayDatastore(YVal);
cdsVal = combine(adsXVal,adsYVal);
```

5) Define LSTM Network Architecture

To create the LSTM network architecture, we will use DNN app. DNN app lets us build, visualize, edit, and train deep learning networks. Launch DND as explained in Chapter 7. | **Deep Network Designer (DND)** section.

6) Selection of Network Type

From the top menu, click on "**New**" button. Doing so will open the main windows of DNN app. From, the left pane (i.e., Layer Library), add the following layer as shown in Figure 9.36. Their properties are also shown on the right side.

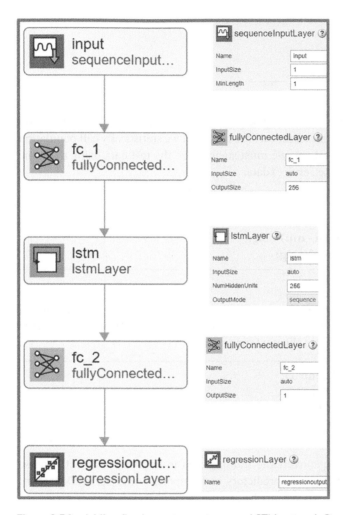

Figure 9.36 Adding five layers to create a new LSTM network. Properties are shown right.

The properties of the layers are adjusted so that they are suitable for the chickenpox data set. This data has a single input feature and a single output feature. Select "**sequenceInputLayer**" and set the "**InputSize**" to *1*. Select the last "**fullyConnectedLayer**" and set the "**OutputSize**" to *1*, as shown in Figure 9.36.

Check your network by clicking on "**Analyze**" button. The network will be ready for training if Deep Learning Network Analyzer reports zero errors, as shown in Figure 9.37.

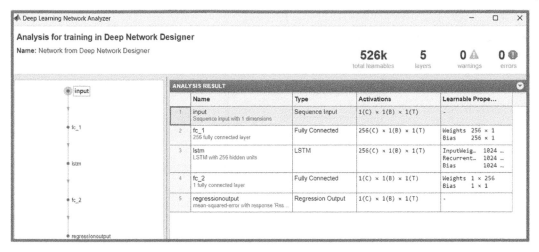

Figure 9.37 Checking for bugs using DND "**Analyze**" button.

7) Import Data

To import the training datastore, select the "**Data**" tab | click on "**Import Data**" button | and select "**Import Datastore**" sub-menu. Select *cdsTrain* as the training data and *cdsVal* as the validation data, as shown in Figure 9.38.

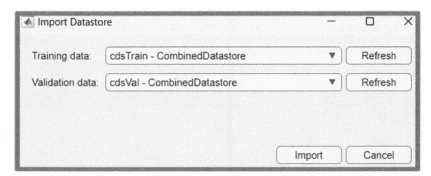

Figure 9.38 Importing combined array data stores for both training and validation.

8) Specify Training Options

On the "**Training**" tab, click on "**Training Options**" button. For example, set "**Solver**" to *adam*, "**InitialLearnRate**" to *0.0008*, "**MiniBatchSize**" to *120*, and "**MaxEpochs**" to *300*. To avoid the gradient blow-up condition, set the "**GradientThreshold**" to *1*. Figure 9.39 shows all changes. Close the "**Training Options**" window.

For more information about setting the training options, see "**trainingOptions**" in offline, built-in MATLAB help.

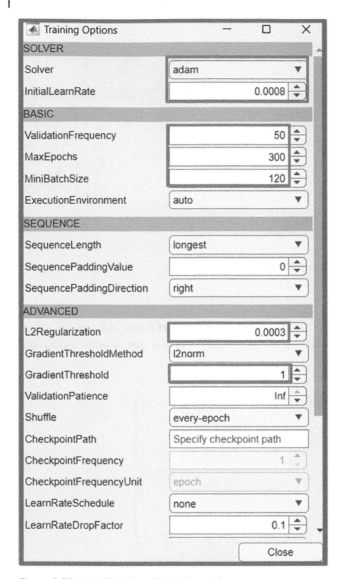

Figure 9.39 Modification of training options.

9) Train Network

Click on "**Train**" button. DNN displays a plot showing the training progress. The plot shows mini-batch loss and accuracy, validation loss and accuracy, and additional information on the training progress, as shown in Figure 9.40.

Once training is complete and under the "**Training**" tab, export the trained network by clicking on "**Export**" tab | "**Export**" button followed by "**Export Trained Network and Results**" sub-menu. The DNN pop-up window tells us that the trained network is saved as the "**trainedNetwork_1**" variable to the workspace. So does the case with its results. At this stage, we can save the trained network to a *.mat file so that it can be re-used later.

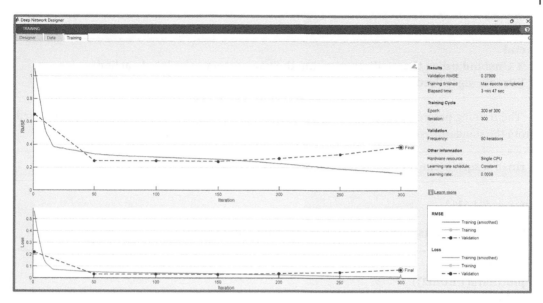

Figure 9.40   LSTM network accuracy and loss for both training and validation step.

10) Forecast Future Time Steps

Test the trained network by forecasting multiple time steps in the future. Use the **predictAndUp-dateState** function to predict time steps one at a time and update the network state at each prediction. For each prediction, use the previous prediction as input to the function.

Standardize the test data using the same parameters as those of the population data.

```
dataTestStandardized = (dataTest − mu) / sig;
XTest = dataTestStandardized(1:end−1);
YTest = dataTest(2:end);
```

To initialize the state of the network, we begin by making predictions on the training data, **XTrain**. Following that, the first prediction is made using the last time step of the training response, **YTrain(end)**. We then proceed to iterate through the remaining predictions, using the previous prediction as input for the **predictAndUpdateState** function.

When dealing with extensive data collections, lengthy sequences, or large networks, computing predictions on a GPU typically offers faster processing compared to using a CPU. However, in cases where the data size is smaller or single time step predictions are required, utilizing the CPU generally results in faster computation. Therefore, it is recommended to employ the CPU for single time step predictions. To use the CPU for prediction, set the '**ExecutionEnvironment**' option of **predictAndUpdateState** to *'cpu'*.

```
net = predictAndUpdateState(trainedNetwork_1,XTrain);
[net,YPred] = predictAndUpdateState(net,YTrain(end));
numTimeStepsTest = numel(XTest);
for i = 2:numTimeStepsTest
   [net,YPred(:,i)] = predictAndUpdateState(net,YPred(:,i-1),'ExecutionEnvironment','cpu');
```

(Continued)

(Continued)
end **% Unstandardize the predictions using the parameters calculated earlier.** **YPred = sig*YPred + mu;**

The training progress plot (Figure 9.40) reports the root-mean-square error (RMSE) calculated from the standardized data. Calculate RMSE from the unstandardized predictions.

rmse = sqrt(mean((YPred-YTest).^2))

rmse = 404.6362

11) Plot the Forecasted Cases with the Training Time Series

figure **plot(data(1:num2))** **hold on** **idx = num2:(num2+numTimeStepsTest);** **plot(idx,[data(num2) YPred],'.--')** **hold off** **xlabel("Month")** **ylabel("Cases")** **title("Forecast")** **legend(["Observed" "Forecast"])**

Figure 9.41 shows the observed and the forecasted number of chickenpox cases. The forecasting is only made for the tested dataset time zone.

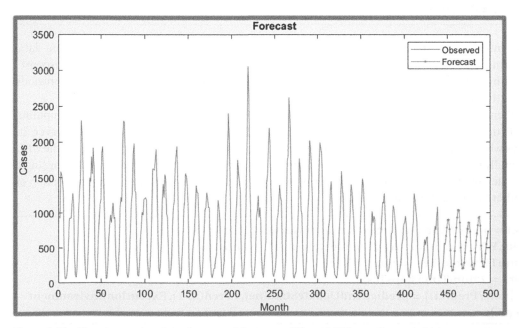

Figure 9.41 The observed and the forecasted (between 450- and 500-month zone) chickenpox cases.

12) Compare the Forecasted Cases with the Test Data

```
figure
subplot(2,1,1)
plot(YTest)
hold on
plot(YPred,'.–')
hold off
legend(["Observed" "Forecast"])
ylabel("Cases")
title("Forecast")
subplot(2,1,2)
stem(YPred – YTest)
xlabel("Month")
ylabel("Error")
title("RMSE = " + rmse)
```

Figure 9.42 shows the observed versus forecasted chickenpox cases as a function of time within the test data time zone (*top*) and the error defined as (YPred –YTest) = (YForecast – YObserved) (*bottom*).

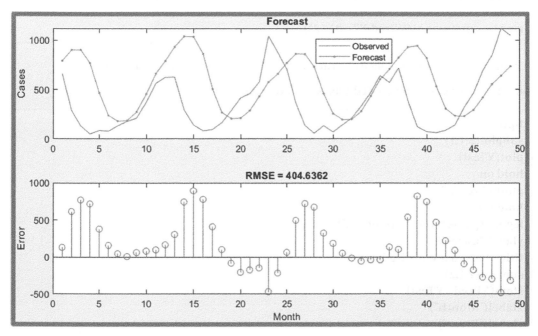

Figure 9.42 The observed versus forecasted chickenpox cases as a function of time within the test data time zone (*top*) and the error defined as (YPred – YTest) = (YForecast – YObserved) (*bottom*).

13) Update Network State with Observed Values

If we have an access to the actual values of time steps between predictions, then we can update the network state with the observed values instead of the predicted values.

First, initialize the network state. To make predictions on a new sequence, reset the network state using **resetState**. Resetting the network state prevents previous predictions from affecting the predictions on the new data. Then initialize the network state by predicting on the training data.

```
net = resetState(net);
net = predictAndUpdateState(net,XTrain);
```

Predict on each time step. For each prediction, predict the next time step using the observed value of the previous time step. Set the '**ExecutionEnvironment**' option of **predictAndUpdateState** to '*cpu*'.

```
YPred = [];
numTimeStepsTest = numel(XTest);
for i = 1:numTimeStepsTest
   [net,YPred(:,i)] = predictAndUpdateState(net,XTest(:,i),'ExecutionEnvironment','cpu');
end
%Unstandardize the predictions using the parameters calculated earlier.
YPred = sig*YPred + mu;
```

Calculate the root-mean-square error (RMSE) for the unstandardized data.

```
rmse = sqrt(mean((YPred-YTest).^2))
```

rmse = 201.4239 (after update) < rmse = 404.6362 (before update; see above)

14. Re-compare the Forecasted Cases with the Test Data

```
figure
subplot(2,1,1)
plot(YTest)
hold on
plot(YPred,'.-')
hold off
legend(["Observed" "Predicted"])
ylabel("Cases")
title("Forecast with Updates")
subplot(2,1,2)
stem(YPred - YTest)
xlabel("Month")
ylabel("Error")
title("RMSE = " + rmse)
```

Figure 9.43 shows the observed and forecasted chickenpox cases but with update status in between; i.e., updating the network state with the observed instead of the predicted values, using **predictAndUpdateState** function. Thus, the network predicts on each time step. For each prediction, the net predicts the next time step using the observed value of the previous time step.

Notice here that the predictions are more accurate when updating the network state with the observed instead of the predicted values.

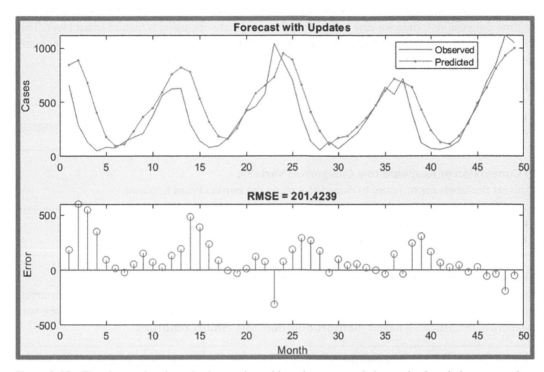

Figure 9.43 The observed and precited cases, but with update status of observed values in between, using **predictAndUpdateState** function (*top*) and the error defined as (YPred – YTest) = (YForecast – YObserved) (*bottom*).

Figure 9.44 shows workspace variables created by running the previous code.

Workspace			Workspace	
Name ▲	Value		Name ▲	Value
adsXTrain	1x1 ArrayDatastore		mu	732.4076
adsXVal	1x1 ArrayDatastore		net	1x1 SeriesNetwork
adsYTrain	1x1 ArrayDatastore		num1	398
adsYVal	1x1 ArrayDatastore		num2	448
cdsTrain	1x1 CombinedDatastore		numTimeStepsTest	49
cdsVal	1x1 CombinedDatastore		rmse	201.4239
data	1x498 double		sig	589.7333
dataTest	1x50 double		trainedNetwork_1	1x1 SeriesNetwork
dataTestStandardized	1x50 double		trainInfoStruct_1	1x1 struct
dataTrain	1x398 double		XTest	1x49 double
dataTrainStandardized	1x398 double		XTrain	1x397 double
dataVal	1x50 double		XVal	1x49 double
dataValStandardized	1x50 double		YPred	1x49 double
i	49		YTest	1x49 double
idx	1x50 double		YTrain	1x397 double
layer1	5x1 Layer		YVal	1x49 double

Figure 9.44 The workspace variables created by running the previous code.

Train Network with Numeric Features

If we have a data set of numeric features (for example a collection of numeric data without spatial or time dimensions), then we can train a deep learning network using a feature input layer. The following data pertain to "Turbofan Engine Degradation Simulation Data Set". For more information, please refer to Saxena et al. [5] and Saxena and Goebel [6].

1) Read the Input Data

```
filename = "transmissionCasingData.csv";
tbl = readtable(filename,'TextType','String');
```

2) Conversion of Response to a Categorical Variable
Convert the labels for response to categorical using the **convertvars** function.

```
GearCond = "GearToothCondition";
tbl = convertvars(tbl,GearCond,'categorical');
```

3) Conversion of Predictors to Categorical Variables
Convert the categorical predictors to categorical using **convertvars** function by specifying a string array containing the names of all the categorical input variables. In this data set, there are two categorical features with names "**SensorCondition**" and "**ShaftCondition**".

```
CatInNames = ["SensorCondition" "ShaftCondition"];
tbl = convertvars(tbl,CatInNames,'categorical');
```

4) Conversion of Categorical to One-hot Vector
Let us loop over the categorical input variables. For each variable:

- We convert the categorical values to one-hot encoded vectors using **onehotencode** function. The outputs of **onehotencode** are categorical variables as binary vectors. Each category is represented by a binary vector where all elements are zero except for the index corresponding to the category, which is set to one.
- Add the one-hot vectors to the table using **addvars** function. Specify to insert the vectors after the column containing the corresponding categorical data.
- Remove the corresponding column containing the categorical data.

```
for i = 1:numel(CatInNames)
  ColName = CatInNames(i);
  OHCode = onehotencode(tbl(:,ColName));
  tbl = addvars(tbl,OHCode,'After',ColName);
  tbl(:,ColName) = [];
end
```

Figure 9.45 shows the results of conversion to one-hot vector and removal of the corresponding column. Notice that the original columns #19 and #20 were replaced by columns #19 (OHCode) and #20 (OHCode_1), respectively. In both columns, the 1×2 mini-table has either *0* or *1* value for the two mini-table columns. For example, column #20 has originally "**No Shaft Wear**" or "**Shaft Wear**" categorical value. The first two rows will assume *1* for "**No Shaft Wear**" and *0* for "**Shaft Wear**" column; on the contrary, the third row will assume *0* for "**No Shaft Wear**" and *1* for "**Shaft Wear**" column.

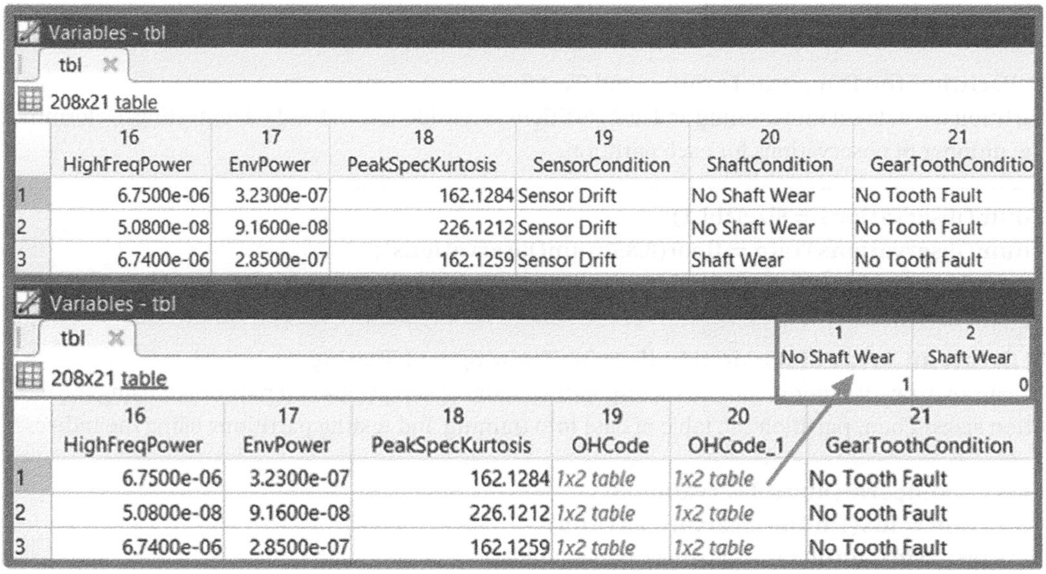

Variables - tbl

tbl

208x21 table

	16 HighFreqPower	17 EnvPower	18 PeakSpecKurtosis	19 SensorCondition	20 ShaftCondition	21 GearToothConditio
1	6.7500e-06	3.2300e-07	162.1284	Sensor Drift	No Shaft Wear	No Tooth Fault
2	5.0800e-08	9.1600e-08	226.1212	Sensor Drift	No Shaft Wear	No Tooth Fault
3	6.7400e-06	2.8500e-07	162.1259	Sensor Drift	Shaft Wear	No Tooth Fault

Variables - tbl

tbl

208x21 table

1 No Shaft Wear	2 Shaft Wear
1	0

	16 HighFreqPower	17 EnvPower	18 PeakSpecKurtosis	19 OHCode	20 OHCode_1	21 GearToothCondition
1	6.7500e-06	3.2300e-07	162.1284	1x2 table	1x2 table	No Tooth Fault
2	5.0800e-08	9.1600e-08	226.1212	1x2 table	1x2 table	No Tooth Fault
3	6.7400e-06	2.8500e-07	162.1259	1x2 table	1x2 table	No Tooth Fault

Figure 9.45 The two predictor columns original present (*top*) were converted to one-hot encoded vectors (*bottom*).

5) Split Pregnant Vectors into Separate Columns

Split the vectors into separate columns using **splitvars** function. Notice that **splitvars** function takes a table or a timetable as input and returns a new table or timetable with each column of the original variable as a separate variable. The original variable is removed from the resulting table or timetable.

```
tbl = splitvars(tbl);
```

6) View of Table Heads

View the first few rows of the table.

head(tbl)

As shown in Figure 9.46, the categorical (pregnant) predictors have been split into multiple (children) columns with the categorical values as the variable names.

PeakSpecKurtosis	No Sensor Drift	Sensor Drift	No Shaft Wear	Shaft Wear	GearToothCondition
162.13	0	1	1	0	No Tooth Fault
226.12	0	1	1	0	No Tooth Fault
162.13	0	1	0	1	No Tooth Fault

Figure 9.46 Split of composite (pregnant) into separate children columns.

7) Partition the Data into Training and Test Set

Partition the data set into training and test partitions. Set aside 15% of the data for testing. Determine the number of observations for each partition.

```
numObservations = size(tbl,1);
numObservationsTrain = floor(0.85*numObservations);
numObservationsTest = numObservations - numObservationsTrain;
```

8) Create an Array of Random Indices for Training and Testing

Create an array of random indices corresponding to the observations and partition it using the partition sizes. Then, partition the table of data into training and testing partitions using the indices.

```
idx = randperm(numObservations);
idxTrain = idx(1:numObservationsTrain);
idxTest = idx(numObservationsTrain+1:end);
tblTrain = tbl(idxTrain,:);
tblTest = tbl(idxTest,:);
```

9) Define a Network with a Feature Input Layer

Define a network with a feature input layer and specify the number of features. Also, configure the input layer to normalize the data using Z-score normalization.

```
classNames = categories(tbl{:,GearCond});
numFeatures = size(tbl,2) - 1;
numClasses = numel(classNames);
 layers = [
   featureInputLayer(numFeatures,'Normalization', 'zscore')
   fullyConnectedLayer(50)
   batchNormalizationLayer
   reluLayer
   fullyConnectedLayer(numClasses)
   softmaxLayer
   classificationLayer];
```

10) Specify the Training Options

```
miniBatchSize = 16;
maxEpochs = 40;
```

```
options = trainingOptions('adam', ...
  'MiniBatchSize',miniBatchSize, ...
  'MaxEpochs',maxEpochs,...
  'Shuffle','every-epoch', ...
  'Plots','training-progress', ...
  'Verbose',false);
```

11) Train the Network

Train the network using the architecture defined by layers, the training data, and the training options. Then, predict the labels of the test data using the trained network and calculate the accuracy. The accuracy is the proportion of the labels that the network correctly predicts.

```
net = trainNetwork(tblTrain,layers,options);
YPred = classify(net,tblTest,'MiniBatchSize',miniBatchSize);
YTest = tblTest{:,GearCond};
Accuracy = (sum(YPred == YTest)/numel(YTest))*100;
DSP = ['Accuracy is ',num2str(Accuracy),' % '];
disp(DSP)
```

Accuracy is 96.875%

Figure 9.47 shows the network model accuracy and loss as a function of number of iterations.

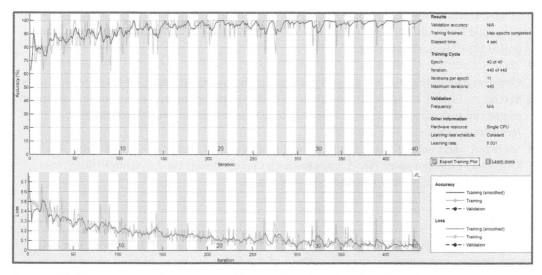

Figure 9.47 The network model accuracy and loss as a function of number of iterations.

Figure 9.48 shows workspace variables, including accuracy %.

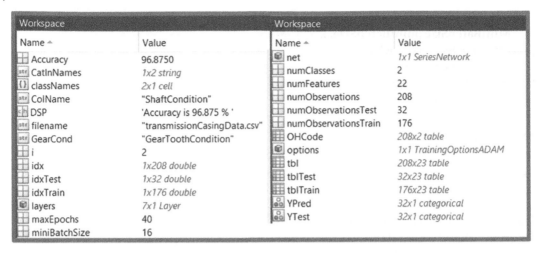

Figure 9.48 Workspace variables, including accuracy %.

References

1 Kudo, M., Toyama, J., and Shimbo, M. (1999) Multidimensional curve classification using passing-through regions. *Pattern Recognition Letters* 20 (11–13): 1103–1111.

2 UCI Machine Learning Repository: Japanese Vowels Dataset. https://archive.ics.uci.edu/ml/datasets/Japanese+Vowels.

3 Iqbalal, T. and Qureshia, Sh. (2022). The survey: text generation models in deep learning. *Journal of King Saud University – Computer and Information Sciences* 34 (6), Part A: 2515–2528. https://doi.org/10.1016/j.jksuci.2020.04.001.

4 Zhou, S. (2020). Research on the application of deep learning in text generation. *Journal of Physics: Conference Series* 1693: 012060. https://iopscience.iop.org/article/10.1088/1742-6596/1693/1/012060.

5 Saxena, A., Goebel, K., Simon, D., and Eklund, N. (2008). Damage propagation modeling for aircraft engine run-to-failure simulation. In *Prognostics and Health Management, 2008. PHM 2008. International Conference on*, 1–9. IEEE.

6 Saxena, A. and Goebel, K. (2008). Turbofan engine degradation simulation data set. *NASA Ames Prognostics Data Repository*, NASA Ames Research Center, Moffett Field, CA. https://ti.arc.nasa.gov/tech/dash/groups/pcoe/prognostic-data-repository.

Quiz #1

We would like to predict the age of children based on X-ray images of their hands. Our data is X-ray images and a numeric response. Which of the following types of deep networks would be most appropriate?

a) CNN   b) YOLO   c) LSTM

Quiz #2

We would like to decide whether to buy or sell based on stock price data. Our data is a numeric ordered sequence. Which of the following types of deep networks would be most appropriate?

a) CNN   b) YOLO   c) LSTM

Quiz #3

Our data consists of 10 sequences of lengths 10, 10, 12, 12, 14, 14, 16, 16, 18, and 18. Our training consists of two mini-batches of size 5. If we sort the data by length before training, what's the most padding a sequence will have?

a) 0   b) 1   c) 2   d) 3   e) 4   f) 5

Quiz #4

Our data consists of 10 sequences of lengths 10, 10, 12, 12, 14, 14, 16, 16, 18, and 18. If we sort the data by length before training, what is the largest mini-batch size that would ensure there is no padding?

a) 1   b) 2   c) 3   d) 4   e) 5   f) 6

Note: In addition to the original source, the data set for each problem is available at Wiley Companion Website.

HCW 9.1 Text Classification: Factory Equipment Failure Text Analysis

Refer to "**Classify Text Data Using Deep Learning: Factory Equipment Failure Text Analysis – 1**" section and carry out deep learning to predict the equipment failure for the following three reports. You can use the same outlined approach in the given section.

a. reportsNew1 = [...
    "The storage tank has water pool underneath."
    "Some products keep falling off the belt."
    "The control panel is off; no screen light."
    "Control panel software freezes more than once."];
b. reportsNew2 = [...
    "Some end products are squeezed or bent."
    "The control panel has power outage."
    "Control panel light is on and off."
    "The storage tank has liquid discharge problem."];
c. reportsNew3 = [...
    "Printer spools get stuck."
    "The cooling water is found outside the drum."
    "Loud sounds are heard from the rotating mixer."
    "The controller program has bugs."];

Show the training performance and the predicted results.

HCW 9.2   Text Classification: Sentiment Labeled Sentences Data Set

The data set information source: 'From Group to Individual Labels using Deep Features'. Dimitrios Kotzias, Misha Denil, Nando de Freitas, Padhraic Smyth. Proceedings of the 21th ACM SIGKDD International Conference on Knowledge Discovery and Data Mining, August 2015, Pages 597–606. https://doi.org/10.1145/2783258.2783380

The data set can be downloaded from:
https://github.com/mr-atharva-kulkarni/UCI-Sentiment-Analysis/tree/master/Dataset
https://data.world/uci/sentiment-labelled-sentences

The data sets contain sentences labelled with positive or negative sentiment. Score is either 1 (for positive) or 0 (for negative). The sentences come from three different websites/fields: imdb.com, amazon.com, and yelp.com.

Sentences have a clearly positive or negative connotation; the goal was for no neutral sentences to be selected. The attributes are text sentences, extracted from reviews of products, movies, and restaurants.

Carry out text classification deep learning. Combine all three data sets into one big data set, as shown below:

XComb=[Amazon_Strng;Yelp_Strng;Imdb_Strng];

YComb=[Amazon_Class;Yelp_Class;Imdb_Class];

Partition the combined data set into 80% for training, 10% for validation, and the last 10% for testing. Preprocess the text, convert document to sequence, and select the proper sequence length. Select the layers configuration, the training options, and contrast the trained model versus the test data set. Two methods are explained in this chapter on how to deal with text classification and carrying out documents to sequence transformation. Carry out training and report the accuracy-based results.

HCW 9.3   Text Classification: Netflix Titles Data Set

Source: https://www.kaggle.com/datasets/shivamb/netflix-shows

Netflix movies and TV shows are rated or classified into different categories (response variable classes), like: PG-13, TV-MA, PG, TV-14, TV-PG, TV-Y, ... etc. There is another column for the description of a movie or TV show title (predictor variable). There are 8800 items or data points. Partition the entire data set into 75% for training, 15% for validation, and the last 10% for testing. Preprocess the text, convert document to sequence, and select the proper sequence length. Select the layers' configuration, the training options, and contrast the trained model versus the test data set. Two methods are explained in this chapter on how to deal with text classification and carrying out documents to sequence transformation. Carry out training and report the accuracy-based results.

HCW 9.4   Text Regression: Video Game Titles Data Set

Video games (https://www.kaggle.com/datasets/gregorut/videogamesales) are listed as Name column which represents the predictor variables. There is another column called Global_Sales which accounts for the sum of sales in North America (NA), European (EU), Japanese (JP), and other market. There are 16600 items or data points. Partition the entire data set into 80% for training,

10% for validation, and the last 10% for testing. Preprocess the text, convert document to sequence, and select the proper sequence length. Select the layers' configuration, the training options, and contrast the trained model versus the test data set. Two methods are explained in this chapter on how to deal with text classification and carrying out documents to sequence transformation. Carry out training and report the accuracy-based results. Keep in mind that this is a regression not classification problem, which means the number of classes of the output layer is 1.

HCW 9.5   Multivariate Classification: Mill Data Set

The data source: (https://www.kaggle.com/datasets/vinayak123tyagi/milling-data-set-prognostic-data)

Acknowledgement: A. Agogino and K. Goebel. BEST lab, UC Berkeley. "Milling Data Set ", NASA Ames Prognostics Data Repository (http://ti.arc.nasa.gov/project/prognostic-data-reposi tory), NASA Ames Research Center, Moffett Field, CA. (2007).

This set represents experiments from runs on a milling machine under various operating conditions. In particular, tool wear was investigated in a regular cut as well as entry cut and exit cut. Data sampled by three different types of sensors (acoustic emission sensor, vibration sensor, current sensor) were acquired at several positions.

There are 16 specific cases which correspond to different combination of Depth of Cut (mm), Feed (mm/revolution), and Material type (cast iron or steel). For each case, varying number of runs were carried out. The number of runs was dependent on the degree of flank wear (VB) that was measured between runs at irregular intervals up to a wear limit (and sometimes beyond). Flank wear was not always measured and at times when no measurements were taken, no entry was made.

There are 146 non-empty, VB entries. Since the data size is small, instead of data partition, use random shuffling of the original 146 data set and create two additional, same-size data sets for validation and testing.

The following code can be used to achieve the random shuffling of the given data-set.

```
r= size(mill2,1);
P = 1.0 ;
idx = randperm(r) ;
idx2 = idx(1:round(P*r));
tblTrain = mill2(idx2,1:end);
YTrain = categorical(mill2.VB(idx2));
idx3 = randperm(r) ;
idx4 = idx3(1:round(P*r));
tblVal = mill2(idx4,1:end);
YVal = categorical(mill2.VB(idx4));
idx5 = randperm(r) ;
idx6 = idx5(1:round(P*r));
tblTest = mill2(idx6,1:end-1);
YTest = categorical(mill2.VB(idx6));
```

Initially, we have 10 predictor variables and one response variable denoted as VB. The ten predictors are: **time, DOC, feed, material, smcAC, smcDC, vib_table, vib_spindle, AE_table,** and **AE_spindle**. Convert and replace DOC, feed, and material variable (or column) into categorical utilizing **onehotencode** function.

Select the layers' configuration, the training options, carry out the model training. Contrast the trained model versus the test data set. Since we have categorical variables, we will carry out a classification case study. The number of classes in the output layer will be equal to the number of sub-categories of the categorical VB variable.

HCW 9.6 Word-by-Word Text Generation Using Deep Learning

Read an HTML page from the world-wide web and carry out word-by-word text generation. See the relevant two sections in this chapter.

10

Image/Video-Based Apps

In this chapter, we will introduce both image labeler and video labeler apps to facilitate the process of image and video labeling. The user can then use such labeled images or videos in his/her training of network classification and regression using CNNs, like: R-CNN, Fast R-CNN, Faster R-CNN, and YOLO. For automated image and video labeling, we can use either built-in or custom automation algorithms. We can synchronize ground truth objects with other time series sensors, like LiDAR or Radar. We can use labeled data outside MATLAB environment, as well. We will also introduce the application called Experiment Manager which is basically used to carry out parametric optimization for a given training network model in one place with the flexibility to see the model goodness results in the form of plots and tables upon completion of simulation trials.

Image Labeler (IL) App

Quoted from offline MATLAB help, the Image Labeler app enables us to label ground truth data in a collection of images. Using the app, we can:

- Define rectangular regions of interest (ROI) labels, polyline ROI labels, pixel ROI labels, polygon ROI labels, and scene labels. We use these labels to interactively label our ground truth data.
- Use built-in detection or tracking algorithms to label our ground truth data.
- Write, import, and use our own custom automation algorithm to automatically label ground truth. See Create Automation Algorithm for Labeling.
- Evaluate the performance of our label automation algorithms using a visual summary. See View Summary of Ground Truth Labels.
- Export the labeled ground truth as a **groundTruth** object. We can then use this object for system verification or for training an object detector or semantic segmentation network. See Training Data for Object Detection and Semantic Segmentation.

The Image Labeler app is compatible with all image file formats that can be read by the **imread** function. In addition, it also supports the Digital Imaging and Communication in Medicine (DICOM) format, including the capability to load **multiframe** data. If you need to read other file formats, you can create an **imageDatastore** and utilize the **ReadFcn** property.

Machine and Deep Learning Using MATLAB: Algorithms and Tools for Scientists and Engineers, First Edition.
Kamal I. M. Al-Malah.
© 2024 Kamal I. M. Al-Malah. Published 2024 by John Wiley & Sons, Inc.
Companion Website: www.wiley.com/go/al-malah/machinelearningmatlab

When images are loaded in the **Image Labeler** app, there is a provision to convert an image into a blocked image if it exceeds 8000 pixels in dimension or if it is a multi-resolution image. A blocked image is essentially a large image divided into smaller blocks that can be accommodated in memory. Once the **Image Labeler** app performs this conversion, we can proceed to process the blocked image within the app, just like any other image. Using blocked images allows us to work with images that would otherwise be difficult to handle. However, it's important to note that there are certain limitations associated with the use of blocked images. For more information, see Label Large Images in the Image Labeler. Here are some worth-mentioning points, as described by MATLAB offline help:

1) **imageLabeler**(*imageFolder*) opens the app and loads all the images from the folder named *imageFolder*. The images in the folder can be unordered and can vary in size. Example:

imageFolder = fullfile(toolboxdir('vision'),'visiondata','bookcovers')
imageLabeler(imageFolder)

2) To label a video, or a set of ordered images that resemble a video, use the **Video Labeler app**, instead.

   imageLabeler(*imageDatastore*) opens the app and reads all of the images from an *image-Datastore* object. The **ReadFcn** property of the *imageDatastore* object specifies how to read the data. For example, to open the app with a collection of stop-sign images:

stopSignsFolder = fullfile(toolboxdir('vision'),'visiondata','stopSignImages');
imds = imageDatastore(stopSignsFolder)
imageLabeler(imds)

3) **imageLabeler**(*sessionFile*) opens the app and loads a saved Image Labeler session, *sessionFile*. The *sessionFile* input contains the path and file name. The MAT-file that sessionFile points to contains the saved session.

4) **imageLabeler**(*gTruth*) opens the app and loads a **groundTruth** object. The ground truth object data source must be an image collection or an **imageDatastore**.

To open the image labeler, at the MATLAB command prompt, enter:

>>**imageLabeler**

Alternatively, from **APPS** menu | **APPS** group | **IMAGE PROCESSING AND COMPUTER VISION** category, click on "**Image Labeler**" icon. The main window of the image labeler shows up as in Figure 10.1. I have modified the default layout of the main window to show its main features. However, the user may click on the "**LAYOUT**" button and revert to the default layout.

Clicking on "**Import**" button will allow the user to import an image from a file or from the workspace environment. The main window shows "**ROI Labels**" and "**Scene Labels**" tab window. An ROI label corresponds to either a rectangular, polyline, pixel, or polygon region of interest. These labels contain two components: the label name, such as *"cars"* and the region we create. On the other hand, a Scene label describes the nature of a scene, such as *"sunny"*. We can associate this label with a frame.

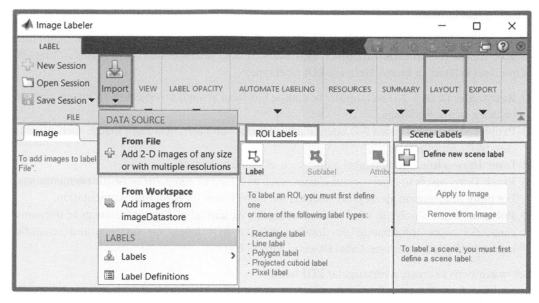

Figure 10.1 Opening the image labeler and importing a 2-D image for labeling purposes.

Using the "**Import**" button let us import a 2-D image from a file named "**boats.png**" found in MATLAB installed directory: **C:\Program Files\MATLAB\R2022a\toolbox\vision\visiondata**. The image will appear in the canvas or working area of the main window, as shown in Figure 10.2.

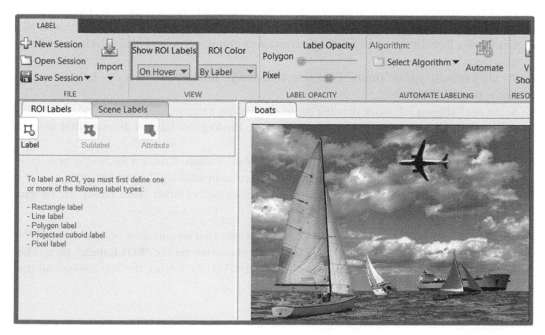

Figure 10.2 Importing **boats.png** image into image labeler environment.

Creating ROI Labels

We can define labels directly within the app. To define labels from the MATLAB® command line instead, we use the **labelDefinitionCreator**. An ROI label is a label that corresponds to a region of interest (ROI) in an image. Here are ROI label types:

1) **Rectangle**: Draw a 2-D rectangular bounding box label around an object in an image, such as vehicles, boats, buildings.
2) **Projected cuboid**: Draw a 3-D bounding box label around an object in an image, such as vehicles, boats, buildings.
3) **Line**: Draw a linear ROI to label a line, such as a lane boundary.
4) **Pixel**: Draw pixels to label various classes, such as a road or a sky, for semantic segmentation. For more information about pixel labeling, see Label Pixels for Semantic Segmentation.
5) **Polygon**: Draw a polygon label around an object. We can label distinct instances of the same class. For more information on drawing polygon ROI labels for instance and semantic segmentation networks, see Label Objects Using Polygons.

Below are steps to create a rectangular **ROI** label:

a) We will define a "**Boats**" group for labeling types of boats, and then create a Rectangle **ROI** label for a sailboat and a tanker. In addition, we will create a Rectangle **ROI** label for an aeroplane.
b) To control showing the **ROI** label names during labeling, select "*On Hover*", "*Always*", or "*Never*" from the "**Show ROI Labels**" drop-down menu (see top of Figure 10.2).
c) In the **ROI** Labels pane on the left, click on "**Label**" button. Create a rectangular label type named "**Sailboat**". Optionally, change the label color by clicking the preview color.
d) From the "**Group**" drop-down menu, select *New Group* ... and name the group "**Boats**".
e) The "**Boats**" group name appears in the **ROI** Labels pane with the label "**Sailboat**" created. We can move a label in the list to a different position or group in the list by left-clicking and dragging the label up or down.
f) To add a second type of "**Boats**" label, select the group "**Boats**", then click on "**Label**" button. Name the label "**Tanker**".
g) Select the **Sailboat** sub-label, then use the mouse to draw a rectangular **ROI** around each of the three sailboats. Again, using the mouse, select the **Tanker** sub-label to draw an **ROI** around the tanker ship.
h) In the **ROI** Labels pane on the left, click on "**Label**" button. Create a Rectangle label type named "**Aeroplane**". Optionally, change the label color by clicking the preview color.
i) To edit or delete a **ROI** label, right-click on the label and select either "**Edit Label**" or "**Delete Label**" option, respectively.

Figure 10.3 shows what we created in previous steps. Notice that we can show or hide the labels or sublabels in a labeled image by using the "**Eye**" icon appearing on the "**ROI Labels**" pane. The "**Eye**" icon appears only after we define a label or sub-label. By default, the app displays all the labels and the sub-labels.

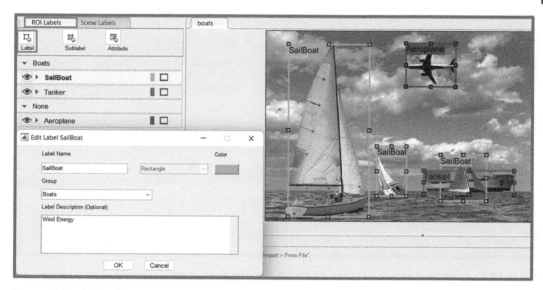

Figure 10.3 Adding ROI labels for both a boat and an aeroplane type.

Creating Scene Labels

A scene label defines additional information for the entire scene. We use scene labels to describe conditions, such as lighting and weather, or events, such as lane changes or point of sail.

a) Create a scene label to apply to an image. In the left pane of the app, select the **"Scene Labels"** next to the **"ROI Labels"** tab. Click on **"Define new scene label"** button and in the **"Label Name"** box, enter a scene label named *"Daytime"*.

b) Change the color of the label definition to light (sky) blue to reflect the nature of the scene label. Under the Color parameter, click the color preview and select the standard light blue colors. Then, click on **"OK"** button to close the color selection window.

c) Leave the Group parameter set to the default of None and click on **"OK"** button. The **"Scene Labels"** pane shows the scene label definition. Click on **"Apply to Image"** to apply the day-time label to the scene. A checkmark appears for the scene label.

d) To edit or delete a scene label, right-click on the label and select either **"Edit Label"** or **"Delete Label"** option, respectively.

Figure 10.4 shows the implementation of the previous steps.

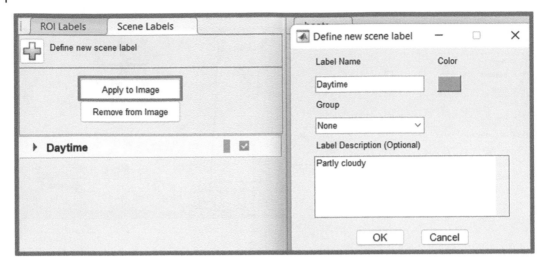

Figure 10.4 Creation of a scene label and applying it unto the given image.

At this stage, save your work in the form of MATLAB session so that it can be opened later by the "**Image Labeler**" app. For example, I saved the file as "**boatSession.mat**".

Label Ground Truth

We can label images manually, use a suitable built-in automation algorithm, create a new algorithm, or import an algorithm.

Label Ground Truth Manually
The ROI and Scene labels are defined for the entire session and all images, but we must draw the labels and sublabels for each image and update the attribute information.

Label Ground Truth Using an Automation Algorithm
To speed up the labeling process, we can use an automation algorithm to label the remainder of the images. Continue on "**boatSession**" session and select one of these types of automation algorithms by going to "**AUTOMATE LABELING**" tab | "**Select Algorithm**" drop-down menu.

- **Use a Built-in Automation Algorithm**: ACF People Detector algorithm can be used to detect people using "**Aggregate Channel Features (ACF)**" option. Under "**AUTOMATE LABELING**" tab, the "**Select Algorithm**" drop-down menu will be reset to "**ACF People Detector**" sub-menu. Follow the instructions that appear in the right pane. If the instructions do not show up, click on "**Automate**" button to show instructions.
- **Add Whole Image Algorithm**: We can create a new automation algorithm or import one. For details on both options, see MATLAB | Built in Help | Create Automation Algorithm for Labeling.
- **Add Blocked Image Algorithm:** We can create a new blocked image automation algorithm or import one. For details on both options, see MATLAB | Built in Help | Label Large Images in the Image Labeler.

After using an automation algorithm, we can manually label the remaining frames with sub-label and attribute information. To further evaluate the created labels, we can view a visual summary of the labeled ground truth. From the app toolstrip, select "**Summary**" tab | "**View Label Summary**" button. Use this summary to compare the frames, frequency of labels, and scene conditions. For more details, see MATLAB | Built in Help | View Summary of Ground Truth Labels. This summary does not show sub-labels or attributes.

Export Labeled Ground Truth

We can export the labeled ground truth to a MAT-file or to a variable in the MATLAB workspace. In both cases, the labeled ground truth is stored as a **groundTruth** object. We can use this object to train a deep-learning-based computer vision algorithm. For more details, see MATLAB | Built in Help | Training Data for Object Detection and Semantic Segmentation.

If we export pixel data, the pixel label data and ground truth data are saved in separate files but in the same folder. For considerations when working with exported pixel labels, see MATLAB | Built in Help | How Labeler Apps Store Exported Pixel Labels.

Let us export the labeled ground truth to the MATLAB workspace. From the app toolstrip, select "**Export**" tab | "**Export Labels**" button | "**To Workspace**" sub-menu. One may give a name other than the default name "**gTruth**". The properties of the exported MATLAB variable, **gTruth**, are shown below.

```
gTruth = groundTruth with properties:
        DataSource: [1×1 groundTruthDataSource]
LabelDefinitions: [4×5 table]
        LabelData: [1×4 table]
```

Figure 10.5 (*top left*) shows properties of "**gTruth**", the groundTruth augmented image (or object); the entries of "**LabelDefintions**" table are shown (*top right*); the pixel location of each **ROI** rectangular label, measured from top left edge of the image, are shown (*bottom left*); and finally, the pixel location, measured from the top left edge of the image, of the three sailboats are shown (*bottom right*). Notice that each bounding rectangle is described by the x, y pixel location of its top left corner and its width and height, respectively. In other words, the bounding rectangle is described by $\begin{bmatrix} x_{top-left} & y_{top-left} & width & height \end{bmatrix}$ pixel units.

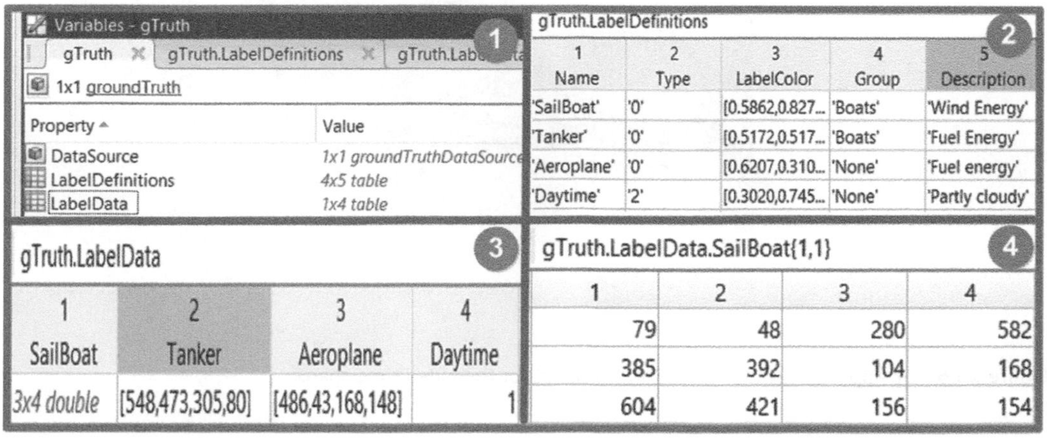

Figure 10.5 Name of labels; their groups, and pixel location of each labeled object as measured from the top left edge of the original image.

Video Labeler (VL) App: Ground Truth Data Creation, Training, and Prediction

The Video Labeler (VL) app provides an easy way to mark rectangular region of interest (**ROI**) labels, polyline ROI labels, pixel **ROI** labels, and scene labels in a video or image sequence.

We can use labeled data to validate or train algorithms such as image classifiers, object detectors, and semantic and instance segmentation networks. Consider the application when choosing a labeling drawing tool to create **ROI** labels. In this section, using Video Labeler app, we will show how to:

- Manually label an image frame from a video.
- Automatically label across image frames using an automation algorithm.
- Export the labeled ground truth data.

For a definition of **ROI** and Scene Label, please, refer to "Image Labeler App" section.

1. Load Unlabeled Data
Programmatically, open the app and load a video. Videos must be in a file format readable by **VideoReader** function.

```
videoLabeler('Dog_Cat.mp4')
```

Figure 10.6 shows the main window of VL app and the loaded video is shown in the main canvas (i.e., working area) as well as the control panel (shown at the bottom), dedicated for the video in terms of time span of the video.

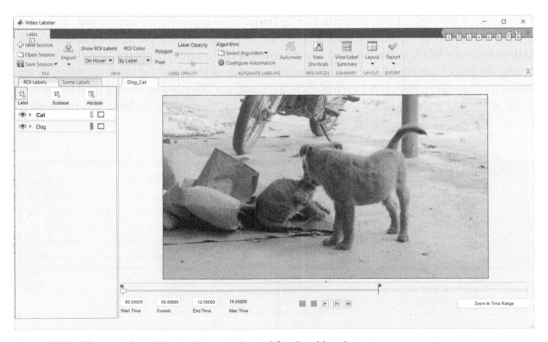

Figure 10.6 VL app main window with control panel for the video time span.

Alternatively, we can also load a video from the app. Simply, open the app from the **Apps** tab, under **Image Processing and Computer Vision**. Click **Import** to load a video or image sequence, or click **Open Session** to load a saved session.

We may explore the video. We can click the "**Play**" button to play the entire video, or use the slider to navigate between frames.

To load an image sequence with corresponding timestamps, select **Import > Image Sequence**. The app supports all image file formats supported by **imread** function. To read additional file formats, we can create an **imageDatastore** and use the **ReadFcn** property.

To load a custom data source that cannot be read by **VideoReader** or **imread** function, see MATLAB | Built in Help | "**Use Custom Image Source Reader for Labeling**".

2. Set Time Interval

We can label the entire video or start with a portion of the video. In this example, we label a twelve-second time range within the loaded video. In the text boxes below the video, enter these times in seconds:

1) In the **Current Time** box, type 0 and press **Enter**.
2) In the **Start Time** box, type 0 so that the slider is at the start of the time range.
3) In the **End Time** box, type 12.

Optionally, to make adjustments to the time range, we can also click and drag the red interval flags, as shown in Figure 10.6.

3. Create ROI Labels

Let us define the labels we intend to draw. In this section, we define labels directly within the app. To define labels from the MATLAB® command line instead, please refer to MATLAB | Built in Help | **labelDefinitionCreator** function. Here, we define two ROI labels called "**Cat**" and "**Dog**", as shown in Figure 10.6. See the previous section on image labeling.

4. Automatic Algorithm for Labeling

From "**Algorithm**" menu, we will select "**Temporal Interpolator**" automatic labeling algorithm and then click on "**Automate**" button to proceed with the selected algorithm type, as shown in Figure 10.7. This algorithm requires us to draw the bounding box for each **ROI** label at different time intervals between zero and the end time 12 seconds. For example, one may select to draw "Cat" ROI label every 1.5 to 2 seconds. To bracket all possible video frames, start at the earliest image frame where the cat appears for the first time and end at the last image frame where the cat no longer appears. Notice that the user may define his/her own custom model which will appear in the algorithm list after refreshing the list. Moreover, the other algorithms, shown in the list of Figure 10.7, can be used for lane detection, vehicle detection, and people detection, respectively. The "**Point Tracker**" algorithm is similar in behavior to that of the "**Temporal Interpolator**" algorithm; it tracks one or more rectangle ROIs over a short interval, using Kanade-Lucas-Tomasi (KLT) algorithm.

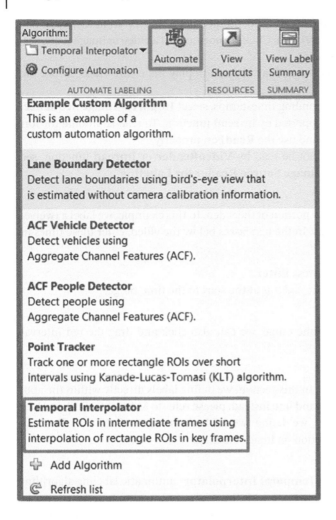

Figure 10.7 The "**Temporal Interpolator**" algorithm is selected and automated after drawing **RO** labels once in a while from the start to the end time.

Figure 10.8 shows the VL platform upon initiating the automation process. Notice the following points:

- The instructions are shown on the right pane for a given automated labeling scenario.
- After clicking on "**Run**" button, the automatic algorithm will start the labeling process throughout the entire time span.
- The user may manually modify any of the automated **ROI** labels, drawn around the "**Cat**" object by resizing the bounding box using the mouse. Moreover, the user may also manually add a missing bounding box or delete any mistakenly drawn bounding box.
- After being satisfied with the drawing/labeling scenario, the user may click on "**Accept**" button to proceed and implement the automated labeling scenario.
- Clicking on "**Undo Run**" button will cancel the automated labeling process and bring the user to the start of the labeling scenario. On the other hand, clicking on "**Cancel**" button will abort the labeling attempt and bring the user back to square number zero with no labeling effect.
- The user may choose to divide the labeling task at different stages. For example, the user may select the time span to be between 0 and 6 seconds, then carry out the automated labeling process, accept the process, and save the VL opened session for later use. At a later time, the user may open the previously saved VL session and continue the process of labeling for the remaining time span, that is, between 6 and 12 seconds.

- It is preferred to carry out the labeling process for one ROI label class. In this section, we finish first the labeling process for "**Cat**" ROI label and upon having satisfactory results, we then move to the labeling process of "**Dog**" ROI label, as shown in Figure 10.9.

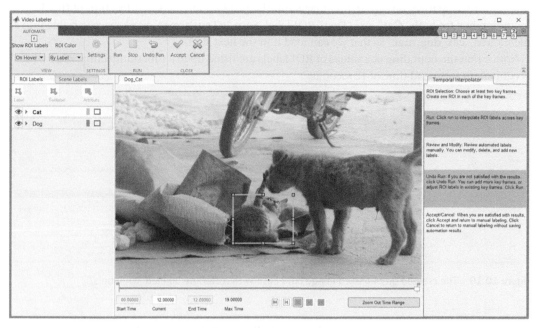

Figure 10.8 Run automatic labeling, check if the labeling algorithm precisely describes the labeled object, and click on "**Accept**" button to implement the labeling/drawing process throughout the selected time span.

Figure 10.9 shows the labeling process for "**Dog**" ROI label.

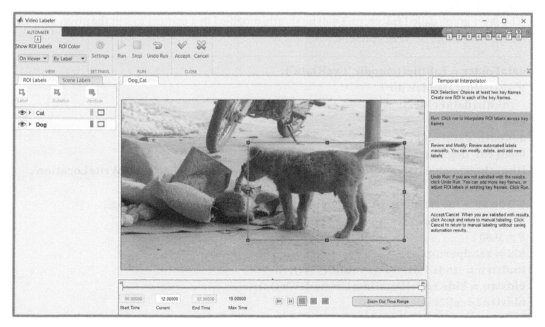

Figure 10.9 The automated "**Temporal Interpolator**" algorithm for labeling "**Dog**" ROI label.

5. Saving the Labeled Set of Video Frames for Future Training and Prediction

Clicking on "**Export**" button will prompt us to save the labeled set of image frames into a workspace 1×1 **groundTruth** class. Figure 10.10 shows the details of this ground truth database where it contains the source of images, a table of labels, and the label data table where each row represents the time span value of a frame image and the pixel dimensions and location of the drawn bounding box for both "**Cat**" and "**Dog**" ROI label. Save "**gTruth.mat**" workspace variable into a file named "**cat_dog.mat**" so that we can load it in the next step.

Notice that the bounding box values of ROI labels are listed up to a timestamp entry #301 which is equivalent to an end time of 12 seconds.

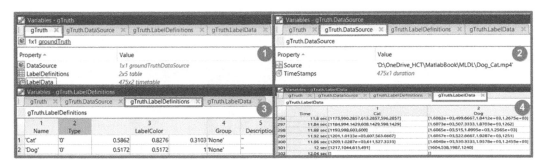

Figure 10.10 The created ground truth image data base with rectangle labeled ROI labels.

6. Creation of Image Data Stores and Box Labeling Data Stores for Training

Below is the code for creating both types of an image and box labeling data store. In addition, the entire data set is further divided into training, validation, and test set. A combined data store is also created for training, validation, and testing phase. Moreover, the layer graph of a previously created faster RCNN will be used. See Chapter 8 | Object Detection Using R-CNN Algorithms | Faster R-CNN section.

The layer graph called **fastercnn3** will be shortly modified in a manner to suit the given data set in terms of number of classes present.

```
dataFolder = fullfile(tempdir,"EvalData");
if ~exist(dataFolder,'dir')
   mkdir(dataFolder);
end
addpath(dataFolder);
load('cat_dog.mat', 'gTruth')
[imds blds]=objectDetectorTrainingData(gTruth,'SamplingFactor',1,'WriteLocation',
dataFolder);
% Top 80 % for training, 10 % for validation, and the last 10 % for testing
r= size(imds.Files,1);
P = 0.80 ;
idx = randperm(r)  ;
imdstrn = imds.Files(idx(1:round(P*r)), :);
bldstrn = blds.LabelData(idx(1:round(P*r)),:);
bldstrn2=cell2table(bldstrn);
```

```
bldstrn3 = boxLabelDatastore(bldstrn2(:,1:2));
imdstrn2 = imageDatastore(imdstrn);
dstrn = combine(imdstrn2, bldstrn3);
dstrn = transform(dstrn,@(data)preprocessData(data,[200 200 3]));
imdsval = imds.Files(idx(round(P*r)+1:round((P+0.1)*r)), :);
bldsval = blds.LabelData(idx(round(P*r)+1:round((P+0.1)*r)), :);
bldsval2=cell2table(bldsval);
bldsval3 = boxLabelDatastore(bldsval2(:,1:2));
imdsval2 = imageDatastore(imdsval);
dsval = combine(imdsval2, bldsval3);
dsval = transform(dsval,@(data)preprocessData(data,[200 200 3]));
imdstst=imds.Files(idx(round((P+0.1)*r)+1:end), :);
bldstst = blds.LabelData(idx(round((P+0.1)*r)+1:end), :);
bldstst2=cell2table(bldstst);
bldstst3 = boxLabelDatastore(bldstst2(:,1:2));
imdstst2 = imageDatastore(imdstst);
dstst = combine(imdstst2, bldstst3);
dstst = transform(dstst,@(data)preprocessData(data,[200 200 3]));
% Any of YOLO or R-CNN object detectors can be used. Please, refer to Chapter 8.
% For simplicity, we will exploit one fasterRCNN layer graph.
```

7. Training Options

The following are training options:

```
miniBatchSize  = 1;
valData=dsval;
options = trainingOptions('adam', ...
'MiniBatchSize',miniBatchSize, ...
'MaxEpochs',1, ...
'InitialLearnRate',1e-3, ...
'LearnRateSchedule','piecewise', ...
'LearnRateDropFactor',0.1, ...
'LearnRateDropPeriod',10, ...
'Shuffle','every-epoch', ...
'ValidationData',valData,...
'ValidationFrequency',30,...
'Plots','training-progress', ...
'Verbose',false);
```

8. Train the Network

Use **trainFasterRCNNObjectDetector** function to train the faster RCNN network.

```
detector = trainFasterRCNNObjectDetector(dstrn, fasterrcnn3, options, ...
'NegativeOverlapRange', [0 0.33], ...
'PositiveOverlapRange', [0.4 1]);
```

The following error is issued:

Error using **trainFasterRCNNObjectDetector**
Invalid network.

Caused by:
Layer '**rcnnBoxDeltas**': The input size must be 1×1×8. This R-CNN box regression layer expects the third input dimension to be four times the number of object classes the network should detect (2 classes). See the documentation for more details about creating Fast or Faster R-CNN networks.
Layer '**rcnnClassification**': The input size must be 1×1×3. The classification layer expects the third input dimension to be the number of object classes the network should detect (2 classes) plus 1. The additional class is required for the "background" class. See the documentation for more details about creating Fast or Faster R-CNN networks.

So, we open DND app, import **fasterrcnn3** from workspace environment, and modify the two indicated layers. Be sure to run the network analysis via clicking on "**Analyze**" button found at the top tool bar of DND app, to assure that there are no errors as a result of mis-entered properties for one or more layers.

After the network analysis is complete with no bugs, export the layer graph back to workspace environment under the name **fasterrcnn4**. Figure 10.11 summarizes changes made to the original configuration where the output size of the two pre-requisite layers is modified to match the given classification problem in hand, with two classes ("**Cat**" and "**Dog**") plus the one class for the background.

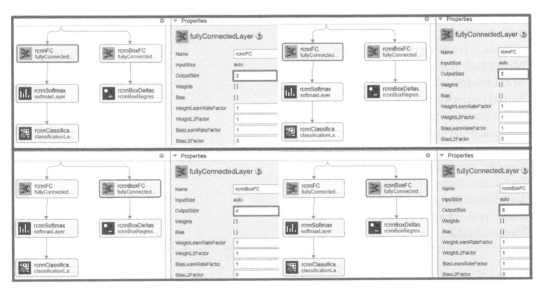

Figure 10.11 The output size of two pre-requisite layers (*left*) are modified to new values (*right*) in line with the given classification problem.

Now, the modified layer graph, **fasterrcnn4**, is ready to go as shown below:

```
detector = trainFasterRCNNObjectDetector(dstrn, fasterrcnn4, options, ...
'NegativeOverlapRange', [0 0.35], ...
'PositiveOverlapRange', [0.5 1]);
```

Figure 10.12 shows the training information in terms of the model accuracy, RMSE, and loss as a function of number of iterations.

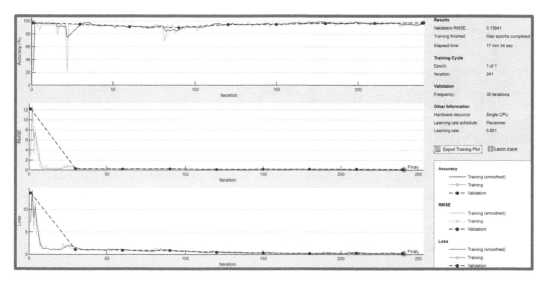

Figure 10.12   The faster R-CNN model performance under training and validation step.

Figure 10.13 shows a sample of six images, out of 30-image test data set, showing the predictability of the faster R-CNN object detector, using the following code:

```
I=imread(imdstst2.Files{7});%# Repeated for the remaining 5 images.
I=imresize3(I,[200 200 3]);
[bbox, score, label] = detect(detector, I);
detectedImg = insertObjectAnnotation(I,'rectangle',bbox,label);
figure (1)
imshow(detectedImg)
```

Figure 10.13 A sample of six images, out of 30-image test data set, showing faster R-CNN model prediction.

Figure 10.14 shows workspace variables created by running the previous code.

Workspace		Workspace	
Name ▲	Value	Name ▲	Value
bbox	3x4 double	I	200x200x3 uint8
blds	1x1 boxLabelDatastore	idx	1x301 double
bldstrn	241x2 cell	imds	1x1 ImageDatastore
bldstrn2	241x2 table	imdstrn	241x1 cell
bldstrn3	1x1 boxLabelDatastore	imdstrn2	1x1 ImageDatastore
bldstst	30x2 cell	imdstst	30x1 cell
bldstst2	30x2 table	imdstst2	1x1 ImageDatastore
bldstst3	1x1 boxLabelDatastore	imdsval	30x1 cell
bldsval	30x2 cell	imdsval2	1x1 ImageDatastore
bldsval2	30x2 table	label	3x1 categorical
bldsval3	1x1 boxLabelDatastore	lgraph_1	1x1 LayerGraph
detectedImg	200x200x3 uint8	miniBatchSize	1
detector	1x1 fasterRCNNObjectDetector	options	1x1 TrainingOptionsADAM
dstrn	1x1 TransformedDatastore	P	0.8000
dstst	1x1 TransformedDatastore	r	301
dsval	1x1 TransformedDatastore	score	[0.7563;0.7811;0.9988]
fasterrcnn4	1x1 LayerGraph	valData	1x1 TransformedDatastore
gTruth	1x1 groundTruth		

Figure 10.14 The code-generated workspace variables.

On the other hand, let us use **trainYOLOv2ObjectDetector** for training purpose. Figure 10.15 shows the creation of an image data store and box labeling data store, partition of data set into training, validation, and test set, creation of yolo v2 layer out of resnet50 DAGNewtork (requires the Deep Learning Toolbox Model for ResNet-50 Network support package), training options, and using **trainYOLOv2ObjectDetector** function.

```
%The DND-labeled gTruth must be loaded to workspace environment, first.
load('cat_dog.mat', 'gTruth')
dataFolder = fullfile(tempdir,"EvalData2");
if ~exist(dataFolder,'dir')
   mkdir(dataFolder);
end
addpath(dataFolder);
[imds blds]=objectDetectorTrainingData(gTruth,'SamplingFactor',1,'WriteLocation',
dataFolder);
% Top 80% for training, 10 % for validation, and the last 10 % for testing
r= size(imds.Files,1);
P = 0.80 ;
idx = randperm(r)  ;
imdstrn = imds.Files(idx(1:round(P*r)), :);
bldstrn = blds.LabelData(idx(1:round(P*r)),:);
bldstrn2=cell2table(bldstrn);
bldstrn3 = boxLabelDatastore(bldstrn2(:,1:2));
imdstrn2 = imageDatastore(imdstrn);
dstrn = combine(imdstrn2, bldstrn3);
dstrn = transform(dstrn,@(data)preprocessData(data,[800 800 3]));
imdsval = imds.Files(idx(round(P*r)+1:round((P+0.1)*r)), :);
bldsval = blds.LabelData(idx(round(P*r)+1:round((P+0.1)*r)), :);
bldsval2=cell2table(bldsval);
bldsval3 = boxLabelDatastore(bldsval2(:,1:2));
imdsval2 = imageDatastore(imdsval);
dsval = combine(imdsval2, bldsval3);
dsval = transform(dsval,@(data)preprocessData(data,[800 800 3]));
imdstst=imds.Files(idx(round((P+0.1)*r)+1:end), :);
bldstst = blds.LabelData(idx(round((P+0.1)*r)+1:end), :);
bldstst2=cell2table(bldstst);
bldstst3 = boxLabelDatastore(bldstst2(:,1:2));
imdstst2 = imageDatastore(imdstst);
dstst = combine(imdstst2, bldstst3);
dstst = transform(dstst,@(data)preprocessData(data,[800 800 3]));
inputSize = [800 800 3];
numClasses = 2;
numAnchors = 4;
[anchorBoxes, meanIoU] = estimateAnchorBoxes(dstrn, numAnchors)
featureExtractionNetwork = RestNet50;
```

Figure 10.15 The code for utilizing **trainYOLOv2ObjectDetector** in creating a trained yolov2ObjectDetector.

```
featureLayer = 'activation_40_relu';
lgraph3 = yolov2Layers(inputSize,numClasses,anchorBoxes,featureExtractionNetwo
rk,featureLayer);
valData=dsval;
options = trainingOptions('adam', ...
'MiniBatchSize',1, ...
'MaxEpochs',1, ...
'InitialLearnRate',1e-3, ...
'LearnRateSchedule','piecewise', ...
'LearnRateDropFactor',0.1, ...
'LearnRateDropPeriod',10, ...
'Shuffle','every-epoch', ...
'ValidationData',valData,...
'ValidationFrequency',40,...
'Plots','training-progress', ...
'Verbose',false);
[detector,info] = trainYOLOv2ObjectDetector(dstrn,lgraph3,options);
```

Figure 10.15 (Cont'd)

Figure 10.16 shows yolov2ObjectDetector performance in terms of the model RMSE and loss.

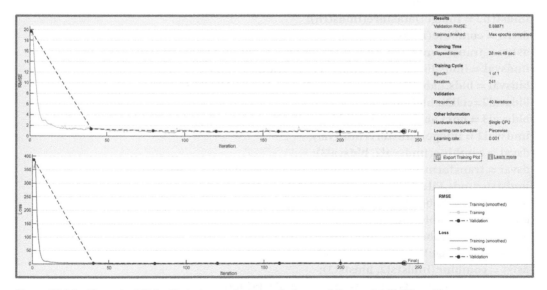

Figure 10.16   The yolov2ObjectDetector performance in terms of the model RMSE and loss.

The following code shows a sample of three images, out of the 30-image test data set, to demonstrate how yolov2ObjectDetector predicts the label box and its tag.

```
I=imread(imdstst2.Files{7});%Repeated twice for image #11 and image #23.
I=imresize3(I,[800 800 3]);
[bbox, score, label] = detect(detector, I);
detectedImg = insertObjectAnnotation(I,'rectangle',bbox,label);
figure
imshow(detectedImg)
```

Figure 10.17 shows the predicted labeled boxes for the three images, using the trained yoloveObjectDetector.

Figure 10.17 Demonstration of the prediction goodness of the trained yoloveObjectDetector.

Ground Truth Labeler (GTL) App

At the command prompt, type in:

```
>>groundTruthLabeler
```

Alternatively, we go to APPS | AUTOMOTIVE menu and click on **"Ground Truth Labeler"** icon. Figure 10.18 shows a portion of the main window (*top*) and the Add/Remove Signal window (*bottom*) where we can browse for a video file and import into the GTL environment. Click on **"OK"** button to close the Add/Remove Signal window.

As pointed out earlier in the previous section, we create two ROI labels: one for the red pole and another for the green pole. The automatic labeling has been explained in the previous section.

First, we need to label the red and green pole at different time steps or intervals, starting from zero up to an end time of 10 seconds, as indicated in Figure 10.19.

Second, one may select one of the last two automated algorithms, shown in the **"Select Algorithm"** menu of Figure 10.19. Either one can be used to detect the red and green pole between the start and end time.

Third, initiate the automatic labeling and follow the instructions. Error messages will pop-up if the user violates any of the requirements imposed by the selected automated labeling algorithm.

Fourth, review each frame between the start and end time. You may modify any of the automatically labeled ROI boxes.

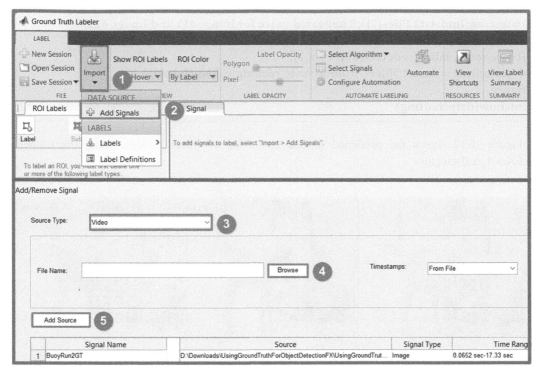

Figure 10.18 The main window (*top*) of the Ground Truth Labeler and the Add/Remove Signal window (*bottom*).

Figure 10.19 The span of the video is set, the automatic labeling algorithm is selected, some frames in between are labeled, and the selected algorithm will do the job of automatic labeling.

Fifth, accept the automatic ROI labeling at the end to let things take into effect.

To export a **groundTruth** object from a labeling app, on the app toolstrip, select "**Export to Workspace**" sub-menu. The app exports the object to MATLAB® workspace. The **groundTruth** object can be saved to a MATLAB *.mat file and be loaded later.

Figure 10.20 shows the complete code for retrieving the ground Truth object which was created earlier by GTL app and saved to a *.mat file, splitting the data into training, validation, and test data set, creating yolov2Layers as a layer graph to be used in training trainYOLOv2ObjectDetector, defining training options, training the network, and evaluating the performance via testing the trained model against the test data set.

```
load('gTruthTraining.mat')
% Treatment below is to convert gTruth from groundTruthMultisignal into gTruth2.
% Otherwise, MATLAB will tell that an error results from using %objectDetectorTrai
ningData>parseInputs
%The value of 'gTruth' is invalid; its type is groundTruthMultisignal.
labelDefinitions = gTruth.LabelDefinitions;
labelDefinitions = removevars(labelDefinitions, 'LabelColor');
labelDefinitions = removevars(labelDefinitions, 'SignalType');
labelDefinitions(2,:) = [];
labelDefinitions(end,:) = [];
labelData = gTruth.ROILabelData.BuoyRun2GT(:,1:2);
gSource = groundTruthDataSource("BuoyRun2GT.mp4");
gTruth2 = groundTruth(gSource,labelDefinitions,labelData);
dataFolder = fullfile(tempdir,"EvalData3");
if ~exist(dataFolder,'dir')
  mkdir(dataFolder);
end
addpath(dataFolder);
[imds blds] = objectDetectorTrainingData(gTruth2,...
 'SamplingFactor',1,'WriteLocation',dataFolder);
% Top 80% for training, 10 % for validation, and the last 10 % for testing
r= size(imds.Files,1);
P = 0.80 ;
idx = randperm(r) ;
imdstrn = imds.Files(idx(1:round(P*r)), :);
bldstrn = blds.LabelData(idx(1:round(P*r)),:);
bldstrn2=cell2table(bldstrn);
bldstrn3 = boxLabelDatastore(bldstrn2(:,1:2));
imdstrn2 = imageDatastore(imdstrn);
dstrn = combine(imdstrn2, bldstrn3);
imdsval = imds.Files(idx(round(P*r)+1:round((P+0.1)*r)), :);
bldsval = blds.LabelData(idx(round(P*r)+1:round((P+0.1)*r)), :);
bldsval2=cell2table(bldsval);
bldsval3 = boxLabelDatastore(bldsval2(:,1:2));
imdsval2 = imageDatastore(imdsval);
dsval = combine(imdsval2, bldsval3);
```

Figure 10.20 The code for training, validating, and testing a YOLOv2 object detector, and evaluating its performance.

```
imdstst=imds.Files(idx(round((P+0.1)*r)+1:end), :);
bldstst = blds.LabelData(idx(round((P+0.1)*r)+1:end), :);
bldstst2=cell2table(bldstst);
bldstst3 = boxLabelDatastore(bldstst2(:,1:2));
bldstst4 = transform(bldstst3,@(data)preprocessData(data,[224 224 3]));
imdstst2 = imageDatastore(imdstst);
dstst = combine(imdstst2, bldstst3);
% Let us create yolov2Layers and use Object Detection Using YOLO v2 Deep Learning
inputSize = [224 224 3];
numClasses = 2;
numAnchors = 4;
[anchorBoxes, meanIoU] = estimateAnchorBoxes(bldstrn3, numAnchors);
%To generatee a copy of CNN, use RestNet50=resnet50;
% Or use a created copy by: load('ReseNet50.mat');
featureExtractionNetwork = RestNet50;
featureLayer = 'activation_40_relu';
lgraph3=yolov2Layers(inputSize,numClasses,anchorBoxes,featureExtractionNetwor
k,featureLayer);
valData=dsval;
options = trainingOptions('adam', ...
'MiniBatchSize',1, ...
'MaxEpochs',15, ...
'InitialLearnRate',1e-5, ...
'LearnRateSchedule','piecewise', ...
'LearnRateDropFactor',0.1, ...
'LearnRateDropPeriod',10, ...
'Shuffle','every-epoch', ...
'ValidationData',valData,...
'ValidationFrequency',40,...
'Plots','training-progress', ...
'Verbose',false);
[detector,info] = trainYOLOv2ObjectDetector(dstrn,lgraph3,options);
results2 = struct;
for i2 = 1: size(imdstst,1)
  I = imread(imdstst2.Files{i2});
  [bboxes, scores, labels] = detect(detector, I);
  I=[];
  results2(i2).Boxes = bboxes;
  results2(i2).Scores = scores;
  results2(i2).Labels = labels;
end
results2 = struct2table(results2);
[am2, fppi2, missRate2] = evaluateDetectionMissRate(results2,dstst);
%Plot log-miss-rate (y) vs. FPPI (x) curve, using a base-10 logarithmic scale for both
axes.
figure;
```

Figure 10.20 (Cont'd)

```
loglog(fppi2{1,1}, missRate2{1,1});
grid on;
loglog(fppi2{2,1}, missRate2{2,1});
xlabel('FPPI');
ylabel('Log Miss Rate')
title(sprintf('Mean log MissRate for TestData 2 classes= %.2f', mean(am2)))
grid off;
[ap2, recall2, precision2] = evaluateDetectionPrecision(results2, dstst);
figure;
plot(recall2{1,1}, precision2{1,1});
xlabel('Recall');
ylabel('Precision')
grid on;
plot(recall2{2,1}, precision2{2,1});
xlabel('Recall');
ylabel('Precision')
grid on;
title(sprintf('Average Precision for TestData 2 classes= %.2f', mean(ap2)))
```

Figure 10.20 (Cont'd)

Figure 10.21 shows the YOLO network RMSE and loss as a function of number of iterations.

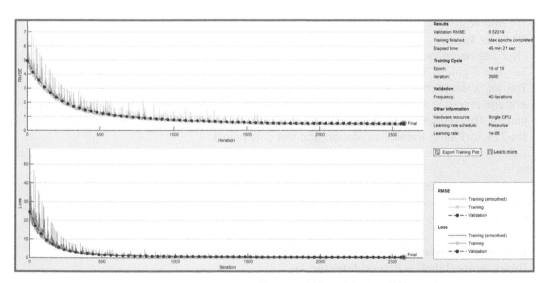

Figure 10.21   The YOLO network performance as a function of number of iterations.

Figure 10.22 (*left*) shows the mean log miss rate for the two classes found in **gTruth** object and (*right*) the average precision for the same classes, using the test data set.

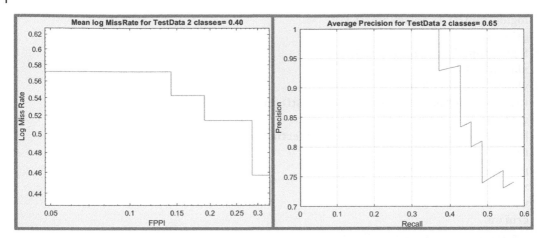

Figure 10.22 The YOLO network performance in terms of the mean log miss rate (*left*) for the two classes found in **gTruth** object and (*right*) the average precision for the same classes, using the test data.

NOTE #1

The terms appearing in Figure 10.22 are defined as:
Precision = TP/(TP+FP)
Recall =   TP/(TP+FN)
Miss Rate = FN/(TP+FN)
Where;
TP: True Positive
FP: False Positive
FN: False Negative
See also Figure 7.76.
The log average miss rate should be as low as possible (close to zero); on the contrary, the average precision should be close to unity. In Figure 10.22, the values are somewhere in the middle. This requires more training data than we have here. In our case the total data points are 215, including, training, validation, and test data points.

The following code reads a sample image and lets the trained YOLO object detector evaluate the bounding box, the objectness score, and the label, then displays the original image embossed by the predicted objects: RPole and GPole.

```
I=imread(imdstst2.Files{3});% Repeated for image #9 and # 13.
[bbox, score, label] = detect(detector, I);
detectedImg = insertObjectAnnotation(I,'rectangle',bbox,label);
figure (1)
imshow(detectedImg)
```

Figure 10.23 shows a sample of three images out of the examined test data set to demonstrate the predictability of YOLO object detector.

Figure 10.23 A sample of three images out of the examined test data set, demonstrating YOLO object detectability.

Figure 10.24 shows workspace variables created upon compiling the code of Figure 10.20.

Workspace		Workspace	
Name ▲	Value	Name ▲	Value
am2	[0.2718;0.5367]	idx	*1x215 double*
anchorBoxes	[130,35;76,27;68,18;47,13]	imds	*1x1 ImageDatastore*
ap2	[0.7536;0.5364]	imdstrn	*172x1 cell*
bbox	*3x4 double*	imdstrn2	*1x1 ImageDatastore*
bboxes	[299,129,12,50]	imdstst	*21x1 cell*
blds	*1x1 boxLabelDatastore*	imdstst2	*1x1 ImageDatastore*
bldstrn	*172x2 cell*	imdsval	*22x1 cell*
bldstrn2	*172x2 table*	imdsval2	*1x1 ImageDatastore*
bldstrn3	*1x1 boxLabelDatastore*	info	*1x1 struct*
bldstst	*21x2 cell*	inputSize	[224,224,3]
bldstst2	*21x2 table*	label	*3x1 categorical*
bldstst3	*1x1 boxLabelDatastore*	labelData	*260x2 timetable*
bldstst4	*1x1 TransformedDatastore*	labelDefinitions	*2x4 table*
bldsval	*22x2 cell*	labels	*1x1 categorical*
bldsval2	*22x2 table*	lgraph3	*1x1 LayerGraph*
bldsval3	*1x1 boxLabelDatastore*	meanIoU	0.7585
detectedImg	*374x558x3 uint8*	missRate2	*2x1 cell*
detector	*1x1 yolov2ObjectDetector*	numAnchors	4
dstrn	*1x1 CombinedDatastore*	numClasses	2
dstst	*1x1 CombinedDatastore*	options	*1x1 TrainingOptionsADAM*
dsval	*1x1 CombinedDatastore*	P	0.8000
featureExtractionNet...	*1x1 DAGNetwork*	precision2	*2x1 cell*
featureLayer	'activation_40_relu'	r	215
fppi2	*2x1 cell*	recall2	*2x1 cell*
gSource	*1x1 groundTruthDataSource*	RestNet50	*1x1 DAGNetwork*
gTruth	*1x1 groundTruthMultisignal*	results2	*21x3 table*
gTruth2	*1x1 groundTruth*	score	[0.8114;0.6138;0.7531]
I	*374x558x3 uint8*	scores	0.7196
i2	21	valData	*1x1 CombinedDatastore*

Figure 10.24 Workspace variables created by running the code in Figure 10.20.

Running/Walking Classification with Video Clips using LSTM

Our approach involves utilizing deep learning techniques for video classification and activity recognition. By employing deep learning, we can effectively categorize the activity or action depicted in a series of images extracted from visual data sources, including video streams. Through the analysis of multiple video frames, our vision-based activity recognition system can accurately predict actions within image sequences, such as walking, swimming, or sitting. The applications of activity recognition from video extend to various domains, including human-computer interaction, anomaly detection, and surveillance, where it proves highly advantageous.

This is a simple example of video classification using LSTM-based RNN with MATLAB. The official [1] example requires down-loading a dataset about 2 GB, nevertheless, this example requires a small amount of data [2], which helps explain the concept of video classification. A pre-trained network, **resnet18**, is used for features extraction using **activations** function, The features extracted from the pre-trained network are fed into LSTM layer, as shown in Figure 10.25. One may choose, however, other networks such as **GoogleNet, resnet50**, and **AlexNet,** when the final accuracy is not high enough.

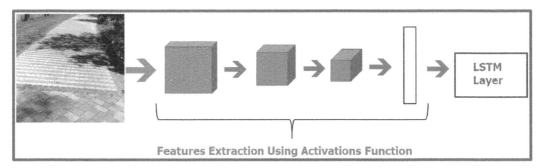

Figure 10.25   Features extraction to be plugged into an LSTM-based RNN.

1. Resnet CNN
Load a copy of resnet18 CNN.

```
netCNN = resnet18;
```

2. Load Video Data
- The video clips used for classification purposes were obtained from recordings of running and walking sessions, with durations of approximately 5 minutes and 10 minutes, respectively.
- Both types of videos were captured along the same path to eliminate any variations in the scene that could affect the analysis.
- In future instances, video scenes featuring diverse locations and moving conditions can be prepared as examples.

```
VF= 'D:\OneDrive_HCT\MatlabBook\MLDL\video_classification_LSTM\';
RunVideo=VideoReader([VF, 'Running.mov']); % load the video taken while running
WalkVideo=VideoReader([VF, 'Walking.mov']);%
f=figure;
```

```
title('Video while Running/Walking');hold on
set(gcf,'Visible','on')
numFrames = 6/(1/RunVideo.FrameRate); %"6" is a duration (second) to show
for i = 1:numFrames
    RunFrame=readFrame(RunVideo);
    WalkFrame=readFrame(WalkVideo);
    imshow([RunFrame,WalkFrame]);
    drawnow
    pause(1/RunVideo.FrameRate)
    hold on
end
hold off
```

Figure 10.26 shows a snapshot for the previously presented videos.

Figure 10.26 A snapshot for the two presented videos for both running and walking scenario.

3. Video Time Reset to Zero

Reset the state of videos back to zero.

```
% reset the state of waking/running video, otherwise the frames already
% read are not to be retrieved
WalkVideo.CurrentTime=0;
RunVideo.CurrentTime=0;
```

4. Features Extraction

Read all frames and extract features to save into variables R and W

- Image features are extracted with the pre-trained network to feed to the LSTM-based RNN.
- The function "**activations**" returns the vector of the extracted features.

```
if (exist('W.mat')==2)&&(exist('R.mat')==2)
  load W.mat
  load R.mat
else
  RFrames=zeros(224,224,3,RunVideo.NumFrames,'uint8');
  WFrames=zeros(224,224,3,WalkVideo.NumFrames,'uint8');
  for i=1:RunVideo.NumFrames
    RFrames(:,:,:,i)=imresize(readFrame(RunVideo),[224 224]);
    % the video frames should be resized into 224 by 224 since the
    % resnet18 only accepts that size.
  end
  for i=1:WalkVideo.NumFrames
    WFrames(:,:,:,i)=imresize(readFrame(WalkVideo),[224 224]);
  end
  R=single(activations(netCNN,RFrames,'pool5','OutputAs','columns'));
  W=single(activations(netCNN,WFrames,'pool5','OutputAs','columns'));
end
```

5. Image Datasets with a Short Time Video Clip
Prepare a set of image feature with a short time video clip

- A video clip whose duration is from **minDuration** to **maxDuration** as defined below was obtained.
- Specify the number of the clips to obtain from each video.

```
minDuration=2;
maxDuration=4;
numData=100;
FrameRate=RunVideo.FrameRate;
RData=cell(numData,1);
WData=cell(numData,1);
for i=1:numData
  ClipDuration=randi((maxDuration-minDuration)*FrameRate,[1 1])
+minDuration*FrameRate;
  StartingFrameNumRun=randi(RunVideo.NumFrames-(maxDuration+minDurati
on)*FrameRate,[1 1])+minDuration*FrameRate;
  StartingFrameNumWalk=randi(WalkVideo.NumFrames-(maxDuration+minDura
tion)*FrameRate,[1 1])+minDuration*FrameRate;
  RData{i}=R(:,StartingFrameNumRun:StartingFrameNumRun+ClipDuration);
  WData{i}=W(:,StartingFrameNumWalk:StartingFrameNumWalk+ClipDuration);
end
```

6. Training and Validation Data Preparation
Partition the data. Assign 70% of the data to the training partition and 30% to the validation partition.

```
idx = randperm(numData);
N = floor(0.7 * numData);
```

```
sequencesTrainRun = {RData{idx(1:N)}};
sequencesTrainWalk = {WData{idx(1:N)}};
sequencesTrain=cat(2,sequencesTrainRun,sequencesTrainWalk);
labelsTrain=categorical([zeros(N,1);ones(N,1)],[0 1],{'Run','Walk'});
sequencesValidRun = {RData{idx(N+1:end)}};
sequencesValidWalk = {WData{idx(N+1:end)}};
labelsValidation=categorical([zeros(numel(sequencesValidWalk),1);ones(numel(seq
uencesValidWalk),1)],[0 1],{'Run','Walk'});
sequencesValidation=cat(2,sequencesValidRun,sequencesValidWalk);
```

7. Create LSTM RNN

Create an LSTM network that can classify the sequences of feature vectors representing the videos. Define the LSTM network architecture. Specify the following network layers:

- A sequence input layer with an input size corresponding to the feature dimension of the feature vectors.
- The dimension of the extracted feature with **resnet18** is 512; hence, **numFeatures** is set to *512*.
- A biLSTM layer with 1600 hidden units followed by a dropout layer. To output only one label for each sequence by setting the '**OutputMode**' option of the biLSTM layer to '*last*'.
- 2 LSTM layers can also be inserted into the "layers" instead of one biLSTM layer to seek a potential improvement in model accuracy.
- A fully connected layer with an output size corresponding to the number of classes (2 for both running and walking scenario), a softmax layer, and a classification layer.
- If we prefer to estimate a numerical value out of the time-series data, we can prepare "**regressionLayer**" with numerical label data instead of the categorical "**labelsTrain**".
- The dropout layer contributes to prevent the network from reaching the saturation or "overfitting" case with the given training data.

```
numFeatures = size(R,1);
numClasses = 2;
layers = [
  sequenceInputLayer(numFeatures,'Name','sequence')
  bilstmLayer(1600,'OutputMode','last','Name','bilstm')
  dropoutLayer(0.4,'Name','drop')
  fullyConnectedLayer(numClasses,'Name','fc')
  softmaxLayer('Name','softmax')
  classificationLayer('Name','classification')];
```

8. Specify Training Options

```
miniBatchSize = 20;
numData = numel(sequencesTrainRun);
numIterationsPerEpoch = floor(numData / miniBatchSize)*3;
options = trainingOptions('adam', ...
  'MiniBatchSize',miniBatchSize, ...
  'MaxEpoch',24, ...
```

```
'InitialLearnRate',1e-4, ...
'GradientThreshold',3, ...
'Shuffle','every-epoch', ...
'ValidationData',{sequencesValidation,labelsValidation}, ...
'ValidationFrequency',numIterationsPerEpoch, ...
'Plots','training-progress', ...
'Verbose',false);
```

9. Train LSTM Network

Train the network with the extracted image features, using the **trainNetwork** function.

```
[netLSTM,info] = trainNetwork(sequencesTrain,labelsTrain,layers,options);
```

Figure 10.27 shows the LSTM RNN training accuracy and loss for both training and validation step.

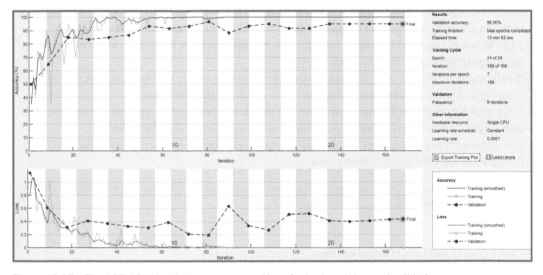

Figure 10.27   The LSTM RNN training accuracy and loss for both training and validation step.

10. The Network Accuracy

Calculate the classification accuracy of the network on the validation set. If the accuracy is quite satisfactory, please prepare the test video clips to explore the feasibility of this LSTM RNN.

```
YPred = classify(netLSTM,sequencesValidation,'MiniBatchSize',miniBatchSize);
accuracy = mean(YPred == labelsValidation)
% please confirm the balance of the classification.
confusionchart(labelsValidation,YPred)
```

Figure 10.28 shows the confusion matrix plot for the true versus predicted class of run/walk scenario.

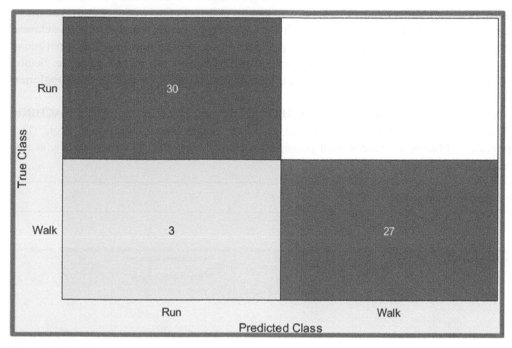

Figure 10.28 The confusion matrix plot for the true versus predicted class of run/walk scenario.

Figure 10.29 shows workspace variables created upon running the previous code.

Name ▲	Value	Name ▲	Value
FrameRate	10	RData	100x1 cell
i	100	RFrames	4-D uint8
idx	1x100 double	RunFrame	250x250x3 uint8
info	1x1 struct	RunVideo	1x1 VideoReader
labelsTrain	140x1 categorical	sequencesTrain	1x140 cell
labelsValidation	60x1 categorical	sequencesTrainRun	1x70 cell
layers	6x1 Layer	sequencesTrainWalk	1x70 cell
maxDuration	4	sequencesValidation	1x60 cell
minDuration	2	sequencesValidRun	1x30 cell
miniBatchSize	20	sequencesValidWalk	1x30 cell
N	70	StartingFrameNumRun	1796
netCNN	1x1 DAGNetwork	StartingFrameNumWalk	3164
netLSTM	1x1 SeriesNetwork	VF	'D:\OneDrive_HCT\MatlabBook\
numClasses	2	W	512x6733 single
numData	70	WalkFrame	250x250x3 uint8
numFeatures	512	WalkVideo	1x1 VideoReader
numFrames	60	WData	100x1 cell
numIterationsPerEpoch	9	WFrames	4-D uint8
options	1x1 TrainingOptionsADAM	YPred	60x1 categorical
R	512x3608 single		

Figure 10.29 Workspace variables created by running the previous code.

Experiment Manager (EM) App

In this section, we show how to train a deep learning network for classification using Experiment Manager app. We train two networks to classify images of flowers into twelve classes. Each network is trained using three algorithms. In each case, a confusion matrix compares the true classes for a set of validation images with the classes predicted by the trained network. This experiment requires the Deep Learning Toolbox™ Model for **GoogleNet** Network support package. Before one runs the experiment, he/she must install this support package by calling the **googlenet** function and clicking the download link.

To open Experiment Manager app, from MATLAB top tool menu, go to APPS | MACHINE LEARNING AND DEEP LEARNING, and click on "**Experiment Manager**" icon. The "**Experiment Manager**" window will pop-up as shown in Figure 10.30. Alternatively, at the command prompt type in

```
>>experimentManager
```

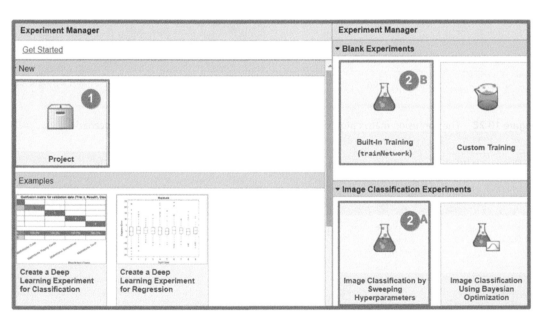

Figure 10.30 The "**Experiment Manager**" window where the user selects the project type.

In this case we select a new project with a type as "**Image Classification by Sweeping Hyperparameters**".

Alternatively, the user may also use "**Built-In Training (trainNetwork)**" template for doing the same job, except the latter one does not populate the hyperparameters table, shown in Figure 10.31. Moreover, "**Built-In Training (trainNetwork)**" template is valid for both classification and regression cases as long as **trainNetwork** function can be used. See previous chapters (5-9) for using **trainNetwork** function both under classification and regression cases.

NOTE #2
For the sake of learning, the user may open one of the given examples, shown in Figure 10.30, with the same project type. Experiment Manager loads a project with a preconfigured experiment that the user can inspect and run.

The "**Specify Project Folder Name**" window will pop-up asking to select a folder and give a name for the project so that the project can be saved there and be opened later by EM app. Figure 10.31 shows the main window of EM app after making the above selection as far as the project type is concerned.

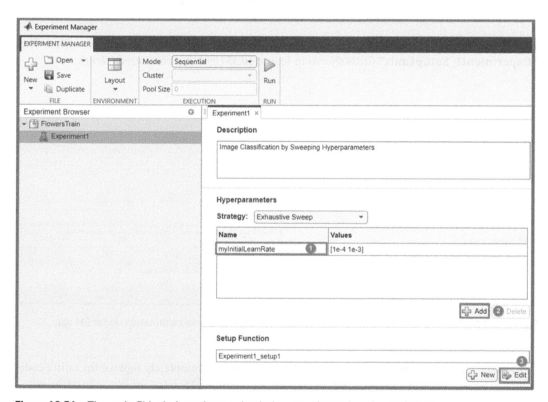

Figure 10.31   The main EM window where a simulation experiment is to be carried out.

The idea here is to select some training settings as pre-defined hyperparameters which will, of course, affect the performance of training model in hand. We will rename the given parameter. Moreover, will add two more hyperparameters: One for the solvent type and another for the network type. To modify the given parameter, shown in Figure 10.31, double click on the name "**myInitialLearnRate**" and change it to "**MyILR**".

To add two more parameters and change their names and assign values, we first click on "**Add**" button twice and EM will add two more parameters but with [0] default values as shown in Figure 10.32 (*left*). Figure 10.32 (*right*) shows the modified names and values of the two added parameters. We will shortly show the code to handle how EM will pick up such three parameters and run the simulation, accordingly.

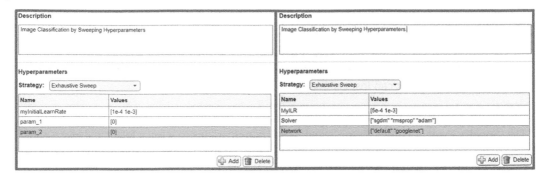

Figure 10.32 Insertion of two parameters, in addition to the initially added parameter.

Click on "**Edit**" button, as shown in Figure 10.31, to open up the MATLAB editor. The "**Experiment1_Setup1.mlx**" file is shown in Figure 10.33 (*left*) and continued (*right*).

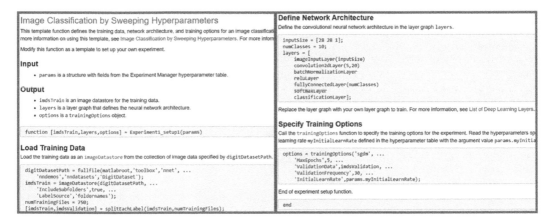

Figure 10.33 The "**Experiment1_Setup1.mlx**" file associated with running simulation under EM app.

To make use of the three parameters found in Figure 10.32, completely replace the entire code found in Figure 10.33 by the following code shown in Figure 10.34.

> %**Experiment Manager uses the outputs of this function to call the trainNetwork**
> %**function. For more information on using this %template, see Image Classification**
> **by %Sweeping Hyperparameters. For more information on built-in experiments, see**
> %**Configure Built-In Training Experiment.**
> %**Modify this function as a template to set up your own experiment.**
> %**Input is params which is a structure with fields from the Experiment Manager**
> %**hyperparameter table.**
> %**Output terms are three:**
> %**1. imdsTrain is an image datastore for the training data.**
> %**2. layers is a layer graph that defines the neural network architecture.**

Figure 10.34 The new code to be fed to EM app to make use of the added hyperparameters.

```
%3. options is a trainingOptions object.
function [imdsTrain,layers,options] = Experiment1_setup1(params)
%Replace the data with your own training and validation data. For more
information, see Data Sets for Deep Learning.
%Define Network Architecture
ImDir=fullfile('D:\OneDrive_HCT\MatlabBook\MLDL\Flowers');
%Create an imageDatastore, specifying the read function as a handle to imread
function.
JPGDS = imageDatastore(ImDir, 'IncludeSubfolders',true,'LabelSource','foldernames',...
'FileExtensions','.jpg','ReadFcn',@(x) imread(x));
imdsTrain=augmentedImageDatastore([227 227],JPGDS);
rng(123);
imds2=shuffle(JPGDS);
imdsValidation=augmentedImageDatastore([227 227],imds2);
rng(123);
switch params.Network
  case "default"
    inputSize = [227 227 3];
    numClasses = 12;
    layers = [
      imageInputLayer(inputSize)
      convolution2dLayer(5,20)
      batchNormalizationLayer
      reluLayer
      fullyConnectedLayer(numClasses)
      softmaxLayer
      classificationLayer];
  case "googlenet"
    inputSize = [224 224];
    numClasses = 12;
    imdsTrain=augmentedImageDatastore(inputSize,JPGDS);
    imdsValidation=augmentedImageDatastore(inputSize,imds2);
    net = googlenet;
    layers = layerGraph(net);
    newLearnableLayer = fullyConnectedLayer(numClasses, ...
    Name="new_fc", ...
    WeightLearnRateFactor=10, ...
    BiasLearnRateFactor=10);
    layers = replaceLayer(layers,"loss3-classifier",newLearnableLayer);
    newClassLayer = classificationLayer(Name="new_classoutput");
    layers = replaceLayer(layers,"output",newClassLayer);
  otherwise
    error("Undefined network selection.");
end
options = trainingOptions(params.Solver, ...
```

Figure 10.34 (Cont'd)

```
    MiniBatchSize=20, ...
    MaxEpochs=6, ...
    InitialLearnRate=params.MyILR, ...
    Shuffle="every-epoch", ...
    ValidationData=imdsValidation, ...
    ValidationFrequency=50, ...
    Verbose=false);
end
```

Figure 10.34 (Cont'd)

NOTE #3

BEFORE RUNNING EM APP, YOU MUST SAVE THE CHANGES MADE IN THE *.MLX FILE SIMPLY BY CLICKING ON "SAVE" BUTTON, FOUND UNDER MATLAB EDITOR. (FIGURE 10.35)

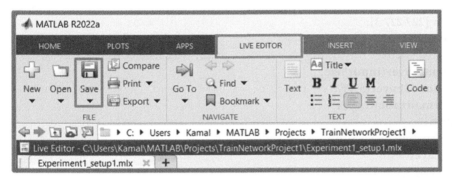

Figure 10.35 Saving the changes made to *.mlx code prior to running the EM simulation.

Click on "**Run**" button, shown at the middle of the top tool menu of Figurer 10.31, to let EM app start the simulation. Figure 10.36 shows the end of simulation where we have (2×2×3) = 12 possible cases or scenarios for the assigned three hyper parameters.

Trial	Status	Actions	Progress	Elapsed Time	MyILR	Solver	Network	Training Accu...	Training Loss	Valid
1	Complete (..		100.0%	0 hr 4 min 26 sec	0.0005	sgdm	default	100.0000	0.0131	
2	Complete (..		100.0%	0 hr 4 min 24 sec	0.0010	sgdm	default	70.0000	1.1127	
3	Complete (..		100.0%	0 hr 4 min 41 sec	0.0005	rmsprop	default	85.0000	6.9203	
4	Complete (..		100.0%	0 hr 4 min 47 sec	0.0010	rmsprop	default	80.0000	19.3115	
5	Complete (..		100.0%	0 hr 4 min 50 sec	0.0005	adam	default	100.0000	0.0000	
6	Complete (..		100.0%	0 hr 4 min 51 sec	0.0010	adam	default	100.0000	0.0000	
7	Complete (..		100.0%	0 hr 17 min 58 sec	0.0005	sgdm	googlenet	100.0000	0.0605	
8	Complete (..		100.0%	0 hr 19 min 45 sec	0.0010	sgdm	googlenet	100.0000	0.0315	
9	Complete (..		100.0%	0 hr 19 min 12 sec	0.0005	rmsprop	googlenet	55.0000	1.2469	
10	Complete (..		100.0%	0 hr 17 min 34 sec	0.0010	rmsprop	googlenet	20.0000	2.0350	
11	Complete (..		100.0%	0 hr 17 min 43 sec	0.0005	adam	googlenet	65.0000	1.3767	
12	Complete (..		100.0%	0 hr 17 min 20 sec	0.0010	adam	googlenet	5.0000	2.4941	

Figure 10.36 The EM simulation results showing the 12 possible scenarios in light of the given hyper parameters.

The user may sort the results by selecting any of the results columns, clicking on the right top corner of the selected column, and then selecting the order theme. Figure 10.37 shows the results, but ascendingly ordered based on column titled: **"Validation Loss"**. This is the first best case (scenario #8) where the initial learning rate is *0.001*, the network type is **GoogleNet**, and the solver method is **"sgdm"**. At the same time, this scenario #8 has also a low training loss and 100% training accuracy. Notice that the second-best case (scenario #7) is quite comparable with the first best case (#8) as both practically have very close indices in terms of training and validation accuracy and loss values. For the second-best case (#7), the initial learning rate, however, is *0.0005*. In general, **GoogleNet** is more accurate than the **default** network as the former has more layers than the latter. However, the CPU time for **GoogleNet** is at least three times longer than that of the **default** network.

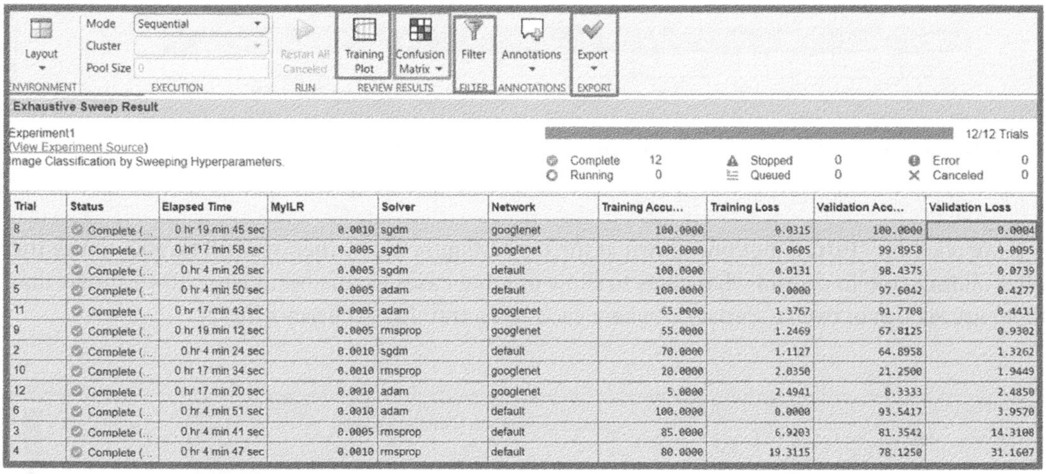

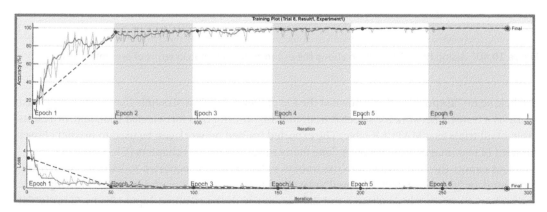

Figure 10.37 The results are ascendingly ordered based on the reported validation loss value.

If we wish to see the training plot and confusion matrix for the first-best case, we have to click on **"Training Plot"** and **"Confusion Matrix"** button shown at the top menu of Figure 10.37. Figure 10.38 shows the training plot for the trained model in terms of accuracy and loss as a function of number of iterations.

Figure 10.38 The training performance of the network in terms of accuracy and loss.

Figure 10.39 shows the confusion matrix plot for the first-best case.

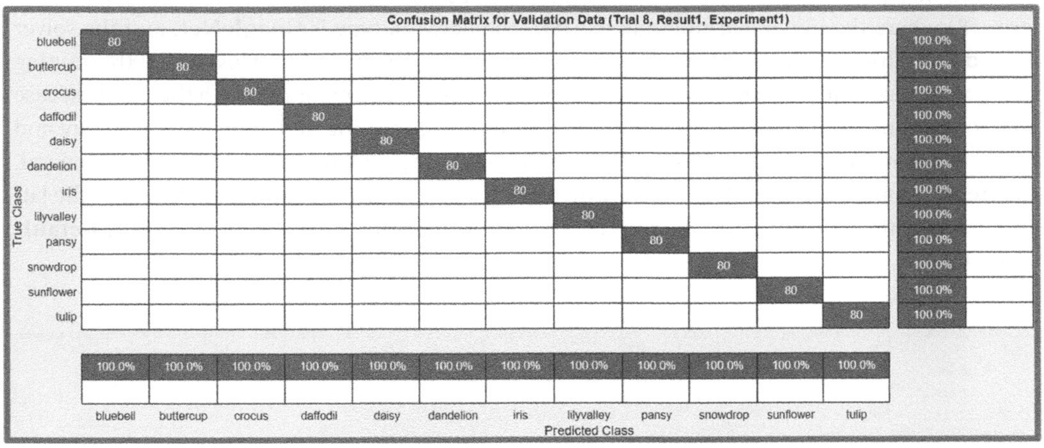

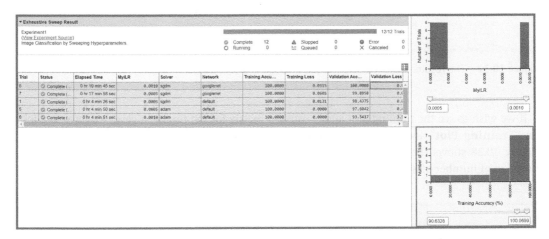

Figure 10.39 The confusion matrix plot for true versus predicted flower type.

If we click on "**Filter**" button, shown at the top of Figure 10.37, then we will be able to use the filter sliding rule to control what cases to show or what cases to rule out. Figure 10.40 shows that there are six out of twelve scenarios, where each has a training accuracy of 100%.

Figure 10.40 Using the "**Filter**" button allows the use to zoom in or out, in terms of presented cases.

Clicking on "**Export**" button, shown at the top of Figure 10.37, will enable the user to export his/ work as a trained network for selected trial, training information for selected trial, and/or results table of all trials as a MATLAB table. The exported trained network can be used to predict classes for a set of test images, as it has been done many times in previous chapters (Ch. 5–9).

Let us export the first case where the trained network will be exported to workspace environment and be given a name: "**trainedNetwork**", a DAGNetwork variable. The following code shows how to randomly generate a test dataset, resize to the required input size by **GoogleNet**

DAG network, predicting the image classes by "**trainedNetwork**" object, and comparing the predicted versus the true flower class.

```
imds3=shuffle(JPGDS);
targetSize = [224,224];
imds4 = transform(imds3,@(x) imresize(x,targetSize));
Ytestpreds= classify(trainedNetwork, imds4);
Ytst= imds3.Labels;
confusionchart(Ytst,Ytestpreds);
numtrue=nnz(Ytst == Ytestpreds);
accuracy=(numtrue/numel(Ytestpreds))*100;
disp(['Accuracy = ', num2str(accuracy),' %'])
```

Copied from Command Window: Accuracy = 100%. The generated confusion matrix plot is exactly the same as that appearing in Figure 10.39.

Figure 10.41 shows workspace variables created by EM simulation and testing data.

Figure 10.41 Workspace variables related to EM simulation and test data prediction.

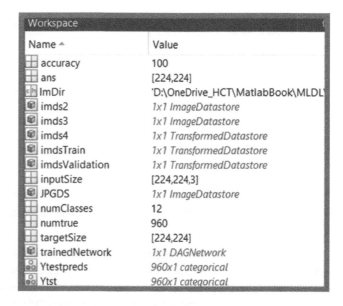

Finally, if the user switches back to MATLAB main window, then he/she will be able to see the folder name and location for EM project, its created resources folders and files, like *.mlx, *.mat, and *.prj file.

Image Batch Processor (IBP) App

To open Image Batch Processor (IBP), from MATLAB top tool menu, go to APPS | IMAGE PROCESSING AND COMPUTER VISION, and click on "**Image Batch Processer**" icon. The "**Image Batch Processor**" window will pop-up as shown in Figure 10.42. Alternatively, at the command prompt type in

>>**imageBatchProcessor**

We also clicked on "**Load Images**" button, selected the folder name or location for the batch of images to be processed, then clicked on "**Load**" button at the middle bottom of the inset figure. The set made of 960 images is ready to be processed. Processing here means applying an action partly or totally on the set of images. This means that we can select a subset or the entire set of images to be proceed by applying a selected function from the "**Function Name**" menu. The function can be viewed or edited by clicking on "**Edit**" button. Furthermore, the user can create a new function which will carry out any image modification or treatment, like image type conversion and image resizing.

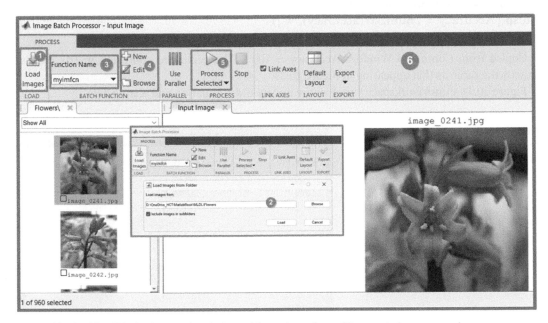

Figure 10.42 The IBP platform main window with a selected set of images to be processed.

The default function named **myimfcn(im)** takes in the image as an input argument and carries out conversion into both grayscale and binary (black-and-white) image. These two types will be the output of the structure called results. Figure 10.43 shows the code for **myimfcn.m** file. For example, the user may remove the last statement and the output will be grayscale images only.

```
function results = myimfcn(im)
%Image Processing Function
% IM    - Input image.
% RESULTS - A scalar structure with the processing results.
%-------------------------------------------------------------
% Auto-generated by imageBatchProcessor App.
% When used by the App, this function will be called for every input image
```

Figure 10.43   **myimfcn.m** code for image conversion into both gray and binary (black-and-white) images

```
% file automatically. IM contains the input image as a matrix. RESULTS is a
% scalar structure containing the results of this processing function.
imgray = im2gray(im);
bw = imbinarize(imgray);
results.imgray = imgray;
results.bw    = bw;
```

Figure 10.43 (Cont'd)

Let us process all images to create a main folder with image-class subfolders but all images will be grayscale images with a size of [227 227 1], unlike the original RGB images which have a size of [227 227 3]. This gives the user the freedom to use either a grayscale-based set of images or RGB-based set of images. From **"Process Selected"** button (see Figure 10.42), select **"Process All"** option and IBP app will start the image conversion process into both grayscale and binary scale images, as shown in Figure 10.44.

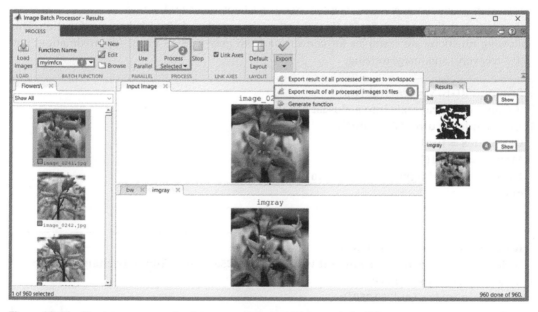

Figure 10.44   The image conversion into a grayscale and binary scale by IBP app.

After the image conversion process for all images is complete, the user will have the chance to save the converted images into a new main folder with image-class subfolders. The name of images as well as the name of sub-folders are all preserved. Figure 10.45 shows the new selected main folder and its image-class sub-folders, similar to the original set of images.

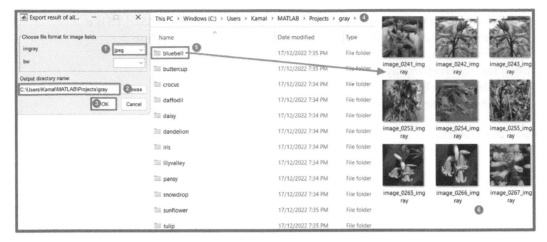

Figure 10.45 The creation of a main folder and image-class sub-folder for grayscale images.

The user will have two separate sets of images: One is RGB-based and another is grayscale-based. Keep in mind that the image conversion can also be done via augmented image data-store conversion, as shown in previous chapters. See also Chapter 6 | NOTE #2.

In addition to the given **myimfcn(im)** function, the user may define his/her own function. The following functions are examples for conversion from gray to RGB image (approximation; not 100% recovery) and image resizing.

```
function RGBIM=gray2rgb(im)
% Converts a grayscale into a closest rgb image.
%GrayIndex = uint8(floor(im));
GrayIndex = uint64(im);
Map      = jet(416);
RGBIM    = ind2rgb(GrayIndex, Map);
end
```

```
function results = imresizecn(im)
% Resize an image from its current values to 250×250 values. You may change to the
%size dimensions you need.
results=imresize(im,[250 250]);
end
```

```
function results = imresizecn(im)
% Convert an RGB into a grayscale image; it does the job of myimfcn(im) function
% See help on imresize3 function on resizing a 3-D volumetric intensity image.
results=imresize3(im,[250 250 1]);
end
```

References

1 *MATLAB Official Documentation: Classify Videos Using Deep Learning.*

2 Kenta. Video classification using LSTM (LSTMによる動画の分類) https://github.com/ KentaItakura/video_classification_LSTM_matlab/releases/tag/v1.4, *GitHub*. Retrieved November 23, 2022.

NOTE #4: End of Chapter 10 Problems

In addition to using **shuffle** and **splitEachLabel** function for creating two additional image datastores for validation and testing step, the user can divide the entire image data store into three image data stores for training, validation, and testing step, using **subset** function. See Appendix A.23 for explaining how to split a given image data store into three sub image data stores.

HCW 10.1 Cat Dog Video Labeling, Training, and Prediction – 1

Use the following video called cat_dog.mp4 (available with Wiley's web companion) as the source of video data. Carry out the following steps:

1) Import to either VL or GTL app.
2) Create two ROI labels: One for cat and another for dog.
3) Carry out automated labeling.
4) Export the ground truth (gTruth) object to workspace environment.
5) Save the variable as *.mat file.
6) Create image data stores and box labeling data stores for training, validation, and testing.
7) Use any R-CNN object detector.
8) Create training options.
9) Train the network.
10) Test the trained network detector to demonstrate its goodness.
11) Repeat steps #7 through #10 for YOLOv2/YOLOv4 object detector.

HCW 10.2 Cat Dog Video Labeling, Training, and Prediction – 2

Use the following video called cat_dog2.mp4 (available with Wiley's web companion) as the source of video data. Carry out the following steps:

1) Import to either VL or GTL app.
2) Create two ROI labels: One for cat and another for dog.
3) Carry out automated labeling.
4) Export the ground truth (gTruth) object to workspace environment.
5) Save the variable as *.mat file.

6) Create image data stores and box labeling data stores for training, validation, and testing.
7) Use any R-CNN object detector.
8) Create training options.
9) Train the network.
10) Test the trained network detector to demonstrate its goodness.
11) Repeat steps #7 through #10 for YOLOv2/YOLOv4 object detector.

HCW 10.3   EM Hyperparameters of CNN Retraining for Merchandise Data Prediction

The MATLAB Merch data set is a small data set containing 75 images of MathWorks merchandise, belonging to five different classes (cap, cube, playing cards, screwdriver, and torch). You can use this data set to try out transfer learning and image classification quickly.

It can be found at one of the following folder locations:

C:\Program Files\MATLAB\R2022a\examples\deeplearning_shared\data\MerchData.zip
C:\Program Files\MATLAB\R2022a\examples\nnet\data\MerchData.zip
C:\Program Files\MATLAB\R2022a\examples\vision\data\MerchData.zip

The images are of size 227×227×3.

Carry out the following steps:

1. Extract MATLAB Merch Data set.

```
filename = 'MerchData.zip';
dataFolder = fullfile(tempdir,'MerchData');
if ~exist(dataFolder,'dir')
    unzip(filename,tempdir);
end
```

2. Load the data as an image datastore using the **imageDatastore** function and specify the folder containing the image data.
3. RestNet-50 network requires an image with 224×224×3 input size. Since the data size is relatively small, use **shuffle** function to create two additional data sets both for validation and testing. See Chapter 7 | **Using shuffle function** section.
4. Build a network by starting with a pretrained **ResNet-50** network (net=resnet50;).
5. Create a fully connected layer with five classes and classification layer. Take RestNet50 CNN layers but replace the last fully connected layer by the newly created FC. The name of the last fully connected layer is **"fc1000"** and the name of the output classification layer is **"ClassificationLayer_fc1000"**. Replace those two layers by your own, but of the same type, layers.
6. Define EM Hyperparameters as shown in Table HCW10.3.

Table HCW10.3   EM Exhaustive Parameters.

Strategy: Exhaustive Sweep	
Name	**Values**
MyILR	[5e-4 1e-3]
MBS	[20 50]
Solver	["adam" "rmsprop" "sgdm"]

Define the image data stores for training, validation, and testing, define the layers variable, and use the following training option in your code:

```
valData=dsval;
miniBatchSize=params.MBS;
options = trainingOptions(params.Solver, ...
      'MiniBatchSize',miniBatchSize, ...
'MaxEpochs',1, ...
'InitialLearnRate',params.MyILR, ...
'LearnRateSchedule','piecewise', ...
'LearnRateDropFactor',0.1, ...
'LearnRateDropPeriod',10, ...
'Shuffle','every-epoch', ...
'ValidationData',valData,...
'ValidationFrequency',30,...
'Plots','training-progress', ...
'Verbose',false);
```

Carry out EM simulation to see which of the above cases will be the best in terms of maximum training accuracy and/or minimum training loss. For the best case, show the training plot, the confusion matrix plot for training and also for testing, and the accuracy of predicting test data classes.

HCW 10.4   EM Hyperparameters of CNN Retraining for Round Worms Alive or Dead Prediction

In this problem, we are given the images of worms (available with Wiley's web companion). Images of roundworms are either alive or dead. (alive worms are round; dead ones are straight).

The images are given in a folder called **WormImages** and the labeling is given in an Excel file where it has the image file name and the worm status.

Carry out the following steps:

1. Get the training images and classes via saving them to an image data store and image data store label.
2. Divide data into training (80%), validation (10%), and testing (10%) sets and prepare the image size that suits **AlexNet** network. See Chapter 6 | NOTE #2 for a grayscale image data store conversion.
3. Build a network by starting with a pretrained **AlexNet** network.

4. Consider all **AlexNet** layers except the last three layers. In addition, add three new layers at the end: **One fullyConnectedLayer**, one **softmaxLayer**, and one **classificationLayer**.

5. Define EM Hyperparameters as shown in Table HCW10.4.

Table HCW10.4   EM Exhaustive Parameters.

Strategy: Exhaustive Sweep	
Name	**Values**
MyILR	[5e-4 1e-3]
MBS	[20 50]
Solver	["adam" "rmsprop" "sgdm"]

Define the images data stores for training, validation, and testing, define the layers variable, and use the following training option in your code:

```
valData=dsval;
miniBatchSize=params.MBS;
options = trainingOptions(params.Solver, ...
    'MiniBatchSize',miniBatchSize, ...
'MaxEpochs',32, ...
'InitialLearnRate',params.MyILR, ...
'LearnRateSchedule','piecewise', ...
'LearnRateDropFactor',0.1, ...
'LearnRateDropPeriod',10, ...
'Shuffle','every-epoch', ...
'ValidationData',valData,...
'ValidationFrequency',30,...
'Plots','training-progress', ...
'Verbose',false);
```

Carry out EM simulation to see which of the above cases will be the best in terms of maximum training accuracy and/or minimum training loss. For the best case, show the training plot, the confusion matrix plot for training and also for testing, and the accuracy of predicting test data classes.

HCW 10.5   EM Hyperparameters of CNN Retraining for Food Images Prediction

The Example Food Images data set (available with Wiley's web companion) contains 978 photographs of food in **nine** classes (caesar-salad, caprese-salad, french-fries, greek-salad, hamburger, hot-dog, pizza, sashimi, and sushi).

Carry out the following steps:

1. Download the Example Food Images data set using the downloadSupportFile function and extract the images using the unzip function. This data set is about 77 MB.

```
fprintf("Downloading Example Food Image data set (77 MB)... ")
filename = matlab.internal.examples.downloadSupportFile('nnet', ...
   'data/ExampleFoodImageDataset.zip');
fprintf("Done.\n")
filepath = fileparts(filename);
dataFolder = fullfile(filepath,'ExampleFoodImageDataset');
unzip(filename,dataFolder);
```

The Zip-format file is usually downloaded to a folder location similar to the following location: C:\Users\Kamal\Documents\MATLAB\Examples\R2022a\supportfiles\nnet\data\ExampleFoodImageDataset.zip

2. Create an images data store.
3. Divide data into training (80%), validation (10%), and testing (10%) sets and prepare the image size that suits **GoogleNet** network.
4. Build a network by starting with a pretrained **GoogleNet** network.
5. Create a fully connected layer with nine classes and classification layer. Take GoogleNet CNN layers but replace the last fully connected layer by the newly created FC. The name of the last fully connected layer is "**loss3-classifier**" and the name of the output classification layer is "**output**". Replace those two layers by your own, but of the same type, layers.
6. Define EM Hyperparameters as shown in Table HCW10.5.

Table HCW10.5   EM Exhaustive Parameters.

Strategy: Exhaustive Sweep	
Name	**Values**
MyILR	[5e-4 1e-3]
MBS	[2 10]
Solver	["adam" "rmsprop" "sgdm"]

Define the images data stores for training, validation, and testing, define the layers variable, and use the following training option in your code:

```
valData=dsval;
miniBatchSize=params.MBS;
options = trainingOptions(params.Solver, ...
'MiniBatchSize',miniBatchSize, ...
'MaxEpochs',1, ...
'InitialLearnRate',params.MyILR, ...
'LearnRateSchedule','piecewise', ...
'LearnRateDropFactor',0.1, ...
```

```
'LearnRateDropPeriod',10, ...
'Shuffle','every-epoch', ...
'ValidationData',valData,...
'ValidationFrequency',30,...
'Plots','training-progress', ...
'Verbose',false);
```

Carry out EM simulation to see which of the above cases will be the best in terms of maximum training accuracy and/or minimum training loss. For the best case, show the training plot, the confusion matrix plot for training and also for testing, and the accuracy of predicting test data classes.

Appendix A

Useful MATLAB Functions

A.1   Data Transfer from an External Source into MATLAB

The first method, by which we bring data from a file named **basketball.xlsx** into MATLAB as a table named **mytbl**, is to use the following built-in **readtable** function.

>> **mytbl =readtable('basketball.xlsx');**

Notice that in MATLAB Workspace window, **mytbl** is created with **table** class type. Each column of the table is referred to as a *variable* in the table, and can be accessed using the dot notation.

A.2   Data Import Wizard

The second method, by which we can import the same Excel file, is to use the import wizard method. Under **"Home"** tab, click on **"Import Data"** button, or go to the respected folder where the Excel file resides, right click on the Excel file, select **"Import Data..."** submenu, and the **"Import"** wizard will pop-up as shown in Figure A.2.1. You may select to create one large table including all columns as part of it, create one column vector for each column appearing in the

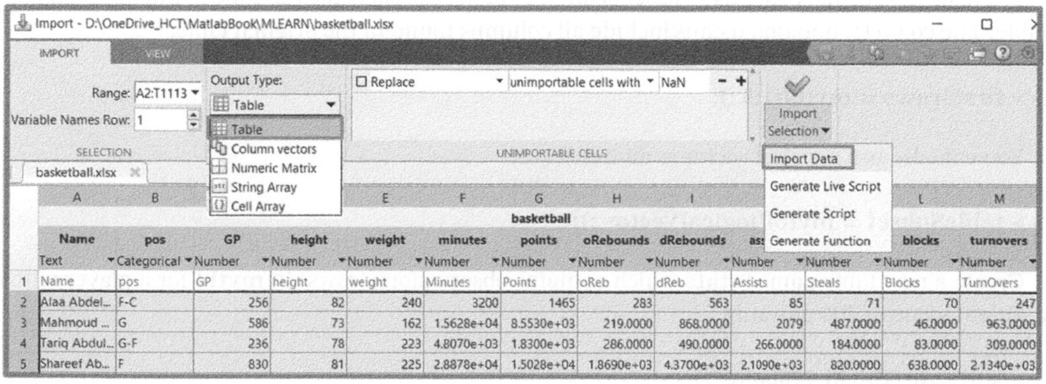

Figure A.2.1 The "**Import**" wizard used to assure a smooth transition between the data source and MATLAB environment.

Machine and Deep Learning Using MATLAB: Algorithms and Tools for Scientists and Engineers, First Edition.
Kamal I. M. Al-Malah.
© 2024 John Wiley & Sons, Inc. Published 2024 by John Wiley & Sons, Inc.
Companion Website: www.wiley.com/go/al-malah/machinelearningmatlab

Excel sheet, a numeric matrix, a string array, or a cell array. I will select the table format for presenting the table under MATLAB environment.

If you look at the "**Workspace**" window, you will notice that MATLAB has already created a table named **basketball** of "**table**" class type. Notice that **basketball** is the same as **mytbl**, except column two is shown without quotation marks in the former case.

A.3   Table Operations

Here are some examples on carrying out some row- and column-related operations:
The following code will create a numeric vector named **h** containing the values of the height variable in **mytbl**.

>> **h=mytbl.height;**

You can create a logical array using relational operators, such as >, <, ==, and ~=, which perform comparisons between two values.

The following code creates a logical array named **isVeryTall** where the values are true when the corresponding **h** values are greater than *85* and false otherwise.

>> **isVeryTall = h > 85;**

The following code creates a numeric vector **w** containing the values of weight variable in **mytbl**.

>>**w=mytbl.weight;**

The following code creates a logical array named **idw200plus** containing a value of true for all elements in **w** that are greater than or equal to *200*.

>> **idw200plus=w>=200;**

To create a table that contains a subset of another table, use parentheses to index into the original table. The colon (:) notation means include all columns found in the original table.

>> **first5rows = mytbl(1:5,:);**

You can also use a logical vector as an index.

>> **tableSubset = mytbl(logicalVector,:);**

Create a new table named **stats** which contains observations (rows) in **mytbl** for a player with a weight value of *200* onwards.

>> **stats=mytbl(idw200plus,:);**

You can delete rows or variables from a table by first selecting those rows or variables and then assigning the output to an empty array.

The following command removes the last two columns from **mytbl2**.

```
>> mytbl2=mytbl;
>> mytbl2(:,end-1:end) = [];
```

Figure A.3.1 shows the result of the previous executed commands and also how the size of columns of **mytbl2** is reduced by two, after carrying out the last command.

Workspace					Workspace			
Name ▲	Value	Size	Class		Name ▲	Value	Size	Class
first5rows	5x20 table	5x20	table		first5rows	5x20 table	5x20	table
h	1112x1 double	1112x1	double		h	1112x1 double	1112x1	double
✓ idw200plus	1112x1 logical	1112x1	logical		✓ idw200plus	1112x1 logical	1112x1	logical
✓ isVeryTall	1112x1 logical	1112x1	logical		✓ isVeryTall	1112x1 logical	1112x1	logical
mytbl	1112x20 table	1112x20	table		mytbl	1112x20 table	1112x20	table
mytbl2	1112x20 table	1112x20	table		mytbl2	1112x18 table	1112x18	table
stats	800x20 table	800x20	table		stats	800x20 table	800x20	table
w	1112x1 double	1112x1	double		w	1112x1 double	1112x1	double

Figure A.3.1 List of Workspace variables and their properties, as generated by the previously executed commands.

Exercise A.3.1: Remove everything after the 16th column in **mytbl2**.
Index into mytbl2 with all rows, :, and columns **17:end** and assign it to **[]**.

```
>> mytbl2(:,17:end) = [];
```

Alternatively, you can index into **mytbl**, by keeping the desired rows and columns and reassign them to a new table called **mytbl3**.

```
>> mytbl3 = mytbl(:,1:16);
```

Assigning a table variable to the empty matrix [] removes the variable from the table.

```
>> table.variable = [];
```

You can create a new table, as a subset of the original table by imposing a restriction, in one line of code.

```
>>hGT80 = mytbl(mytbl.height>=80,1:20);
```

When text labels (i.e., quotation marks) are intended to represent a finite set of possibilities, a cell array of strings is unnecessary and utilizes more memory. Instead, you can use a categorical array.

The following command converts the values in **mytbl2.pos** to categorical values by dropping the quotation marks.

```
>> mytbl2.pos = categorical(mytbl2.pos);
```

The following command shows a list of the possible categories in **mytbl2.pos**.

>> categories(mytbl2.pos)

We can save the categories to a cell array named **catags** containing all the categories in **mytbl2. pos**.

>>catags=categories(mytbl2.pos);

You may want to define the categories for your categorical classes. For example, if you pull data from multiple sources and not all the categories are represented in that particular set, then this becomes important when you want to merge the sets together. It also means that any data not specified by a category becomes **<undefined>**. The following code converts the data in **mytbl4. pos** to categorical that is represented by **'C','C-F','F','F-C','F-G','G'**, or **<undefined>**. I deliberately removed **'G-F'** category to let MATLAB assign **<undefined>** value for any **pos** value originally holding **'G-F'**. From the **Workspace** window, double click on **mytbl4** and MATLAB will edit the matrix and you can see the second column values. Figure A.3.2 shows a portion of both **mytbl4** and the original **mytbl** values.

>> mytbl4=mytbl;
>>mytbl4.pos = categorical(mytbl4.pos,{'C','C-F','F','F-C','F-G','G'});

	Variables - mytbl4		Variables - mytbl	
	mytbl4 ✕		mytbl ✕	
	1112x20 table		1112x20 table	
	1	2	1	2
	Name	pos	Name	pos
1	'Alaa Abdelnaby'	F-C	'Alaa Abdelnaby'	'F-C'
2	'Mahmoud Abdul-Rauf'	G	'Mahmoud Abdul-Rauf'	'G'
3	'Tariq Abdul-Wahad'	<undefined>	'Tariq Abdul-Wahad'	'G-F'

Figure A.3.2 The entry at (3,2) shows **<undefined>** value as it is originally assigned **'G-F'** value.

You can use the categories defined in the cell array **catags** and convert the data in **mytbl5.pos** into a categorical.

>>mytbl5=mytbl;
>>mytbl5.pos = categorical(mytbl5.pos, catags);

From the **Workspace** window, if you double click on **mytbl5**, you will see that the quotation marks are dropped from each entry of the second column values.

A.4   Table Statistical Analysis

Let us carry out some statistical analysis for data arrays. Unfortunately, this table contains only one column category, that is, column #2 with title **pos**. Let us first create a dataset array for either table.

```
>> dsa = mytbl(:,{'pos','GP','height','weight','Minutes','Points'});
>> dsa2 = basketball(:,{'pos','GP','height','weight','Minutes','Points'});
```

The following command

```
>>mystat = grpstats(dsa, 'pos', {'mean', 'min', 'max'})
```

will give the mean, minimum, and maximum for each of the five selected numeric columns defined in the dataset **dsa** and at the same time grouped by **pos** category. Other stats are: **'std'**, **'var'**, **'range'** **'meanci'**, **'predci'**, **'numel'**, and **'sem'**. See MATLAB built-in help under **grpstats** topic.

The following command calculates the standard deviation of the data in **dsa**, grouped by '**pos**'.

```
>>Varstd = grpstats(dsa,'pos', @std)
```

The following command calculates the sum of the data in **dsa2**, grouped by '**pos**'.

```
>>Varsum = grpstats(dsa2,'pos', @sum)
```

A.5   Access to Table Variables (Column Titles)

You can access the variable names in a table using the **VariableNames** property of **Properties** of the table. The following code returns the variables in **mytbl**.

```
>> mytbl.Properties.VariableNames
```

Save the variable names in **mytbl** to a cell array named **Vars**.

```
>>Vars=mytbl.Properties.VariableNames;
```

A.6   Merging Tables with Mixed Columns and Rows

The following command combines two tables by matching up rows that have the same key variables.

```
>> totalData = innerjoin(tableA,tableB);
>>Tleft = table({'a' 'b' 'c' 'e' 'h'}',[1 2 3 11 17]', 'VariableNames',{'Key1' 'Var1'});
>>Tright = table({'a' 'b' 'd' 'e'}',[4 5 6 7]', 'VariableNames',{'Key1' 'Var2'});
>> [T] = innerjoin(Tleft,Tright);
```

Figure A.6.1 shows the **innerjoin** function on merging two tables, where only rows are kept whose values in the variable **Key1** match.

Tleft =		Tright =		T =		
5×2 **table**		4×2 **table**		3×3 **table**		
Key1	**Var1**	**Key1**	**Var2**	**Key1**	**Var1**	**Var2**
{'a'}	1	{'a'}	4	{'a'}	1	4
{'b'}	2	{'b'}	5	{'b'}	2	5
{'c'}	3	{'d'}	6	{'e'}	11	7
{'e'}	11	{'e'}	7			
{'h'}	17					

Figure A.6.1   The **innerjoin** function between two tables with a common variable for both.

A.7   Data Plotting

Use the **boxplot** function

>>boxplot(mytbl.height,mytbl.pos)

to create a box plot of the heights for each position. On each box, the central mark indicates the median, and the bottom and top edges of the box indicate the 25th and 75th percentiles, respectively. The whiskers extend to the most extreme data points not considered outliers, and the outliers are plotted individually using the '+' marker symbol, as shown in Figure A.7.1.

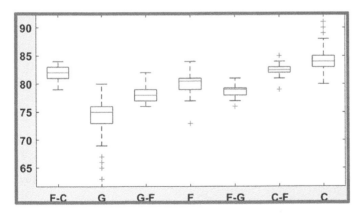

Figure A.7.1   The plot of height using the boxplot function where it shows the scatter of the variable around its mean or median.

You can use **gscatter** to explore the relationship between points per game and rebounds per game, grouped by position. Notice that **gscatter(x,y,g)** creates a scatter plot of **x** and **y**, grouped by **g**. The inputs **x** and **y** are vectors of the same size.

>>**gscatter(mytbl.dReb./mytbl.GP, mytbl.Points./mytbl.GP,mytbl.pos)**

Use **gname** to label the name of the player that corresponds to a chosen data point.

>>**gname(mytbl.Name)**

Click on any colored point within the dedicated plot area and MATLAB will indicate the player's name, as shown In Figure A.7.2. Use the "**Esc**" button to stop interacting with the plot.

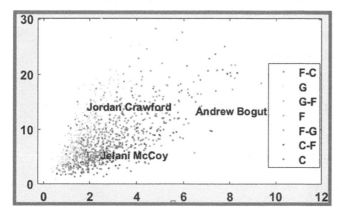

Figure A.7.2   gscatter plot for points per game and rebounds per game, grouped by position.

A.8   Data Normalization

You can use the normalize function to perform several methods of normalization. By default, this function uses the z-score method.

>>**dataNorm = normalize(Xdata)**

This normalizes the data to have a mean of *0* and a standard deviation of *1*. This applies only for a numeric **Xdata**. All columns must be numeric vectors.

A.9   How to Scale Numeric Data Columns to Vary Between 0 and 1

Figure A.9.1 shows how to scale numeric data for each column to vary between 0 and 1.

```
Atable= readtable('Atable.xlsx','Format', 'auto');
% Extract numeric data
origData = Atable{:,1:end-1};
MaxVal=max(origData,[], 1);
MinVal=min(origData,[], 1);
for j=1:size(origData,2)
numData(:,j)=(origData(:,j)-MinVal(j))/(MaxVal(j)-MinVal(j));
end
```

Figure A.9.1   The code to scale a numeric column data to vary between 0 and 1.

Thus, **numData** matrix will be scaled between zero and one for each numeric column. This is different from data normalization or standardization.

A.10 Random Split of a Matrix into a Training and Test Set

Consider a matrix **A [r, c]** made of **r** rows and **c** columns and we would like to split it into two sets: **XTrain [p*r, c]** and **XTest [(1-p)*r, c]**. Figure A.10.1 shows how to split a given matrix into a training and test data set.

```
% A fraction of A matrix dedicated for training
[r,c] = size(A) ;
P = 0.70 ;
%Returns a row vector containing a random permutation of the integers from 1 to r
%without repeating elements.
idx = randperm(r);
% Allocation of 70 % for training and the remaining 30 % for testing.
ATrain = A(idx(1:round(P*r)),:) ;
ATest = A(idx(round(P*r)+1:end),:) ;
```

Figure A.10.1 Splitting a given matrix into a training and test dataset.

A.11 Removal of NaN Values from a Matrix

Consider a matrix **A [r, c]** made of **r** rows and **c** columns and we would like to remove rows that contain any not a number, NAN, value. Figure A.11.1 shows the code to remove rows with any NAN value. To remove the row, at least one NAN value must appear anywhere in that row.

```
temp=rmmissing(A);
Xnum=temp(:,1:end);
```

Figure A.11.1 Removal of rows containing any NAN value.

A.12 How to Calculate the Percent of Truly Judged Class Type Cases for a Binary Class Response

Figure A.12.1 shows how to estimate the truly judged classification type for a binary class response out of all tested/predicted observations.

```
[cm,cl] = confusionmat(ytest, predictquality);
figure(11);
heatmap(cl,cl,cm);
trueclass=0.0;
for k=1:height(cm)
trueclass = trueclass+sum(cm(k,k));
end
TC=100*trueclass/height(ytest);
disp(['Percentage of true classification: ',num2str(TC),'%'])
```

Figure A.12.1 The code to estimate the per cent of truly predicted class type cases for a binary class response.

A.13 Error Function m-file

Notice that you do not have to create the training or test data. The sequential feature selection function internally creates the training and test data and calls this function (saved under **errorFun.m**). Figure A.13.1 shows the code for **errorFun.m** file. Save it anywhere but within MATLAB path folders. Moreover, **fitcknn** can be replaced by any **fitc*** function for a classification case, or even can be replaced by any **fitr*** function for regression case.

```
function error = errorFun(Xtrain,ytrain,Xtest,ytest)
% Create the model with the learning method of your choice
mdl = fitcknn(Xtrain,ytrain); %It can be changed; you may use any fitc* model.
% Calculate the number of test observations misclassified
ypred = predict(mdl,Xtest);
error = nnz(ypred ~= ytest);
```

Figure A.13.1 The code for the m-file **errorFun.m**, where the user may use any **fitc*** model.

An alternative to writing the error function in a separate file is to create an anonymous function. To create an anonymous function that calculates the number of misclassifications given training and test data, use the code shown in Figure A.13.2:

```
af = @(Xtrain,ytrain,Xtest,ytest)...
nnz(ytest ~= predict(...
fitc*(Xtrain,ytrain),Xtest))
```

Figure A.13.2 The code for writing an anonymous error function instead of m-file.

A.14   Conversion of Categorical into Numeric Dummy Matrix

Figure A.14.1 shows the m-file code for **cattbl2mat** function which converts a categorical predictor column into a number of dummy predictors columns depending on the number of subcategories present in the original predictor column.

```
function [data,vars]=cattbl2mat(data)
% Makes a matrix from a table, with categ. variables replaced by
% numeric dummy variables
vars=string(data.Properties.VariableNames);
idxCat=varfun(@iscategorical,data,"OutputFormat","uniform");
for k=find(idxCat)
    % get list of categories
    c=categories(data.(vars(k)));
    % replace variable with matrix of dummy variables
    data=convertvars(data,vars(k),@dummyvar);
    % split dummy variable and make new variable names (by appending
    % the category value to the categorical variable name).
    varnames=vars(k) + "_" + replace(c, " ", "_");
    data=splitvars(data, vars(k),"NewVariableNames",varnames);
end
vars=string(data.Properties.VariableNames);
```

Figure A.14.1   The m-file code for **cattbl2mat** function for creation of dummy out of categorical predictor columns.

Conversion of a table into array:

Xnum = table2array(Table);

Conversion of an array into table:

Tbl=array2table(Xnum,'VariableNames',{'Column1','Column2','Column3', 'Column4'});

Here is an example shown in Figure A.14.2:

```
% Create a table by reading any MATLAB compatible file.
Mushroom=readtable('Mushroom.csv','Format', 'auto');
T1=Mushroom;
% Extract Column titles or names found in the table
Vars=T1. Properties.VariableNames;
% Create a categorical table called T2
T2 = convertvars(T1,Vars,'categorical');
```

Figure A.14.2   Handling a table with mixed datatypes to be finally converted into a numeric matrix.

```
% The last column is excluded as it represents the response variable
temp= T2(:,1:end-1);
% Use cattbl2mat(data) defined in Figure A.14-1 to convert table of predictor data
[xdata, xvar] = cattbl2mat(temp);
% Convert the predictor numeric table into numeric array
Xnum= table2array(xdata);
% Use cattbl2mat(data) defined in Figure A.14-1 to convert vector of response data
[ydata, yvar] = cattbl2mat(T2(:,end));
% Convert the response numeric table into numeric array
Ynum = table2array(ydata);
```

Figure A.14.2 (Cont'd)

A.15 evaluateFit2 Function

evaluateFit2 function (Figure A.15.1) creates four plots. This function requires four input arguments: The actual response values, the predicted response results, the type of the regression model as a text, and the name of the response variable as a text. The data is sorted by the actual response. The first plot shows the actual values (in blue) and the predicted values (in red) against the observed data. The second plot shows the actual value on the *x*-axis against the predicted value on the *y*-axis. The third plot is a histogram showing the distribution of the errors between the actual and predicted values. The last plot is another histogram showing the per cent relative error (error as a percentage of the actual value). MAPE stands for mean absolute percent error.

```
function evaluateFit2(y,ypred,name,yname)
figure
% Plot against observation number
subplot(2,2,1)
plot(y,'.')
hold on
plot(ypred,'.')
hold off
title(name)
legend('Actual','Predict')
xlabel('Observation number')
ylabel(yname)
% Plot predicted and actual against each other
subplot(2,2,2)
scatter(y,ypred,'.')
% Add 45-degree line
xl = xlim;
%xlim is [0 1] double
hold on
```

Figure A.15.1 The m-file for **evaluateFit2** function which shows the model goodness by comparing the actual versus the predicted response value.

```
plot(xl,xl,'k:')
hold off
title(name)
xlabel(['Actual ', yname]);
ylabel(['Predicted ', yname]);
% Distribution of errors
subplot(2,2,3)
err = y-ypred;
MSE = mean(err.^2,'omitnan');
histogram(err)
title(['MSE = ',num2str(MSE,4)])
xlabel('Prediction error')
% Distribution of percentage errors
subplot(2,2,4)
err = 100*err./y;
MAPE = mean(abs(err),'omitnan');
histogram(err)
title(['MAPE = ',num2str(MAPE,4)])
xlabel('Prediction percentage error')
end
```

Figure A.15.1 (Cont'd)

A.16 showActivationsForChannel Function

Figure A.16.1 shows the code for **showActivationsForChannel** function where it accepts three input arguments: the virgin image, the layer activation, and the number of activation channel.

```
function showActivationsForChannel(origImage,ac,ch)
%show the image features of specific layer activation input:
% origImage= Original Image
%ac= layer Activation
% ch= number of channel of the activation
imageSize = size(origImage);
imageSize = imageSize([1 2]);
act_ch = ac(:,:,ch);
act_ch = imresize(mat2gray(act_ch),imageSize);
imshowpair(origImage,act_ch,'montage')
title(['Compare Channel ',num2str(ch)])
end
```

Figure A.16.1 Definition of **showActivationsForChannel** function.

A.17   upsampLowRes Function

Figure A.17.1 shows the code for **upsampLowRes** function where it accepts one low-resolution image argument and outputs a high-resolution counterpart of the same image.

```
function dataOut = upsampLowRes(dataIn)
    temp = dataIn;
    temp = imresize(temp,[7,7],'method','bilinear');
    dataOut = {imresize(temp,[28,28],'method','bilinear')};
end
```

Figure A.17.1   **upsampLowRes** function to upgrade a low- to high-resolution image.

A.18A   preprocessData function

Figure A.18A.1 shows the code for **preprocessData** function where it accepts an image as an input and the target size to be scaled by. Also, any bounding box will be scaled in proportion with the same target size.

```
function data = preprocessData(data,targetSize)
% Resize image and bounding boxes to the targetSize.
scale = targetSize(1:2)./size(data{1},[1 2]);
data{1} = imresize(data{1},targetSize(1:2));
bboxes = round(data{2});
data{2} = bboxresize(bboxes,scale);
end
```

Figure A.18A.1   **preprocessData** function for scaling a given image as well as its bounding boxes.

A.18B   preprocessData2 function

Figure A.18B.1 shows another code called **preprocessData2** function where it accepts an image as an input and the target size to be scaled by. Also, any bounding box will be scaled in proportion with the same target size.

```
function data = preprocessData2(data,targetSize)
for num = 1:size(data,1)
  I = data{num,1};
  imgSize = size(I);
  bboxes = data{num,2};
```

Figure A.18B.1   **preprocessData2** function for scaling a given image as well as its bounding boxes.

```
   I = im2single(imresize(I,targetSize(1:2)));
   scale = targetSize(1:2)./imgSize(1:2);
   bboxes = bboxresize(bboxes,scale);
   data(num,1:2) = {I,bboxes};
end
end
```

Figure A.18B.1   (Cont'd)

A.19   processTurboFanDataTrain function

The function definition is explained by the comment statements, at the beginning of Figure A.19.1.

```
function [predictors,responses] = processTurboFanDataTrain(filenamePredictors)
% The function processTurboFanDataTrain extracts the data from
% filenamePredictors and returns the cell arrays predictors and responses
% which contain the predictor and response sequences, respectively. The
% data contains zip-compressed text files with 26 columns of numbers,
% separated by spaces. Each row is a snapshot of data taken during a single
% operational cycle, and each column is a different variable. The columns
% correspond to the following:
%    1:   Unit number
%    2:   Time in cycles
%    3-5: Operational settings
%    6-26: Sensor measurements 1-17
dataTrain = readmatrix(filenamePredictors);
numObservations = max(dataTrain(:,1));
predictors = cell(numObservations,1);
responses = cell(numObservations,1);
for i = 1:numObservations
   idx = dataTrain(:,1) == i;
   predictors{i} = dataTrain(idx,3:end)';
   timeSteps = dataTrain(idx,2)';
   responses{i} = fliplr(timeSteps);
end
end
```

Figure A.19.1   processTurboFanDataTrain function extracts the data from a file and returns the cell arrays predictors and responses.

A.20   processTurboFanDataTest Function

The function definition is explained by the comment statements, at the beginning of Figure A.20.1.

```
function [predictors,responses] = processTurboFanDataTest(filenamePredictors,file
nameResponses)
% The function processTurboFanDataTest extracts the data from filenamePredictors
% and filenameResponses and returns the cell arrays predictors and
% responses, which contain the test predictor and response sequences. In
% filenamePredictors, the time series ends some time prior to system
% failure. The data in filenameResponses provides a vector of true RUL
% values for the test data.
predictors = processTurboFanDataTrain(filenamePredictors);
RULTest = dlmread(filenameResponses);
numObservations = numel(RULTest);
responses = cell(numObservations,1);
for i = 1:numObservations
  X = predictors{i};
  sequenceLength = size(X,2);
  rul = RULTest(i);
  responses{i} = rul+sequenceLength-1:-1:rul;
end
end
```

Figure A.20.1 processTurboFanDataTest function extracts the data from a file and returns the cell arrays predictors and responses.

A.21   preprocessText Function

A function that tokenizes, converts to lower case, and removes the punctuation of the text data, as shown in Figure A.21.1.

```
function documents = preprocessText(textData)
% Tokenize the text.
documents = tokenizedDocument(textData);
% Convert to lowercase.
documents = lower(documents);
% Erase punctuation.
documents = erasePunctuation(documents);
end
```

Figure A.21.1 preprocessText function that tokenizes, converts to lower case, and removes the punctuation of the text data.

A.22   documentGenerationDatastore Function

A mini-batch datastore to input the documents to the network as mini-batches of sequence data. The datastore converts documents to sequences of numeric word indices, as shown in Figure A.22.1.

```
classdef documentGenerationDatastore < matlab.io.Datastore & ...
    matlab.io.datastore.MiniBatchable
  properties
    Documents
    Encoding
    MiniBatchSize
  end
  properties(SetAccess = protected)
    NumObservations
  end
  properties(Access = private)
    CurrentFileIndex
  end
  methods
    function ds = documentGenerationDatastore(documents)
      % ds = documentGenerationDatastore(documents) creates a
      % document mini-batch datastore from an array of tokenized
      % documents.
      % Add startOfText token to documents
      startTokens = repmat(tokenizedDocument("startOfText"), size(documents));
      documents = startTokens + documents;
      % Set Documents and MiniBatchSize properties.
      ds.Documents = documents;
      ds.MiniBatchSize = 128;
      % Create word encoding.
      ds.Encoding = wordEncoding(documents);
      % Datastore properties.
      numObservations = numel(documents);
      ds.NumObservations = numObservations;
      ds.CurrentFileIndex = 1;
    end
    function tf = hasdata(ds)
      % tf = hasdata(ds) returns true if more data is available.
      tf = ds.CurrentFileIndex + ds.MiniBatchSize - 1 ...
        <= ds.NumObservations;
    end
    function [data,info] = read(ds)
      % [data,info] = read(ds) read one mini-batch of data.
      miniBatchSize = ds.MiniBatchSize;
      enc = ds.Encoding;
      info = struct;
      % Read batch of documents.
      startPos = ds.CurrentFileIndex;
      endPos = ds.CurrentFileIndex + miniBatchSize - 1;
```

Figure A.22.1 A mini-batch datastore to input the documents to the network as mini-batches of sequence data. It converts documents to sequences of numeric word indices.

```matlab
            documents = ds.Documents(startPos:endPos);
            % Convert documents to sequences.
            paddingValue = 0;
            predictors = doc2sequence(enc,documents, ...
                'PaddingDirection','right', ...
                'PaddingValue',paddingValue);
            % Create categorical sequences of responses.
            classNames = [enc.Vocabulary "EndOfText"];
            for i = 1:miniBatchSize
                X = predictors{i};
                % Remove start token.
                X(1) = [];
                % Count padding values.
                nz = sum(X==paddingValue);
                % Convert words to indices and pad with "EndOfText". Append
                % additional "EndOfText" token to account for unpadded
                % sequences.
                words = [ind2word(enc,X(X~=0)) repmat("EndOfText",[1 nz+1])];
                responses{i,1} = categorical(words,classNames);
            end
            % Update file index
            ds.CurrentFileIndex = ds.CurrentFileIndex + miniBatchSize;
            % Convert data to table.
            data = table(predictors,responses);
        end
        function reset(ds)
            % reset(ds) resets the datastore to the start of the data.
            ds.CurrentFileIndex = 1;
        end
        function dsNew = sort(ds)
            % dsNew = sort(ds) sorts the observations in the datastore by
            % sequence length.
            dsNew = copy(ds);
            documents = dsNew.Documents;
            documentLengths = doclength(documents);
            [~,idx] = sort(documentLengths);
            dsNew.Documents = documents(idx);
        end
    end
    methods (Hidden = true)
        function frac = progress(ds)
            % frac = progress(ds) returns the percentage of observations
            % read in the datastore.
            frac = (ds.CurrentFileIndex - 1) / ds.NumObservations;
        end
    end
end
```

Figure A.22.1 (Cont'd)

A.23   subset Function for an Image Data Store Partition

In addition to **splitEachLabel** function (refer to Chapter 7 | **Validation Data** section; the user can also divide the entire data set into three subsets for training, validation, and testing steps, using subset function. See Figure A.23.1 for explaining how to split a given image data store into three subsets.

```
ImDir=fullfile(DirFolder);
imds=imageDatastore(ImDir,'IncludeSubfolders',true,'LabelSource','foldernames');
% Top 80 % for training, 10 % for validation, and the last 10 % for testing
r= size(imds.Files,1);
P = 0.80 ;
idx = randperm(r)  ;
idx2 = idx(1:round(P*r));
dstrn = subset(imds,idx2);
trainingData=augmentedImageDatastore([224 224],dstrn);
idx3=idx(round(P*r)+1:round((P+0.1)*r));
dsval = subset(imds,idx3);
dsval=augmentedImageDatastore([224 224],dsval);
idx4=idx(round((P+0.1)*r)+1:end);
dstst = subset(imds,idx4);
dstst2=augmentedImageDatastore([224 224],dstst);
```

Figure A.23.1   The code for creating three subsets using subset function, out of a given image data store.

Index

a

activations function 302–303
  AlexNet CNN layers 263
  extraction and visualization 261–263
  2-D layer features 258–261
activity recognition 520
adam algorithm 299
Adult Census Income Data (1994) 233–234
aggregate channel features (ACF) 500
AlexNet
  CNN 263, 361–362, 364–370
  filters 308–311
  network 539–540
  training 251–253
anchor boxes 408
augmentedImageDatastore function 244, 322
autocorrelation method 437
automated labeling algorithm 513, 514
autonomous vehicles 198
average precision (AP) 427–429

b

BiLSTM network 430–436
  see also long short-term memory (LSTM)
binary decision tree model 68–73
binary linear classifier (fitclinear)
  breast cancer diagnosis 102
  cardiac arrhythmia data 102
  mushroom edibility data 98–100
  1994 adult census income data 100–101
  wine quality dataset 101–102
Boston House Price 235–236
breast cancer diagnosis 234

c

Cat Dog video labeling, training and
  prediction 537–538
cautious/conservative jump 296
cepstral deconvolution 437
channel-wise separable convolution 264
Chat Generative Pre-trained Transformer
  (ChatGPT) 465
classical multidimensional scaling 4–6
Classification Learner app 57–68
classification models, supervised learning
  1, 42–43
  binary linear classifier 98–100
  Classification Learner app 57–68
  customizing classifier 43
  decision tree model 68–73
  discriminant analysis (DA) classifier
    for 79–84
  evaluating accuracy 45–47
  KNN model 47–57
  multiclass support vector machine (fitcecoc)
    model 92–98
  Naïve Bayes classifier 74–79
  predicting response 45
  support vector machine 84–92
  training and test datasets 43–44, 104
    see also Cross-validation model
clustering techniques 1
  Gaussian mixture model (GMM)
    clustering 15–17
  hierarchical clustering 23–27
  k-means clustering 13–15
  PCA and 27–34

Machine and Deep Learning Using MATLAB: Algorithms and Tools for Scientists and Engineers, First Edition.
Kamal I. M. Al-Malah.
© 2024 Kamal I. M. Al-Malah. Published 2024 by John Wiley & Sons, Inc.
Companion Website: www.wiley.com/go/al-malah/machinelearningmatlab

quality evaluation   21–23
CNN-based yolov4ObjectDetectors   424–425
CNNs *see* convolutional neural networks (CNNs)
command-based feed-forward neural
      networks   210–214
   regression   223–225
confusion matrix   204
Continental United States (CONUS)   291
convolution2dLayer image   307
convolutional neural networks (CNNs)   240,
      361, 384, 399
   algorithms   298–299
   architecture   290
   channels and activations   256–267
   chest X-ray prediction   359
   confusion matrix plot   302
   DAG   329–333
   DND   333–342
   fetching response categories   300
   filters   307–311
   food images prediction   286–287, 357
   fruit/vegetable varieties prediction   289
   hand-written synthetic digit images   364–374
   image augmentation   322–328
   land satellite images   291–294
   landcover dataset   299–302
   layers and filters   302–306
   merchandise data prediction   287–288, 358
   modification   361–364
   musical instrument spectrograms
      prediction   288, 358–359
   network performance   319–322
   neurons   290
   original and red version image   303, 304
   root-mean-square error   364
   round worms alive/dead prediction   286, 356–357
   semantic segmentation   342–356
   training options   294–299
   validation data   311–319
covariance method   437
crossentropy function   203
cross-validation model   229–231
   cvpartition options   105–106
   model evaluation   104
   partition K-fold creation   105–106
customer service chatbots   198

d
DAG *see* directed acyclic graph (DAG)
data normalization   29, 30
data scaling   29
DBN *see* deep belief network (DBN)
decision tree model   68
   data types   72–74
   numeric predictors   70–71
   properties of   69–70
deep belief network (DBN)   240
deep learning   386
   image networks   237–241
   MATLAB tools and description   238
   network analysis   405
   sequence-to-sequence regression   446–453
   techniques   520
   text data   453–464
   word-by-word text generation   465–475, 494
deep network designer (DND)
   checking bugs   479
   chickenpox cases   476
   CNNs   333–342
   creating network architecture   376–378
   data partition   477
   data standardization   477
   forecast future time steps   481–482
   generating training data   375–376
   image data loading   375
   importing data   378, 479
   load sequence data   476
   LSTM network architecture   477
   network type selection   478–479
   prepare predictors and responses   477
   spatial resolution   374
   test data   483
   testing network   383–384
   training network   378–382, 480–481
   training options   479–480
   training time series   482
   update network state, observed
      values   483–484
depth-wise separable convolution   264
Digital Imaging and Communication in Medicine
      (DICOM)   495
digital ortho quarter quad tiles (DOQQs)   291
directed acyclic graph (DAG)   320, 329–333

discriminant analysis classification model   79–80
  computational performance of   80
  numeric predictors   82–84
  properties of   81–82
DND *see* deep network designer (DND)
Document Object Model (DOM)   466
documentGenerationDatastore
     function   557–559
DOQQs *see* digital ortho quarter quad tiles
     (DOQQs)
dummy predictors   200, 446
dummy variables   200–201

e

edge detecting kernel   305–306
EM app *see* Experiment Manager (EM) app
EM hyperparameters of CNN retraining
  exhaustive parameters   539–541
  food images prediction   540–542
  merchandise data prediction   538–539
  round worms alive/dead prediction   539–540
ensemble learning   126
  creating data   130–131
  fitensemble   126–129
  quality classification   131
  weak learners and   127
epoch training   202
evaluateDetectionMissRate function   417–421
evaluateDetectionPrecision function   413–417
evaluateFit2 function   553–554
Experiment Manager (EM) app   495
  command prompt   526
  hyperparameters   527–530
  parameters insertion   528
  project type   526
  running simulation   528
  saving changes   530
  simulation experiment   527
  training performance   531
  trainNetwork function   526, 532–533
  true *vs.* predicted flower type   532
  using filter button   532
  validation loss   531
  workspace variables   533
export network function
  MATLAB Coder   209
  MATLAB compiler   208, 209

f

factor analysis   110–112
factory equipment failure text analysis
  alternative coding   462–463
  categorical classes, transformation to   455
  class distribution, histogram of   455
  classify new sequences   461
  convert document to sequences   457–458
  convert test text data to sequences   461
  create and train LSTM network   459
  data partition   455–456
  import data   454
  LSTM network   453–454
  preprocess text data   456–457
  sequence length and padding   458–459
  specify training options   459–460
  test text data, preprocess   461
  text classification   491
  tokenized train data   457
  train LSTM network   460
  using new data   460
  vector arithmetic equation   454
  word cloud creation   456
  workspace variables   461, 464
faster RCNN-based car objects
     prediction   427–429
fasterRCNN object detector   424
feature extraction   256, 269–271, 275–277
feature selection, predictors   115–116
  with categorical data   122–126
  fitensemble predictive model with   134–135
  sequential   118–121
  using neighborhood component analysis   146–148
  using predictorImportance function   116–117
  using Regression Learner app   145
feature transformation, predictors   108–110
  and factor analysis   113–115
  fitensemble predictive model with   135–136
  using regression learner app   145
feed-forward neural networks   240
  big car data, command-based regression   223–225
  classification   199–200
  command prompt-based code   212
  fully connected regression   198, 199
  heart data, command-based
     classification   210–214

regression   200–201
two-layer   198
filters
  AlexNet   308–311
  convolution layers   307–308
financial forecasting   197
fitclinear classification model   98–100
fitensemble predictive model   126–127
  description of   127
  with feature selection   134–135
  with feature transformation   135–136
  improvement of   132–134
  selection of   127
  value and method description   128–129
fitnet function   210
fitrnet function   226–229
gaming   198
gated recurrent unit (GRU)   240

g
Gaussian mixture model (GMM) clustering
    15–17, 29
Gausslan process regression (GPR)   172
  automatic hyper parameter optimization   175
  fitting and prediction using   172–173
  nonparametric regression model   173–175
GoogleNet
  fast R-CNN object detector   425–426
  faster R-CNN object detector   426, 427
  network function   244, 253–255, 362–364
  yolov4 object detector   427
GPR *see* Gaussian process regression (GPR)
GPU *see* Graphics Processing Unit (GPU)
gradCAM features explainers   284–285
gradient clipping   297
gradients-based critical issues   439–440
gradient-weighted class activation mapping
    (Grad-CAM)   284–285
Graphics Processing Unit (GPU)   267
ground sample distance (GSD)   291
Ground Truth Labeler (GTL) app
  add/remove signal window   514
  automatic labeling   513, 514
  command prompt   513
  ROI labels   513
  test data set   519
  workspace variables   519

YOLO network performance   517–518
YOLOv2 object detector   515–517
groundTruth object   501
GRU *see* gated recurrent unit (GRU)
GSD *see* ground sample distance (GSD)
GTL app *see* Ground Truth Labeler (GTL) app

h
heartData table   201
HeartDisease variable   201
hyperparameter optimization   231–233

i
IBP app *see* Image Batch Processor (IBP) app
IKConv-treated image   304–305
IL app *see* Image Labeler (IL) app
ILSVRC *see* ImageNet Large-Scale Visual
    Recognition Challenge (ILSVRC)
image augmentation   322–328
Image Batch Processor (IBP) app
  command prompt   533–534
  grayscale and binary scale   535
  main folder and image-class
    sub-folder   535–536
  main window   534
  myimfcn.m file   534–535
  RGB-based and grayscale-based   536
  type conversion and resizing   534
Image Labeler (IL) app
  automation algorithm, label ground
    truth   500–501
  blocked image   496
  boats.png image   497
  components   496
  creating ROI labels   498–499
  creating scene labels   499–500
  export labeled ground truth   501
  label ground truth   500
  MATLAB command prompt   496
  2-D image labeling purposes   497
  usage   495
image networks, deep learning   237–241
image recognition   197, 247–255
image sequences   431
imageDataAugmenter function   322–327
imageDatastore function   243
imageLIME features explainer   282–284

ImageNet Large-Scale Visual Recognition
    Challenge (ILSVRC)   238, 329
image/video-based apps
    Experiment Manager (EM) app   526–533
    Ground Truth Labeler (GTL) app   513–519
    Image Batch Processor (IBP) app   533–536
    Image Labeler (IL) app   495–501
    LSTM, running/walking
        classification   520–525
    Video Labeler (VL) app   502–513
imfilter function   306
imshow function   310
intersection over union (IoU)   353–354, 356
Iris flower database   37–38

k

Kanade-Lucas-Tomasi (KLT) algorithm   503
k-means clustering   13, 29
    distance metrics   13–14
    replications   14–15
KNN classification model   47–48
    cost penalty   52
    heart disease numeric data   48–50
    mushroom edibility data   52–55
    number of neighbors and weighting factor   51
    properties   50–51
    red wine data   55–57

l

labeled ground truth   500
    automation algorithm   500–501
    export   501
    manually   500
land satellite images
    bands   291
    categorisation   291
    channels   292
    displaying   291–294
Laplacian (fsulaplacian) for unsupervised
        learning   35–37
lasso regularized regression method   183–186
learning rates   296–297
least absolute shrinkage and selection operator
        (lasso)   183–186
Levenberg-Marquardt algorithm   216
LIME *see* locally interpretable model-agnostic
        explanation (LIME)

linear classification model   93–94
linear predictive coding (LPC)   437
linear predictive coding cepstral (LPCC)   437
linear regression models   154–155, 198
    boston house price dataset   192
    forest fires data   193
    fuel economy data   195–196
    nonparametric *see* nonparametric regression
        models
    organizing data   155–156
    polynomial model   158–159
    regularized parametric   176–186
    robust polynomial model   160–162
    stepwise parametric   186–192
    telemonitoring data   194–195
    using fitlm function   155, 164–166
    Wilkinson–Rogers notation   156
locally interpretable model-agnostic explanation
        (LIME)   282
long short-term memory (LSTM)   240
    classification   434
    components   433
    data flow   432
    data preparation, training and
        validation   522–523
    features extraction   520, 521–522
    hidden and cell state   431–433
    image datasets, short time video clip   522
    layers explanation   435–436
    learnable weights   432
    load video data   520–521
    lstmLayer function   433
    network architecture   438–439
    network layers   523
    performance   441–445
    pre-trained network   520
    regression   434
    resnet CNN   520
    running and walking scenario   521
    sentences   433
    sigmoid gate and tanh function   433–434
    tasks   431
    train network   524
    training options   523–524
    training results   440
    types   430–431
    video classification   434–435, 520

video time reset to zero   521

word embeddings   454

workspace variables   525

see also LSTM RNNs

LPC *see* linear predictive coding (LPC)

LPCC *see* linear predictive coding cepstral
    (LPCC)

LSTM *see* long short-term memory (LSTM)

LSTM RNNs

accuracy   441

MiniBatchSize   442–445

performance   441–445

sequence classification   437–440

m

machine learning (ML)   1

algorithms   1–2

classical multidimensional scaling   4–6

classification   268–269

feature extraction   269–271, 275–277

flower features   267–268

hyper parameter optimization   271

image features   268

pattern recognition network
    generation   271–275

unsupervised/supervised   1–2

MAPE *see* mean absolute percent error (MAPE)

MATLAB functions

boxplot function   548

cattbl2mat function   552

data import wizard   543–544

data normalization   549

data plotting   548–549

documentGenerationDatastore
    function   557–559

error function m-file   551

evaluateFit2 function   553–554

external source, data transfer   543

gscatter plot   549

innerjoin function   548

merging tables, mixed columns and
    rows   547–548

mini-batch datastore   558

NaN values removal   550

numeric dummy matrix   552–553

preprocessData function   555

preprocessData2 function   555–556

preprocessText function   557

processTurboFanDataTest function   556–557

processTurboFanDataTrain function   556

random split, training and test set   550

scale numeric data columns   549–550

showActivationsForChannel function   554

subset function, image data store
    partition   560

table operations   544–546

table statistical analysis   547

table variables (column titles)   547

truly judged class type cases, binary class
    response   550–551

upsampLowRes function   555

workspace variables and properties   545

MATLAB regression learner   136–145

MATLAB workspace   208, 370

max function   211

mean absolute percent error (MAPE)   553

mean squared errors (MSE)   226

medical diagnosis, neural networks   198

Mill data set, multivariate classification   493–494

MIMO *see* multiple input multiple output
    (MIMO)

mini batches   295–296

MISO *see* multiple input single output (MISO)

miss rate   417–421, 427

ML *see* machine learning (ML)

MSE *see* mean squared errors (MSE)

multiclass support vector machine (fitcecoc)
    model   92–93

linear classification model   93–94

red wine data   95–98

multiple input multiple output (MIMO)   241

multiple input single output (MISO)   241

mushroom edibility data   233

n

NAIP *see* National Agriculture Imagery Program
    (NAIP)

Naïve Bayes classification model   74, 276

computational performance of   74

data types   77–79

Gaussian/kernel distribution   76–77

numeric data to   75–76

properties of   75

National Agriculture Imagery Program (NAIP)   291

natural language processing (NLP)
    197, 431, 465

net, neural network   210
Netflix titles data set   492
network depth   238
network layers architecture   370–374
network object prediction explainers
   gradCAM features   284–285
   imageLIME features   282–284
   occlusion sensitivity   278–282
network performance
   architecture   320–321
   training algorithm options   319
   training data   319
network training algorithms   216
neural net pattern recognition   198
neural network fitting (nftool)   214
neural network pattern recognition
   (nprtool)   199, 201
   application   201–210
   HDDSCG.m file   205
   nntraintool   206
   performance and model error   203
   window of   202
neural network regression (nftool)   198
   absolute error histogram   218
   data fitting problems   214
   MATLAB toolstrip   215
   MPGNR.m file   221
   network training algorithms   216
   nntraintool pops up   220
   output and target response value   218
   predictors   215
   regression model   216
   training, validation, and test data subsets   217
   training script/MATLAB m-code   219
   workspace window   215
   X and Y data   215
neural networks
   confusion matrix   214
   cross entropy, *f(epoch)*   207
   custom hyperparameter
      optimization   231–233
   examples   197–198
   Export Model button   207
   feed-forward *see* Feed-forward neural
      networks
   fitnet function, regression model   226–229
   nntraintool, workspace variables   213

optimum regularization strength,
   cross-validation   229–231
pattern recognition (nprtool)
   application   201–210
patterns   197
problems   239–240
regression (nftool)   214–222
simulation block   208
types and properties   241
see also pretrained neural networks
neurons   240
NLP *see* natural language processing (NLP)
non-maximum suppression (NMS)   384
nonparametric regression models   166–167
   fitrtree   167–170
   Gaussian process regression   172–176
   support vector machine   170–171
numeric features
   categorical to one-hot vector   486–487
   feature input layer   488
   predictors to categorical variables   486
   random indices, training and testing   488
   read input data   486
   response to categorical variable   486
   split pregnant vectors into columns   487
   train network   489
   training and test set partition   488
   training options   488–489
   view table heads   487–488
   workspace variables   489–490

o

observations and clusters visualization   18–21
occlusion sensitivity   278–282
ONEover9 kernel   305
onosphere database   38–39
outputSize argument   199

p

PAN *see* path aggregation network (PAN)
parametric regression model   166–167, 176
   lasso   183–186
   ridge linear   176–183
   stepwise linear regression model   186–191
Pareto chart   6
Path aggregation network (PAN)   385
pattern recognition network algorithms   271–275

patterns   197
PCA *see* principal component analysis (PCA)
pixelLabelDatastore function   344
Point Tracker algorithm   503
polynomial model   158–159
prediction error   452, 453
predictive model
  accuracy and robustness of   103–104
  cross-validation   104–106
  dummy variables   121–126
  ensemble learning   126–136
  factor analysis   110–112
  ionosphere data   148–149
  MATLAB regression learner   136–145
  multiple classification algorithms   103
  neighborhood component analysis   146–148
  partition K-fold creation   106–108
  quality dataset   150–152
  selection techniques 115–117.   134–135,
    145–148
  sequential feature selection   118–121
  small car data   152
  Sonar Dataset   149–150
  transformation   106–115, 145
predictors   201, 215
  classical multidimensional scaling   4–6
  reduction of   108–117
preprocessData function   555
preprocessData2 function   555–556
preprocessText function   557
pretrained neural networks
  accessing image file   246–247
  CNN   256–267
  data stores, MATLAB   241–243
  deep learning   237–241
  image and augmented image
    datastores   243–246
  machine learning, features extraction
    for   267–278
  network object prediction explainers   278–285
  RGB 3-D image processing   239
  transfer learning   247–255
principal component analysis (PCA)   6–8
  algorithms in   8
  and clustering   27–34
  feature transformation   108–109
  NaN values   9

parallel coordinates plot   109–110
pareto plot creation   9–11
pc-squared plot   12
priori boxes   386
processTurboFanDataTest function   556–557
processTurboFanDataTrain function   556

q
quefrency   437

r
RBMs *see* restricted Boltzmann machines
    (RBMs)
R-CNN *see* regions with convolutional neural
    networks (R-CNN)
receiver operating characteristic (ROC)   204
recommender systems   197
rectified linear unit (ReLU)   226, 290, 293
recurrent neural network (RNN)   240
  accuracy and loss function   470, 474
  BiLSTM   430–436
  categorical sequences   445–446
  deep learning   446–453
  DND *see* deep network designer (DND)
  encoder-decoder framework   465
  LSTM *see* long short-term memory (LSTM)
  numeric features, train network with   486–490
  sequential/time-dependent data   430
  training model   463–464
  vanishing gradient problem   430
recurrent neuron/cell   240
region of interest (ROI)   397, 398, 498–499, 502
region proposal network (RPN)   398, 408
regions with convolutional neural networks
    (R-CNN)
  algorithms, object detection   396–399
  components   397
  creation and training   399–413
  workspace variables   412
regression
  data preparation   361–374
  DND   374–384
  evaluateDetectionMissRate, miss rate
    metric   417–421
  evaluateDetectionPrecision function, precision
    metric   413–417
  R-CNN algorithms   396–399

transfer learning   399
YOLO object detectors   384–396
see also convolutional neural networks (CNNs)
regression learner   136–145
regression techniques, supervised learning   1–2
see also linear regression models
regularized parametric linear
    regression   176–183
Remaining Useful Life (RUL)   446
residual network (ResNet)   320
ResNet-50 model   329–331
response matrices   213
restricted Boltzmann machines (RBMs)   240
ridge linear regression   176–178
    big car data   179–183
    parameter, lambda   179
    predicting response using   178
rmsprop algorithm   298–299
RNN *see* recurrent neural network (RNN)
robust polynomial model   160–162
ROC *see* receiver operating characteristic (ROC)
ROI *see* region of interest (ROI)
root-mean-square error (RMSE)   364
RPN *see* region proposal network (RPN)
RUL *see* Remaining Useful Life (RUL)

s

scaled conjugate gradient   202, 205
semantic segmentation
    architecture   349
    CamVid dataset   359–360
    creating network   345–350
    five-image data store   343
    IoU   353–354
    parameters   353
    SemSegNet, command-created   349
    training and testing network   350–356
    training data analysis   343–345
sensor data sequences   431
sentiment labeled sentences data set   492
sequence length   441–444
sequential feature selection (SFS)   118–121
shallow neural network *see* two-layer feed-
    forward network
showActivationsForChannel function   554
shuffle function   316–319
Simulink environment   208

single input multiple output (SIMO)   240
single input single output (SISO)   240
small car data (regression case)   234–235
small car database   39–40
sparse gradients/noisy data   299
spatial pyramid pooling (SPP)   385
spectrograms   275–277, 288
splitEachLabel function   314, 316
SPP *see* spatial pyramid pooling (SPP)
standard schnauzer   285
stepwise linear regression model   186–191
stochastic gradient descent   296
supervised ML techniques   1
    binary decision tree model for multiclass
        classification   68–74
    binary linear classifier (fitclinear)   98–100
    Classification Learner app   57–68
    classification models   42–47
    discriminant analysis   79–84
    KNN model   47–57
    multiclass support vector machine
        model   92–98
    Naïve Bayes classification model   74–79
    predictor data   42
    response variable   42
    support vector machine classification
        model   84–92
support vector machine (SVM)   84–85, 239, 397
    all data types   90–92
    numeric data types   87–89
    properties of   85–87

t

temporal interpolator algorithm   503–505
time series data   430
trainFasterRCNNObjectDetector
    function   507–508
trainYOLOv2ObjectDetector function   511–512
transfer learning   237, 245, 247–255, 399
Turbofan Engine Degradation Simulation
    clip responses   448
    create directory   447
    download data   446–447
    network architecture   449
    network testing   450–451
    normalize training predictors   448
    predict RUL   451–453

prepare data padding   448–449
prepare training data   447
remove constant values   447, 451
training network   450
training options   449–450
2-D/2-D grouped convolution layer   264–267
two-layer feed-forward network   198

u
United States Geological Survey (USGS)   291
unsupervised ML techniques   1–2
    Gaussian mixture model (GMM)
        clustering   15–17
    hierarchical clustering   23–27
    ionosphere database   38–39
    Iris flower database   37–38
    k-means clustering   13–15
    Laplacian (fsulaplacian) for   35–37
    observations and clusters visualization   18–21
    small car database.   39–40
    wheat seeds dataset   40–41
upsampLowRes function   555

v
vanishing gradient problem   240
video game titles data set   492–493
Video Labeler (VL) app
    automatic algorithm   503–505
    code-generated workspace variables   510
    control panel, video time span   502
    create ROI Labels   502, 503
    faster R-CNN model performance   509
    frames, training and prediction   506
    image and box labeling data stores
        506–507
    load unlabeled data   502–503
    pre-requisite layers   508
    set time interval   503
    30-image test data set   510
    training network   507–513

training options   507
video streams   520
vision-based activity recognition system   520
VL app *see* Video Labeler (VL) app

w
wheat seeds dataset   40–41
word-by-word text generation   494
    ChatGPT   465
    computational techniques   465
    create and train LSTM network   469–470
    data extraction   466, 467
    extracting and generating code   473–474
    load training data   465–466
    new text generation   470–471
    parse HTML code   466
    prepare data training   468–469
    refinement   471
    regeneration   472
    remove empty paragraphs   466–467
    text data visualization   467
    workspace variables   472, 475

y
YOLO object detectors
    attributes   386
    COCO-based creation   387–389
    components   384–385
    detector.Network command   390
    evaluation   394–396
    fine-tuning   389–394
    object detection   386–387
    overview   384
    workspace variables   396
Yolo v4 architecture   385
yolov2ObjectDetector   512–513, 515–517
yolov4ObjectDetector   424

z
Z-score normalization   488

Printed and bound by CPI Group (UK) Ltd, Croydon, CR0 4YY

27/10/2024

14580678-0005